Nanotechnologies for the
Life Sciences
Volume 9
Tissue, Cell and
Organ Engineering

Edited by
Challa S. S. R. Kumar

Related Titles

zur Hausen, H.

Infections Causing Human Cancer

2006
Hardcover
ISBN 3-527-31056-8

Kelsall, R., Hamley, I. W., Geoghegan, M. (eds.)

Nanoscale Science and Technology

2005
Hardcover
ISBN 0-470-85086-8

Willner, I., Katz, E. (eds.)

Bioelectronics

From Theory to Applications

2005
Hardcover
ISBN 3-527-30690-0

Minuth, W. W., Strehl, R., Schumacher, K.

Tissue Engineering

Essentials for Daily Laboratory Work

2005
Hardcover
ISBN 3-527-31186-6

Kumar, C. S. S. R., Hormes, J., Leuschner, C. (eds.)

Nanofabrication Towards Biomedical Applications

Techniques, Tools, Applications, and Impact

2005
Hardcover
ISBN 3-527-31115-70

Goodsell, D. S.

Bionanotechnology

Lessons from Nature

2004
Hardcover
ISBN 0-471-41719-X

Niemeyer, C. M., Mirkin, C. A. (eds.)

Nanobiotechnology

Concepts, Applications and Perspectives

2004
Hardcover
ISBN 3-527-30658-7

Nanotechnologies for the Life Sciences
Volume 9

Tissue, Cell and Organ Engineering

Edited by
Challa S. S. R. Kumar

WILEY-VCH Verlag GmbH & Co. KGaA

The Editor

Dr. Challa S. S. R. Kumar
The Center for Advanced
Microstructures and Devices
(CAMD)
Louisiana State University
6980 Jefferson Highway
Baton Rouge, LA 70806
USA

Cover

Cover design by G. Schulz based on
micrograph courtesy of Peter X. Ma,
Department of Biologic and Materials
Sciences, University of Michigan, USA.

Library of Congress Card No.: applied for
British Library Cataloguing-in-Publication Data:
A catalogue record for this book is available
from the British Library.

**Bibliographic information published by
the Deutsche Nationalbibliothek**
Die Deutsche Nationalbibliothek lists this
publication in the Deutsche Nationalbibliografie;
detailed bibliographic data are available in the
Internet at ⟨http://dnb.d-nb.de⟩.

© 2006 WILEY-VCH Verlag GmbH & Co.
KGaA, Weinheim

Printed in the Federal Republic of Germany.
Printed on acid-free paper.

Typesetting   Asco Typesetters, Hong Kong
Printing   Strauss GmbH, Mörlenbach
Binding   Litges & Dopf GmbH, Heppenheim

ISBN-13: 978-3-527-31389-1
ISBN-10: 3-527-31389-3

Contents of the Series

Contents

Preface

Tissue Cell and Organ Engineering is the title of the ninth volume of the series on Nanotechnologies for the Life Sciences. As the title indicates the book can be considered as an encyclopedia on nanotechnological approaches to tissue, cell and organ engineering. The publication of the book is timely for a number of reasons. First, with the increase in life expectancy there is an unfulfilled demand for replacing non functional tissues and organs. Second, life time for implants and vascular grafts made from conventional synthetic materials (with constituent dimensions greater than 1 micron) is limited ranging from 5 to 15 years. Third, utilization of living tissue/organ for replacement is not a practical solution as evidenced by the fact that millions of surgeries are performed annually to treat tissue and organ failure and the number of people waiting for such transplantations are more than 80,000 at any time. Fourth, there is the ever-increasing cost of health care related to tissue loss or end-stage organ failure exceeding 400 billion US dollars. Furthermore, laboratory experiments to date suggest that nanomaterials have great potential for tissue, cell and organ engineering. Until the publication of this book, there is no single source of information about how nanotechnologies are impacting various facets of tissue engineering. With close to five hundred pages, the book has turned out to be a comprehensive source of information covering different types of nanomaterials being investigated for tissue, cell and organ engineering in addition to innovative strategies for assembling functional and structural artificial tissues.

I am extremely pleased with the final product, which is a tribute to harmonious integration of tissue engineering and nanomaterials science. Such an integration is only possible due to the scholarly presentations from some thirty researchers around the world, spanning twelve chapters. My special thanks to all of them for making this project a reality. I am always grateful to my employer, family, friends and Wiley-VCH publishers, who are part and parcel of this long journey into the union of nanomaterials and life sciences. I am thankful to you, the reader, who has taken time to join this journey. I do hope you will enjoy reading the book and will find it useful in your future endeavors. Let me first give you a bird's eye view of the various facets of tissue engineering being touched by nanomaterials prior to your plunging into the more specific topics.

In comparison with nanomaterials such as nanoparticles and nanotubes, nano-fibers are more suitable for use as scaffolds in tissue engineering as they are

Nanotechnologies for the Life Sciences Vol. 9
Tissue, Cell and Organ Engineering. Edited by Challa S. S. R. Kumar
Copyright © 2006 WILEY-VCH Verlag GmbH & Co. KGaA, Weinheim
ISBN: 3-527-31389-3

morphologically similar to natural scaffolds and possess suitable properties such as high porosity, various pore-size distributions, and high surface-to-volume ratio. The majority of the investigations being carried out by tissue engineering researchers centers on utilization of nanofibers for development of artificial tissues. Therefore, the first four chapters in the book have been dedicated to provide an understanding of various aspects of nanofibers with respect to tissue engineering. The book begins with a chapter on *Nanotechnology and Tissue Engineering: The Scaffold Based Approach* providing an overview of the importance of nanotechnologies in developing scaffolds that closely mimic the structure and functions of the extra cellular matrix (ECM). Lakshmi S. Nair, Subhabrata Bhattacharyya and Cato T. Laurencin from the University of Virginia, USA, have demonstrated the importance of the scaffold approach for utilizing nanostructured materials in engineering ECM replacements. Of the various types of nanofibers currently under investigation for tissue engineering, polymeric nanofibers have been attracting a great deal of attention especially for applications in the area of ophthalmology, hepatic biology, nerve, skin, bone and cartilage regeneration, heart and vascular grafts, and stem cell research. Researchers Seow Hoon Saw, Karen Wang, Thomas Yong and Seeram Ramakrishna from the National University of Singapore have contributed the second chapter by bringing out an exhaustive review on *Polymeric Nanofibers for Ttissue Engineering* covering different types of nanofibers, fabrication methodologies, degradation kinetics, biocompatibility and their applications in regeneration of several types of tissues. Focusing more specifically on electro-spinining technology, researchers Wan-Ju Li, Rabie M. Shanti and Rocky S. Tuan from the National Institute of Arthritis, and Musculoskeletal and Skin Diseases in Bethseda, USA, demonstrate that this technology is a useful, economical, and easily set-up method for the fabrication of three-dimensional, highly porous and nano-fibrous scaffolds which have been shown to support cellular activities and tissue formation. Their contribution in the third chapter, *Electrospinning Technology in Tissue Engineering*, is a thorough review on not only electrospinning technology per se but also on various chemical, physical and biological properties of nano-fibers prepared using this technology. In the fourth chapter, electrospinning technology is compared with two other well known approaches, namely self-assembly and phase separation for fabrication of nano-fibrous scaffolds. Researchers from the University of Michigan in Ann Arbor, USA, systematically analyze the differences and unique characteristics of the three approaches for the development of scaffolds and their applications in tissue engineering. This chapter, *Nano-fibrous Scaffolds and their Biological Effects*, reviewed by Laura A. Smith, Jonathan A. Beck and Peter X. Ma provides some useful insights into achieving the dream of engineering three-dimensional tissue formations.

The fifth chapter in the book, *Nanophase Biomaterials for Tissue Engineering*, contributed by R. Murugan and S. Ramakrishna from the National University of Singapore is a source of information on nanoscale biomaterials in particular, ceramic and polymer-based materials that are being investigated for tissue engineering applications. In addition to the fabrication techniques for nanobiomaterials, the chapter also provides information related to their influence on cells and cell growth. In the next chapter, the sixth, reader's attention is brought to nanotechnologies that are bringing solutions to long-standing problems with current orthopedic implants. The chapter, *Orthopedic Tissue Engineering using Nanomaterials*, written by Michiko Sato and Thomas J. Webster from Purdue University and Brown University respectively in West Lafayette, USA, discusses the effect of ceramic, metallic, polymeric, and composite nanomaterials on cellular functions particularly related to the bone. Norberto Roveri and Barbara Palazzo from the University of Bologna in Italy provide an up to date review on hydroxyapatite nanoparticles, which are being thoroughly investigated as materials for bone substitution. In the seventh chapter, entitled *Hydroxyapatite Nanocrystals as Bone Tissue Substitute*, they describe morphological, structural, chemical-physical, and surface characteristics of hydroxyapatite nanocrystals, followed by research investigations leading to understanding their high bioactivity and ability to induce bone regeneration and remodeling.

Moving away from nanofibers, ceramic nanomaterials and hydroxyapatite nanocrystals, the eighth chapter, *Magnetic Nanoparticles for Tissue Engineering*, authored by Akira Ito and Hiroyuki Honda from Kyushu University and Nagoya University respectively in Japan contains research investigations about application of the magnetic force-based tissue engineering technique (Mag-TE) to cell-seeding (termed "Mag-seeding") and its effectiveness in enhancing cell-seeding efficiency, leading to three-dimensional porous scaffolds for tissue engineering. In addition to magnetic nanomaterials, carbon nanotubes (CNTs) have been attracting the attention of life scientists. Due to their extraordinary physical, chemical, mechanical and electrical properties, in parrticular single-walled carbon nanotubes (SWNTs) are being touted as materials that have immense potential to interface directly with biological systems leading to innovative applications. In the ninth chapter, Peter S. McFetridge and Matthias U. Nollert from the University of Oklahoma in Norman, USA, delve deeply into the applications of SWNTs in the field of tissue engineering. The chapter, *Applications and Implications of Single-Walled Carbon Nanotubes in Tissue Engineering*, provides the history behind the electrical stimulation of cells and the reasoning behind the use of SWNTs as a conductive material to support or promote organ regeneration with specific examples of applications of SWNT in tissue engineering from the literature.

Cellular engineering is conceptually more fundamental to tissue engineering and nanotechnologies are helping in modulating cellular functions as well. In the tenth chapter, a more specific case of the effect of nanomaterials on cellular engineering related to free radical mediated oxidative stress is examined. The chapter, *Nanoparticles for Cell Engineering—A Radical Concept*, is a contribution from the laboratories of Beverly A. Rzigalinski at Virginia College of Osteopathic Medicine and Virginia Polytechnic & State University in Blacksburg, USA. The authors describe in detail potential application of three nanoparticle redox reagents viz rare earth oxide nanoparticles (particularly cerium), fullerenes and their derivatives and carbon nanotubes in preservation of cellular redox status and treatment of disease. Adding to the specifc case of cellular engineering described in the tenth chapter, Jessica Winter from Ohio State University in Columbus, USA, presents in the eleventh chapter an exhaustive review on engineering of cells using nanotechnologies. The chapter aptly entitled *Nanoparticles and Nanowires for Cellular Engineering* is a one-stop source of information on applications of nanostructures to cellular engineering in general and for manipulating specific cellular components. The chapter deals with a gamut of issues on nanomaterial-based cellular engineering related to a wide spectrum of biomedical applications ranging from tissue engineering, intracellular tracking, biosensing and drug delivery. The final chapter of the book is an excellent review on *Nanoengineering of Biomaterial Surfaces* contributed by Ashwath Jayagopal and V. Prasad Shastri from Vanderbilt University in Nashville, USA. In this chapter, a variety of surface engineering techniques are presented for achieving micro- and nanoscale surface features, in addition to their applicability in the construction of hard and soft materials and three-dimensional geometries relevant to cellular and tissue engineering. This chapter brings to close the first comprehensive treatise on nanotechnological approaches for tissue, cell and organ engineering and I am very confident that the book will be a knowledge base for further advances that are bound to take place in the near future.

June 2006   *Challa S. S. R. Kumar*
Baton Rouge, USA

List of Authors

Jonathan A. Beck
University of Michigan
Department of Biomedical Engineering
Ann Arbor, MI 48019
USA

Subhabrata Bhattacharyya
Department of Chemistry
University of Virginia
Charlottesville, VA 22903
USA

Courtney A. Cohen
Virginia College of Ostheopathic Medicine
NanoEuroLab
Research II Bldg – 1861 Pratt Drive
Blacksburg, VA 24060
USA

Igor Danelisen
Virginia College of Ostheopathic Medicine
NanoEuroLab
Research II Bldg – 1861 Pratt Drive
Blacksburg, VA 24060
USA

Hiroyuki Honda
Nagoya University
Department of Biotechnology
School of Engineering
Furo-cho, Chikusa-ku
Nagoya 464-8603
Japan

Akira Ito
Kyshu University
Department of Chemical Engineering
744 Motooka, Nishi-ku
Fukuoka 819-0395
Japan

Ashwath Jayagopal
Biomaterials, Drug Delivery, and
Tissue Engineering Laboratory
Department of Biomedical Engineering
Vanderbilt University
5924 Stevenson Center
Nashville
Tennessee 37232
USA

Cato T. Laurencin
Departments of Orthopaedic Surgery
Biomedical Engineering
and Chemical Engineering
University of Virginia
Charlottesville, VA 22903
USA

Wan-Ju Li
National Institutes of Health, National
Institute of Arthritis, and Musculoskeletal and
Skin Diseases
Cartilage Biology and Orthopaedics Branch
50 South Drive, Bldg. 50, Room 1314
Bethesda, MD 20892
USA

Chengya Liang
Virginia College of Osteopathic Medicine
NanoEuroLab
Research II Bldg – 1861 Pratt Drive
Blacksburg, VA 24060
USA

Peter X. Ma
The University of Michigan
Department of Biologic and Materials Sciences
Department of Biomedical Engineering
Macromolecular Science and Engineering
Center
1011 North University Ave., Room 2211
Ann Arbor, Mi 48019-1078
USA

Peter S. McFetridge
School of Chemical, Biological and
Materials Engineering
and the University of Oklahoma
Bioengineering Center
University of Oklahoma
Norman
Oklahoma 73019-1004
USA

R. Murugan
Division of Bioengineering
National University of Singapore
9 Engineering Drive 1
Singapore 117576

Lakshmi S. Nair
Department of Orthopaedic Surgery
University of Virginia
Charlottesville, VA 22903
USA

Matthias U. Nollert
School of Chemical, Biological and
Materials Engineering
and the University of Oklahoma
Bioengineering Center
University of Oklahoma
Norman
Oklahoma 73019-1004
USA

Barbara Palazzo
Department of Chemistry "G. Ciamician"
Alma Mater Studiorum
University of Bologna
via Selmi 2
Bologna 40126
Italy

Seeram Ramakrishna
Division of Bioengineering
Block E3-05-29
9 Engineering Drive 1
National University of Singapore
Singapore 117576

Norberto Roveri
Department of Chemistry "G. Ciamician"
Alma Mater Studiorum
University of Bologna
via Selmi 2
Bologna 40126
Italy

Beverly A. Rzigalinski
Nanobiology Laboratories
Virginia College of Osteopathic Medicine
and Virginia Polytechnic & State University
Virginia-Maryland Regional College of
Veterinary Medicine
Blacksburg
Virginia 24060
USA

Michiko Sato
Purdue University
School of Materials Engineering
600 Central Drive
West Lafayette, IN 47907
USA

Seow Hoon Saw
Division of Bioengineering
Block E3-05-29
2 Engineering Drive 3
National University of Singapore
Singapore 117576

Rabie M. Shanti
HHMI-NIH Research Scholar
National Institutes of Health, National
Institute of Arthritis, and Musculoskeletal and
Skin Diseases
Cartilage Biology and Orthopaedics Branch
Bethesda, MD 20892
USA

V. Prasad Shastri
Biomaterials, Drug Delivery, and
Tissue Engineering Laboratory
Department of Biomedical Engineering
Vanderbilt University
5924 Stevenson Center
Nashville
Tennessee 37232
USA

Laura A. Smith
University of Michigan
Department of Biomedical Engineering
Ann Arbor, MI 48109
USA

Elizabeth T. Strawn
Virginia College of Osteopathic Medicine
NanoEuroLab
Research II Bldg – 1861 Pratt Drive
Blacksburg, VA 24060
USA

Rocky S. Tuan
National Institutes of Health, National
Institute of Arthritis and Musculoskeletal and
Skin Diseases
Cartilage Biology and Orthopaedics Branch
Bethesda, MD 20892
USA

Karen Wang
Nanoscience and Nanotechnology Initiative
Block E3-05-29
2 Engineering Drive 3
National University of Singapore
Singapore 117576

Thomas J. Webster
Brown University
Division of Engineering and Orthopaedics

184 Hore Street
Providence, RI 02912
USA

Jessica O. Winter
The Ohio State University
Department of Chemical Engineering
335 Kottolt Laboratories
140 W 19th Avenue
Columbus, OH 43210
USA

Thomas Yong
Division of Bioengineering
Block E3-05-29
2 Engineering Drive 3
National University of Singapore
Singapore 117576

1
Nanotechnology and Tissue Engineering:
The Scaffold Based Approach

Lakshmi S. Nair, Subhabrata Bhattacharyya,
and Cato T. Laurencin

1.1
Overview

Biodegradable porous three-dimensional (3D) structures have been extensively used as scaffolds for tissue engineering to temporarily mimic the structure and functions of the natural extracellular matrix (ECM). The ECM functions to provide 3D structure with mechanical and biochemical cues to support and control cell organization and functions. Even though macro- and micro-fabrication techniques enabled the development of highly porous 3D scaffolds that could support the adhesion and proliferation of cells, their ability to closely mimic the complex nanostructured topography and biochemical functions of the ECM is far from optimal. However, recent developments in nanofabrication techniques have afforded various nanostructured bioactive scaffolds. These include top-down approaches such as electrospinning and phase separation to develop nanofibrous scaffolds from polymer solutions or bottom-up approaches such as self-assembly to develop nanofibrous scaffolds from specifically designed bioactive peptide motifs. Although significant improvements are needed for these nanofabrication processes to produce scaffolds that could precisely mimic the structure and functions of the ECM, the developments so far have significantly enhanced our ability to recreate the natural cellular environment for regenerating tissues.

1.2
Introduction

Tissue engineering has now emerged from the stage of infancy demonstrating proof of principle and developing various functional tissues using different approaches to the stage of an established scientific discipline capable of developing viable products for clinical applications. Tissue engineered skin can be considered as one of the first commercialized products developed using the principles of tissue engineering. In addition to clinical applications, tissue engineering has also

Nanotechnologies for the Life Sciences Vol. 9
Tissue, Cell and Organ Engineering. Edited by Challa S. S. R. Kumar
Copyright © 2006 WILEY-VCH Verlag GmbH & Co. KGaA, Weinheim
ISBN: 3-527-31389-3

raised significant interest as a novel tool for investigating cell and developmental biology and developing novel drugs using tissues grown in 3D environments [1].

The ultimate goal of tissue engineering is to address the current organ shortage problem, i.e., development of an alternative therapeutic strategy to autografting and allografting, two common approaches currently used to repair or reconstruct damaged tissues or organs. Autografts and allografts have several shortcomings that significantly limit their applications. These include limited availability and donor site morbidity associated with autografts and risk of infection and immunogenicity associated with allografts [2]. Conversely, regeneration or repair of tissue using tissue engineering approaches attempts to recreate functional tissue using bioresorbable synthetic materials and other required components that can be routinely assembled and reliably integrated into the body without any of the above-mentioned adverse side effects. Tissue engineering thus holds promise to revolutionize current health care approaches to improve the quality of human life in a practical and affordable way.

The term "tissue engineering" was coined in 1987 during a National Science Foundation (NSF) Meeting inspired by a concept presented by Dr. Y.C. Fung of the University of California at San Diego [3]. At a subsequent workshop held by NSF in 1988, tissue engineering was defined as "the application of principles and methods of engineering and life sciences to obtain a fundamental understanding of structure–function relationships in novel and pathological mammalian tissues and the development of biological substitutes to restore, maintain and improve tissue functions" [4]. However, widespread interest of the scientific community in tissue engineering was triggered by two phenomenal reviews: one by Nerem [5] on cellular engineering and another by Langer and Vacanti on tissue engineering [6]. These reviews discuss in depth, for the first time, the possibilities of tissue engineering and presented some of the preliminary studies demonstrating proof of the concept. Figure 1.1 shows the process of tissue engineering [7]. The field of tissue engineering has now developed into a highly interdisciplinary science and has attempted to recreate or regenerate almost every type of human tissue and organ [8]. This was possible within a short time due to the highly multidisciplinary nature of the tissue engineering approach, which makes use of the combined efforts of basic and material scientists, cell biologists, engineers and clinicians. Several different definitions for tissue engineering followed the NSF consensus definition due to the interdisciplinary approach and our laboratory defines tissue engineering as "the application of biological, chemical and engineering principles towards the repair, restoration or regeneration of living tissues using biomaterials, cells and factors, alone or in combination", describing the different possible approaches for tissue engineering [9]. Thus, three or more approaches are currently used to regenerate tissues using the principles of tissue engineering. One approach is the guided tissue engineering that uses a biomaterial membrane to guide the regeneration of new tissue; another approach called cell transplantation uses the application of isolated cells, manipulated cells (gene therapy) or cell substitutes to promote tissue regeneration. A third approach uses biomaterial in combination with bioactive molecules called growth factors to induce and guide tissue regenera-

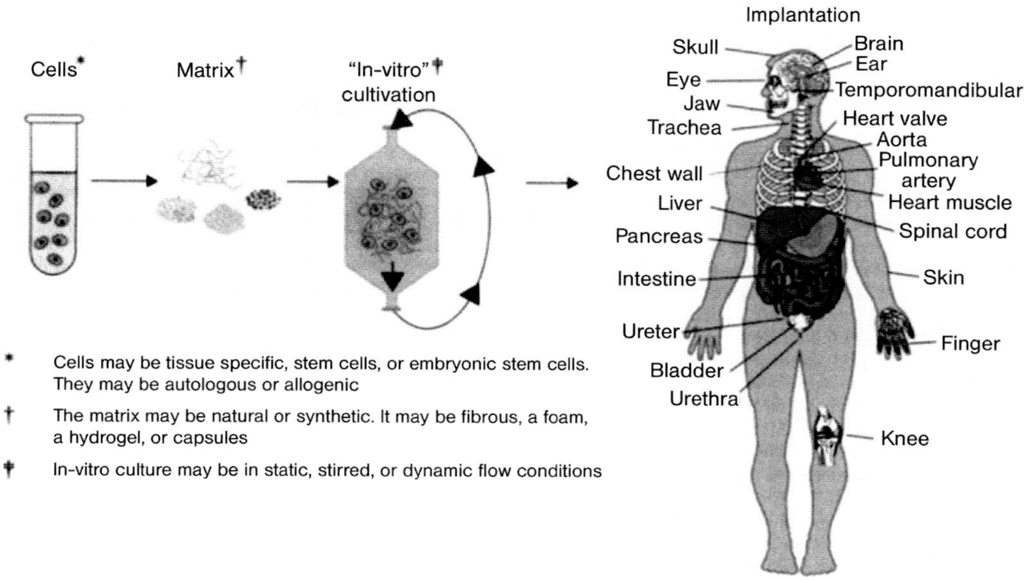

Fig. 1.1. Scheme showing the process of tissue engineering.
(Adapted from Ref. [7] with permission from Elsevier.)

tion and a fourth, the most extensively investigated approach, uses biomaterials in combination with cells (with and without biological factors). Within the cell–biomaterial combination approach two different methods are used, a closed system and an open system. In a closed system cells are protected from the immune response of the body by encapsulating in a semi-permeable membrane that can allow nutrient and waste transport to keep them functional. In an open system, the cell–biomaterial construct is developed *in vitro* and is directly implanted in the body. In the open system, biomaterials are used to develop supporting matrices or scaffolds for cell implantation. Bioresorbable polymers (both synthetic and natural polymers) are commonly used for fabricating scaffolds. Several fabrication techniques are used to develop porous 3D scaffolds from these biomaterials. The function of the scaffold is to guide the regeneration of new tissue and to provide appropriate structural support, i.e., to mimic the structure and functions of natural extracellular matrix (ECM). The exogenous cells delivered through the scaffolds along with endogenous cells are used to regenerate or remodel the damaged tissue. During this process the bioresorbable scaffold will degrade and disappear resulting in the formation of remodeled native tissue [8]. Research to date has identified different cell sources, including stem cells that, when combined with degradable, matrices can form 3D living structures. The technique has led to the development of many tissues in the laboratory scale such as bone, ligament, tendon, heart valves, blood vessels, myocardium, esophagus, and trachea. However, several engineering and biological challenges still remain for successful clinical translation of the laboratory research to make tissue engineering a reliable route for organ/tissue regeneration. These include mimicking the complex structure and biology of the ECM using synthetic materials, controlling cell interactions using artificial scaffolds, vascularization of cell–scaffold constructs, development of efficient bioreactors for *in vitro* culture, storage and translation [10]. The present chapter reviews progress made in tissue engineering to overcome some of the engineering and biological challenges in developing ideal 3D synthetic scaffolds by harnessing nanotechnology and material science.

This chapter also overviews the importance of mimicking the structure and functions of the ECM when developing ideal scaffolds for tissue engineering and the recent developments and advantages of nanotechnology assisted techniques to fabricate scaffolds that closely mimic the ECM.

After the present section, which gives a brief introduction to tissue engineering, Section 1.3 lays out the importance of scaffolds in tissue engineering and the need for mimicking the structure and functions of the ECM. Section 1.4 reviews the important aspects of the structure and functions of the ECM that need to be mimicked to develop ideal scaffolds for tissue engineering. Section 1.5 includes in-depth examination of the applications of nanotechnology in developing ECM mimic nanostructured scaffolds for tissue engineering. Section 1.6 reviews some recent studies, demonstrating the advantages of nanostructured scaffolds for tissue engineering, and Section 1.7 overviews some of the current applications of nanostructured scaffolds for engineering different tissues.

1.3
The Importance of Scaffolds in Tissue Engineering

The importance of the extra-cellular matrix (ECM) in cellular assembly and tissue regeneration was demonstrated by the pioneering works of Mina Bissell along with others [11]. The cells in mammalian tissues are connected to the ECM which provide three-dimensionality, organize cell–cell communications and provide various biochemical and biophysical cues for cellular adhesion, migration, proliferation, differentiation and matrix deposition. Their studies have shown the significant differences in behavior of cells when grown in two-dimensional (2D) and 3D environments [11].

Even though 2D cell culture techniques have been extensively used by cell biologists to derive valuable information regarding cellular processes and cell behavior, in the light of recent studies it is evident that *in vivo* tissue response can be simulated only through 3D cell culture techniques [12]. Considering the complex biomechanical and biochemical interplay between cells and the ECM, it is apparent that tissue engineers will be unable to address the biological subtleties if the cells are grown on 2D biomaterials before implantation in the body.

The strategy of using bioresorbable porous synthetic scaffolds as artificial ECM was introduced by Langer and Vacanti in 1988 [13]. This seminal paper significantly influenced investigators throughout the world in the practical area of scaffold based tissue engineering and has led to hundreds of research articles and patents to date.

A bioresorbable α-hydroxyester was used as the candidate polymer in the first study by Langer's group for developing the scaffolds. The α-hydroxyesters being aliphatic polyesters have the ability to undergo hydrolytic degradation *in vivo* and therefore could resorb and disappear once regeneration is complete. Studies that followed have shown that the properties of the biomaterial play a crucial role in the success of the tissue engineered construct. Since the dynamics of different tissues vary significantly, appropriate materials need to be carefully chosen to satisfy the properties required. This knowledge has led to the design and development of several bioresorbable polymeric biomaterials to fabricate scaffolds for engineering different types of tissues [14]. These include synthetic polymers such as α-hydroxyesters, polyanhydrides, polyphosphazenes, polyphosphoesters and natural polymers such as collagen, gelatin, chitosan and hyaluronic acid [15, 16]. Among these, synthetic polymers are mostly preferred for developing tissue engineering scaffolds due to immunogenic problems and batch by batch variations associated with many of the natural polymers.

Apart from the properties of the materials, the 3D architecture of the scaffold is very important when attempting to mimic the structure and functions of the natural ECM. Several unique fabrication processes have been developed to form 3D porous structures from bioresorbable materials as scaffolds for tissue engineering [9, 17]. These 3D structures have been primarily designed to direct tissue growth by allowing cell attachment, proliferation and differentiation. Most of these fabrication processes have been designed based on a set of criteria that have been identified as crucial to promote cellular infiltration and tissue organization. Some of the basic requirements of scaffolds for tissue engineering, summarized by Agarwal and Ray [18], are that they should be:

- Biocompatible.
- Bioresorbable and hence capable of being remodeled.
- Degrade in tune with the tissue repair or regeneration process.
- Highly porous to allow cell infiltration.
- Highly porous and permeable to allow proper nutrient and gas diffusion.
- Have the appropriate pore sizes for the cell type used.
- Possess the appropriate mechanical properties to provide the correct micro-stress environment for cells.
- Have a surface conducive for cell attachment.
- Encourage the deposition of ECM by promoting cellular functions.
- Able to carry and present biomolecular signals for favorable cellular interactions.

Various studies have been performed so far, using macro- and micro-fabrication techniques, to form 3D scaffolds that could address the requirements listed above to develop ideal synthetic scaffolds with some success. The results of these studies have been extensively reviewed [17, 19–23]. Particulate leaching can be considered as one of the first techniques widely used to develop micro-porous matrices from

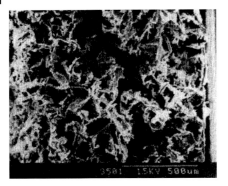

Fig. 1.2. Porous bioresorbable poly(L-lactic acid) foams developed by particulate leaching. (Adapted from Ref. [24] with permission from Elsevier.)

biodegradable polymers for tissue engineering applications (Fig. 1.2) [24–26]. The technique has several advantages such as ease of processing, ability to develop foams from wide range of polymers, and the ability to control the pore size by varying the size of the porogen. However, the porogen leaching process has some serious limitations to fabricate scaffolds for tissue engineering, such as the inability to completely remove the porogen from the porous matrix and to control the pore shape and maintain interconnectivity between pores. Consequently, several modifications to the particulate leaching method as well as new fabrication techniques were developed. Some of the newer processes include sintered microsphere process and rapid prototyping.

Sintered microsphere matrix fabrication technique of Laurencin was developed as a robust technique to fabricate 3D porous structures with reproducible porosities and interconnected pore structure [27, 28]. Sintered microsphere matrices are developed by heat sintering bioresorbable polymeric microspheres (Fig. 1.3)

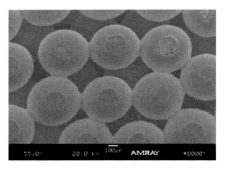

Fig. 1.3. SEM showing the 3D porous structure of a PLAGA scaffold formed by the sintered microsphere fabrication process. (Adapted from Ref. [29] with permission from Elsevier.)

Fig. 1.4. SEM showing human osteoblast attachment and infiltration in porous PLAGA sintered microsphere matrix. (Adapted from Ref. [29] with permission from Elsevier.)

[27, 28]. Poly(lactide-*co*-glycolide)s (PLAGA) having different ratios of lactic acid (LA) and glycolic acid (GA) were used as the polymers to develop sintered matrices. Polymeric microspheres are prepared by the commonly used solvent evaporation technique [27, 28]. The sintered microsphere matrices demonstrated controllable pore size and interconnectivity depending on the size of the microspheres used to fabricate the matrices. Thus, the pore size of the scaffolds could be varied from 100 to 300 μm, depending on the size of the microspheres used. The 3D porous sintered microsphere scaffolds were investigated as potential candidates for bone tissue engineering and showed appropriate mechanical properties for orthopedic applications. The osteoconductivity of the porous 3D matrices were evaluated using human osteoblast cells and showed good osteoblast attachment and infiltration (Fig. 1.4) [28, 29]. An *in vivo* evaluation demonstrated the efficacy of the bioresorbable sintered microsphere matrix in healing a critical segmental bone defect in a rabbit model [30]. Figure 1.5 shows the X-ray of a bone defect site implanted with a sintered microsphere matrix after eight weeks of implantation. The study showed the formation of new bone throughout the entire structure of the implant indicating significant bone regeneration at the defect site. The fabrication process led to the development of porous scaffolds having high interconnectivity and good mechanical integrity, with the percentage pore volume of the matrices equal to ~40%.

Recently, different types of computer-assisted design and manufacturing processes (CAD/CAM) were investigated as potential methods to develop scaffolds

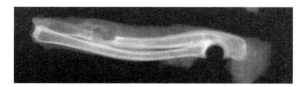

Fig. 1.5. Radiograph of a defect site implanted with sintered microsphere matrix, bone marrow cells and BMP-7 after 8 weeks of implantation, showing significant bone regeneration.

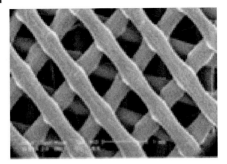

Fig. 1.6. SEM showing the porous structure of a polymer scaffold developed by FDM. (Adapted from Ref. [23] with permission from Elsevier.)

having controllable pore size, shape and porosity. One of the first developed computer assisted techniques for scaffold fabrication was solid free form fabrication or 3D printing. In this process a complex 3D structure is first designed using CAD software. An inkjet printing of a binder on appropriate polymer powder layers is then used to fabricate the porous structure based on the computer model. Even though complex structures can be designed and fabricated using this automated process, the preciseness of the technology has various limitations imparted by the size of the polymer particle, size of the binder drop and the type of the nozzle tip [31]. Another rapid prototyping technique extensively investigated for developing porous scaffolds for tissue engineering is fused deposition model (FDM) developed by Hutmacher [22, 23]. The FDM can be used to develop 3D structures from a CAD or an image source such as computer tomography (CT) or magnetic resonance imaging (MRI) of the object. The computer design is then imported into software that mathematically slices the model into different horizontal layers. The FDM extrusion head and the platform are then synchronized to deposit fused polymeric melt based on the computer model, one layer at a time. Figure 1.6 shows a porous 3D structure developed from a bioresorbable polymer poly(caprolactone) (PCL). The FDM process has several advantages, such as the ability to precisely control the pore size, pore morphology and pore interconnectivity. The process enables also the development of multiple-layer designs and different localized pore morphologies needed for multiple tissue types or interfaces. Another advantage of FDM is the good mechanical properties and structural integrity of the scaffolds due to the use of mechanically stable designs and proper fusion between individual material layers. However, the fabrication process has some limitations that make it less than an optimal method for developing porous matrices for tissue engineering applications. These include the limitations associated with the processing technique such as the requirement of temperature, the need for materials that are appropriate for fused deposition, the necessity of supporting structures to construct complex structures, and variable pore openings observed along different axis [32].

A   B

Fig. 1.7. (A) Pro/Engineer rendered CAD image of prototype PPF construct. The series of slots and projections test the interslice PPF registration (50 × 4 mm). (B) Three-dimensional structure developed from CAD model data from PPF. (Adapted from Ref. [33].)

Several stereolithographic techniques were investigated to develop porous 3D matrices from polymers to overcome the problems associated with temperature-assisted fabrication methods. Stereolithographic techniques have been extensively used to develop 3D structures from photopolymerizable polymer solutions such as poly(propylene fumarate) (PPF) in presence of photoinitiators (Fig. 1.7A and B) [33]. Preliminary studies showed the feasibility of developing structures having controlled pore sizes (50–300 μm) and different layer thicknesses using a highly controlled laser light source.

Most of the techniques described above are used to develop 3D structures from synthetic hydrophobic polymers. However, a wide range of techniques using hydrophilic polymers have also been investigated to develop novel structures as cell delivery vehicles. Hydrophilic polymers are good candidates to develop tissue engineering scaffolds due to their high water content and ability to mimic the properties of various tissues. One such technique is the use of photolithography to pattern hydrogel films with hydrophilic porous structures [34].

The fabrication techniques discussed so far have been developed to fabricate acellular scaffolds that are populated with appropriate cells after fabrication for tissue engineering applications. However, this process has the limitation of obtaining uniform cell distribution throughout the scaffold even with the use of bioreactors during *in vitro* culture. Therefore some studies have also focused to develop materials and fabrication processes to form cellular scaffolds. These studies have led to the development of different types of stimuli sensitive hydrophilic polymers that can be used to encapsulate cells under mild conditions to form cellular scaffolds [35, 36]. Cells can be uniformly distributed in the aqueous stimuli sensitive polymer solutions before the gelling process (Fig. 1.8) [37]. Attempts are currently underway to combine this process with the lithographic techniques to form cell incorporated 3D structures under very mild conditions.

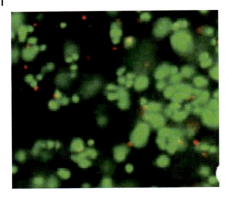

Fig. 1.8. Photomicrograph showing chondrocytes encapsulated within a hydrogel after 21 days in culture stained using Live dead stain (green shows live cells and red shows dead cells). (Adapted from Ref. [37] with permission from Elsevier.)

Another strategy recently developed to form structures with uniform distribution of cells throughout the scaffold is cell printing [38]. This approach combines rapid prototyping procedures with microencapsulation to print viable free form structures using bio-ink with custom-modified ink-jet printers. One advantage is the feasibility of placing quickly and precisely various cells layer by layer to develop multi-cell systems. However, the process is still in its infancy and further research is necessary with regards to developing appropriate bio-ink, optimizing the rheologic and surface properties of the inks, and designing printers optimized for these properties [39]. Another strategy is organ printing, which makes use of nature's ability to assemble many tissue forms such as blood vessels. The technology is based on the hypothesis that when cell aggregates are placed in close approximation they can assemble to form a disc or tube of tissue (Fig. 1.9) [40–42]. This process is also still in its infancy, has various scaling up limitations and further studies are needed to demonstrate the potential of the approach.

The previous discussion demonstrates the importance of the ECM in tissue repair and regeneration and serves as a brief overview of the attempts made to mimic the structure of the ECM using polymeric biomaterials and various macro/micro fabrication techniques to develop interconnected porous structures having porosities in the micron range. These studies have led to the design and synthesis of novel bioresorbable materials with unique chemistries, fabrication of 3D structures having different properties and demonstrated the feasibility of growing cells in appropriate 3D forms *in vitro* and *in vivo* with the help of these scaffolds. Figure 1.10(A and B) shows the feasibility of developing an artificial ear on the back of a mouse using a bioresorbable PCL scaffold having the macroscopic shape of an ear seeded with chondrocytes [43].

Even though these materials and fabricated 3D structures showed the feasibility of 3D organization of cells into tissue, they are far from being ideal for develop-

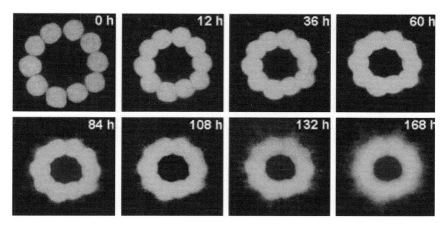

Fig. 1.9. Time evolution of the fusion of aggregates of Chinese Hamster Ovary (CHO) cells encapsulated in collagen gel. The nuclei of the cells are fluorescently labeled. Cell before fusion (top left) and the final disc-like configuration after fusion (bottom right). (Adapted from Ref. [37] with permission from the National Academy of Sciences, USA.)

ing fully functional tissues and organs *in vivo* in a reproducible way under clinical setting.

So far, most of the biomaterial design has focused on developing materials that are capable of degrading at a rate that matches tissue regeneration, have the ability to degrade into non-toxic degradation products and can support the adhesion and proliferation of cells without placing much emphasis on the bioactivity of the materials. Conversely, most fabrication techniques are focused on developing scaffolds with macroscale properties, such as the ability to provide sufficient transport prop-

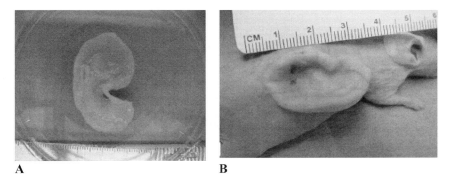

A   B

Fig. 1.10. (A) Photomicrograph of tissue engineered ear construct developed from chondrocyte-PCL composite after 8 weeks *in vitro* culture. (B) The regenerated ear on the back of athymic mice. (Adapted from Ref. [43] with permission from Elsevier.)

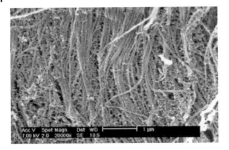

Fig. 1.11. High magnification picture of collagen fibrils in human aortic valve. Individual fibrils are separated by a narrow space crossed by interfibrillar bridges formed by small proteoglycans interconnecting adjoining fibrils. (Adapted from Ref. [44] with permission from Elsevier.)

erties (interconnected microporous structure), and adequate mechanical properties (to match the properties of the tissue to be replaced or repaired).

However, the organization of the cells, and hence the properties of the tissue, are highly dependent on the structure of the ECM, which has a hierarchical structure with nano-sized features. Figure 1.11 shows the ultrastructure of collagen fibrils in human aortic valves, illustrating the nanoscale topographic features of the native tissue [44]. Thus, successful fabrication of a fully functional tissue is a far more complex and involved process that requires the creation of an appropriate environment at both a micro- and nano-scale level to allow for cell viability and function along with macroscopic properties [45]. In fact, just as important as these structural features are the biological principles that govern cell–cell and cell–matrix interactions. These interactions form the basis of cellular performance and appropriate tissue organization and are controlled by various biochemical cues present in the natural ECM. The recreation of this process requires the incorporation of various bioactive molecules in synthetic porous scaffolds with molecular precision to mimic the functions of the ECM. Recently, a paradigm shift has been observed from developing macro/micro-structured scaffolds to nanostructured bioactive scaffolds in an attempt to improve tissue design and reconstruction in reparative medicine [46, 47].

1.4
Structure and Functions of Natural Extracellular Matrix

Since the ultimate goal of tissue engineering is to develop tissue substitutes that could temporarily mimic the structure and functions of damaged tissue to be replaced, it is crucial that the engineered substitutes mimic the natural tissue structurally and functionally for successful regeneration. Extensive research performed in different areas such as tissue and organ development during embryogenesis, the normal tissue healing process, tissue structure and functions and development of various characterization techniques at the micro- and nano-scale levels have significantly enhanced our ability to mimic native tissue. The human body is a very complex structure that is organized in a hierarchical way with body composed of systems, systems composed of organs, organs composed of tissues and tissues composed of cells, vasculature and extracellular matrix. In tissues, the ECM provides the structured environment with mechanical and biochemical cues that enable the cells to interact with each other and with the ECM to allow for control of growth, proliferation, differentiation and gene expression.

The ECM is composed of a physical and chemical crosslinked network of fibrous proteins and hydrated proteoglycans with glycosaminoglycan side chains (collectively called the physical signals) in which other small molecules (such as growth factors, chemokines and cytokines) and ions are bound. Figure 1.12 shows the ultrastructural features of an ECM with a condensed basement membrane and the stromal tissue [48]. The ECM proteins are mainly composed of more than 20 different types of collagens as well as elastin, fibrillin, fibronectin, and laminin [49]. In the natural environment these macromolecular ECM components are secreted by the cells and then modified and assembled to form the matrix during the tissue development and repair process. Among the ECM proteins, type I collagen is mainly involved in the formation of the fibrillar and microfibrillar structure of the ECM. Type I collagen molecules (~300 nm long and ~1.5 nm in diameter) are packed to form collagen fibrils. Each collagen fibril displays a characteristic

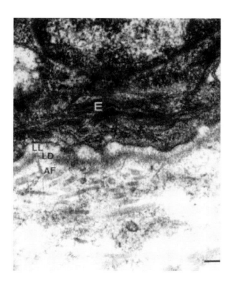

Fig. 1.12. Ultrastructure of ECM matrix. Adjacent to an epithelial cell (E) is the basement membrane with its amina lucida (LL) and lamina densa (LD). The interstitial matrix contains collagen fibrils and is close to the basement membrane anchoring fibrils (AF), composed of type VII collagen fibrils. (Adapted from Ref. [48].)

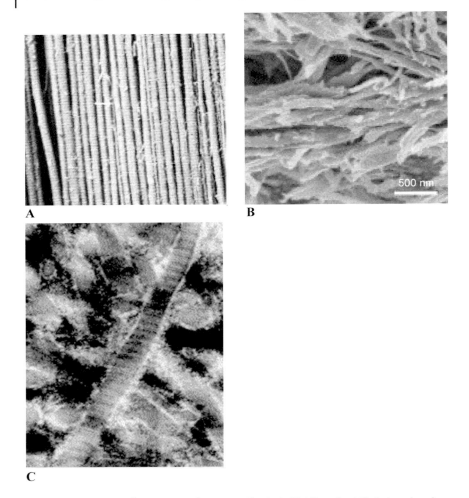

Fig. 1.13. Structure and orientation of collagen fibrils of various tissues. (A) Mature rat ligament-collagen fibrils are primarily aligned along the long axis of the ligament. (Adapted from Ref. [51] with permission from Elsevier.) (B) Mineralized fibrils in trabecular bone without the non-fibrillar matrix. (Adapted from Ref. [52] with permission from Elsevier.) (C) Articular cartilage. (Adapted from Ref. [53] with permission from Elsevier.)

~67 nm D-repeat with uniform or multi-model diameter distribution varying from ~25 to 500 nm and several micrometers in length, depending on the nature of the tissue. All of these molecules are arranged in a unique tissue specific 3D architecture [50]. Figure 1.13 shows the different arrangement patterns of collagen fibrils as observed in three different types of tissues (A: ligament; B: bone; and C: articular cartilage) [51–53]. Fibrils show varied orientation in different tissue types that give the appropriate physical and mechanical properties to the tissue. Collagen fibrils are further bundled together to form collagen fibers. The hierarchical structure of the ECM has length scales, varying from a few nanometer (nm) to millimeter (mm) that control the cellular functions and corresponding tissue properties. The fact that cells are highly sensitive to the environmental structural features has been demonstrated using *in vitro* cell culture studies on nanopatterned surfaces fabricated by electron beam lithographic techniques. These studies revealed that the cells are sensitive to nanoscale dimensions and could react to objects as small as 5 nm [54]. This can be attributed to the structural details ECM presents to the cells *in vivo*. Thus, the 3D hierarchical structure of the ECM significantly affects cellular behavior and hence tissue functions through topographical cues.

However, the function of the ECM is not just to provide an inert support for cellular adhesion. Almost all of the molecules present in ECM have both structural and functional roles. The ECM serves mainly to organize cells in space to give them form, provide them with environmental signals, to direct site-specific cellular regulation, and separate one tissue space from another. Thus the orientation and position of cells with respect to each other is dictated by the ECM and the orientation varies with different tissues. This is achieved by providing chemical cues such as insoluble signals or factors of the ECM which could interact with the soluble signals of cells along with the structural features to promote adherence, migration, division and differentiation of cells. In a natural tissue the ultimate decision of cellular processes such as adhesion, proliferation, differentiation, migration and matrix production takes place as a result of this continuous cross-talk between cells and ECM effectors [55]. Figure 1.14 shows a schematic representation of various interactions taking place between cells and ECM during tissue organization and function [50].

At least three mechanisms have been identified though which the ECM can regulate cell behavior. The first mechanism is through the composition of the ECM such as various proteins and glycosaminoglycans, which is highly tissue and cell specific. The second mechanism is through synergistic interactions between growth factors and matrix molecules. The growth factors are found to bind with the ECM through the glycosaminoglycan side chains or protein cores and this increases the stability of growth factors and creates the appropriate cellular environment or niche to regulate cell proliferation and differentiation. The third mechanism is through the cell surface receptors or integrins that mediate cell adhesion to extracellular matrix components [56]. The integrin–ECM ligand interactions play a major role in anchoring cells to the ECM. An integrin is ~280 Å long and consists of one α (150–180 kDa) and one β (~90 kDa) subunit, both of which are type I membrane proteins [57]. About 18α and 8β subunits that can form 24 different heterodimers have been identified so far and the ligand specificity of the integrin is determined by the specific $\alpha\beta$ subunit combination [58]. Integrin-mediated cell adhesion to the ECM occurs through a cascade of processes. First cell attachment occurs where cell attach to the surface with ligand binding through integrins to withstand gentle shear forces followed by cell spreading. Next organization of actin into microfilament bundles or stress fibers occurs. In the last stage the formation of focal adhesion occurs, which links the ECM to molecules of the actin

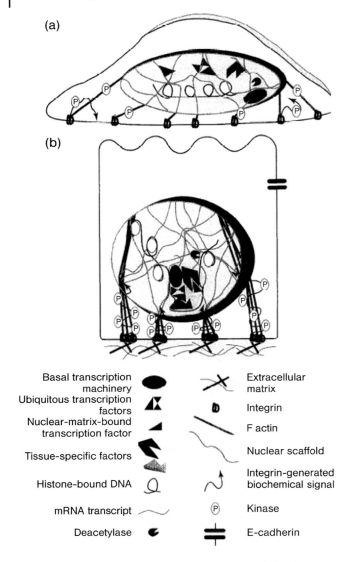

(a)

(b)

Basal transcription machinery	●	Extracellular matrix	✕
Ubiquitous transcription factors	▲	Integrin	◗
Nuclear-matrix-bound transcription factor	◢	F actin	—
Tissue-specific factors	◤	Nuclear scaffold	∿
Histone-bound DNA	◠	Integrin-generated biochemical signal	⤴
mRNA transcript	∿	Kinase	Ⓟ
Deacetylase	◖	E-cadherin	≡

Fig. 1.14. Scheme showing the various interactions between cells and the ECM during tissue organization and repair. (a) Flattened cells in the absence of ECM. Due to incompatible cytoskeletal organization ro signals originating from integrin receptors can be properly propagated. (b) In the presence of ECM, binding of ECM components to integrin receptors induces integrin clustering and generates biochemical signals. Cytoskeleton filaments intimately associated with the cytoplasmic domains of the integrins are modified and reorganized to facilitate interaction of the incoming signals with downstream mediators. This reorganized cytoskeleton can evoke further architectural changes via its association with the nuclear matrix. The subsequent nuclear reorganization brings together incoming signaling molecules, transcriptional activators, histone deacetylases, and the basal transcriptional machinery to promote the assembly of a functional transcriptional complex on the gene. (Adapted from Ref. [50] with permission from Elsevier.)

cytoskeleton. The focal adhesion is mainly composed of clustered integrins and other transmembrane molecules. In this process, the integrins have two-fold activities, anchoring the cells to the ECM and signal transduction through the cell membrane [59]. This allows for continuous cross-talk between ECM and cells which is highly crucial for proper tissue functioning [60].

Several studies have been performed to characterize and analyze the largest and most stable types of contacts between the ECM and cells. These include focal adhesions, focal contacts or adhesion plaques, fibrillar adhesions and hemidesmosomes (Table 1.1) [61]. Several morphological criteria have been used to characterize the

Tab. 1.1. Characterization of cell–matrix contact structures. (Reprinted from Ref. [61] with permission from Birkhäuser Verlag, Basel.)

Contact type	Dimensions	IRM Image separation from substratum	Characteristic associations
Close contact	(Associated with lamellipodium)	Grey in IRM 30–50 nm from substratum	Submembranous densities parallel to F-actin meshwork at plasma membrane
Filopodium	20–200 μm long 0.2–0.5 μm diameter	grey in IRM	core bundle of F-actin, integrins, syndecans
Focal contact/focal adhesion/	0.25 μm wide 1.5 μm long	Black in IRM 10–15 nm from substratum	At termini of microfilaments, contain integrins, syndecan-4, low tensin content
Hemidesmosome	Plaque ca. 0.15 μm by 0.04 μm	–	Connect to intermediate filaments, contain $\alpha6\beta4$ integrin, plectin, BP230
Matrix assembly sites/fibronexus/ fibrillar adhesions	ca. 3–5 μm long	White in IRM 100 nm from substratum	ECM cables align parallel with microfilaments, contain $\alpha5\beta1$ and tensin
Podosomes	0.2–0.4 μm diameter	Dark in IRM	Core bundle of act in perpendicular to substratum, in macrophages contain $\beta2$ integrins, fimbrin
Spike or microspike	2–10 μm long 0.2–0.5 μm diameter	Grey in IRM	Core bundle of F-actin, contain fascin

contact type and size of the contact including evaluation of phase-dark structures detected by phase contrast or interference reflexion microscopy (IRM), electron-dense and organized structures detected by transmission electron microscopy and cell surface topography detected by scanning electron microscopy (SEM) [61]. Table 1.1 shows that even though the size of a cell is ∼10 μm, the activities leading to cell adhesion and the following processes take place mainly at the nanometer level.

Detailed studies on integrin-mediated cell adhesion to the ECM clearly point to the importance of ECM ligands on cell behavior. In addition to the structure and size of the ECM–cell contact points, several studies have been performed to elucidate the biological molecules involved in the interactions. The results of these studies have significantly influenced tissue engineers and have become a great tool in their attempts to recreate a natural cellular environment using synthetic scaffolds. Many of the identified biological molecules have been utilized to decorate synthetic scaffolds to form ligand-functionalized matrices to increase their bioactivity. Various surface modifications or one-dimensional nanotechnological modifications are used to develop ligand functionalized scaffolds [62].

Several cell recognition motifs such as fibronectin, vitronectin, collagen and laminin present in the natural ECM have been used to modify the surface of biomaterial scaffolds to increase their bioactivity [63–65]. Even though preliminary studies show significant promise in developing biomimetic scaffolds, the modification of matrices using bioactive proteins has several limitations. Proteins are bioactive molecules and can elicit an immunological response as they are mostly isolated from different sources and also possess the risk of associated infections. Another serious limitation associated with protein surface modification is that the surface topography and chemistry of the synthetic matrix could influence the orientation and conformation of the attached or adsorbed protein, thereby affecting its functionality. Owing to the low stability of the proteins, the immobilization process as well as subsequent storage could also affect its patency.

The breakthrough research that revolutionized the biomimetic surface modification approach towards biomaterials development is the finding that low molecular peptides from ECM proteins such as the tetrapeptide "arginine-glycine-aspartate-serine" (GRGDS) sequence could significantly modulate cellular behavior [66, 67]. Following this, several RGD-containing sequences were found in other ECM proteins and several other short linear adhesive sequence motifs have also been identified as active molecules to promote cell adhesion, proliferation and migration. Several studies have been performed to elucidate the mechanisms by which these sequences could interact with cells. The tetrapeptide and tripeptide sequences such as arginine-glycine-aspartate can bind to members of the integrin family of the transmembrane receptors, thereby activating a series of signaling events within the bound cells favorably affecting their functions [67–72]. The RGD sequence has been quickly identified as a potential candidate to develop biomimetic scaffolds and extensive research has followed. Because the RGD sequence is present in multiple ECM proteins such as fibronectin, laminin, collagen and vitronectin, a broad range of cell types could respond to this peptide sequence. Furthermore, small peptide sequences are highly stable compared to the corresponding proteins [73], they are cost effective [74], can be packed densely on surfaces due to their small size, and can selectively address one type of cell adhesion receptors for controlled cell adhesion during multicellular tissue development [68].

Various techniques have been attempted to immobilize these biological motifs on synthetic biomaterial surfaces to increase their bioactivity [75]. Stable immobilization of these ligands to the surface is crucial for proper functioning, as the peptide sequence should be able to withstand the cells contractile forces during initial attachment and prevent internalization by cells [76, 77]. The most extensively investigated approach to covalently immobilize the RGD sequence on surfaces is by using active functional groups such as hydroxyl, carboxyl or amino groups on the RGD and polymer surfaces, involving carbodiimide chemistry [75]. For polymers devoid of these functional groups, several approaches were attempted, such as coating the surface with a polymer having such active groups such as polylysine [75, 78] and coating with RGD modified pluronics via hydrophobic interactions [75, 79]. Another approach to incorporate active groups is by copolymerizing with a monomer having active groups such as acrylic acid [75, 80] or lysine in the case of poly(lactic acid-*co*-lysine) [75, 81]. Another extensively investigated approach is chemical or physical surface modification of biomaterials such as alkaline hydrolysis [75, 82], oxidation [75, 83], reduction [75, 84], etching [75, 85] or plasma deposition [75, 86].

Even though immobilization via carbodiimide chemistry is a versatile approach to covalently immobilize RGDs on various surfaces, it is not a highly selective process as RGD has two reactive groups (amino and carboxyl) and therefore can lead to various un-wanted side reactions. A recent study has demonstrated the feasibility of incorporating RGDs on the surface of polymers without the various functionalization routes described above. The approach is called chemoselective ligation. Under mild conditions, selected pairs of functional groups are used to form stable bonds with RGD without interfering with other functional groups [87]. Thus thiol-functionalized surfaces can be modified using bromoacetyl containing RGD cyclopeptides [88] or a thiol-containing RGD can be linked to maleinimide-functionalized surfaces under mild conditions [89]. Benzophenone or aromatic azide functionalized RGD has been developed as a versatile technique to immobilize RGDs on the surface [90–92] by streptavidin–biotin capture [93].

In addition to direct linking, attachment of RGD to surfaces using spacers significantly increases the activity of immobilized RGDs. This increased activity has been attributed to the ability of RGD peptide binding site to reach the hollow globular head of an integrin. Several studies have confirmed a spacer length of 35–40 Å is optimal for maximum activity [75, 94]. However, recent studies on the crystal structure of the ligand bound extracellular domain of the $\alpha V\beta 3$ integrin show the RGD binding site on the surface region of the head of the $\alpha V\beta 3$ integrin, suggesting that it is only a few angstroms deep [71]. This indicates that spacers may not be needed for the ligand–integrin interaction. The experimental improvement in activity of RGDs with spacers found in some studies has been attributed to the spacer presumably contributing to the surface roughness of the substrates [75].

All of these approaches have shown the feasibility of covalent attachment of

RGDs on the surface of polymeric biomaterials and several studies were also performed to demonstrate the bioactivity of RGD immobilized surfaces. Numerous polymers, various immobilization techniques, different RGD peptides and different cell types were used to investigate the biological activity of biomimetic surfaces. The cell behavior towards RGD modified surfaces has been found to depend on various parameters such as the structure and conformation of RGD as well as the density and arrangement of RGD on the surface. Some of the RGD peptides investigated include RGD, RGDS, GRGD, YRGDS, YRGDG, YGRGD, GRGDSP, GRGDSG, GRGDSY, GRGDSPK, CGRGDSY, GCGYGRGDSPG, and RGDSPASSKP peptides [75]. One study systematically investigated the cell attachment activity of different types of RGDs including RGD, RGDS (from fibronectin), RGDV (from vitronectin) and RGDT (from collagen) immobilized on polymeric surfaces. The study demonstrated that tetrapeptides show distinct increases in cell attachment compared to tripeptides indicating that peptides with higher integrin affinity bear higher cell attachment [80]. No significant differences in cell attachment between the tetrapeptides were observed [80]. Another study showed that cyclic RGD peptides on surfaces can show higher activity than linear molecules, which has been attributed to their higher stability and increased $\alpha v \beta 3$ binding of cyclic peptide compared with linear molecules [95].

Another unique application of RGD modified biomaterials is the development of materials that can promote selective adhesion of various cell types. Since it has been found that each cell type has its own typical pattern of different integrins, RGD peptides could be used to promote selective cell adhesion on a surface by modifying the surface with an appropriate RGD [69]. Some *in vitro* studies have demonstrated the feasibility of integrin specificity of RGD leading to selective cell adhesion on RGD modified surfaces [69]. The study showed that fibroblasts rather than endothelial cells preferably adhered to a RGDSPASSKP (which is selective to $\alpha 5 \beta 1$) modified surface [96]. Similarly, enhanced fibroblast attachment was observed to an $\alpha 5 \beta 1$ integrin selective GRGDSP peptide functionalized surface where as $\alpha 5 \beta 3$ selective cyclic G*PenGRGDSPC*A supported higher smooth muscle cell and endothelial cell densities [97]. However, no data is currently available to show if such modification holds for more complex *in vivo* environments since cells could express more than one type of integrin and also because the integrin expression pattern of a cell is a highly dynamic phenomenon.

The surface density of RGD on the material also has a profound effect on the number of cells attached as well as cell spreading, cell survival, focal contact formation and to some extent proliferation. Studies have shown a sigmoidal increase in cell attachment with RGD concentration on the surface, indicating a critical minimum density for cell response [94]. Thus Neff et al. demonstrated that maximum proliferation of fibroblast occurred on surfaces with intermediate surface concentration (~ 1.33 pmol cm^{-2}) [79]. Another study by Massia and Hubbell using RGD functionalized glycophase glass surface has shown that a minimal amount, as low as 1 fmol RGD peptide cm^{-2}, is sufficient for cell spreading on the surface and as low as 10 fmol cm^{-2} sufficient for formation of focal contacts and stress fibers [98].

However, a higher RGD peptide concentration requirement has been reported for polymers and has been attributed to the entropic penalty that results from attachment of a peptide to flexible polymer chain compared to a rigid glass surface as well as the inefficient transmission of forces through polymer surfaces [75, 99]. Studies have shown that in addition to surface concentration, the mode of presentation of ligands also could affect integrin behavior [100]. One study by Maheshwari et al. evaluated surfaces with controlled overall peptide density and controlled nanoscale spatial ligand distribution with an overall RGD distribution of 0.15–20.50 nmol cm^{-2} [101]. The results demonstrated that a significantly higher fraction of fibroblasts showed higher shear stress resistance and exhibited well-formed stress fibers and focal contacts when the ligand was presented in a clustered versus a random individual format. The use of a higher affinity peptide GRGDSPK afforded a lower RGD density of 0.06–0.88 nmol, showing that activity of the RGD is also very important [102].

Nanoscale RGD clustering on the surface of biomaterials seems to be a promising approach to elicit favorable cell responses with minimal amounts of RGD peptides. Studies are ongoing to determine the technique that could be used to create nanoscale clustering on the surface as well as to determine which arrangement elicits a particular cell response.

Thus the foremost challenge in developing a tissue engineered construct is the development of a resorbable synthetic microenvironment that could closely mimic the complex hierarchical micro-nano architecture of the ECM along with the molecular level spatial organization of biological cues found in native tissue *in vivo*.

1.5
Applications of Nanotechnology in Developing Scaffolds for Tissue Engineering

Nanotechnology has been defined as "research and technology development at the atomic, molecular and macromolecular levels in the length scales of approximately 1–100 nm range, to provide a fundamental understanding of phenomena and materials at the nanoscale and to create and use structures, devices and systems that have novel properties and functions because of their small and/or intermediate size" [103]. Nanotechnology has emerged as an exciting field that deals with both the design and fabrication of structures with molecular precision. Nanotechnology enables the control and manipulation of individual constituent molecules/atoms to have them arranged to form the bulk macroscopic substrate. The uniqueness of the nanotechnological approach is that it considers spatial and temporal scales at the same time, thereby forming an excellent technique to develop hierarchical structures. The biological milieu that tissue engineers attempt to mimic using synthetic materials and techniques is a highly complex system with spatial and temporal levels of organization that span several orders of magnitude, with different levels nested within higher order levels (nm to cm scale). To study and mimic this complex system, highly sophisticated technology is required. For instance, the visualization and characterization of these biological structures, processes, and their manipulation require sophisticated imaging and quantitative techniques with spatial and temporal control at or below the molecular level.

Recent developments in nanotechnology have revolutionized the visualization and characterization of biological processes in various ways. The capability of imaging living cells after implantation is very crucial in studying cell behavior and processes *in vivo*. The recent developments in nanotechnology assisted fluorescent probes such as quantum dots (QD) have significantly improved our capability of *in vivo* imaging. QDs are nanocrystals or nanoparticles with size ranging from 1 to 10 nm with unique photophysical and photochemical properties not available with conventional organic fluorophores [104].

Similarly, the scanning probe microscopic techniques (SPM) provide a great tool to investigate atomic and molecular level biological phenomena even though its potential in biology is yet to be realized. One of the most extensively investigated SPM techniques for tissue engineering application is atomic force microscopy (AFM). AFM has provided various strategies to investigate the interactions of living cells with the ECM [105]. AFM has also enabled the visualization of nano-scale biomolecules and significantly contributed to the in-depth understanding of their structure and role in biological process [106].

The developments in current nanoscale fabrication techniques have also significantly increased our understanding of nanoscale features on cellular behavior and tissue organization. Several nanoprinting/etching/electron beam lithographic techniques have been developed to form substrates with large areas of controlled nanoscale features. *In vitro* studies using these substrates confirmed the importance of nanoscale topography of scaffolds for developing tissue *in vitro* [107–110]. One study examined the interaction of fibroblasts with nanoscale islands having heights varying from 10 to 95 nm on polymer films. The fibroblasts underwent rapid organization of cytoskeleton and improved adhesion during initial reaction to the islands with concomitant cell spreading. The lamellae of the cells on the islands also showed many filopodia showing better interaction with the islands. Another study by Dalby et al., using nano- and micro-patterned surfaces has demonstrated the importance of nanoscale features in modulating human mesenchymal bone marrow stromal cell (HBMSC) adhesion [111]. HBMSCs were found to be well-spread and attained normal morphologies on polymer thin films similar to the morphology cells attained on flat topographies. However, the cells on nanofeatured surfaces were found to respond to the nanofeatures. This included cells conforming to the shape of the nanosized pits (Fig. 1.15A – 310 nm deep and 30 μm wide), filopodia production, contact guidance and production of endogenous extracellular matrix. On nanometer depth grooves, the cells were found to be highly aligned along the groove direction showing pronounced contact guidance (Fig. 1.15B – 500 nm deep and 5 μm wide). The study demonstrated that the nanoscale features of the substrates could elicit significant control over cell adhesion, cytoskeletal organization, cell-growth, and production of the osteoblastic markers osteocalcin and osteopontin [111].

To recreate structures having features at the nanoscale level, novel nanotech-

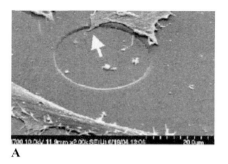 

A   **B**

Fig. 1.15. Scanning electron micrographs of HBMSCs cultured on polymer surface with pits and grooves having nanometer depth. (A) Cells conforming to a groove edge of a nanopit (arrow). (B) Contact guidance of cells and their filopodia on the narrow grooves. (Adapted from Ref. [111] with permission from Elsevier.)

niques that enable the conversion of existing macromolecules into nanostructured forms or development of novel structures from atomic or molecular constituents with spatial organization of biofunctionality are needed. It is presumed that these developed nanostructures, due to their ability to interact with cells and tissues at a molecular (subcellular) level with a high degree of functional specificity, would allow a greater extent of integration than previously attainable. Thus, research in this direction is ongoing to develop structures that could temporarily mimic the structure and functions of the ECM as ideal scaffolds for tissue engineering using various nanofabrication processes.

Nanofabrication techniques have shown the feasibility to develop nanostructured scaffolds that better mimic the structure of the ECM compared to the structures developed by macro/micro fabrication techniques. Two different approaches are currently under investigation to develop synthetic nanostructured scaffolds that could resemble the structure of nanoscale collagen fibrils of the ECM as scaffolds for tissue engineering. The first approach can be considered as a "top-down approach" which uses synthetic polymeric materials to develop nanostructures using various nanofabrication processes. The second approach can be considered as a "bottom-up approach" and is based on short peptides or block polymers that can assemble into nanofibers by a self-assembly process.

1.5.1
Polymeric Nanofiber Scaffolds

As discussed earlier, collagen fibrils are the major building blocks of the natural ECM and they have diameters in the range 50–500 nm and orientation in different directions depending on the tissues. A logical method to develop scaffolds for tissue engineering is to mimic the structure of collagen fibrils, i.e., by using synthetic polymeric nanofiber matrices. Developments in nanofabrication techniques have enabled the fabrication of synthetic nanofiber matrices from a wide range of polymers. Polymeric nanofibers have been defined as fibers having diameters less than 1 μm and are developed from synthetic and natural polymers [112]. Fibers with diameters ranging from 1–1000 nm and a very high surface area can be developed by the nanofabrication processes. Thus a nanofiber with a diameter of 100 nm has a specific surface area of 1000 m^2 g^{-1} [113]. Porous matrices developed using polymeric nanofibers have excellent structural and mechanical properties, high axial strength combined with extreme flexibility, high surface to volume ratio, high porosity (>70%), and variable pore sizes – all of these properties are highly beneficial for cell adhesion, migration and proliferation.

1.5.1.1 Top-down Approaches in Developing Scaffolds for Nano-based Tissue Engineering

The top-down approach is considered as a classical approach used to size down macrostructures to smaller sizes using various fabrication techniques. Several top-down techniques have been developed to form polymeric nanofibers from pre-formed macromolecules such as electrospinning, phase separation and templating [112, 114, 115].

Polymeric Nanofibers by Electrospinning Electrospinning has developed into a promising, versatile and economical technique to produce nanostructured scaffolds for tissue engineering [112, 116]. Figure 1.16 shows the schematic of the electrospinning process. Briefly in an electrospinning process an electric field is applied to a pendant droplet of polymer solution at the tip of a needle or capillary attached to a syringe or pipette. The polymer solution feed to the needle/capillary is controlled using a syringe pump or allowed to flow under gravity. The electrode can be either inserted in the polymer solution or connected to the tip of the needle. When an electric potential is applied to the droplet, the droplet will be subjected to couple of mutually opposing forces. One set of forces (surface tension and viscoelastic forces) tend to retain the hemispherical shape of the droplet and another set

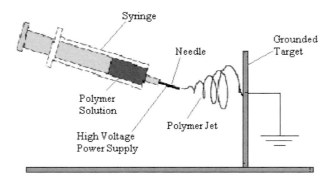

Fig. 1.16. Scheme of the electrospinning process.

of forces (due to the applied electric field) tend to deform the droplet to form a conical shaped "Taylor cone". Beyond a threshold voltage, the electric forces in the droplet predominate and at that point a narrow charged polymer jet will be ejected from the tip of the Taylor cone. However, the viscosity of the polymer solution plays a crucial role in maintaining the ejected jet. If the viscosity of the polymer solution is low, the ejected jet break into droplets by a process called "electrospraying". For solutions with higher viscosities, the ejected jet travels in a nearly straight line towards the grounded collector for some time due to the stabilization imparted by the longitudinal stress of the external electrical field on the charge carried by the jet. However, at some point along the course, the jet reaches a point of instability due to the repulsive forces arising from the opposite charges in the jet. The unstable jet then passes through a series of bending instabilities and it tends to bend back and forth following a bending, winding, spiraling and looping path in three dimensions. This bending instability of the jet has been demonstrated using high speed videography. During this process the polymer jet is continuously stretched resulting in significant reduction of the fiber diameter. This, along with the rapid evaporation of the solvent from the ultrathin jets results in the formation of ultrathin fibers that are deposited on a grounded collector surface [117–122].

Extensive studies have been performed to investigate the fundamental aspects of the process of electrospinning to determine the parameters that modulate the morphology and diameter of the electrospun fibers and for determining appropriate conditions for developing fibers from a wide range of polymers [112, 123–126]. These studies have clearly demonstrated the flexibility of the electrospinning process. Electrospun nanofiber scaffolds can be developed from a wide range of polymers with varying physical, chemical, and mechanical properties, thereby creating scaffolds with varying strength, surface chemistry, degradation patterns (in the case of matrix developed from bioresorbable polymers) and physical properties. The electrospinning process also enables co-spinning two or three different polymers, which further extends the ability to control the properties of the resulting scaffolds/matrices. Another advantage of electrospinning process is the feasibility of developing composite nanofiber scaffolds/matrices by incorporating small insoluble particles such as drugs or bioactive particles within polymeric nanofibers. Since the shape of the nanofiber scaffold/matrix depends on the properties of the collector, complex and seamless 3D structures can be developed using the appropriate collectors.

Parameters that Affect the Electrospinning Process Extensive studies have been undertaken to determine the parameters/variables that affect the electrospinning process. These include system parameters, solution properties and processing variables. The system parameters include the nature (chemistry and structure) of the polymer, molecular weight of the polymer and molecular weight distribution of the polymer. Solution properties include viscosity, elasticity, conductivity and surface tension of the polymer solution. The processing variables in the electrospinning process are electric potential, flow rate of the polymer solution, concentration of the polymer solution, the distance between the tip and the target, ambient parameters such as solution temperature, humidity, air velocity in the electrospinning chamber, and motion of the target screen [112, 124, 126].

Most studies correlated the electrospinning parameters/variables to fiber diameter and/morphology. The effect of molecular weight of polymer on the process of electrospinning was evaluated using poly(ethylene oxide)s (PEO) of different molecular weights electrospun under identical conditions and by following the morphology of the fibers [127]. Figure 1.17(A–C) shows the effect of polymer molecular weight on the morphology of resultant nanofibers. In this study viscosity, surface tension and conductivity of all the solutions were kept constant to correlate the morphology of fibers to the molecular weight. Electrospinning of the low molecular weight polymer (20 000) resulted in the formation of mostly beads rather than fibers (Fig. 1.17A). Increasing the molecular weight to 500 000 resulted in the formation of fibers, however, with spindle shaped defect structures or beads (Fig. 1.17B). A further increase in molecular weight to 4×10^6 resulted in the formation of bead-free fibers (Fig. 1.17C). The formation of bead-free structures with high molecular weight PEO has been attributed to the increasing entanglement of the polymer chains with high molecular weight polymer.

The effect of electric potential and polymer concentration on the morphology and diameter of electrospun polymer fibers were demonstrated by Katti et al. using a bioresorbable polymer poly(lactide-*co*-glycolide) (PLAGA) [128]. PLAGA dissolved in a dimethyl formamide–tetrahydrofuran (1:3) mixture was used for electrospinning. The study demonstrated that the concentration of the polymer solution has a significant effect on the diameter and morphology of the electrospun fibers. Figure 1.18(A–C) shows the morphologies of PLAGA fibers formed from polymer solutions having different concentrations. A lower polymer concentration (0.15 g mL^{-1}) resulted in the formation of beaded nanofibers (Fig. 1.18A). Increasing the polymer concentration (0.2 g mL^{-1}) significantly reduced the probability of bead formation (Fig. 1.18B). The concentration also showed significant effects on the diameter of the resulting fibers. At low concentration (0.15 g mL^{-1}), fibers having diameters $\sim$270 nm with beads were formed and increasing the concentration to 0.2 g mL^{-1} increased the diameter of the fibers to $\sim$340 nm with minimal amount of beads. A further increase in concentration to 0.25 g mL^{-1} resulted in the formation of fibers having diameters $\sim$1000 nm with apparently no bead formation (Fig. 1.18C). Increase in fiber diameter with increasing solution concentration followed a power law relationship. The same behavior has been observed for different polymer systems by other investigators [123, 129]. Demir et al., using polyurethane solution, have shown that fiber diameter can be correlated to polymer concentration as proportional to the cube of the polymer concentration [130].

Similarly the electrospinning voltage has a profound effect on the morphology and diameter of the fibers. An increase in electric voltage decreases the diameter of electrospun fibers up to a certain voltage and above that tends to increase the fiber diameter [126, 128]. Spinning voltage has been found to strongly correlate with the formation of beads. Deitzel et al. have shown that an increase in electrical potential increases the feasibility of formation of beads along the fibers [131].

In addition to varying the polymer concentration and the electric potential the

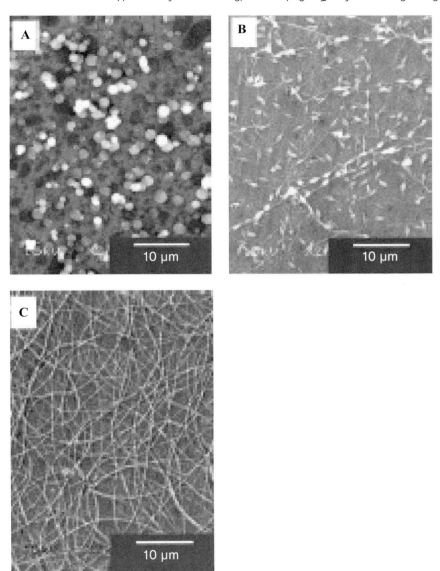

Fig. 1.17. Morphology of fibers formed by the electrospinning process using poly(ethylene oxide)s of different molecular weights: (A) 20 000, (B) 500 000, and (C) 4×10^6. (Adapted from Ref. [127] with permission from Elsevier.)

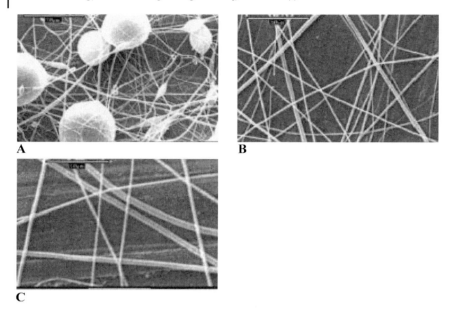

Fig. 1.18. Morphology of fibers formed by the electrospinning of PLAGA solutions having varying polymer concentrations (g mL^{-1}): (A) 0.15, (B) 0.2, and (C) 0.25. (Adapted from Ref. [128].)

diameter and morphology of electrospun polymer nanofibers can be modulated by the addition of various additives to the spinning solution. One study by Zong et al. has shown that addition of ionic salts to the polymer solution can significantly reduce the bead formation and could result in thinner fibers. This has been attributed to the higher charge density of the jet due to the presence of ionic salts. The higher the charge carried by the jet, the greater will be the pull or elongation force the jet will experience under the electrical field, resulting in fewer beads and thinner fiber [132].

The electrospinning process can also result in the formation of fibers having various cross-sectional features in addition to circular fibers, as demonstrated in the case of various polymers. Fibers having varying shapes have been created such as flat ribbon, bent ribbon, ribbons with other shapes, branched fibers and fibers that were split longitudinally from larger fibers from different polymers and polymer–solvent systems [133]. Figure 1.19 shows the SEM of fibers having varying cross-sectional shapes. The occurrence of skin on the polymer jet accounts for a number of these observations. The phenomenon has been attributed to various causes, including contribution of fluid mechanical effects, electrical charge carried by the jet, and evaporation of solvent from the jet. In addition to varied cross-sectional features, fibers have been found to form with surfaces having varying nanotopographies such as nanopores or ridges. The formation of these structures

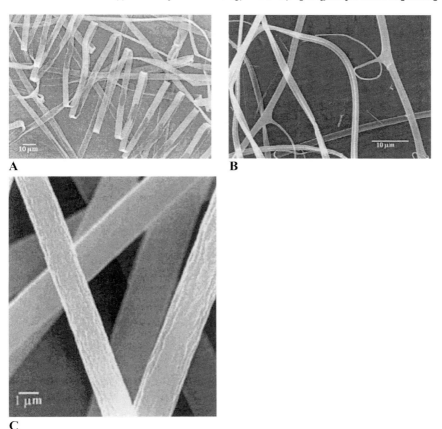

Fig. 1.19. SEM showing the fibers having varying cross-sectional shapes. (A) Flat ribbon formed from 10% solution of poly(ether imide). (B) Branched fibers from 16% HEMA. (C) Round fibers with skin collapsed to form longitudinal wringles. (Adapted from Ref. [133].)

have been attributed to various parameters such as the nature of the solvent, glass transition temperature of the polymer, solvent–polymer interactions, and environmental parameters such as humidity and temperature [134–136].

Nanofibers deposited by electrospinning using a static target result in the formation of a non-woven matrix composed of randomly oriented fibers. However, properties of fiber matrices can evidently be improved, if the fibers could be aligned in appropriate directions. This is particularly important for developing scaffolds for tissue regeneration as this would enable the development of scaffolds with specific orientation and architecture. Aligning the fibers formed by the process of electrospinning, however, is very difficult to achieve because during electrospinning process the jet trajectory follows a complex 3D whipping and bending path towards

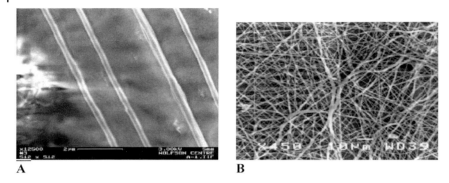

A   B

Fig. 1.20. SEM showing PEO nanofibers: (A) Aligned fibers
and (B) randomly deposited fibers. (Reprinted from Ref. [139]
with permission from Institute of Physics Publishing.)

the target rather than a straight line. Several attempts have been performed to develop aligned electrospun polymeric nanofibers. Earlier attempts were performed using a high speed rotating cylinder collector [137]. However, fiber alignment could be achieved only to a certain extent using this process, presumably due to the low control that can be achieved over a polymer jet that undergoes chaotic motion. In another attempt, an auxiliary electrical field was applied to align the fibers which substantially improved the fiber alignment [138]. Another successful approach to align electrospun nanofibers was developed by Theron et al. using a thin wheel with sharp edge device. The thin edge of the wheel helped to concentrate the electrical field so that almost all of the spun nanofibers were attracted and wound to the bobbin edge of the rotating wheel [139]. Figure 1.20(A and B) shows the SEMs of aligned PEO nanofibers developed using the thin wheel with sharp edge collector and randomly deposited fiber matrix using conventional static collector. Another approach investigated is the use of a frame collector and has been found to significantly improve the alignment of nanofibers [140]. In this process, however, the extent of alignment depends significantly on the frame material. Figure 1.21 shows the SEM of aligned PLLA-CL copolymer fibers formed by the frame method.

In addition to the above techniques, several processing techniques were also attempted to increase the versatility of the electrospinning process. These include electrospinning the mixture of polymer with sol–gel solution [141], electrospinning blend polymer solutions [142], electrospinning polymer solution containing nanomaterials to form composite matrices [126], core–shell nanofiber spinning [143] and side by side/multijet electrospinning of different polymers (to increase the rate of fiber deposition and develop matrices having unique properties such as biohybrid matrices) [144–146].

The electrospinning process is a very mild fabrication process. This makes it very attractive for developing structures for biomedical applications. Studies have shown the ability of electrospun fibers to preserve the biological activity of highly sensitive biomolecules encapsulated within the fibers during the electrospinning

Fig. 1.21. Aligned PLLA-CL nanofibers formed by the frame method. (Adapted from Ref. [140] with permission from Elsevier.)

process. Hamdan et al. have encapsulated RNase and trypsin in poly(2-ethyl-2-oxazoline) nanofibers by electrospinning [147]. The enzymes were found to preserve their biological activity after being encapsulated within the nanofibers. Another study by Jiang et al. demonstrated the feasibility of incorporating model proteins such as bovine serum albumin and lysozyme within PCL nanofibers with preservation of their biological activity [148]. A novel electrospinning fabrication process called coaxial spinning was recently developed to encapsulate water-soluble macromolecules within hydrophobic nanofibers [143]. The fibers developed are called core–shell nanofibers where the aqueous phase containing the protein solution forms the core of the fiber surrounded by the hydrophobic polymer layer. The thickness of the core and the shell can be adjusted by the feed rate of the inner dope. Circular dichroism and SDS-PAGE studies on the released lysozyme and BSA encapsulated in the core–shell fibers revealed that both the proteins maintained their structure and bioactivity after encapsulation. The core–shell structures could have several advantages, such as improved protection of the encapsulated molecules and feasibility of achieving their controlled delivery when used as a macromolecular delivery vehicle for biomedical applications, including tissue engineering.

The mechanical properties of non-woven nanofiber matrices developed from bioresorbable polymers have been investigated. Thus, nanofiber matrices developed from PLAGA with LA:GA ratio of 85:15 showed a tensile strength similar to that of natural skin [149]. Ding et al. have demonstrated that the mechanical properties of non-woven nanofiber matrices could be further modulated by developing blend nanofibrous matrices using multi-jet electrospinning [144]. Another study by He et al. has demonstrated the low stiffness of non-woven polymeric nanofiber matrices compared to large diameter dacron grafts using poly(L-lactic acid-co-caprolactone) copolymer (PLLA-CL) nanofiber matrices. The low stiffness of the matrix makes it a suitable candidate for vascular graft applications [150]. The nanofiber matrix developed from PLLA-CL showed an ultimate strain of 175 ± 49%. This high distension property has been attributed to the ability of the randomly oriented fibers to rearrange themselves in the direction of the stress. Another study compared the mechanical properties of nanofiber matrices with microfiber matrices developed from poly(L-lactide-*co*-caprolactone) [PLCL] by the electrospinning process [151]. Three matrices composed of fibers with diameters of ~0.3, ~1.2, and ~7 μm were investigated. The differences in fiber diameter could lead to differences in specific density of the resultant matrices. The matrix composed of the smallest diameter fibers (~0.3 μm) gave the densest matrix. Also, the Young's modulus of the densest matrix (~0.3 μm) was the highest followed by the matrix composed of fibers having diameter 1.2 μm. The matrix composed of fibers with diameter ~7 μm showed the lowest Young's modulus. This indicates that nanofiber matrices could show better mechanical performance than microfibrous matrices of the same material presumably due to the increase in fiber density. In terms of the mechanical properties of aligned nanofiber matrices developed by electrospinning, the nanofiber matrices showed different properties along different directions [152]. The ultimate strength of the aligned polyurethane fiber matrices (3520 ± 30 kPa) was significantly higher than randomly deposited fiber matrices (1130 ± 21 kPa) [153].

In summary, these studies demonstrate the feasibility of developing polymeric nanofibers with diameters ranging from 1 to 1000 μm from a wide range of polymers using the process of electrospinning. The diameter and morphology of the nanofibers can be controlled to a great extent by varying the process parameters/variables that govern the electrospinning process. The properties of the matrices fabricated using polymeric nanofibers can be modulated by varying the properties of the polymer, co-spinning or multiple spinning polymer mixtures, incorporating nanoparticles or fillers, varying the rate of deposition as well as varying the properties of the target. The feasibility of aligning the fibers significantly increases the ability to modulate the properties of nanofiber matrices for biomedical applications. Another notable advantage of the electrospinning process is the cost effectiveness compared to other nanofabrication techniques.

Polymeric Nanofibers by Phase Separation   Phase separation is another type of top-down approach used to develop polymeric nanofiber matrices from different polymer solutions. This technique has been found to be effective in developing nanofibrous matrices having high porosities (up to 98.5%) from biodegradable polymers and has been investigated for tissue engineering applications [154].

Liquid–liquid phase separation can be achieved by lowering the temperature of a polymer solution having an upper critical solution temperature. The phase separation at low temperature could lead to the formation of a continuous polymer-rich and polymer-lean solvent phases. The removal of solvent from the phase separated system at low temperature affords a scaffold with an open porous structure [154, 155]. Thus the development of nanofibrous porous matrices using phase separation of polymer solution takes place in five steps. Polymer dissolution in a solvent system, phase separation and gelation, solvent extraction from the gel with water, freezing and freeze drying under vacuum [114]. Similar to the electrospinning process, various processing variables can be controlled to modulate the properties of the nanofibrous matrices formed by phase separation process. These include type of solvent and polymer, polymer concentration, solvent exchange, thermal treatment and order of procedures [114].

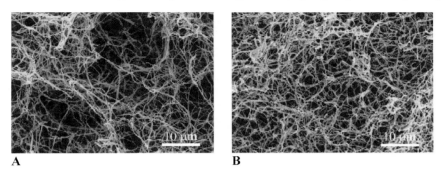

Fig. 1.22. SEM micrographs of PLLA fibrous matrices prepared from PLLA/THF solution with different PLLA concentrations at a gelation temperature of 8 °C: (A) 1% and (B) 5% (w/v). (Adapted from Ref. [154].)

Ma and Zhang have developed 3D continuous nanofibrous structures that could mimic the structure of the ECM using biodegradable poly(L-lactic acid) [PLLA]. Fibers of the nanofiber matrix exhibited diameters in the range 50–500 nm. Figure 1.22(A and B) shows the SEM of PLLA fibrous matrices prepared from 1% (w/v) and 5% (w/v) PLLA/THF solution. The figures demonstrate the feasibility of varying the porosity of matrices by varying the concentration of the polymer solution. The higher the concentration of the solution, the lower was the porosity of the matrices formed.

Another advantage of fabricating nanofibrous matrices using the phase separation technique is that it allows for incorporation of macropores along with nanopores in the matrices by adding various porogens such as sugars, inorganic salts or paraffin spheres to the mold with the polymer solution during phase separation [156]. Figure 1.23(A and B) shows macro-nano porous matrices of PLLA fabricated by the combined porogen leaching phase separation method. Thus the phase separation process is a mild processing technique that provides the flexibility to control the properties of the nanofiber matrices such as fiber diameter, interconnectivity, porosity and size of the pores.

1.5.1.2 Bottom-up Approaches in Developing Scaffolds for Nano-based Tissue Engineering

Bottom-up approaches are based on self-assembly, a ubiquitous natural phenomenon that harnesses the physical and chemical forces operating at the nanoscale to assemble small building blocks into larger structures. Thus, the basic principle of the bottom-up approach is molecular self-assembly, which is the spontaneous organization of molecules under near thermodynamic equilibrium conditions into structurally well-defined and stable arrangements through non-covalent interactions [157]. These interactions include weak non-covalent bonds, such as hydrogen bonds, ionic bonds, hydrophobic interactions, van der Waals interactions and water-mediated hydrogen bonds [158]. The development of a self-assembling sys-

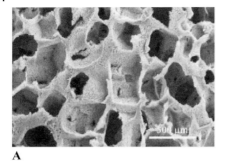
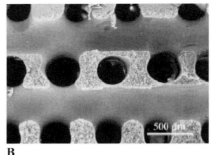

A   **B**

Fig. 1.23. (A) SEM micrograph of PLLA nano-fibrous matrix with particulate macropores prepared from PLLA/THF solution and sugar particles; particle size 250–500 mm. (B) SEM micrograph of PLLA nano-fibrous matrix with an orthogonal tubular macropore network prepared from PLLA/THF solution and an orthogonal sugar fiber assembly. (Adapted from Ref. [156].)

tem requires the design and development of small building blocks that can spontaneously self-assemble and be stabilized to form functional nano/microstructures. Recently, there has been significant interest in using self-assembly to develop nanostructured scaffolds for tissue engineering.

The most extensively investigated self-assembled nanostructured scaffolds for tissue engineering application are developed from peptide molecules. In 1993, Zhang et al. demonstrated the feasibility of an aqueous solution of a 16-residue ionic self-complementary peptide to spontaneously associate to form a macroscopic membrane [159]. The ionic complementary oligopeptides used in the study have regular repeating units of positively charged residues (lysine or arginine) and negatively charged residues (aspartate or glutamate) separated by hydrophobic residues (alanine or leucine). The membrane was highly stable to varying pHs and temperatures [159]. SEM of the self-assembled membrane revealed that the structure is composed of nanofibers with diameters ranging from 10 to 15 nm. The study raised interest in self-assembling peptide motifs and led to the development of different self-assembling peptide structures capable of assembling into unique nanostructures. Thus different types of self-assembling peptides have been identified to form different self-assembled structures such as nanofibers, nanotubes, nanowires and nanocoatings. Among these the Type I peptides also called molecular Legos developed by Zhang et al. have been identified as a potential peptide motif for developing self-assembled scaffolds for tissue engineering. These peptide motifs are called molecular Legos because at the nanometer scale they resemble the Lego bricks that have pegs and legs in a precisely determined organization. These peptides form β-sheet structures in aqueous solution resulting in distinct hydrophilic and hydrophobic surfaces [158–160]. The hydrophobic surface shields the peptide motif from water, thereby enabling them to self-assemble as in the case of protein folding *in vivo*. Then complementary ionic bonds will be formed with regular repeats on the hydrophilic surface. The complementary ionic sides have been classified into different moduli based on the chemistry of the hydrophilic surface, i.e., having alternating positive and negative charged amino acid residues with different intervals. Depending on the moduli, these molecules could under go ordered self-assembly to form nanofibers. These nanofibers, due to their high aspect ratio, in turn associate to form nanofiber scaffolds that closely mimic the porosity and gross structure of the ECM, making them potential candidates as tissue engineering scaffolds [161]. Figure 1.24(A) illustrates the formation of nanofibrous structures using molecular Lego. "PuraMatrix" a commercially developed ECM mimic self-assembling molecular Lego system has been found to be suitable for performing 3D tissue culture *in vitro*. The nanofiber scaffolds formed from these peptide motifs are formed of interwoven nanofibers with diameter of ~10 nm and pores of ~5–500 nm with very high water content (>99.5%) (Fig. 1.24B).

Naturally occurring amino acids were used in developing peptide motifs in these studies. However, studies by Stupp et al. demonstrated the feasibility of using building blocks other than natural amino acids to create amphiphilic peptides. The peptide amphiphiles were developed from appropriate amino acids using solid phase peptide chemistry and the NH terminus of the peptide sequence was then alkylated to form the amphiphilic molecule. The amphiphilic peptides (PA) molecules are composed of a peptide segment containing 6–12 amino acids coupled via an amide bond to a fatty acid chain that varies in length from 10 to 22 carbon atoms. Even at very low concentrations of 0.25% (w/v) these molecules can self-assemble to form a gel structure composed of a network of cylindrical nanofibers with diameter ranging from 5 to 8 nm, depending on the length of the self-assembling molecules. The matrices were highly hydrated (>99.5%) and the mechanical integrity of such a highly hydrated matrix has been attributed to the high aspect ratio of the nanofibers composing the matrix. These molecules are custom developed so that they can self-assemble to form nanostructured scaffolds that could structurally and biologically mimic the structure of the ECM of specific tissue type. Thus, a composite nanostructured matrix has been developed as potential scaffolds for bone tissue engineering. The scaffold was designed to mimic the ECM of natural bone by self-assembling peptide motifs with appropriate amino acids and mineralizing the matrix *in vitro* [162]. In this study the peptide amphiphile was designed as follows. To make robust nanofibers four consecutive cysteine amino acids were incorporated in the sequence. The cysteine residues were incorporated as they could form disulfide linkages between adjacent molecules upon oxidation to stabilize the supramolecular structure. The formation of the disulfide linkage, however, is a reversible process. A phosphoserine residue was incorporated into the peptide sequence, so that after self-assembly the resulting fibers will have highly phosphorylated surface. These groups are specifically incorporated in the PA to increase the mineralizing capacity (nucleation and deposition of hydroxyapatite an essential inorganic component of natural bone) of the nanostructured scaffold. Anionic groups are known to promote nucleation and deposition of hydroxyapatite on synthetic materials and phosphorylated groups are particularly important in the formation of calcium phosphate minerals. Thus, the phosphorylated surface of self-assembled nanofibers could promote the nucleation and deposition

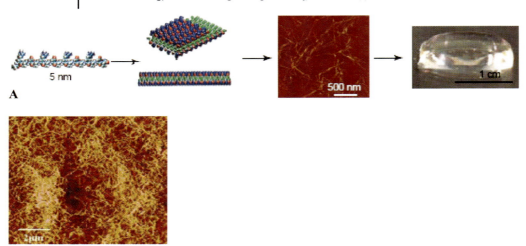

A

B

Fig. 1.24. Fabrication of various peptide materials. (A) Peptide Lego, also called ionic self-complementary peptide, has 16 amino acids, ~5 nm in size, with an alternating polar and nonpolar pattern. The peptide motifs could form stable β-strand and β-sheet structures. The peptide motifs undergo self-assembly to form nanofibers with the nonpolar residues inside (green) and positive (blue) and negative (red) charged residues forming complementary ionic interactions, like a checkerboard. These nanofibers form interwoven matrices that produce a scaffold hydrogel with very high water content (~99.5%). (Adapted from Ref. [161] with permission from Elsevier.) (B) AFM image of the nanofiber scaffold (PuraMatrix). (Adapted from Ref. [161] with permission from Elsevier.)

of hydroxyapatite (HA). To improve the cell adhesivity of the self-assembled nanofiber matrix, an RGD sequence was also incorporated in the peptide. Figure 1.25 shows the structure of the PA molecule designed to self-assemble to form nanofiber scaffold for bone tissue engineering.

The nanofiber scaffold was developed from the PA as follows. The cysteine residues of the PA were first reduced to thiol groups at higher pH. The resulting PA was found to be highly soluble in water. The pH of the aqueous solution was then reduced to 4.0, at that point the material rapidly became insoluble due to the formation of network structure. Cryo-transmission electron microscopy (Cryo-TEM) showed that the gel is composed of a network of fibers that are 7.6 ± 1 nm in diameter and up to several micrometers long. Figure 1.26(A and B) shows the ultrastructure of the gels formed by self-assembly with and without covalent stabilization. The ability of the nanofiber matrix formed from the PA to nucleate hydroxyapatite (HA) along the fiber axis was also demonstrated by incubating the nanofiber matrix in appropriate salt solution (Fig. 1.26C). Another interesting property of the self-assembled gel is its reversibility. The self-assembled matrix could disassemble at higher pHs. Even though the study demonstrated for the first time the feasibility of designing and developing bioactive self-assembled system to mimic the properties of the ECM, it has certain disadvantages. The significant

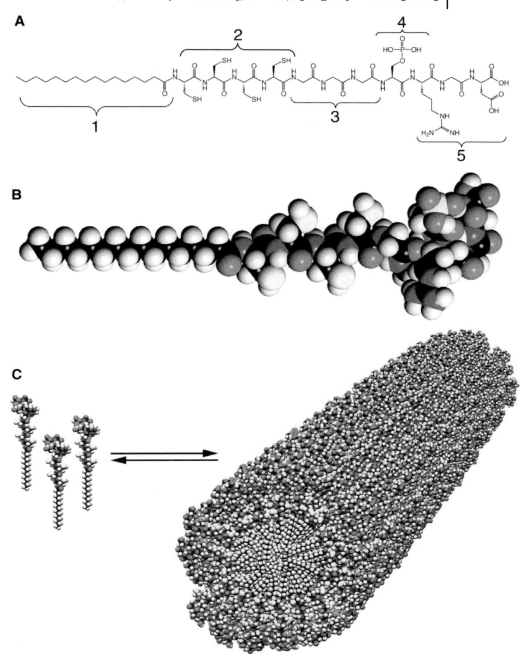

Fig. 1.25. (A) Chemical structure of the peptide amphiphile. Region 1 is a long alkyl tail to make the peptide motif amphiphilic. Region 2 is composed of four consecutive cysteine resides for disulfide linkages. Region 3, a flexible linker region of three glycine residues, provides hydrophilic head group flexibility from the rigid crosslinked region. Region 4 is a single phosphorylated serine residue. Region 5 is a cell adhesion RGD ligand. (B) Molecular model of PA. (C) Scheme showing the self-assembly. (Reprinted with permission from Ref. [162].)

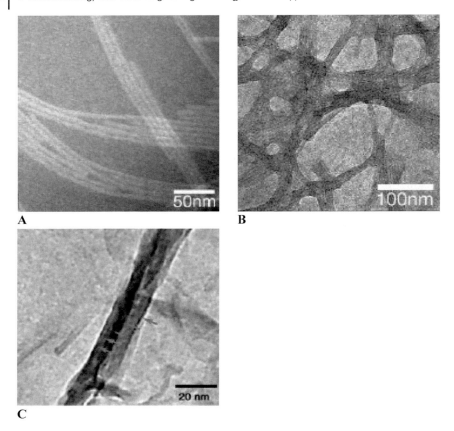

Fig. 1.26. TEM of self-assembled nanofibers (A) before and
(B) after covalent stabilization. (C) TEM showing PA nanofibers
completely covered by mature hydroxyapatite crystals.
(Reprinted with permission from Ref. [162].)

disadvantage of this system is the low stability of the self-assembled structure at physiological pH unless internally crosslinked by covalent bonds. Another study was therefore performed to demonstrate the feasibility of developing self-assembled structures that are stable at physiological pH using PAs with opposite charges based on the electrostatic attraction of the opposite charge [163]. Mixed systems having oppositely charged PAs were used to develop self-assembled systems capable of assembling at physiological pH due to electrostatic attraction. This ability of these materials to undergo mild self-assembly and gelation at physiological conditions and the ability to decorate them with bioactive motifs makes them potential candidates for various biomedical applications.

In addition to peptide motifs, synthetic proteins were also investigated to develop self-assembled matrices. Thus Petka et al. used a recombinant DNA method to create artificial proteins that can undergo reversible gelation in response to pH or temperature. The developed proteins consist of terminal leucine zipper domains flanking a central flexible water-soluble polyelectrolyte segment [164]. In near neutral solution, a 3D network can be formed by coiled-coil aggregates of the terminal domains and the polyelectrolyte segment prevent precipitation of the chain and retain the solvent. An elevation of pH or temperature leads to dissolution of the gel, resulting in the formation of viscous polymer solution. Similarly Nowak et al. developed a diblock copolypeptide amphiphile containing charged and hydrophobic segments, which were found to form hydrogels with high temperature stability [165]. The ability of this system to form gel under mild conditions makes them potential candidates for biomedical applications.

Both top-down and bottom-up approaches have significantly contributed to the development of nanostructured matrices as scaffolds for tissue engineering. The top-down approaches currently used to develop nanofiber matrices such as electrospinning or phase separation are highly economical and easily scalable processes. Furthermore, nanofiber scaffolds with finely controlled physical and mechanical properties and complex structures can be developed due to the versatility of the fabrication processes. The bottom-up approach even though is a more involved process that requires highly specific building blocks for spontaneous self-assembly has several advantages from a biomaterials point of view. The matrix formation via a mild self-assembling process makes it a very attractive process for *in vivo* applications. Another advantage is the ease of incorporating specific bioactive motifs as the building blocks that when combined with the nanostructured topography of the resulting matrix could better mimic the structure and functions of the extracellular matrix. The major disadvantage of the self-assembly process, however, is its relative inability to generate complex patterns for biological devices due to its homogeneous character.

1.6
Cell Behavior Towards Nano-based Matrices

Several studies have confirmed the fact that cells prefer to live in a complex nanostructured environment composed of pores, ridges and fibers of the polymeric nanofiber matrices that mimics the structure of the ECM compared to 2D matrices or microfiber matrices. Kwon et al. have evaluated the adhesion and proliferation potential of human umbilical vein endothelial cells (HUVEC) on three different types of PLL-CL fiber matrices composed of fibers having diameters ~0.3, ~1.2, and 7 μm [151]. Figure 1.27 shows the SEMs of HUVECs cultured for 1 and 7 days on electrospun matrices having fibers of different diameters (0.3, 1.2, and 7 μm). The matrix composed of the smallest diameter fibers (0.3 μm) and the medium diameter fibers (1.2 μm) showed higher cell adhesion and proliferation than matrix composed of fibers having diameter 7 μm. The morphology of the cells on 0.3 and 1.2 μm fiber matrices were comparable but differed significantly from that on the matrix composed of 7 μm diameter nanofibers. Thus the matrices com-

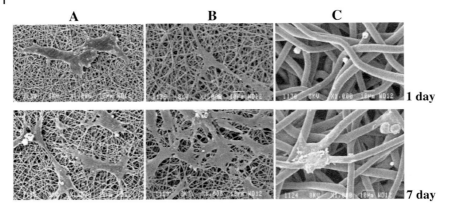

Fig. 1.27. SEM showing the morphologies of HUVECs cultured for 1 and 7 days on PLCL (50/50) electro-spun fibers having different diameters (μm): (A) 0.3, (B) 1.2, and (C) 7. (Adapted from Ref. [151] with permission from Elsevier.)

posed of 0.3 and 1.2 μm diameter fibers promoted the adhesion, spreading and proliferation of cells and the cells were found to be anchored on many fibers on the surface of the matrices. The quantitative determination of cell adhesion on these two matrices did not show any statistically significant differences in cell proliferation. However, the cells on the 7 μm fiber matrix were found to be rounded and showed significant decrease in cell proliferation compared to the other two matrices. This low cell adhesion and proliferation on microfiber matrix has been attributed to the large interfiber distance or a very low surface density of fibers that could not permit cell adhesion between the neighboring fibers.

Schindler et al. have investigated the ability of nanofibrillar matrix to promote *in vivo*-like organization and morphogenesis of cells in culture [166]. The synthetic nanofibrillar matrices were prepared by electrospinning a polymer solution of polyamide onto glass cover slips. The matrices were found to be composed of fibers of diameters ∼180 nm and a pore diameter of ∼700 nm. The surface smoothness of the matrix was found to be within 5 nm over a length of 1.5 μm, which is similar to the 3D organization of fibers in the basement membranes. NIH 3T3 fibroblasts, normal rat kidney (NRK) cells and breast epithelial cells were used for the *in vitro* evaluation. The organizational and structural changes of the intracellular components (actin and focal adhesion components) of the cells were measured as a function of adhesion when cultured on nanofibrous matrices and compared with the responses to cells on glass substrate. Fibroblasts plated on the glass substrate were well spread with an elaborate checkerboard pattern of stress fibers (Fig. 1.28A). Cells on the nanofiber matrix, however, showed significant changes in the morphology and shape. Compared to cells on the glass substrates, the cells on the nanofiber matrices were more elongated and bipolar with thinner actin fibers arranged parallel to the long axis of the cell. Notable increases in the formation of actin-rich lamellipodia, membrane ruffles and cortical actin were also observed

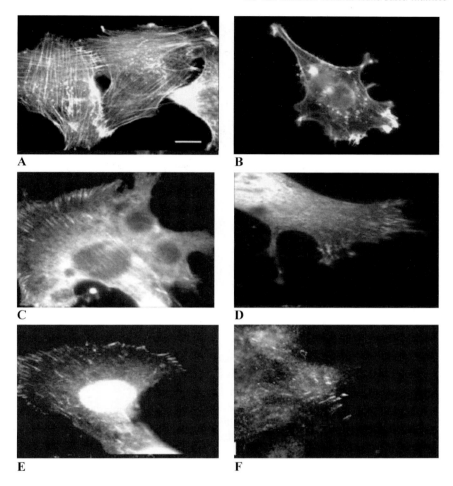

Fig. 1.28. Comparison of the F-actin network, focal adhesion components, fibronectin organization and integrin antibodies for NIH 3T3 fibroblasts cultured on glass substrates and nanofiber matrices. (A, C, E, G, I) are cells on glass substrates and (B, D, F, H and J) are cells on nanofiber matrices (see text for details). (Adapted from Ref. [166] with permission from Elsevier.)

(Fig. 1.28B). Staining of vinculin (a prominent component of focal complexes and focal adhesions that links cytoskeleton, plasma membrane and the ECM) of fibroblasts cultured on glass substrate showed a parallel streaked structure (Fig. 1.28C). However, the streaked staining for vinculin within cells on nanostructured matrix was limited to the edge of the lamellipodia with a more diffuse staining throughout the cell cytoplasm (Fig. 1.28D). Similar to actin distribution, such pattern of vinculin labeling correlates with cellular differentiation and morphogenesis *in vivo*. The cells were also stained for focal adhesion kinase (FAK) which functions as a central mechano-sensing transducer in cells. Cells cultured on glass demon-

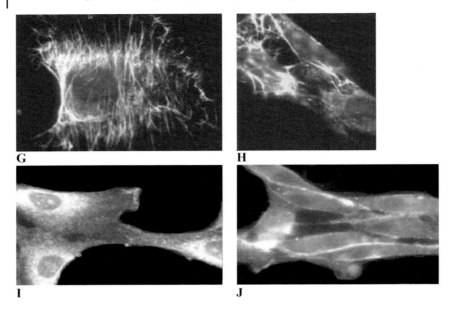

Fig. 1.28 (*continued*)

strated a streaky pattern of FAK labeling similar to the pattern obtained for vinculin (Fig. 1.28E). However, for cells on nanofiber matrix, the localization of FAK was found to be more punctuated and less well defined (Fig. 1.28F). Previous studies using breast epithelial cells have correlated this loss of FAK localization at focal adhesions to morphogenesis and differentiation [167]. The distribution of fibronectin on the cell surface cultured on glass substrate for 24 h revealed a classic linear pattern of fibrils (Fig. 1.28G). The cells on the nanofiber matrix, however, showed a thicker network of more randomly deposited apically localized fibrils indicating that they are permissive for the assembly of a matrix that can promote the formation of 3D-matrix adhesions (Fig. 1.28H). Staining for β1 integrin for NPK cells was punctuated when cultured on glass substrate for 24 h (Fig. 1.28I). However, the cells on nanofibrous matrix showed an organized long slender aggregate staining pattern, indicating the localization of β1 integrin in focal adhesions (Fig. 1.28J).

The ability of nanofiber matrices to promote morphogenesis was demonstrated by culturing T47D epithelial cells on glass and nanofiber coated glass matrices [166]. The T47D cells were used in the present study as they have been demonstrated to form duct like tubular structures and spheroids under conditions that promote morphogenesis. After 5 days in culture, a mixed population of multicellular structures comprised of tubules and spheroids were found on nanofiber matrix. At day 10, multicellular spheroids were dominant compared to tubules. Figure 1.29(A) shows a confocal series through a multicellular spheroid showing a lumen formed on nanofiber matrix. The figure shows the ability of T47D cells to grow into a complex multilayer structure on a nanofiber matrix. The cells cultured on

(a)

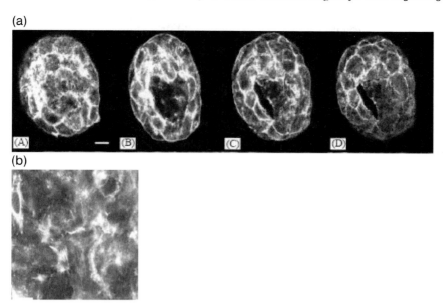

(b)

Fig. 1.29. (a) A series of confocal sections of a multicellular spheroid composed of T47D breast epithelial cells grown on nanofibers and stained with phalloidin-Alexa Fluor (A–D). (b) Cells after 10 days of culture on glass substrate. (Adapted from Ref. [166] with permission from Elsevier.)

glass surface, in contrast, formed a monolayer with groups of F-actin fibers (Fig. 1.29B). These studies demonstrated the advantages of using nanofiber matrices compared to microfiber matrices or 2D surfaces in developing 3D tissues, demonstrating their potential as ideal scaffolds for tissue engineering application.

1.7
Applications of Nano-based Matrices as Scaffolds for Tissue Engineering

Due to the unique properties and favorable cell behavior towards nanofiber matrices, different nanofiber matrices developed by top-down and by bottom-up approaches have been investigated as potential scaffolds for developing various tissues. The following section overviews nanofiber-based matrices as scaffolds for tissue regeneration.

1.7.1
Stem Cell Adhesion and Differentiation

As stem cells have been identified as one of the most appropriate cells for tissue engineering, the interaction of nanofiber matrices with stem cells raise significant interest. The unique ability of nanofiber matrices to support the growth and differentiation of stem cells into appropriate lineages have been demonstrated using self-assembled protein nanofiber matrices. Silva et al. have demonstrated the feasibility of encapsulating neural progenitor cells (NPCs) in self-assembled bioactive peptide amphiphiles and the ability of the cells to differentiate to appropriate lineage [168]. NPCs were selected as they are extensively used to replace lost central nervous system cells after degenerative or traumatic insults. Owing to the design flexibility of peptide amphiphiles as described earlier, unique PA was developed for form self-assembled scaffolds that could provide favorable environment for NPCs. The peptide motif designed to develop the scaffold was composed of a pentapeptide epitope isoleucine-lysine-valine-alanine-valine (IKVAV). The epitope has been selected because it is present in laminin, a cell adhesive protein present in the ECM and known to promote neurite sprouting and direct neurite growth. In addition to the neurite sprouting epitope, a GLu residue was also incorporated in the peptide that could give the peptide a net negative charge at pH 7.4. The rest of the molecule is composed of four Ala and three Gly followed by an alkyl tail of 16 carbon atoms.

The peptide was soluble in aqueous media and upon addition of cell suspension the cations present in the cell culture media screened the electrostatic repulsion, allowing the molecules to self-assemble due to hydrogen bond formation. Upon self-assembly the bioactive motifs were found to be placed on the surface. Figure 1.30 shows the SEM of nanofiber matrix formed by the self-assembly of peptide amphiphiles. A control amphiphile was also developed with a non-physiological sequence of glutamic acid-glutamine-serine (EQS) to compare the cell response to the bioactive self-assembling PA. Even though the control matrices allowed the encapsulation of the progenitor cells by self-assembly, the encapsulated cells did not sprout neuritis or differentiate morphologically or histologically.

The NPCs encapsulated within the peptide amphiphile self-assembled matrix was found to be viable throughout the period of study of 22 days. No differences in cell viability between the encapsulated cells and cells cultured on polylysine 2D

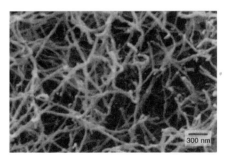

300 nm

Fig. 1.30. SEM showing the morphology of scaffold developed from an IKVAV nanofiber network by adding cell culture media to the peptide amphiphile. The sample is dehydrated and critical point dried caged in a metal grid to prevent network collapse during sample preparation for SEM. (Reprinted with permission from Ref. [168].)

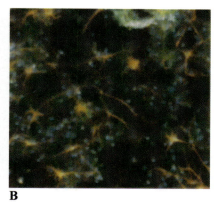

Fig. 1.31. Immunohistochemistry showing β-tubulin III of neurons and astrocytes of cells encapsulated in self-assembled bioactive gels (A) (after 1 day in culture) and lamine-coated cover slips (B) (after 7 days in culture). β-Tubulin III is stained green, differentiated astrocytes (glial cells) are labeled orange. (Reprinted with permission from Ref. [168].)

films were found demonstrating that sufficient diffusion of nutrients, bioactive factors and oxygen is taking place through the highly hydrated nanofiber matrix. Immunocytochemistry demonstrated the differentiation of NPCs encapsulated in the bioactive self-assembled PA matrix to neurons. Furthermore, the results were found to be statistically significant when compared to cells grown on laminin or lysine coated surface. Differentiated neurons were labeled for β-tubulin III and glial fibrillary acidic protein to detect neurons and astrocytes. After only 1 day in culture, 35% of the cells encapsulated within the bioactive nanofiber scaffold stain positive for β-tubulin (Fig. 1.31A). At the same time only less than 5% of the cells in the bioactive scaffold showed astrocyte differentiation even after 7 days in culture (Fig. 1.31A). This is a positive observation since inhibition of astrocyte proliferation is important in the prevention of the glial scar a known barrier to axon elongation following CNS trauma. Cells cultured on 2D laminine coated substrate did not show such differentiation and at the same time showed significant astrocyte proliferation (Fig. 1.31B). Encapsulation in the nanofiber scaffold led to the formation of large neurites after only 1 day (57 ± 26 μm) where as cells grown on laminin or lysine did not form neurites at that time. TEM evaluation of cells encapsulated within the gels after 7 day showed a healthy and normal ultrastructural morphology. The high migrating ability of the cells encapsulated within the gels was also demonstrated by tracking the distance between the center of each neurosphere and cell bodies as a function of time. The study demonstrated the potential of bioactive self-assembling nanofiber scaffolds as stem cell delivery vehicles.

The differentiation ability of progenitor cells into different lineages on polymeric nanofiber matrices developed by electrospinning was demonstrated recently by Li et al., using multipotent human mesenchymal stem cells (MSC). The study demonstrated the feasibility of nanofiber matrices to support the adhesion and differen-

Adipogenesis

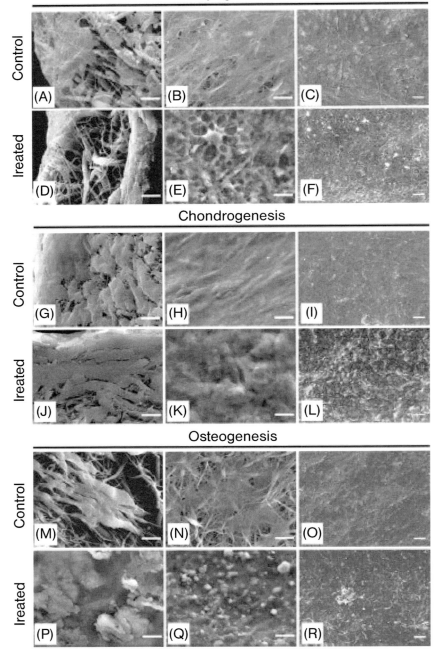

Chondrogenesis

Osteogenesis

tiation of these cells into different lineages [169]. PCL nanofiber matrices were used for the study. The matrix was found to be composed of randomly oriented nanofibers having diameters ~700 nm. The MSCs seeded on the nanofiber matrices were found to attach and remained viable. For multi-lineage differentiation of the cells the MSC seeded nanofiber matrices where placed under specific differentiation promoting culture conditions (Fig. 1.32). Thus, under adipogenic conditions (media with dexamethazone, 3-isobutyl-1-methylxanthine and insulin), the cell–polymer constructs developed into an adipose like tissue with the cells expressing appropriate gene expression. The incubation of the cell seeded nanofiber matrices in chondrogenic media resulted in the formation of a cartilaginous tissue composed of cells showing characteristic chondrocyte phenotypes. The authors observed that the chondrogenic differentiation of MSC on nanofiber matrices takes place even at low cell population compared to 2D cultures or cells encapsulated in hydrogels presumably due to the unique interactions of the cells with the nanostructured topography of the matrix. Incubation of the cell seeded nanofiber matrix in osteogenic media (β-glycerophosphate, ascorbic acid and dexamethasone) resulted in the formation of a dense bone-like tissue with the cells showing characteristic osteoblast phenotypes. The study thus demonstrated that electrospun nanofiber scaffolds can support chondrogenic, osteogenic and adipogenic differentiation and, therefore, are candidate scaffolds for the fabrication of multi-component tissue constructs.

1.7.2
Neural Tissue Engineering

Nerve tissue engineering can be considered as one of the most promising approaches to restore central nervous system function. Several studies have evaluated the efficacy of nanofibrous matrices as scaffolds for neural tissue engineering. In one study Yang et al. determined the efficacy of PLLA nanofiber matrices developed by phase separation process as scaffold for neutral tissue regeneration [170]. The fiber diameters were ~196 nm with a matrix porosity of ~85%. Neural stem cells (NSC) were used in the study. Upon NSC seeding and culturing for a day, the cells were found to randomly spread over the surface of the polymer scaffold without much differentiation. By day 2 the cells had progressively grown throughout the scaffold, with a neurite length twice that of the cell body, and migration of cells

Fig. 1.32. SEM showing nanofiber–MSC constructs maintained with and without differentiation media (A–F adipogenic; G–L chondrogenic; and M–R osteogenic). (A, D, G, J, M, P) Cross sections; (B, C, E, F, H, I, K, L, N, O, Q, R) top views. In adipogenic cultures (D–F), globular, round cells were evident, while fibroblast-like cells were found in the control cultures (A–C). In chondrogenic cultures (J–L), round chondrocyte-like cells were embedded in a thick layer of ECM that was not found in the control group (G–I). In osteogenic cultures (P–R), mineralized nodules were formed in the constructs. In contrast, control cultures (M–O) contained primarily fibroblast-like cells, and mineralization was not seen [169] with permission from Elsevier.

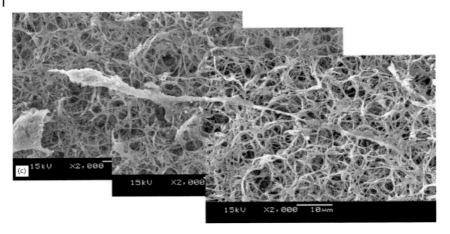

Fig. 1.33. SEM showing the magnified view of a differentiated cell with long neurite cultured on a nanofiber matrix developed by phase separation. (Adapted from Ref. [170] with permission from Elsevier.)

into the porous matrix also occurred (Fig. 1.33). The study showed the feasibility of nanofibrous scaffold to act as a positive guidance cue to guide neurite outgrowth.

Efficacy of aligned polymeric nanofiber matrices as a scaffold for the growth and differentiation of neural NSCs was further evaluated by Yang et al. using PLLA nano/microscaffold [171]. The cell adhesion and differentiation pattern on the aligned nanofiber matrices were compared to aligned microfiber matrices formed by electrospinning as well as random micro and nanofiber matrices. The aligned nanofiber matrices were composed of fibers having diameter $\sim$300 nm and microfiber matrices were composed of fibers having diameter $\sim$1.5 μm. The average fiber diameters of the random nano and microfiber matrices were $\sim$250 nm and 1.25 μm respectively. NSC attached and formed an elongated spindle-like shape on all the surfaces. The direction of NSC elongation and neurite outgrowth was found to be aligned with the direction of aligned fibers and showed classical contact guidance. The cells on random fiber matrices in contrast showed significantly different phenotype. In terms of differentiation, more cells were found to be differentiated on aligned nanofiber matrix (80%) than on aligned microfiber matrix (40%) demonstrating that nanofiber alignment has profound effect on cell differentiation. Since successful nerve regeneration rely on the extensive growth of axonal process, the neurite length of cells on the matrices was evaluated. The neurite length of NSCs on aligned nanofiber matrix was significantly higher than aligned micromatrix or random matrices. Figure 1.34 shows the SEM micrograph of NSC on aligned nanofiber matrix. The cells body shows an apparent bipolar elongated morphology with the outgrowing neuritis. Both cell elongation and neurite outgrowth followed the same direction of PLLA nanofibers. The figure also shows significant interaction between NSCs and the aligned fibers. Some filament-like

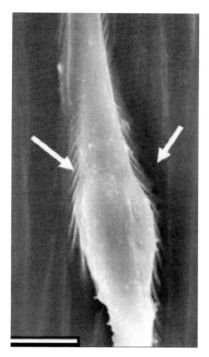

Fig. 1.34. SEM showing the interaction of NSC on an aligned PLLA nanofiber matrix. Bar = 5 μm. (Adapted from Ref. [171] with permission from Elsevier.)

structures, presumably focal adhesions, extend out from the NSC cell body and neurite and attach to the nanofiber. This study thus demonstrated that aligned nanofiber matrix could improve NSC differentiation and support neurite outgrowth compared to other matrices evaluated.

1.7.3
Cardiac and Blood Vessel Tissue Engineering

One of the major reasons for the failure of synthetic small diameter vascular grafts is adverse blood biomaterials interactions resulting in an acute occlusion followed by a chronic intimal hyperplasia. The ability to engineer vascular grafts using biodegradable scaffolds could lead to the development of synthetic grafts with long-term patency. One of the most extensively investigated approaches to prevent graft occlusion is to improve the antithrombogenicity of graft materials by seeding them with endothelial cells. But the development of a stable endothelial covering on the surface of synthetic materials is difficult due to the high sensitivity of the endothelial cells.

Mo et al. have evaluated the efficacy of polymeric nanofiber matrices as scaffolds for growing endothelical and smooth muscle cells [172]. P(LLA-CL) nanofiber matrices developed by the process of electrospinning were used as the 3D scaffold in the study. Endothelial cells and smooth muscle cells were cultured on the nano-fiber matrices for 7 days. Both SMC and EC adhered and spread on nanofiber matrices with the cell number, showing a significant increase from day 1 to day 7. This indicates the ability of the cell to attach and proliferate on the nanofiber matrix. Immunohistochemical evaluation of adhered ECs on the nanofiber matrix showed appropriate EC phenotype expression, indicating favorable interaction of the cells with the matrix.

Even though self-assembly of bioactive peptide motifs can be considered as the direct method to develop bioactive nanofiber matrices, He et al. have recently demonstrated the feasibility of improving the bioactivity of electrospun nanofibers by surface modification. They modified the surface of a P(LLA-CL)nanofiber matrix to improve the efficacy of the matrix towards endothelial cell attachment and proliferation [150]. The surface of p(LLA-CL) nanofiber matrix was functionalized by plasma modification and the nanofiber matrix surface was then coated with collagen. Human coronary artery endothelial cells (HCAECs) were used to evaluate the efficacy of the coated nanofiber scaffold compared to uncoated scaffold. The cells on uncoated scaffold were rounded in shape without significant spreading (Fig. 1.35A). More cells were found to be attached on the collagen-coated p(LLA-CL) nanofibers than uncoated nanofiber matrix since collagen is the main structural and functional protein present in ECM (Fig. 1.35B). Immunohistochemical evaluation of the cultured cells showed the preservation of endothelial phenotype by the cells on the coated nanofiber matrix.

Similarly a collagen coated PCL nanofiber matrix was evaluated for improve SMC adhesion and interaction [173]. Coating the nanofibers with collagen improved the cell adhesion on the fibers, with cells preserving their characteristic phenotype.

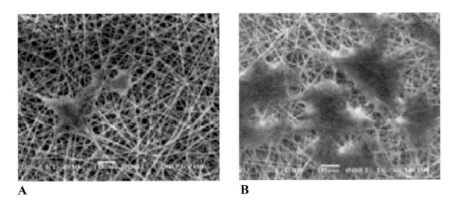

A **B**

Fig. 1.35. SEM images of HCAECs cultured on (A) uncoated P(LLA-CL) nanofiber matrix and (B) collagen-coated p(LLA-CL) nanofiber matrix after 3 days in culture. (Adapted from Ref. [150] with permission from Elsevier.)

Another related study was performed by the same research group to evaluate the effect of cell behavior towards scaffolds that combine nanostructure and bioactivity. Poly(ethylene terephthalate) a non-degradable polymer extensively investigated for developing vascular graft was fabricated into a nanofibrous structure using the process of electrospinning [174]. The nanofibers were then surface modified to attach a bioactive protein gelatin. The surface-modified fiber scaffolds were then seeded with HCAECs and cultured for 7 days. The surface-modified nanofiber matrix showed improved cell adhesion and maintenance of phenotypic activity. These results showed that combining the nanostructure of the scaffold with the bioactive molecule could positively promote cell–matrix and cell–cell interactions, inducing them to express the phenotypic shape.

The effect of fiber orientation on SMC attachment and proliferation of nanofiber matrices was evaluated using aligned nanofiber scaffolds of P(LLA-CL) [140]. Aligned fibers were developed by the process of electrospinning using the wheel with a sharp edge collector. SMCs were found to adhere on aligned nanofiber scaffold as early as within 1 h of seeding and the cells tend to elongate along the direction of the nanofibers. After 3 days in culture, SMCs proliferated approximately along the longitudinal direction of the nanofiber length, which formed an oriented pattern similar to those in native artery. The cell number significantly increased at day 7 and the surface of the nanofibrous scaffold was covered with a continuous SMC monolayer with a regular direction from left to right along the nanofiber alignment. The cell adhesion and proliferation on aligned nanofiber matrix was significantly higher than on 2D film of the same polymer. Figure 1.36 shows the SEM of aligned SMC on aligned nanofiber matrix. Further, the distribution and organization of cytoskeleton proteins inside SMCs were parallel to the direction of the nanofibers (Fig. 1.37). The study demonstrated for the first time the feasibility of using an aligned nanostructured scaffold to mimic natural vessel architecture and the ability of cells to orient along the fiber axis to mimic the natural orientation.

One elegant method to incorporate biofunctionality into nanofiber scaffolds is the use of self-assembling peptides decorated with appropriate protein motifs. Gen-

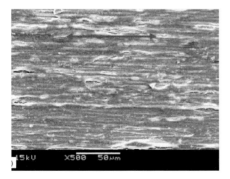

Fig. 1.36. SEM showing the alignment of SMCs along the aligned nanofiber matrix. (Adapted from Ref. [140] with permission from Elsevier.)

Fig. 1.37. LSCM micrograph of immunostained α-actin filaments in SMC after 1 day of culture on aligned nanofiber matrix. (Adapted from Ref. [140] with permission from Elsevier.)

ove et al. recently developed biomimetic self-assembled peptide scaffolds and evaluated the function of human aortic endothelial cells seeded on the scaffolds [175]. The self-assembling protein motifs were developed based on the functionality of basement membrane, which is composed mainly of laminin and collagen. Two peptide sequences present in laminin (YIGSR and RYVVLPR) are known to promote cell adhesion, cell migration and endothelial cell tubular formation, and a peptide sequence in collagen Type IV (TAGSCLRKFSTM) shown to promote the adhesion and spreading of bovine aortic endothelial cells, were selected to develop novel functionalized peptide motifs for self-assembly. The functionalized peptides were developed by solid-phase synthesis and found to self-assemble to form hydrogel under physiological conditions (Fig. 1.38). Four different scaffolds were investigated in the study, unmodified self-assembled peptide scaffolds, composed of RAD 16-1 (AcN-RADARADARADARADA-CONH2), YIG (10%) modified scaffold, RYV (10%) modified scaffold and TAH (10%) modified scaffold. Cell numbers increased about two-fold in modified peptide scaffolds compared to unmodified scaffolds, indicating that cell could sense and respond to the functionalized material. In addition, the matrix could modulate endothelial cell growth only when the sequence was physically attached to the nanofiber matrix, showing the importance of spatial distribution of the bioactive molecules on the nanostructured scaffolds. The peptide scaffolds in general also enhanced the endothelial cell phenotype, such as NO synthesis and deposition of basement membrane components (laminin I and collagen IV), suggesting the potential of these scaffolds in recreating the endothelial microenvironment.

Another application for which biodegradable nanofiber matrices have been evaluated is for developing scaffolds for engineering myocardium. Shin et al. have evaluated the potential of PCL nanofiber matrix as a cardiac graft by assessing the interaction of rat cardiomyocytes with the nanofiber scaffold [176]. The average di-

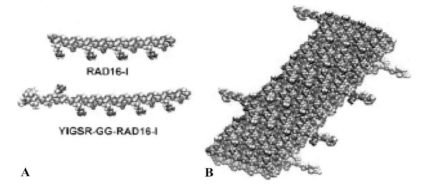

Fig. 1.38. (A) Model representing the peptide RAD16-1 (unfunctionalized peptide) and peptide YIG (functionalized). (B) Model representing the double β-sheet tape of a self- assembled peptide nanofiber of a mixture composed of RAD16-1 and YIG (9:1). (Adapted from Ref. [175] with permission from Elsevier.)

ameter of the fibers used to develop the scaffold was ∼250 nm. The cardiomyocytes attached well on the scaffold and after three days in culture the cardiomyocytes started to contract (Fig. 1.39). The contractions were ubiquitous and synchronized. The mechanical property (softness) of the scaffold was found to be highly appropriate to allow the spontaneous contraction of the cardiomyocytes. The cardiomyocytes were also found to form a tight arrangement and intercellular contacts throughout the entire mesh and stained positive for cardiotypical proteins. However, the system has some limitations for *in vivo* application. The limited thickness of the scaffold might not be able to provide sufficient function when used alone to cover an infracted area. Therefore, a modified multilayered nanostructured nano-fibrous graft was developed [177]. PCL nanofiber matrices composed of fibers having diameters of ∼100 nm were developed and cardiomyocytes were cultured on nanofiber matrices for 5–7 days. After that layering of the individual grafts (five

Fig. 1.39. SEM of the cross-section of a cardiac nanofibrous mesh, showing complete coverage of the mesh with cardiomyocytes. (Adapted from Ref. [176] with permission from Elsevier.)

layers) were performed by gently placing layers on top of each other. The layered constructs were incubated for 2 h at 37 °C without media for interlayer attachment and then cultured for 14 days in culture media. Multilayered scaffolds initially showed weak and unsynchronized contractions but became stronger and synchronized with time. H&E staining of the constructs demonstrated the interconnections between the layers. The immunohistochemistry study showed the presence of connexin43 in multilayered graft, indicating that the cells are rebuilding gap junctions, and synchronized contractions in multilayered grafts. *In vivo* studies using these grafts are under way.

1.7.4
Bone, Ligament and Cartilage Tissue Engineering

Nanofiber matrices were also investigated for developing scaffolds for bone, ligament and cartilage tissue engineering. Li et al. used PLAGA nanofiber matrices to culture MSCs and demonstrated that nanofibrous matrices could support the adhesion and proliferation of these cells [149]. Yoshimoto et al. used PCL nanofiber matrices to culture MSCs under dynamic conditions in osteogenic media to evaluate the feasibility of developing bone *in vitro* [178]. The average fiber diameter of the scaffold was ~400 nm. The MSCs were seeded on the nanofiber matrix and cultured for 4 weeks under dynamic conditions. MSCs were found to attach and proliferate throughout the nanofiber matrix. Furthermore, the cells migrated inside the scaffold and produced an extracellular matrix of collagen throughout the scaffold (Fig. 1.40). After 4 weeks in culture the cell polymer construct was noticeably harder and the extracellular matrix was calcified throughout the matrix, as evidenced from histology.

Polyurethane nanofiber scaffold was used to evaluate the effect of nanofiber alignment on the cellular response of human ligament fibroblasts (HLF) and to evaluate the influence of HLF alignment and strain direction on mechanotransduction [153]. A rotating collecting target was used to develop aligned nanofibers. After 3 days in culture, the HLFs on the aligned nanofibers were spindle shaped

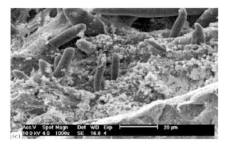

Fig. 1.40. SEM showing a cell nanofiber construct after 4 weeks in culture. Globular accretions, abundant calcification and collagen bundles can be seen. (Adapted from Ref. [178] with permission from Elsevier.)

and oriented (similar to *in vivo* ligament fibroblast morphology) along the nano-fiber direction and the aligned nanofiber formed tissue-like oriented bundles that were confluent over the entire surface within 7 days. The HLFs on randomly oriented nanofiber scaffold were not oriented, but when the scaffold was subjected to uniaxial strain the cells reorganized their spindle shape and became organized. There were significant differences in the ECM production by oriented and un-oriented cells on the matrices, indicating that cell morphology plays a significant role in ECM production. Furthermore, the aligned HLFs on the nanofiber structure were more sensitive to strain in a longitudinal direction. The study demonstrated that aligned nanofiber scaffold forms a promising structure for developing tissue engineered ligament due to its biomimetic structure and the ability to provide a mechanical environment ligament cells encounter *in vivo*.

Li et al. using PCL nanofiber scaffold demonstrated the efficacy of nanofiber scaffold to support the differentiation of MSCs to chondrocytes as a viable method to engineer cartilage *in vitro* [179]. The three-dimensional MSC seeded constructs display a cartilage-like morphology containing chondrocyte like cells surrounded by abundant cartilaginous matrix (Fig. 1.41). The level of MSC chondrogenesis using nanofiber scaffold was found be higher than in high density pellet cultures commonly used for MSCs. This study this demonstrated that biodegradable nano-fiber scaffold due to its microstructure resemblance to a native ECM effectively supported MSC chondrogenesis.

A self-assembling peptide hydrogel scaffold has also been investigated as scaffolds for cartilage repair and regeneration [180]. A KLD12 peptide was developed, the aqueous solution of which was found to form a hydrogel when exposed to salt solution or cell culture media. The encapsulated chondrocytes showed a round morphology with cell viability of 89% immediately after encapsulation. Cell division of encapsulated chondrocytes in peptide hydrogel was found to be much higher than agarose control cultures. Histological evaluation of the encapsulated cell–hydrogel construct showed the formation of cartilage-like ECM rich in proteoglycans and type II collagen. Time dependent accumulation of this ECM was paralleled by increase in stiffness of the material showing the deposition of mechanically functional neo-tissue. This study demonstrated the potential of self-assembling peptide hydrogel as a scaffold for the synthesis and accumulation of a true cartilage-like ECM within a 3D cell culture for cartilage tissue repair.

1.8
Conclusions

Nanotechnology assisted fabrication processes are changing the way bioresorbable scaffolds are being developed for tissue engineering. Novel nanofabrication processes have enabled the development of nanostructured scaffolds that could closely resemble the structure of the ECM, and studies using bioactive nanostructured scaffolds have demonstrated the importance of nanostructure in cell–matrix and cell–cell interaction. Even though top-down and bottom-up approaches developed

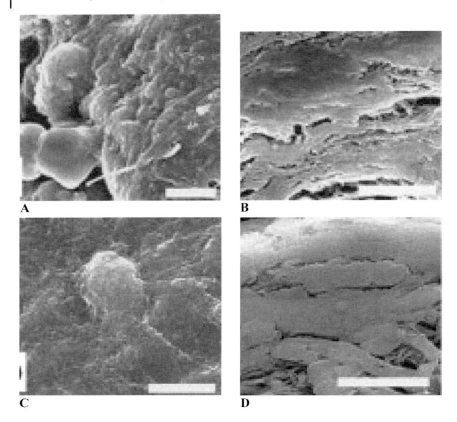

Fig. 1.41. Morphology of differentiated chondrocytes in cell pellet (CP) culture and using PCL nanofiber matrix. (A) Top view of CP with round chondrocyte-like cells on the surface. (B) Cross-sectional view of CP, showing thick ECM. (C) Top view of nanofiber matrix, showing the presence of round, ECM-embedded chondrocyte-like cells. (D) Cross-sectional view of nanofiber matrix, showing a thick, dense ECM-rich layer. (Adapted from Ref. 179] with permission from Elsevier.)

to fabricate nanostructured scaffolds have several advantages of their own, several limitations still hinder the translation of many of these technologies for clinical applications. Future studies will rely on a combination of both approaches as necessary as well as combining these with lithographic techniques to develop hierarchical nanostructures with spatially presented biological cues for developing optimal scaffolds for tissue engineering.

References

1 LEVENBERG S, LANGER R. *Curr. Top. Develop. Biol.* 61, 113–134, **2004**.

2 BAUER TW, MUSCHLER GF. Bone graft materials An overview of the basic

science. *Clin. Orthop.* 371, 10–27, **2000**.

3 HEINEKEN FG, SKALAK R. Tissue engineering: A brief overview. *J. Biomech. Eng.* 113, 111, **1991**.

4 SKALAK R, FOX CF (eds.). *Tissue Engineering*, Proceedings for a Workshop held at Granlibakken, Lake Tahoe, California, February 26–29, Alan Liss, New York, **1988**.

5 NEREM RM, Cellular engineering, *Ann. Biomed. Eng.* 19, 529–545, **1991**.

6 LANGER R, VACANTI JP, Tissue engineering. *Science* 260, 920–926, **1993**.

7 VACANTI JP, LANGER R, Tissue engineering: the design and fabrication of living replacement devices for surgical reconstruction and transplantation. *Lancet* 354 (Suppl 1), 32–34, **1999**.

8 FUCHS JR, NASSERI BA, VACANTI JP. Tissue engineering: A 21st century solution to surgical reconstruction. *Ann. Thorac. Surg.* 72, 577–591, **2001**.

9 LAURENCIN CT, AMBORSIO AMA, BORDEN MD, COOPER JA. Tissue engineering: orthopedic application. *Ann. Rev. Biomed. Eng.* 1, 19–46, **1999**.

10 SIPE JD. Tissue engineering and reparative medicine. *Ann. New York Acad. Sci.* 961, 1–9, **2002**.

11 WEAVER VM, PETERSEN OW, WANG F, LARABELL CA, BRIAND P, DAMSKY C, BISSELL MJ. Reversion of the malignant phenotype of human breast cells in three-dimensional culture and in vivo by integrin blocking antibodies. *J. Cell Biol.* 137, 231–245, **1997**.

12 CUKIERMAN E, PANKOV R, STEVENS DR, YAMADA KM. Taking cell-matrix adhesions to third dimensions. *Science* 294, 1708–1712, **2001**.

13 VACANTI JP, MORSE MA, SALTZMAN WM, DOMB AJ, PEREZATYDE A, LANGER R. Selective cell transplantation using bioabsorbable artificial polymers as matrices. *J. Pediar. Surg.* 23, 3–9, **1988**.

14 LANGER R, TIRRELL DA, Designing materials for biology and medicine. *Nature* 428, 487–492, **2004**.

15 ROSSO F, MARIO G, GIORDANO A, BARBARISI M, PARMEGGIANI D,

BARBARUSU A. Smart materials as scaffolds for tissue engineering. *J. Cell Physiol.* 203, 465–470, **2005**.

16 MALLAPRAGADA S, NARASIMHAN B (Eds.), *Handbook of Biodegradable Polymeric Materials and their Applications*. American Scientific Publishers, North Lewis Way, **2006**.

17 HOLLISTER SJ. Porous scaffold design for tissue engineering. *Nat. Mater.* 4, 518–524, **2005**.

18 AGARWAL CM, RAY RB. Biodegradable polymeric scaffolds for musculoskeletal regeneration. *J. Biomed. Mater. Res.* 55, 141–150, **1988**.

19 LU L, MIKOS AG. The importance of new processing techniques in tissue engineering. *MRS Bull.* 11, 28, **1996**.

20 THOMSON RC, YASZEMSKI MJ, MIKOS AG. Polymer scaffold processing. In: LANZA RP, LANGER R, CHICK WL (Eds.), *Principles of Tissue Engineering*. R.G. Landes Co., Austin, TX, p. 263, **1997**.

21 WIDMER MS, MIKOS AG. Fabrication of biodegradable polymer scaffolds for tissue engineering. In: PATRICK JR CW, MIKOS AG, McINTIRE LV (Eds.), *Frontiers in Tissue Engineering*. Elsevier Science, New York, p. 107, **1998**.

22 HUTMACHER DW, ZEIN I, TEOH SH, NG KW, SCHANTZ JT, LEAHY JC. Design and fabrication of a 3D scaffold for tissue engineering bone. In: AGARWAL CM, PARR JE, LIN ST (Eds.), *Synthetic Bioabsorbable Polymers for Implants*, STP 1396. American Society for Testing and Materials, West Conshohocken, PA, p. 152, **2000**.

23 HUTMACHER DW, SITTINGER M, RISBUD MV. Scaffold-based tissue engineering: rationale for computer aided design and solid free form fabrication systems. *Trends Biotechnol.* 22, 354–362, **2004**.

24 MIKOS AG, THORSEN AJ, CZERWONKA LA, BAO Y, LANGER R, WINSLOW DN, VACANTI JP. Preparation and characterization of poly(L-lactic acid) foams. *Polymer* 35, 1068–1077, **1994**.

25 MA, PX, LANGER R. Fabrication of biodegradable polymer foams for cell transplantation and tissue engineering. In: MORGAN J, YARMUSH M (Eds.), *Tissue Engineering Methods and Protocols*, Humana Press, NJ, p. 47, **1999**.

26 LU L, PETER SJ, LYMAN MD, LAI H, LEITE SM, TAMADA JA, VACANTI JP, LANGER R, MIKOS AJ. *In vitro* degradation of porous poly(L-lactic acid) foams. *Biomaterials* 21, 1595–1605, **2000**.

27 BORDEN MD, KHAN Y, ATTAWIA M, LAURENCIN CT. Tissue engineered microsphere-based matrices for bone repair: Design, evaluation, and optimization. *Biomaterials* 23, 551–9, **2002**.

28 BORDEN M, ATTAWIA M, LAURENCIN CT. The sintered microsphere matrix for bone tissue engineering: In vitro osteoconductivity studies. *J. Biomed. Mater. Res.* 61, 421–429, **2002**.

29 BORDEN MD, EL-AMIN SF, ATTAWIA M, LAURENCIN CT. Structural and human cellular assessment of a novel microsphere based tissue engineered scaffold for bone repair. *Biomaterials*, 24, 597–609, **2003**.

30 BORDEN MD, ATTAWIA M, KHAN Y, EL-AMIN SF, LAURENCIN CT. Tissue-engineered bone formation *in vivo* using a novel sintered polymeric microsphere matrix. *J. Bone Joint Surg. (B)*. 86-B, 1200–1208, **2004**.

31 YANG SF, LEONG KF, DU ZH, CHUA CK. The design of scaffolds for use in tissue engineering: part 1 – Traditional factors. *Tissue Eng.* 7, 679–689, **2001**.

32 LEONG KF, CHEAH CM, CHUA CK. Solid freeform fabrication of three-dimensional scaffolds for engineering replacement tissues and organs, *Biomaterials* 24, 2363–2378, **2003**.

33 COOKE MN, FISHER JP, DEAN D, RIMNAC C, MIKOS AG. Use of stereolithography to manufacture critical-sized 3D biodegradable scaffolds for bone ingrowth, *J. Biomed. Mater. Res.* 64B, 65–69, **2003**.

34 YU T, CHIELLINI F, SCHMALJOHANN D, SOLARO R, OBER CK. Microfabrication of hydrogels as polymer scaffolds for tissue engineering applications. *Polym. Prepr.* 41, 1699–1700, **2000**.

35 GARIEPY E, LEROUX J. In situ forming hydrogels – Review of temperature

sensitive systems. *Eur. J. Pharm. Biopharm.* 58, 409–426, **2004**.

36 GIL ES, HUDSON SM. Stimuli-responsive polymers and their bioconjugates. *Progr. Polym. Sci.* 29, 1173–1222, **2004**.

37 KIM TM, SHARMA B, WILLIAMS CG, RUFFNER MA, MALIK A, MCFARLAND EG, ELISSEEFF JH. Experimental model for cartilage tissue engineering to regenerate the zonal organization of articular cartilage. *Osteoarthritis Cartilage* 11, 653–664, **2003**.

38 VARGHESE D, DESHPANDE M, XU T, KESARI P, OHRI S, BOLAND T. Advances in tissue engineering: cell printing. *J. Thor. Card. Surg.* 129, 470–472, **2005**.

39 BOLAND T, MIRONOV V, GUTOWSKA A, ROTH EA, MARKWALD RR. Cell and organ printing 2: Fusion of cell aggregates in three-dimensional gels. *Anat. Rec.* 272A, 497–502, **2003**.

40 NEAGU A, FORGACS G. Fusion of cell aggregates: a mathematical model. In: VOSSOUGHI J (Ed.), *Biomedical Engineering. Recent Development*, Medical and Engineering Publishers, Inc., Washington DC, USA, pp. 241–242, **2002**.

41 MIRONOV V, BOLAND T, TRUSK T, FORGACS G, MARKWALD RR. Organ printing: Computer-aided jet-based 3D tissue engineering. *Trends Biotechnol.* 21, 157–161, **2003**.

42 JAKAB K, NEAGU A, MIRONOV V, MARKWALD RR, FORGACS G. Engineering biological structures of prescribed shape using self-assembling multicellular systems, *Proc. Natl. Acad. Sci. U.S.A.* 10, 2864–2869, **2004**.

43 SHIEH S, TERADA S, VACANTI JP. Tissue engineering auricular regeneration: *In vitro* and *in vivo* studies. *Biomaterials*. 25, 1545–1557, **2004**.

44 RASPANTI M, PROTASONI M, MANELLI A, GUIZZARDI S, MANTOVANI V, SALA A. The extracellular matrix of the human aortic wall: Ultrastructural observations by FEG-SEM and by tapping-mode AFM. *Micron.* 37, 81–86, **2006**.

45 GRIFFITH LG. Emerging design

principles in biomaterials and scaffolds in tissue engineering. *Ann. New York Acad. Sci.* 961, 83–95, **2002**.

46 SALTZMANN WM, OLBRICHT WL. Building drug delivery into tissue engineering. *Nat. Rev. Drug Discov.* 1, 177–186, **2002**.

47 TIRRELL M, KOKKOLI E, BIESALSKI M. The role of surface science in bioengineered materials. *Surf. Sci.* 500, 61–83, **2002**.

48 BOSMAN FT, STAMENKOVIC I. Functional structure and composition of the extracellular matrix. *J. Pathol.* 200, 432–428, **2003**.

49 OLSEN BR. Matrix molecules and their ligands. In: LANZA RP, LANGER R, CHICK WL (Eds.), *Principles of Tissue Engineering*, 1st edition, Academic Press, New York, pp. 47–65, **1997**.

50 BOUDREAU N, MYERS C, BISSELL MJ. From laminin to lamin: regulation of tissue-specific gene expression by the ECM. *Trends Cell Biol.* 5, 1–4, **1995**.

51 PROVENZANE PP, VANDERBY R. Collagen fibril morphology and organization: Implications for force transmission in ligament and tendon. *Matrix Biol.* 25, 71–84, **2006**.

52 FANTNER GE, RABINOVYCH O, SCHITTER G, THURNER P, KINDT JH, FINCH MM, WEAVER JC, GOLDE LS, MORSE DE, LIPMAN EA, RANGELOW IW, HANSMA PK. Hierarchical interconnections in the nano-composite material bone: Fibrillar cross-links resist fracture on several length scales. *Composites Sci. Technol.* 66, 1205–1211, **2006**.

53 GELSE K, POSCHL E, AIGNER T. Collagen Structure, function and biosynthesis. *Adv. Drug Deliv. Rev.* 55, 1531–1546, **2003**.

54 CURTIS A, WILKINSON C. Nano-techniques and approaches in biotechnology. *Trends Biotechnol.* 19, 97–101, **2001**.

55 LUTOLF MP, HUBBELL JA. Synthetic biomaterials as instructive extra-cellular microenvironments for morphogenesis in tissue engineering. *Nat. Biotechnol.* 23, 47–55, **2005**.

56 ADAMS JC, WATT FM. Regulation of development and differentiation by the extracellular matrix. *Development* 117, 1183–1198, **1993**.

57 XIONG J, STEHLE T, DIEFENBACH B, ZHANG R, DUNKER R, SCOTT DL, JOACHIMIAK A, GOODMAN SL, ARNAOUTI MA, Crystal structure of the extracellular segment of integrin $\alpha V \beta 3$. *Science* 294, 339–345, **2001**.

58 GARCIA AJ. Get a grip. Integrins in cell–biomaterial interactions *Biomaterials* 26, 7525–7529, **2005**.

59 BERSHADSKY AD, BALLESTREM C, CARRAMUSA L, ZILBERMAN Y, GILQUIN B, KHOCHBIN S, ALEXANDROVA AY, VERKHOVSKY AB, SHEMESH T, KOZLOV MM. Assembly and mechanosensory function of focal adhesions: experiments and models *Eur. J. Cell Biol.* 85, 165–173, **2006**.

60 ROSS RS. Molecular and mechanical synergy: Cross-talk between integrins and growth factor receptors. *Cardiovasc. Res.* 63, 381– 390, **2004**.

61 ADAMS JC, Cell matrix contact structure. *Cell Mol. Life Sci.* 58, 371–392, **2001**.

62 HE L, DEXTER AF, MIDDLEBERG APJ. Biomolecular engineering at interfaces. *Chem. Eng. Sci.* 61, 989–1003, **2006**.

63 LI JM, MENCONI MJ, WHEELER HB, ROHRER MJ, KLASSEN VA, ANSELL JE, APPEL MC. Precoating expanded polytetrafluoroethylene grafts alters production of endothelial cell-derived thrombomodulators. *J. Vasc. Surg.* 15, 1010–1017, **1992**.

64 KAEHLER J, ZILLA P, FASOL R, DEUTSCH M, KADLETZ M. Precoating substrate and surface configuration determine adherence and spreading of seeded endothelial cells on polytetrafluoroethylene grafts. *J. Vasc. Surg.* 9, 535–41, **1989**.

65 SEEGER JM, KLINGMAN N. Improved endothelial cell seeding with cultured cells and fibronectin-coated grafts. *J. Surg. Res.* 38, 641–7, **1985**.

66 PIERSCHBACHER MD, RUOSLAHTI E. Cell attachment activity of fibronectin can be duplicated by small synthetic fragments of the molecule. *Nature* 309, 30–33, **1984**.

67 RUOSLAHTI E, PIERSCHBACHER MD. New perspectives in cell adhesion:

RGD and integrins. *Science* 238, 491–497, **1987**.

68 RUOSLAHTI E. RGD and other recognition sequences for integrins. *Annu. Rev. Cell Dev. Biol.* 12, 697–715, **1996**.

69 RUOSLAHTI E. The RGD story. A personal account. *Matrix Biol.* 22, 459–465, **2003**.

70 HYNES RO. Integrins – Versatility, modulation, and signaling in cell-adhesion. *Cell* 69, 11–25, **1992**.

71 XIONG JP, STEHLE T, ZHANG R, JOACHIMIAK A, FRECH M, GOODMAN SL, ARNAOUT MA. Crystal structure of the extracellular segment of integrin avb3 in complex with an Arg–Gly Asp ligand. *Science* 296, 151–155, **2002**.

72 XIONG J, STEHLE T, GOODMAN SL, ARNAOUTI MA. A novel adaptation of the integrin PSI domain revealed from its crystal structure. *J. Biol. Chem.* 279, 40 252–40 254, **2004**.

73 ITO Y, KAJIHARA M, IMANISHI Y. Materials for enhancing cell adhesion by immobilization of cell-adhesive peptide. *J. Biomed. Mater. Res.* 25, 1325–37, **1991**.

74 NEFF JA, CALDWELL KD, TRESCO PA. A novel method for surface modification to promote cell attachment to hydrophobic substrates. *J. Biomed. Mater. Res.* 40, 511–519, **1998**.

75 HERSEL U, DAHMEN C, KESSLER H. RGD modified polymers: Biomaterials for simulated cell adhesion and beyond. *Biomaterials* 24, 4385–4415, **2003**.

76 PELHAM JR RJ, WANG YL. Cell locomotion and focal adhesions are regulated by the mechanical properties of the substrate. *Biol. Bull.* 194, 348–50, **1998**.

77 CASTEL S, PAGAN R, MITJANS F, PIULATS J, GOODMAN S, JONCZYK A, HUBER F, VILARO S, REINA M. RGD peptides and monoclonal antibodies, antagonists of avb3-integrin, enter the cells by independent endocytic pathways. *Lab. Invest.* 81, 1615–1626, **2001**.

78 YANG XB, ROACH HI, CLARKE NM, HOWDLE SM, QUIRK R, SHAKESHEFF KM, OREFFO RO. Human osteoprogenitor growth and differentiation on synthetic biodegradable structures after surface modification. *Bone* 29, 523–531, **2001**.

79 NEFF JA, TRESCO PA, CALDWELL KD. Surface modification for controlled studies of cell–ligand interactions. *Biomaterials* 20, 2377–2393, **1999**.

80 HIRANO Y, OKUNO M, HAYASHI T, GOTO K, NAKAJIMA A. Cell attachment activities of surface immobilized oligopeptides RGD, RGDS, RGDV, RGDT, and YIGSR toward five cell lines. *J. Biomater. Sci. Polym. Ed.* 4, 235–243, **1993**.

81 COOK AD, HRKACH JS, GAO NN, JOHNSON IM, PAJVANI UB, CANNIZZARO SM, LANGER R. Characterization and development of RGD-peptide-modified poly(lactic acid-co-lysine) as an interactive, resorbable biomaterial. *J. Biomed. Mater. Res.* 35, 513–523, **1997**.

82 BREUERS W, KLEE D, HOCKER H, MITTERMAYER C. Immobilization of a fibronectin fragment at the surface of a polyetherurethane film. *J. Mater. Sci. Mater. Med.* 2, 106–109, **1991**.

83 TONG YW, SHOICHET MS. Peptide surface modification of poly(tetrafluoro-ethylene-co-hexafluoropropylene) enhances its interaction with central nervous system neurons. *J. Biomed. Mater. Res.* 42, 85–95, **1998**.

84 PORTE-DURRIEU MC, LABRUGERE C, VILLARS F, LEFEBVRE F, DUTOYA S, GUETTE A, BORDENAVE L, BAQUEY C. Development of RGD peptides grafted onto silica surfaces: XPS characterization and human endothelial cell interactions. *J. Biomed. Mater. Res.* 46, 368–375, **1999**.

85 MARCHAND-BRYNAERT J. Surface modifications and reactivity assays of polymer films and membranes by selective wet chemistry. *Recent Res. Polym. Sci.* 2, 335–361, **1998**.

86 CARLISLE ES, MARIAPPAN MR, NELSON KD, THOMES BE, TIMMONS RB, CONSTANTINESCU A, EBERHART RC, BANKEY PE. Enhancing hepatocyte adhesion by pulsed plasma deposition and polyethylene glycol coupling. *Tissue Eng.* 6, 45–52, **2000**.

87 TAM JP, YU Q, MIAO Z. Orthogonal ligation strategies for peptide and protein. *Biopolymers* 51, 311–332, **1999**.

88 IVANOV B, GRZESIK W, ROBEY FA. Synthesis and use of a new bromoacetyl-derivatized heterotrifunctional amino acid for conjugation of cyclic RGD-containing peptides derived from human bone sialoprotein. *Bioconj. Chem.* 6, 269–277, **1995**.

89 HOUSEMAN BT, GAWALT ES, MRKSICH M. Maleimide functionalized self-assembled monolayers for the preparation of peptide and carbohydrate biochips. *Langmuir* 19, 1522–1531, **2003**.

90 HERBERT CB, MCLERNON TL, HYPOLITE CL, ADAMS DN, PIKUS L, HUANG CC, FIELDS GB, LETOURNEAU PC, DISTEFANO MD, HU WS. Micropatterning gradients and controlling surface densities of photoactivatable biomolecules on self-assembled monolayers of oligo(ethylene glycol) alkanethiolates. *Chem. Biol.* 4, 731–737, **1997**.

91 LIN YS, WANG SS, CHUNG TW, WANG YH, CHIOU SH, HSU JJ, CHOU NK, HSIEH KH, CHU SH. Growth of endothelial cells on different concentrations of Gly Arg-Gly Asp photochemically grafted in poly-ethylene glycol modified polyurethane. *Artif. Organs* 25, 617–621, **2001**.

92 SUGAWARA T, MATSUDA T. Photo-chemical surface derivatization of a peptide containing Arg–Gly Asp (RGD). *J. Biomed. Mater. Res.* 29, 1047–1052, **1995**.

93 REYES CD, GARCIA AJ. Engineering integrin-specific surfaces with a triple helical collagen mimetic peptide. *J. Biomed. Mater. Res. Part A* 65, 511–523, **2003**.

94 KANTLEHNER M, SCHAFFNER P, FINSINGER D, MEYER J, JONCZYK A, DIEFENBACH B, NIES B, HOLZEMANN G, GOODMAN SL, KESSLER H. Surface coating with cyclic RGD peptides stimulates osteoblast adhesion and proliferation as well as bone formation. *Chem-BioChem* 1, 107–114, **2000**.

95 DELFORGE D, GILLON B, ART M, DEWELLE J, RAES M, REMACLE J. Design of a synthetic adhesion protein by grafting RGD tailed cyclic peptides on bovine serum albumin. *Lett. Pept Sci.* 5, 87–91, **1998**.

96 KAO WJ, HUBBELL JA, ANDERSON JM. Protein-mediated macrophage adhesion and activation on bio-materials: A model for modulating cell behavior. *J. Mater. Sci. Mater. Med.* 10, 601–605, **1999**.

97 MASSIA SP, STARK J. Immobilized RGD peptides on surface grafted dextran promotes biospecific cell attachment. *J. Biomed. Mater. Res.* 56, 390–9, **2001**.

98 MASSIA SP, HUBBELL JA. An RGD spacing of 440 nm is sufficient for integrin avb3-mediated fibroblast spreading and 140 nm for focal contact fiber formation. *J. Cell. Biol.* 114, 1089–1100, **1991**.

99 ELBERT DL, HUBBELL JA. Conjugate addition reactions combined with free-radical cross-linking for the design of materials for tissue engineering. *Biomacromolecules* 2, 430–441, **2001**.

100 IRVINE DJ, HUE KA, MAYES AM, GRIFFITH LG. Simulations of cell-surface integrin binding to nanoscale-clustered adhesion ligands. *Biophys. J.* 82, 120–132, **2002**.

101 MAHESHWARI G, BROWN G, LAUFFENBURGER DA, WELLS A, GRIFFITH LG. Cell adhesion and motility depend on nanoscale RGD clustering. *J. Cell Sci.* 113, 1677–1686, **2000**.

102 IRVINE DJ, RUZETTE A-VG, MAYES AM, GRIFFITH LG. Nanoscale clustering of RGD peptides at surfaces using comb polymers – 2. Surface segregation of comb polymers in polylactide. *Biomacromolecules* 2, 545–56, **2001**.

103 Interagency Working Groups on Nanoscience, and Technology, National Nanotechnlgy Initiative: Leading to the Next Industrial Revolution. Washington D.C., Committee on Technology, National Science and Technology Council, **2000**.

104 ALIVISATOS AP. Semiconductor

clustors, nanocrystals, and quantum dots. *Science* 271, 933–937, **1996**.

105 SIMON A, DURRIEU M. Strategies and results of atomic force microscopy in the study of cellular adhesion. *Micron* 37, 1–13, **2006**.

106 WOODCOCK SE, JOHNSON WC, CHEN Z. Collagen adsorption and structure on polymer surfaces observed by atomic force microscopy. *J. Colloid Interface Sci.* 292, 99–107, **2005**.

107 CURTIS AS, CASEY B, GALLAGHER JO, PASQUI D, WOOD MA, WILKINSON CD. Substratum nanotopography and the adhesion of biological cells. Are symmetry or regularity of nanotopography important? *Biophys. Chem.* 94, 275–283, **2001**.

108 DALBY MJ, YARWOOD SJ, RIEHLE MO, JOSHSTONE HJH, AFFROSSMAN S, CURTIS ASG. Increasing fibroblast response to materials using nano-topography: morphological and genetic measurements of cells response to 13 mm-high polymer demixed islands. *Exp. Cell Res.* 276, 1–9, **2002**.

109 DALBY MJ, CHILDS S, RIEHLE MO, JOHNSTONE HJH, AFFROSSMAN S, CURTIS ASG. Fibroblast reaction to island topography; changes in cytoskeleton and morphology with time. *Biomaterials* 24, 927–935, **2003**.

110 DALBY MJ, GIANNARAS D, RIEHLE MO, GADEGAARD N, AFFROSSMAN S, CURTIS ASG. Rapid fibroblast adhesion to 27 nm high polymer demixed nano-topography. *Biomaterials* 25, 77–83, **2004**.

111 DALBY MJ, McCLOY D, ROBERTSON M, WILKINSON CDW, OREFFO ROC. Osteoprogenitor response to defined topographies with nanoscale depths. *Biomaterials* 27, 1306–1315, **2006**.

112 NAIR LS, BHATTACHARYYA S, LAURENCIN CT. Development of novel tissue engineering scaffolds via electrospinning. *Expert Opin. Biol. Ther.* 4, 1–10, **2004**.

113 CHRONAKIS IS. Novel nanocomposites and nonoceramics based on polymer nanofibers using electrospinning process – A review. *J. Mater. Process Tec.* 167, 283–293, **2005**.

114 SMITH LA, MA PX, Nanofibrous scaffolds for tissue engineering. *Collids Surf. B: Biointerfaces* 39, 125–131, **2004**.

115 MARTIN CR. Membrane based synthesis of nanomaterials. *Chem. Mater.* 8, 1739–1746, **1996**.

116 JAYARAMAN K, KOTAKI M, ZHANG Y, MO X, RAMAKRISHNA S. Recent advances in polymer nanofibers. *J. Nanosci. Nanotechnol.* 4, 52–65, **2004**.

117 TAYLOR GI. *Proc. Royal Soc. London* A313, 453, **1969**.

118 RENEKER JD, Electrospinning process and applications of electrospun fibers. *J. Electrostatics* 35, 151, **1995**.

119 RENEKER DH, YARIN AL, FONG H, KOOMBHONGSE S, *J. Appl. Phys.* 87, 4531, **2000**.

120 RENEKER DH, YARIN A, EVANS EA, KATAPHINAN W, RANGKUPAN R, LIU W. *Electrospinning and Nanofibers*, Book of Abstracts. In: New Frontiers in Fiber Science, Spring Meeting 2001. Available from: http://www.tx.ncsu.edu/jtatm/volume1specialissue/presentations/pres_part1.doc.

121 YARIN AL, RENEKER DH. Taylor cone and jetting from liquid droplets in electrospinning of nanofibers. *J. Appl. Phys.* 90, 4836–4846, **2001**.

122 SHIN MY, HOHMAN MM, BRENNER M, RUTELDGE GC. Experimental characterization of electrospinning: The electrically forced jet and instabilities. *Polymer* 42, 9955–9967, **2001**.

123 NAIR LS, BHATTACHARYYA S, BENDER JD, GREISH YE, BROWN PW, ALLCOCK HR, LAURENCIN CT. Fabrication and optimization of methylphenoxy substituted polyphosphazene nanofibers for biomedical applications. *Biomacromolecules*, 5, 2212–2220, **2004**.

124 FRENOT A, CHRONAKIS IS. Polymer nanofibers assembled by electrospinning. *Curr. Opin. Colloid Inter. Sci.* 8, 64–75, **2003**.

125 RENEKER DH, CHUN I. Nanometric diameter fibers of polymer produced by electrospinning. *Nanotechnology*, 7, 216–223, **1996**.

126 HUANG Z, ZANG YZ, KOTAKI M,

Ramakrishana S. A review on polymer nanofibers by electrospinning and their applications in nanocomposites. *Composites Sci. Technol.* 63, 2223–2253, **2003**.

127 Morota K, Matsumoto H, Mizukoshi T, Konosu Y, Minagawa M, Tanioka A, Ya, agata Y, Inoue K. Poly(ethylene oxide) thin films produced by electrospray deposition: Morphology control and additive effects of alcohols on nanostructure. *J. Colloid Interface Sci.* 279, 484–492, **2004**.

128 Katti DS, Robinson KW, Ko FK, Laurenicn CT. Bioresorbable nanofiber-based systems for wound healing and drug delivery: Optimization of fabrication parameters. *J. Biomed. Mater. Res. Part B, Appl. Biomater.* 70, 286–296, **2004**.

129 Deitzel JM, Kleinmeyer D, Hirvanen JK, Tan NCB. Controlled deposition of electrospun poly(ethylene oxide) fibers. *Polymer* 42, 8163–8170, **2001**.

130 Demir MM, Yilgor I, Yilgor E, Erman B. Electrospinning of polyurethane fibers. *Polymer* 43, 3303–3309, **2002**.

131 Deitzel JM, Kleinmeyer J, Harris D, Tan NCB. The effect of processing variables on the morphology of electrospun nanofibers and textiles. *Polymer* 42, 261–272, **2001**.

132 Zong X, Kim K, Fang D, Ran S, Hsiao BS, Chu B. Structure and process relationship of electrospun bioabsorbable nanofiber membranes. *Polymer* 43, 4403–4412, **2002**.

133 Koombhongse S, Liu W, Reneker DH. Flat ribbons and other shapes by electrospinning. *J. Polym. Sci., Polym. Phys. Ed.* 39, 2598–2606, **2001**.

134 Bognitzki M, Czado W, Frese T, Schaper A, Hellweg M, Steinhart M, Greiner A, Wendorff JH. *Adv. Mater.* 13, 70, **2001**.

135 Megelski S, Stephens JS, Chase DB, Rabolt JF. Micro- and nanostructured surface morphology on electrospun polymer fibers. *Macromolecules* 35, 8456–8466, **2002**.

136 Casper CL, Stephens JS, Tassi NG, Chase DB, Rabolt JF. Controlling surface morphology of electrospun polystyrene fibers: Effect of humidity and molecular weight in the electrospinning process. *Macromolecules* 37, 573–578, **2004**.

137 Matthews JA, Wnek GE, Simpson DG, Bowlin GL. Electrospinning of collagen nanofibers. *Biomacromolecules* 3, 232–238, **2002**.

138 Bornat A. Production of electrostatically spun products. *US Patent 4689186*, **1987**.

139 Theron A, Zussman E, Yarin AL. Electrostatic field-assisted alignment of electrospun nanofibres. *Nanotechnology* 12, 384–390, **2001**.

140 Xu CY, Inai R, Kotaki M, Ramakrishna S. Aligned biodegradable nanofibrous structure: A potential scaffold for blood vessel engineering. *Biomaterials* 25, 877–886, **2004**.

141 Ding B, Kim HY, Kim CK, Khil MS, Park SJ. *Nanotechnology* 14, 532–537, **2003**.

142 Kenawy ER, Bowlin GL, Mansfieled K, Laman J, Simpson DG, Sander EH, Wnek GE. *J. Controlled Release* 81, 57–64, **2002**.

143 Sun ZC, Zussman E, Yarin AL, Wendorff JH, Greiner A. *Adv. Mater.* 15, 1929, **2003**.

144 Ding B, Kimura E, Sato T, Fujita S, Shiratori S. Fabrication of blend biodegradable nanofibrous nonwoven mats via multijet electrospinning. *Polymer* 45, 1895–1902, **2004**.

145 Gupta P, Wilkes GL. Some investigations on the fiber formation by utilizing a side-by-side bicomponent electrospinning approach. *Polymer* 44, 6353–6359, **2003**.

146 Stankus JJ, Guan J, Wagner WR. Fabrication of biodegradable elastomeric scaffolds with sub-micron morphologies. *J. Biomed. Mater. Res.* 70, 603–614, **2004**.

147 Hamdan AL, Daniel SJ, Laura HM, Reneker D. Preservation of enzymes in electrospun nanofibers. Technical Papers – American Chemical Society, Rubber Division, Spring Technical Meeting, 163rd, San Francisco, 28–30 April 2003.

148 JIANG H, HU Y, LI Y, ZHAO P, ZHU K, CHEN W. A facile technique to prepare biodegradable coaxial electrospun nanofibers for controlled release of bioactive agents. *J. Controlled Release* 108, 237–243, **2005**.

149 LI WJ, LAURENCIN CT, CATERSON EJ, TUAN RS, KO FK. Electrospun nanofibrous structure: A novel scaffold for tissue engineering. *J. Biomed. Mater. Res.* 60, 613–621, **2002**.

150 HE W, MA ZW, YONG T, TEO WE, RAMAKRISHNA S. Fabrication of collagen-coated biodegradable polymer nanfiber mesh and its potential for endothelial cells growth. *Biomaterials* 26, 7606–7615, **2005**.

151 KWON K, KODOAKI S, MATSUDA T. Electrospun nano- to microfiber fabrics made of biodegradable copolyesters: Structural characteristics, mechanical properties and cell adhesion potential. *Biomaterials* 26, 3929–3939, **2005**.

152 LEE SH, KU BC, WANG X, SAMUELSON LA, KUMAR J. Design, synthesis and electrospinning of a novel fluorescent polymer for optical sensor applications. *Mater. Res. Soc. Symp. Proc.* 708, 403–408, **2002**.

153 LEE CH, SHIN HJ, CHO IH, KANG YM, KIM IA, PARK KD, SHIN JW. Nanofiber alignment and direction of mechanical strain affect the ECM production of human ACL fibroblast. *Biomaterials* 26, 1261–1270, **2005**.

154 MA PX, ZHANG R. Synthetic nano-scale fibrous. *J. Biomed. Mater. Res.* 46, 60–72, **1999**.

155 WEI G, MA PX. Structure and properties of nano-hydroxyapatite/polymer composite scaffolds for bone tissue engineering *Biomaterials* 25, 4749–4757, **2004**.

156 ZHANG R, MA PX. Synthetic nano-fibrillar extracellular matrices with predesigned macroporous architectures. *J. Biomed. Mater. Res.* 52, 430–438, **2000**.

157 WHITESIDES GA, MATHIAS JP, SETO CT. *Science* 254, 1312, **1991**.

158 ZHANG S. Building from bottom-up. *Mater. Today* 6, 20–27, **2003**.

159 ZHANG S, HOLMES T, LOCKSHIN L, RICH A. Spontaneous assembly of a self-complementary oligopeptide to form a stable macroscopic membrane. *Proc. Natl. Acad. Sci. U.S.A.* 90, 3334–3338, **1993**.

160 ZHANG S. Molecular self-assembly. In: BUSCHON KH et al. (Eds.), *The Encyclopdia of Materials Science and Technology*, Elsevier Science, Oxford, pp. 5822–5829, **2001**.

161 ZHAO Z, ZHANG S. Fabrication of molecular materials using peptide construction motifs. *Trends Biotechnol.* 22, 470–476, **2004**.

162 HARTGERINK JD, BENIASH E, STUPP SI. Self assembly and mineralization of peptide amphiphile nanofibers. *Science* 294, 1684–1688, **2001**.

163 NIECE KL, HARTGERINK JD, DONNORS JJM, STUPP SI. Self assembly combining two bioactive peptide-amphiphile molecules into nanofibers by electrostatic attraction. *J. Am. Chem. Soc.* 125, 7146–7147, **2003**.

164 PETKA WA, HARDEN JL, MCGRATH KP, WIRTZ D, TIRRELL DA. Reversible hydrogels from self-assembling artificial proteins. *Science* 281, 389–392, **1998**.

165 NOWAK AP, BREEDVELD V, PAKSTIS L, OZBAS B, PINE DJ, POCHAN D, DEMING TJ. Rapidly recovering hydrogel scaffolds from self assembling diblock copolypeptide amphiphiles. *Nature* 417, 424–428, **2002**.

166 SCHINDLER M, AHMED I, KAMAL J, NUR-E-KAMAL A, GRAFE TH, CHUNG HY, MEINERS S. A synthetic nanofibrillar matrix to promote *in vivo*-like organization and morphogenesis for cell in culture. *Biomaterials* 26, 5624–5631, **2005**.

167 WOZNIAK MA, DESAI R, SOLSKI PA, DER CJ, KEELY PJ, ROCK generated contractility regulates breast epithelial cell differentiationin response to the physical properties of a three-dimensional collagen matrix. *J. Cell. Biol.* 163, 583–595, **2003**.

168 SILVA GA, CZEISLER C, NIECE KL, BENIASH E, HARRINGTON DA, KESSLER JA, STUPP SI. Selective differentiation of neural progenitor cells by high-epitope density nanofibers. *Science* 303, 1352–1355, **2004**.

169 Li W, Tuli R, Huang X, Laquerriere P, Tuan RS. Multilineage differentiation of human mesenchymal stem cells in a three-dimensional nanofibrous scaffold. *Biomaterials* 26, 5158–5166, **2005**.

170 Yang F, Murugan R, Ramakrishna S, Wang X, Ma YX, Wang S. Fabrication of nano-structured porous PLLA scaffold intended for nerve tissue engineering. *Biomaterials* 25, 1891–1900, **2004**.

171 Yang F, Murugan R, Wang S, Ramakrishana S. Electrospinning of nano/micro scale poly(L-lactic acid) aligned fibers and their potential in neural tissue engineering. *Biomaterials* 26, 2603–2610, **2005**.

172 Mo ZM, Xu CY, Kotaki M, Ramakrishna S. Electrospin (P(LLA-CL) nanofiber: A biomimetic extracellular matrix for smooth muscle cell and endothelial cell proliferation. *Biomaterials* 25, 1883–1890, **2004**.

173 Venugopal J, Ma LL, Yong T. Ramakrishna S. In vitro study of smooth muscle cells on polycaprolactorn and collagen nanofibrous matrices. *Cell Biol. Int.* 29, 861–867, **2005**.

174 Ma Z, Kotaki M, Yong T, He W, Ramakrishnan S. Surface engineering of electrospun polyethylene terephthalate (PET) nanofibers towards development of a new material for blood vessel engineering. *Biomaterials* 26, 2527–2536, **2005**.

175 Genove E, Shen C, Zhang S, Semino CE. The effect of functionalized self-assembling peptide scaffolds on human aortic endothelial cell function. *Biomaterials* 26, 3341–3351, **2005**.

176 Shin M, Ishii O, Sueda T, Vacanti JP. Contractile cardiac grafts using a novel nanofibrous mesh. *Biomaterials* 25, 3717–3723, **2004**.

177 Ishii O, Shin M, Sueda T, Vacanti JP. In vitro tissue engineering of a cardiac graft using a degradable scaffold with an extracellular matrix-like topography. *J. Thor. Cardiovasc. Surg.* 130, 1358–1363, **2005**.

178 Yoshimoto H, Shin YM, Terai H, Vacanti JP. A biodegradable nanofiber scaffold by electrospinning and its potential for bone tissue engineering. *Biomaterials*, 24, 2077–2082, **2003**.

179 Li W, Tuli R, Huang X, Okafor C, Derfoul A, Danielson KG, Hall DJ, Tuan RS. A three-dimensional nanofibrous scaffold for cartilage tissue engineering using human mesenchymal stem cells. *Biomaterials* 26, 599–609, **2005**.

180 Kisiday J, Jin M, Kurz B, Hung H, Semino C, Zhang S, Grodzinsky AJ. Self assembling peptide hydrogel fosters chondrocyte extracellular matrix production and cell division: implications for cartilage tissue repair. *Proc. Natl. Acad. Sci. U.S.A.* 99, 9996–10 001, **2002**.

2
Polymeric Nanofibers in Tissue Engineering

Seow Hoon Saw, Karen Wang, Thomas Yong,
and Seeram Ramakrishna

2.1
Overview

Advances in tissue engineering in recent decades has sparked interest in fabricating polymeric fibers, in particular nanofibers, as scaffolds for biomedical applications. The unique properties of nanofibers, such as high surface area, remarkable surface properties and superior mechanical properties, have great potential for a wide variety of applications in medicine, biotechnology and engineering. Among the many features of nanofibers, biocompatibility is the most crucial aspect of tissue engineering. The materials that are mainly used are biopolymer and biodegradable synthetic polymers. Researchers have shown that polymeric nanofibers can be fabricated using various processing techniques such as self-assembly, phase separation and electrospinning.

This chapter reviews the applications of nanofibers in different aspects of tissue engineering. Studies have been carried out to optimize the properties of nanofiber scaffolds for applications in ophthalmology, hepatic biology, nerve, skin, bone and cartilage regeneration, heart and vascular grafts, and stem cell research. Numerous studies have shown that nanofibers that closely mimic the extracellular matrix in natural tissues have potential as scaffolds for tissue regeneration. They provide interconnected pores that facilitate transport of nutrients and growth factors to the cells as well as providing a stable structural support, which are essential for effective tissue engineering.

In nerve regeneration, nanofibers have been shown to effectively control the neurite outgrowth rate and orientation that are critical for the repair of the damaged nervous system. Similarly, the ability to control the morphology, proliferation, adhesion and alignment of smooth muscles cells and endothelial cells cultured on nanofibrous scaffolds provides a possible avenue for successful tissue engineering of blood vessels. Suitable bioactive nanofibers for proliferation of osteoblasts in guided bone regeneration, and dermal fibroblasts for burns and chronic wound treatment have also been reported. Studies on nanofibers for stem cell proliferation

Nanotechnologies for the Life Sciences Vol. 9
Tissue, Cell and Organ Engineering. Edited by Challa S. S. R. Kumar
Copyright © 2006 WILEY-VCH Verlag GmbH & Co. KGaA, Weinheim
ISBN: 3-527-31389-3

and differentiation are still in their infancy but the potential for biomedical applications is tremendous.

Although most of these studies are at the laboratory level, researchers foresee that polymeric nanofibers have good prospects in the near future. Advances in material science, manufacturing technology and design aspects of polymer nanofibers are necessary to realize their full potential in medicine, biology and engineering.

2.2
Introduction

Nanotechnology is a rapidly expanding field and the products resulting from research have increasingly entered into our daily lives. This chapter reviews potential applications of nanofibers, fiber fabrication techniques, and illustrates some of the current studies carried out to optimize the properties of nanofiber scaffolds for applications in ophthalmology, hepatic biology, nerve, skin, bone and cartilage regeneration, heart and vascular grafts, and stem cell research. The advantages and disadvantages of the nanofiber scaffolds are discussed and compared with conventional scaffolds in terms of their physical properties, performances and functionalities for tissue engineering applications. This chapter will contribute to the understanding of nano-scale polymeric fibers, which will facilitate the development of nanofibers with defined properties for specific applications to accelerate advanced research in tissue engineering.

2.2.1
History of Tissue Engineering and Nanofibers

Tissue engineering is a discipline that involves the application of the principles and methods of engineering and life sciences towards a fundamental understanding of structure–function relationships in normal and pathological tissues and the development of biological substitutes to restore, maintain or improve functions [1]. It is a relatively novel and emerging field that involves knowledge of bioengineering, the life sciences, and the clinical sciences with the aim of solving the serious medical problems of tissue loss and organ failure [2]. A new functional living tissue is created with the help of the growth and interaction of cells with a matrix or scaffolding to guide tissue development.

Previously, this area of science was called reconstructive surgery and, later, renamed as tissue engineering when attention was focused on fabricating living replacement parts for the body in the laboratory. Tissue engineering has developed at a logarithmic rate, and with researchers realizing the realistic potentials of this new discipline, many concepts and techniques that can be applied to saving lives

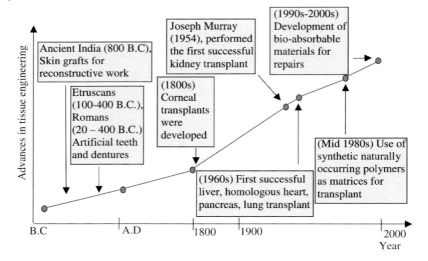

Fig. 2.1. Advances in tissue engineering.

are becoming a reality. Figure 2.1 shows a comprehensive summary of the advances made in tissue engineering.

Scaffolds for tissue engineering are an important aspect as they provide the optimized environment for cell survival, proliferation and differentiation, and also for the formation of tissues for desired tissue engineering applications. Constructing scaffolds for tissue engineering purposes is a challenging task as there are many factors that need to be taken into consideration, e.g., the implants must not be able to induce an immune response that can cause inflammation or rejection, a proper substrate is needed for cell survival and differentiation, and an appropriate environmental condition for tissue maintenance is required [3]. Research in the last decade has aimed to design and develop optimized scaffolds for tissue regeneration.

The field of nanotechnology was introduced by Drexler (1986) [4], and subsequently this has become a highly researched topic and promising area for applications in various industries, in particular biomedical applications. Fibers of nanoscale diameter, called nanofibers, were developed and found to be good candidates in the areas of molecular filters, tissue regeneration, drug delivery system, protective clothing, nanosensors and many more. Researchers have realized that polymeric nanofibers have unique properties such as high surface area, remarkable surface properties and superior mechanical properties, making them good candidates as tissue regeneration scaffolds. Thus, the potential of scaffolds in tissue engineering is enormous and studies have been carried out to gain a better understanding of the interactions between cells and the polymeric nanofibers in order to optimize scaffold designs for tissue engineering applications.

2.3
Classification of Nanofibers

Nanofibers are classified based on the materials of which they are fabricated: bio-degradable synthetic polymers, biopolymers, copolymers and composite polymers, which will impart unique properties to the nanofibers for specific applications.

2.3.1
Synthetic Polymers

A wide variety of synthetic polymers have been commonly used for fabricating nanofibers. In the area of tissue engineering, using synthetic polymers to fabricate nanofibers provides many advantages. Many studies have been performed to evaluate the potential of using these polymers in different tissue engineering applications. Some synthetic polymers such as aliphatic polyesters, poly(L-lactic acid) (PLLA), poly(glycolic acid) (PGA) and polycaprolactone (PCL) have biodegradable properties that are important for the development of biodegradable implants, scaffolds for tissue engineering, and controlled drug delivery. On the other hand, non-biodegradable synthetic polymers are frequently used for permanent or temporary prostheses in the area of biomedical applications, and are not commonly used as biological grafts. As they are not biodegradable, they satisfy the requirement of being sufficiently strong, flexible and hard to withstand the fatigue due to constant motion and friction. Some examples of commonly used non-biodegradable synthetic polymers are poly(vinyl chloride) (PVC), poly(tetrafluoroethylene) (PTFE) and poly(dimethyl siloxane) (PDMS).

Generally, synthetic polymer nanofibers have a major advantage of being mechanically stronger as well as greater flexibility, therefore making them highly durable scaffolds for replacement of desired tissues. In addition, synthetic polymers have another advantage, i.e., they can be produced relatively easily in large quantity with reproducible properties. This allows a consistency in the quality of scaffolds or devices fabricated for tissue engineering applications where obtaining reproducible results is important for determining future applications of the polymer. The degradation rate of the synthetic polymer would also be easier to determine and refine, making them more specific for desired applications. However, the main disadvantage is that synthetic polymers lack biocompatibility, which is a critical issue in tissue engineering.

2.3.2
Biopolymers

In recent years, researches have started using natural biopolymers to fabricate nanofibers. Some of these successfully electrospun biopolymers are silk [5, 6], fibrinogen [7], collagen [8], gelatin [9], chitosan [10] and chitin [11]. Natural biopolymers used in tissue engineering are often components in the extracellular matrix (ECM) produced by cells, giving them a better biocompatibility and providing a better environment for tissue regeneration. This in turn improves the cell attachment and growth on these nanofibers. However, variation in the quality of natural biopolymers is inevitable as they are extracted from different sources, and batch variations are unavoidable even when the biopolymers are extracted from the same source, thus affecting the reproducibility of studies. In addition, it is generally more difficult to obtain submicron or nano-sized fibers from natural biopolymers than from synthetic polymer nanofibers because of the relatively low molecular weight of the polymer and the limited solvent availability to dissolve biopolymers.

2.3.3
Copolymers

To overcome certain limitations of synthetic and natural biopolymer, researchers have developed copolymers and blends to fabricate nanofibers. With their unique and favorable properties combined, this enhances their usefulness as scaffolds in tissue engineering, examples are collagen-blended poly(L-lactide-*co*-ε-caprolactone) (PLLA-CL) [12], collagen/PCL [13], and gelatin/PCL [9]. Besides copolymers and blended nanofibers, surface modification is another means of improving the biocompatibility of nanofibers for specific tissue engineering applications, some examples are gelatin-grafted poly(ethylene terephthalate) (PET) nanofiber mats [14], collagen type I or gelatin grafted PLLA films [15], collagen-coated PLLA-CL nanofibers [16], concentrated hydrochloric acid treated PGA nanofibers [17], and galactosylated poly(ε-caprolactone-*co*-ethyl ethylene phosphate) (PCLEEP) nanofiber scaffolds [18].

2.3.4
Composite Polymers

In general, most synthetic and natural biopolymers lack certain properties such as material strength and biocompatibility. Researchers have included additives to the nanofibers to confer beneficial properties, making them specific for tissue engineering applications. Studies have revealed that composite nanofibers have better material strength and performance than synthetic and natural biopolymers nanofibers, and this is mostly due to the high surface area to volume ratio, thus providing a greater interaction between the additives and the nanofibers. In tissue engineering, examples of some composite polymers used to fabricate nanofibers are poly(L-lactide-*co*-glycolide) (PLGA)/chitin [19], polycarbonate urethane (PCU)/carbon [20], and gelatin/PCL [9].

2.4
Nanofiber Fabrication

Polymeric nanofibers can be fabricated by several techniques, including drawing, template synthesis, phase separation, self-assembly and electrospinning.

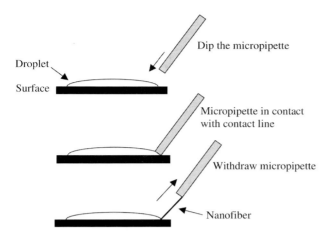

Fig. 2.2. Schematic diagram of the drawing technique.

2.4.1
Drawing

Drawing is a process whereby nanofibers are produced when a micropipette with a tip few micrometers in diameter is dipped into a droplet near the contact line using a micromanipulator and then withdrawn from the liquid at about 1×10^{-4} m s^{-1} (Fig. 2.2). When the end of the micropipette touches the surface, the drawn nanofiber is then deposited. To draw fibers, a viscoelastic material is required so that it can endure the strong deformation and at the same time it has to be cohesive enough to support the stress that build up in the pulling process. This method can be considered as dry spinning at a molecular level [21]. Nanofibers have been fabricated by the drawing method with citrate molecules [22]. The droplet of solution is deposited on a surface to allow evaporation to take place. After a few minutes of evaporation, due to the capillary flow, the edge of the droplet is more concentrated and the drawing process can begin.

2.4.2
Template Synthesis

Template synthesis involves utilizing a template to achieve nanofibers. Aligned polyacrylonitrile nanofibers have been fabricated by this method, using aluminum oxide membrane as the template [23]. In this particular process, when the pressure of the water is increased, it forces the polymer solution through the cylindrical nanopores of the membrane (Fig. 2.3). Once the extruded polymer comes into contact with the solidification solution, it solidifies and forms nanofibers. The diameter of the nanofibers depends on the diameter of the pores of the membrane. This method has been used to obtain nanotubules and nanofibrils of polymers, metals, semiconductors and carbons.

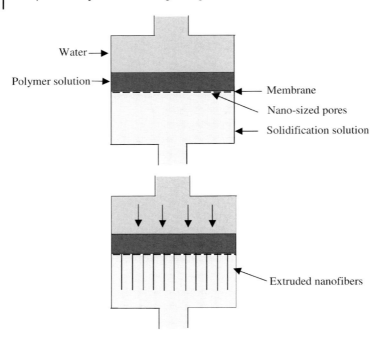

Fig. 2.3. Schematic diagram of template synthesis.

2.4.3
Phase Separation

Phase separation of a homogenous polymer–solvent solution can be induced in several ways, e.g., thermal induced phase separation (TIPS). When the polymer solution undergoes a decrease in temperature, two phases would form, a polymer-rich phase and a polymer-lean phase. Gradually, gelation would occur causing the polymer-rich phase to form a network structure when the solvent is removed via extraction, sublimation or evaporation. The spaces that were occupied by the solvent in the polymer-lean phase become pores [24] (Fig. 2.4). Generally, matrices made by solid–liquid phase separation have a greater pore size than by liquid–

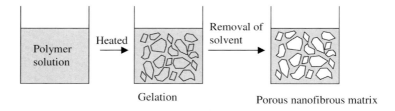

Fig. 2.4. Diagram showing nanofibrous matrix obtained by phase separation.

liquid phase separation. Common materials used for this process are PLLA and PLLA-CL. Fiber dimensions achievable by this process range from 50 to 500 nm [25]. The mechanical properties of nanofibrous mats so-fabricated can be customized by altering the polymer concentration. Various porous structures can be achieved with this method by varying the thermodynamic and kinetic parameters [26]. Although the phase separation method is an easy process, it can only be applied to a few polymers.

2.4.4
Self-assembly

Self-assembly is a complex process that involves individual, pre-existing components to organizing themselves into desired patterns and functions. Several factors significantly affect this process, including the peptide sequence if the self-assembly is based on protein–ligand interactions, concentration, pH, presence of salt, and time (or kinetics) [27]. By varying these parameters, nano-scale or macro-scale structures with great stability and functionality can be constructed. Polymeric nanofibers have been constructed by self-assembly through multiple processes of polymer chain annealing, crosslinking, dissolution and cleaving [25]. In the annealing process, polymer chains are annealed end to end, and are crosslinked chemically, thermally or by UV irradiation. The crosslinked polymers are immersed in chemicals to dissolve the continuous non-crosslinked polymer phase to disentangle the nanofibers. Further modifications involve cleaving and loading the nanofibers with chemicals to yield desired nanofiber morphology and chemical structures. By this method, nanofibers ranging from a few to 100 nm in diameter have been fabricated.

2.4.5
Electrospinning

Electrospinning is a technique used to fabricate polymeric nanofibers by means of an electrostatic force [21]. This cost effective and simple method was invented back in 1934 by Formhals [28]. The principle involves applying a high voltage to the surface of the polymer solution, and when the surface tension is overcome, an electrically charged jet is ejected. The electrical forces cause the discharged jet to stretch, forming a long and thin jet. At the same time, the solvent evaporates before the nanofiber is collected on a grounded collector. Figure 2.5 illustrates schematically the electrospinning process. To date, a great variety of polymer nanofibers have been successfully prepared by electrospinning when compared with other methods mentioned above that can fabricate nanofibers from only a limited number of polymers. With further advancement in this technology, electrospinning has the greatest potential for the mass production and industrial processing of nanofibers.

Electrospinning is influenced by many factors: the most significant is the properties of the polymer solution, e.g., viscosity, conductivity, and surface tension. Controlled variables such as distance between the needle tip and the collector, voltage

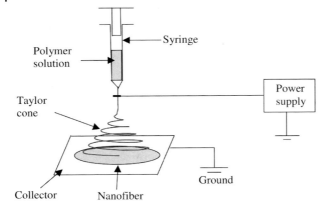

Fig. 2.5. Schematic diagram of the electrospinning technique.

applied, type of collector and feed rate have to be considered, too. Ambient parameters such as humidity, pressure and the type of atmosphere in the electrospinning environment also play a role [29]. With this knowledge, different morphology, structures and arrangements of nanofibers can be achieved by varying the parameters.

As most electrospun nanofibers are in the non-woven form, researchers have spent tremendous time and effort in controlling the alignment of the nanofibers. Several set-ups have been designed, such as the cylinder collector with high rotating speed [8, 30], auxiliary electrode/electrical field [31], parallel conducting collector [32], knife edge disk [33], frame collector [34] and many more. Currently, the thinnest electrospun nanofibers achievable are 3 nm in diameter [25]. Huang et al. (2003) have provided a more comprehensive review of electrospun nanofibers [35].

Comparing all of the above techniques, only drawing, template synthesis, phase separation and electrospinning are simple to perform. However, these processes have a common limitation, i.e., most of the polymeric nanofiber production process is currently only possible at the laboratory level, with the exception for electrospinning, which may be the only technique that has the potential for mass production.

2.5
Degradation and Absorption Kinetics of Nanofiber Scaffolds Compared with Conventional Scaffolds

One of the factors required of a tissue regeneration scaffold is that it needs to be biodegradable. As the tissue grows into the porous structure of the scaffold, it is difficult to remove the scaffold after the tissue has regenerated. Therefore, the scaffold is only required to maintain its volume, structure and mechanical stability just long enough for adequate tissue formation. The rate of the biodegradation needs to be adjusted so that it is able to match the rate of tissue formation. When degradation occurs, the breakdown products released should not induce any inflammation or cause toxicity to the body [36]. The biodegradation rate of the scaffold is influenced by many factors. One common method used to determine the rate is by placing the scaffold in phosphate-buffered saline (PBS) at 37 °C in an incubator (*in vitro*) or implanted into animals (*in vivo*), and biodegradability can be measured by the amount of mass loss, strength loss and the changes in morphology.

Conventional scaffolds such as hydrogels, sheets, meshes, and foams have been widely used in tissue engineering. Many studies were carried out to evaluate their degradation, e.g., Lee et al. (2004) [37] have shown that poly(aldehyde guluronate) (PAG) hydrogel crosslinked with poly(acrylamide-*co*-hydrazide) (PAH) had a weight loss of about 8% at the end of 60 days in Dulbecco's Modified Eagle Medium (DMEM). Generally, nanofibers have a very high surface area to volume ratio, and this will affect the degradation kinetics when compared to more conventional fibers. Studies showed that PGA nanofibers exhibited a rapid degradation rate and was left with a residual weight of approximately 40% at 20 days in PBS. Even after only 1 day of degradation, some PGA nanofibers have already started to undergo hydrolytic degradation and, after 4 days a significant amount of fibers was degraded. At the end of 12 days of degradation, the nanofiber matrix had become chunks of shorter fiber fragments [38]. In contrast, PGA microfiber non-woven matrix, which has been commercialized as a tissue engineering scaffold, only started to degrade at day 3, as shown by surface defects or microcracks. After several days, the surface defect degraded further to form a rupture in the microfiber [39]. The difference in the degradation rate between the microfiber and the nanofiber is probably due to the fiber diameter, which relates to the surface area. This property of rapid degradation is especially beneficial for temporary tissue regeneration scaffolds. However, studies have shown that the breakdown product from PGA was able to cause dedifferentiation as well as reduce proliferation rate of vascular smooth muscle cells. Therefore, PGA would not be the optimal scaffold for vascular tissue engineering [40]. As PGA nanofibers have a faster degradation rate, more breakdown product would be produced in a shorter period and this would cause more dedifferentiation and reduction of the proliferation rate when compared to conventional scaffolds.

Other degradation kinetic studies include that of Zong et al. (2005) [41], which showed that non-woven PLLA scaffold of an average fiber diameter of 1 μm had 90% remaining weight after about 20 days of degradation. The thickness of the scaffold also plays an important role in the degradation rate. With a reduction in the thickness of PLGA films, the degradation rate significantly decreased as there is faster diffusion of the degraded products, thus minimizing autocatalyzed hydrolysis [42].

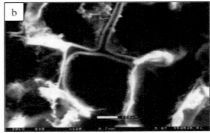

Fig. 2.6. (a) Scanning electron micrograph of a PGA non-woven scaffold with a fiber diameter of approximately 15 μm [54]. (b) Scanning electron micrograph of a PLLA foam fabricated using the salt-leaching technique [54].

2.6
Advantages and Disadvantages of Nanofiber Scaffolds Compared with Other Conventional Scaffolds

Textile technologies have been utilized to produce the earlier tissue engineering scaffolds consisting of fibrous biodegradable polymer fabrics with an average diameter of 15 μm. It has been used alone or in combination with other non-woven biodegradable polymers, such as PGA (Fig. 2.6a), PGA/PDLLA and PGA/PLLA for the engineering of cartilage [43, 44–46], tendon [47], ureter [48], intestine [49], blood vessels [50, 51], heart valves [46, 52] and other tissues [46]. However, the limitations of these scaffolds are a lack of structural stability to withstand biomechanical loading, low mechanical strength, fast degradation rate, difficulty in controlling pore shape and limited fiber diameter variations [53, 54].

Other conventional scaffolds developed using particulate leaching techniques [55, 56] and phase separation techniques have diameters of the order of a few to tens of microns. The fibers are not uniformly distributed and are not suitable for tissue engineering applications [55]. In the particulate leaching method, a polymer solution is cast into the salt-filled mold with the desired size. After the solvent is allowed to evaporate, the salt crystals are leached away using water to form the pores in the scaffold (Fig. 2.6b). It is relatively easy to prepare, the pore size can be controlled by the size of the salt crystals, and the porosity by the salt/polymer ratio. However, this method of scaffold preparation is not able to control the pore shape and interpore opening of the scaffolds. To improve these aspects, the three-dimensional (3D) printing (free-form fabrication) technique was adopted to fabricate scaffolds. In this process, a binder is ink-jet printed onto sequentially laid polymer powder layers [57] where computer-assisted design and manufacture (CAD/CAM) is used to design complex-shaped objects. This method was initially explored at Massachusetts Institute of Technology, USA [58–60]. The precision of this process is severely limited due to the smallest possible powder particle size achievable and pixel size of the binder drops (usually a few hundred μm), and positioning of the printer nozzle [61]. Membrane lamination is another technique that can be used to fabricate scaffolds by laminating membranes and introducing the peptides and proteins layer by layer during the fabrication process. This method will produce porous 3D structures [62]. The limitations of this process are less interconnected pore networks in the scaffold and the lack in mechanical properties. Solvent cast polymer–salt composites have also been extruded into a tubular geometry [63]. The disadvantages of producing the scaffold using this method are extensive use of highly toxic solvents, long duration for solvent evaporation (days to weeks), labor intensive fabrication process, limitation to thin structures, residual particles in the polymer matrix, irregularly shaped pores, and insufficient interconnectivity.

All the above-mentioned methods do not allow tissue engineers to design and fabricate scaffolds with a completely interconnected pore network, and a highly regular and reproducible scaffold morphology. However, many researchers have successfully developed high-performance polymer fibers with a high degree of fiber orientation using conventional fiber spinning techniques such as wet spinning, dry spinning, melt spinning, and gel spinning. Wet spinning is the oldest of these processes and it involves submerging the polymer solvent into a chemical bath, followed by precipitation and solidification of the polymer fiber. Dry spinning is achieved by evaporating the polymer solution in a stream of air or inert gas to solidify the fibers. In melt spinning, the molten polymer is forced through a spinneret and the fibers are directly solidified by cooling. Gel spinning is achieved by binding the polymer chains at various points in the liquid crystal state and further cooled in liquid bath after passing through air to produce the fibers [64–67]. With these methods, many different high-performance and functional polymer fibers have been produced and commercialized for many applications in the various industries.

Several complex molecular processes are involved in conventional fiber fabrication processes such as the spinning and drawing techniques. The ultimate state of molecular order in the polymer fibers depends on process variables such as stress, strain, temperature, time, length and length-distribution of the molecules [65]. Polymer fibers developed by these techniques have diameters in the range of microns ($10–100 \ \mu m$) and are classified as microfibers. The advantages of polymer microfiber are its flexibility which is easily controlled during structure formation, and the superior mechanical properties [68]. These fibers have been used in making ropes, satellite tethers, and high-performance sails. Polymer composite microfibers have found applications in aircrafts, boats, automobiles, sporting goods and biomedical implants. Several significant fiber properties such as ultraviolet resistance, electrical conductivity and biodegradability have been brought to good use.

The specific features of microfiber are fully exploited in industry, which exceeded previous performance level, and this led to the research, development and production of nano-diameter fibers or nanofibers with remarkable properties. When a fiber diameter is reduced to the nanometer scale, the surface area to volume ratio is significantly increased and this property has been exploited by many researchers for applications in various industries such as medicine, biotechnology and engineering [36]. There are more than 70 research groups worldwide investigating electrospun polymer nanofibers [69] and this is due to the unique properties of nanofibers when compared with the conventional macromaterials and polymer microfibers. Being as thin as less than 10 nm in diameter [35], nanofibers exhibit a larger surface area per unit mass of the order of 10^3 m^2 g^{-1} [70], permitting easier addition of surface functionalities as well as providing better mechanical properties. As a result, mats of these nanofibers are being tested for use in filter media, scaffolds for tissue engineering, protective clothing, reinforcement in composite materials and sensors [71]. The increased surface area, micro-scale interstitial space with high interconnectivity, good morphological stability, controlled fiber diameter, and sheet thickness make polymeric nanofibers an ideal scaffold for tissue engineering applications as it allows a significant amount of cell–surface interaction [69]. Polymeric nanofibers as well as other nano-structured materials such as nanotubes, nanowires, nanocrystals, nanorods and nanospheres are actively studied to uncover the many potentially advanced technological applications [72]. Several fabrication techniques are currently being investigated to develop nanofiber matrices from polymers: template synthesis, phase separation, drawing, self-assembly, and electrospinning (Section 2.4). Among these, electrospinning [33] is an efficient method to develop polymeric fibers with diameters in the nanometer range, and has attracted a great deal of attention recently. In addition to the small fiber diameter and pore size, and lightweight scaffolds that can be fabricated, electrospinning is an inexpensive and easy way to produce nanofibers from many types of polymers.

One of most crucial features of nanofibers is the huge availability of surface area per unit mass (Fig. 2.7). It provides a remarkable capacity for the attachment or release of functional groups, absorbed molecules, ions, catalytic moieties and

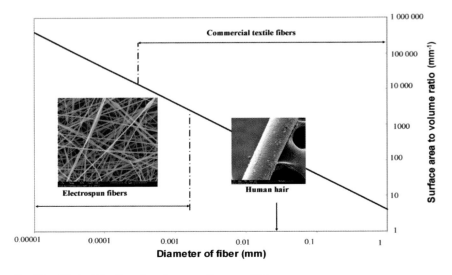

Fig. 2.7. Effect of the fiber diameter on surface area [73].

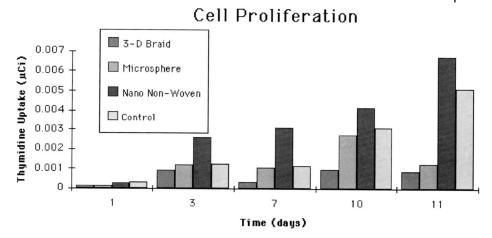

Fig. 2.8. Proliferation of fibroblast cells on four types of scaffolds: (red) 3D braided structure consisting of 20 bundles of 20 μm filaments; (green) 150–300 μm PLGA sintered spheres; (blue) non-woven, consisting of PLGA nanofibrils; (yellow) unidirectional bundles of 20 μm filaments. Proliferation was determined by the thymidine uptake of the cell as a function of time. Clearly, the PLGA nanofiber scaffold is the most favorable for cell growth [73].

nanometer-scale particles of many kinds [73]. The high surface cohesion of nanofibers is capable of trapping tiny particles that are less than 0.5 μm in size. The porosity and pore size of nanofiber membranes are important for application in tissue engineering. Nano-scale fibrous scaffolds can provide an optimal template for cells to attach, migrate, and grow. Mammalian cells can attach and organize well around fibers with diameters smaller than the diameter of the cells [74] and this is mainly due to the nano-scale interaction between the cells and the many biologically functional molecules and extracellular matrix components [54]. In a study of osteoblast bioactivity cultured on matrices consisting of 3D braided filaments, microspheres, non-woven nanofibrils, and microfilaments (Fig. 2.8), nanofibrils had the best support for cell growth and proliferation as determined by thymidine uptake [75]. This study also showed that a small fiber diameter facilitates greater available surfaces for the cell adhesion and migration. Non-woven mats formed from nanofibers have additional advantages of controllable pore size, high porosity, and permeability [76]. These remarkable properties motivate extensive interests in using these materials for applications in industries, especially in the biomedical fields. Composites nanofibers have been identified as potential candidates for high technology applications [35] due to its unique physical and mechanical properties. Hence, nano-featured synthetic scaffold design is one of the exciting new areas in tissue engineering.

Some recent studies showed that electrospun nanofibers can be collected in the form of highly oriented arrays that are able to elicit favorable biological responses due to their ability to mimic the biological microarchitecture and act to support and guide cell growth [33, 35, 67, 77]. Converting biopolymers into fibers and networks that mimic native structures will ultimately enhance the utility of these materials, as large diameter fibers do not mimic the morphological characteristics of the native fibrils [65]. The various properties of electrospun nanofibers such as mechanical strength, fiber diameter, porosity, and biocompatibility are dependent on the properties of the polymer type. Electrospun nanofibers fabricated from polyesters, polyurethanes, silicones, and poly(ethylene-*co*-vinyl acetate) (PEVA) have been identified as potential candidates for medical applications and have been fabricated as drug delivery devices, wound dressings and prosthetic devices such as vascular grafts [78–80] (Section 2.8). In the area of wound healing, electrospun nanofibrous membrane showed good and immediate adherence to wet wound surface where it attained uniform adherence to the wound surface without any fluid accumulation. The rate of epithelialization was increased and the dermis was well organized in the nanofibrous membrane, providing a good support for wound healing. The porosity of the nanofibrous membrane promotes evaporative water loss, excellent oxygen permeability and fluid drainage from the wound [81, 82]. Nanomaterial scaffolds have also been observed to enhance the functions of osteoblasts, such as adhesion, synthesis of alkaline phosphatase, and deposition of calcium-containing mineral compared with conventional scaffolds [83], and promote greater than 1.7× the osteoblastic cell attachment of conventional scaffolds. Nanofibrous scaffolds have also shown to adsorb four times more serum proteins than conventional scaffolds. More interestingly, the nanofibrous architecture selectively enhanced protein adsorption including fibronectin and vitronectin, even though both scaffolds were made from the same PLLA material. These results demonstrate that the biomimetic nanofibrous architecture serves as superior scaffolding for tissue engineering [84]. However, as-spun nanofibrous matrix can quickly lose its structural integrity or biological functionality in an aqueous environment such as in the human body [85]. Bhattarai et al. (2005) [85] have investigated the cellular compatibility of electrospun chitosan nanofibers incorporated with 40 wt.% poly(ethylene oxide) (PEO) but the nanofibrous matrix swelled readily in water and completely lost its fibrous structure in a few days. This would limit its application where a prolonged material functionality *in vivo* is required. Thus, processing conditions and mechanisms that would yield the desirable nanofibrous structure and material properties need to be investigated in order to construct nanofibrous matrix that exhibit good structural integrity to promote cell attachment and potentially serve as scaffold materials for tissue engineering.

Nanomaterials are unique as they are stronger than bulk materials. As fiber diameter decreases, the strength of the fiber (in this particular case glass fiber) increases exponentially [86] and this is probably due to the reduction in structural flaws of the glass fibers (Fig. 2.9a). Dresslhaus et al. (2000) [87] have described that, in the case of nanotubes, as the radius of matter gets smaller the strain energy per atom increases exponentially (Fig. 2.9b), contributing to over 30 GPa of increased strength. These unique characteristics in addition to their light weight, high permeability and the ability to absorb toxic materials enable these materials to be developed into high-performance filters, catalysis systems as well as for novel military fabrics [35]. Nano-sized fibers that can conduct electrical current and respond to electronic stimuli over metal contacts have also been developed. Norris

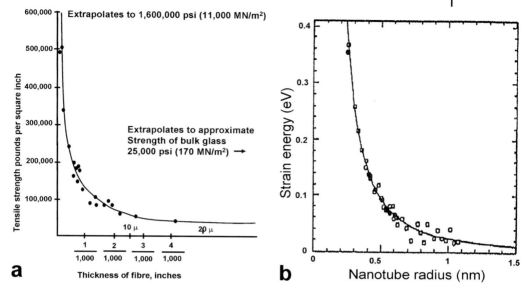

Fig. 2.9. Effect of fiber size on strength. (a) Glass fiber diameter versus the tensile strength [86]. (b) Strain energy as a function of nanotube radius [87].

and colleagues (2000) [88] found that polyaniline/PEO sub-micron fibrils have a response time an order of magnitude faster than bulk polyaniline/PEO in a doping–de-doping experiment. El-Aufy et al. (2003) [89] also demonstrated a significant increase in conductivity using sub-micron polyethylene dioxythiophene (PEDT) conductive fiber mat or when the fiber diameter decreases (Fig. 2.10). This effect is probably attributed to the intrinsic fiber conductivity effect or the geometric surface and packing density effect. However, nanofibers of smaller diameter have also been reported to exhibit lower and less stable conductivities. Zhou et al. (2003) [90] fabricated electrospun polyaniline/PEO nanofibers with diameters below 15 nm and found that the nanofibers were in fact electrically insulating. Many factors play important roles in polyaniline-based conductometric sensors, and one has to balance these factors to fully realize the potential of nano-structured conducting polymers. In some instances, "old-fashioned" materials may provide more desirable sensing characteristics [91]. Although the nanofiber films demonstrated faster responses, they exhibited lower sensitivity than conventional thin films. Researchers have shown that high surface area nanofibril structures do not effectively enhance sensor sensitivity due to the relatively open structure inherent in the many polymeric materials, and the adverse contribution from the interfibrillar contact resistance associated with nanofiber films [92, 93].

Current research focuses on the exploitation of these unique properties of nanofibers to improve their performance and functionalities for various applications. Improvement in the electrospinning technique has also been a focus of many research groups to refine and improve the properties and features of nanofibers. This

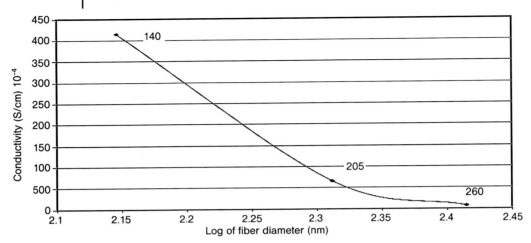

Fig. 2.10. Log of fiber diameter (nm) versus conductivity
(S cm^{-1}). As fiber diameter decreases from 260 to 140 nm, the
electrical conductivity of PEDT nanofibers showed almost two
orders of magnitude increase [89].

is because electrospinning is a simple process to fabricate polymer nanofibrous
materials for high-performance applications. Nano-scale fibers have the potential
to revolutionize many industries. During the past decade, rapid technological advances in the areas of surface microscopy, silicon fabrication, biochemistry, molecular biology, physical chemistry, and computational engineering have enhanced
and enabled research and implementation of nanotechnology in academia and industry. Figure 2.11 summarizes the advantages of nanofibrous scaffolds.

2.7
Biocompatibility of Nano-structured Tissue Engineered Implants

The objective of tissue engineering is to develop reproducible and biocompatible
3D scaffolds that act as bio-matrix composites to support cell in-growth in tissue
repair and replacement procedures [65]. Researchers have focused on making
such scaffolds with synthetic biopolymers and/or biodegradable polymer nanofibers [94–96]. Polymeric nanofibers have been widely studied as potential tissue
engineering materials for biomedical applications. As a matrix, nanofibers mimic
the structure and certain functions of the natural extracellular matrix to restore,
maintain or improve the function of human tissues. Being biocompatible, nanofibers have been used for the replacement of structurally or physiologically deficient tissues and organs in human [35].

Biocompatibility is described as the ability of an implant (biomaterial, biotextile,
prosthesis, artificial organ, biomedical device) to be accepted by the cellular and
biological response from the host environment and does not have the potential to

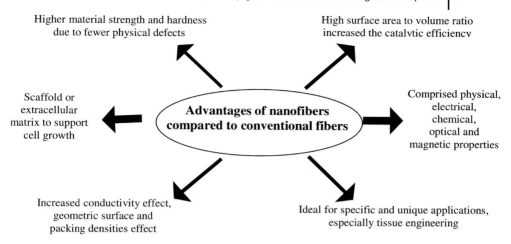

Fig. 2.11. The many advantages of nanofibers.

elicit an immunological or clinically detectable primary or secondary foreign body reaction [53, 97, 98]. It should be compatible with tissues in terms of mechanical, chemical, surface and pharmacological properties. *In vivo* biocompatibility testing of the biomaterials requires introduction of the implant into the biological environment to evaluate adverse reactions [99], such as alterations in homeostatic mechanisms, magnitude and duration of these altered homeostatic mechanisms if any, to determine the host biocompatibility response to the implant. If the implants are not rejected by the body, the following phases of recovery will be initiated: resolution (response of the biological environment to the injury and the presence of the implant), restitution (the return of the tissue environment to its normal structure and with the implant) and reorganization (the result of the wound healing response initiated by the injury and the presence of the implant). In addition, the rate of polymer biodegradation will also affect the biocompatibility of the implant in the biological system. Jayaraman and colleagues (2004) [71] have demonstrated that mats made of nanofibers from biodegradable polymers may be helpful in adjusting the degradation rate of specified biomaterials in the *in vivo* environment. Although no reports have been established on the influence of nanofiber diameter on the degradation behavior of polymers in the *in vivo* environment, there is evidence that the diameter of fibers do affect the degradation features and related mechanical properties of the materials [100]. In an *in vivo* study carried out in our laboratory on electrospun PLGA nanofibers tubes [101] that act as nerve guidance conduits, we observed that there was no inflammatory response after a month's implantation of the nanofiber nerve guidance conduit to the right sciatic nerve of the rat. Nerve growth factors or Schwann cells [102, 103] could be introduced into the nanofiber conduits to increase the biocompatibility of the nanofibers and hence improve the quality of the nerve regeneration and lead to clinical applications.

Polymeric nanofibers have been fabricated from various natural and synthetic biodegradable nanofibers [71] and most of the research on nanofibers for tissue engineering applications has centered on *in vitro* cell culturing to evaluate cell adhesion, proliferation, gene expression and extracellular matrix secretion [104]. Natural polymers are generally preferred over synthetic polymers in tissue engineering due to their tissue compatibility and biodegradation products which are readily resorbed into the body. In addition, nanofibers made from natural materials usually have better cell interactions than synthetic polymer nanofibers. Thus, the surface properties of the scaffold have been modified by adding surface coatings of collagen, gelatin, heparin, albumin or pyrolytic carbon to enhance the biocompatibility of the scaffold [69].

The most commonly used naturally derived polymers as tissue engineering scaffolds are collagen and chitosan [105]. Scientists are able to fabricate these nanofibers with diameters in the range of 100 nm to a few microns [8]. Collagen is a natural substrate for cell attachment, growth and differentiation. It provides considerable strength in its natural polymeric state. Collagen is also important in the design of tissue engineered devices [106] and it meets the requirements in wound healing since it has higher gas permeation and is able to protect the wound from infection and dehydration [81]. Venugopal et al. (2005) [107] fabricated nanofiber matrices from biodegradable PCL incorporated with collagen and showed that human dermal fibroblasts were able to grow, proliferate and migrate inside the matrices. These scaffolds support the attachment and proliferation of the cells and can be used as dermal substitutes for skin regeneration. In another study, Matthews et al. (2001) observed that smooth muscle cells cultured on electrospun collagen (calfskin) nanofiber matrix were able to grow into the nanofiber network [8]. Other studies showing infiltration of cells into the nanofiber networks include chondrocyte cultured on electrospun collagen type II scaffolds [108], human bone marrow stromal cell (BMSC) cultured on electrospun silk fibroin based-nanofibers [6], and human keratinocytes and fibroblasts proliferation on silk fibroin-based nanofibers [5].

Chitosan, a biologically renewable, biodegradable, non-antigenic and biocompatible natural polymer [105], is widely studied in wound dressing, wound healing [109], drug delivery systems [110], bone tissue engineering [111–113] and in various tissue engineering applications [114–117]. Electrospun chitosan nanofibers incorporated with synthetic polymers such as chitosan/poly(vinyl alcohol) and ultrathin hybrid nanofibers containing chitosan and PEO have been prepared by Ohkawa et al. (2004) [118], Duan et al. (2004) [119] and Bhattarai et al. (2005) [85]. Chitosan nanofibers with controllable size and alignment to form non-woven mats or 3D porous structures will provide an unlimited source for the development of natural scaffolds for tissue engineering [85]. Elastin, another structural protein of the extracellular matrix, has been successfully electrospun into elastin-mimetic peptide polymer nanofibers. The elastic property of elastin nanofibers has been capitalized on for the engineering of arterial blood vessels [96, 120].

In bone tissue engineering, nano-structured composites consisting of nano-hydroxyapatite and collagen (nHAC) have been developed by co-precipitation of hydroxyapatite (HA) and collagen [121], and mineralizing type I collagen sheet with hydroxyapatite [122, 123]. Preliminary *in vitro* and *in vivo* studies indicate that nHAC is bioactive and biodegradable [124]. However, its mechanical properties were too weak for practical application. To improve the mechanical strength, a nano-HA/collagen/PLA (nHAC/PLA) scaffold has been developed [124] which has high biocompatibility and strength, and serve as a promising scaffold in traditional bone-defect repair and in bone tissue engineering. Osteoblast cultured on the nHAC/PLA nano-composites performed well in adhesion, proliferation and maturation that accompanied the morphological change from spindle shaped to predominantly polygonal morphology. To further enhance nHAC mechanical properties, Zhang et al. (2003) [125] integrated nHAC composite with a small fraction of Ca-crosslinked to alginate to produce a porous structure that mimicked the composition of natural bone as well as to increase the osteoconductive activity of the scaffold. Polysaccharide alginate possesses good biocompatibility and provides satisfying mechanical support. Fibroblasts and osteoblasts co-cultured with nHAC/ Ca-alginate composite exhibited good cell morphology, adhesion and proliferation, inferring the composite as a good scaffold material for bone tissue engineering. Although the efforts in fabricating natural polymer-based fibrous structures are encouraging, much remains to be explored and improved, especially in *in vivo* applications.

The uses of biocompatible synthetic polymer nanofibers have been widely investigated to evaluate their feasibility as scaffolds for applications in tissue engineering. Materials that have been used are polyesters, polycaprolactone, poly(amino acids), polycarbonates, poly(ethylene glycol) and many more. Polyesters such as PLA, PGA and copolymers of lactide/glycolide are among the most commonly used biomaterials for drug delivery [126] and tissue engineering. They break down to naturally occurring metabolites and degradation requires only water. Fine fibers produced from PLA and PGA by electrospinning for tissue engineering applications have been intensively investigated [30, 127]. ECM-like scaffolds fabricated from PLLA nanofibers by phase separation promote cell growth, proliferation and migration into the interstices of the nanofiber network [128, 129]. A nano-3D PLGA scaffold constructed by solvent casting and salt leaching processes had been shown to be biocompatible and support human bladder smooth muscle cells adhesion, growth and proliferation [130]. Cells cultured on nano-dimensional scaffolds showed a corresponding increase in cellular protein production when compared to cells cultured on conventional, micro-dimensional scaffolds. In addition, evidence in pressure experiments showed that, in general, scaffolds and resident cells experiencing a sustained pressure stimulus of 10 cm-H_2O functioned similarly to those experiencing atmospheric (control) pressures. These results suggest that the novel porous, nano-dimensional PLGA scaffold is promising for *in vivo* replacements of the urinary bladder wall. Webster et al. (2005) [83] observed that the functions of osteoblasts such as adhesion, synthesis of alkaline phosphatase, and deposition of calcium-containing mineral showed an increase on nanofibrous PLGA scaffolds compared with conventional ceramics. Since both the nanofibrous PLGA and conventional ceramics share the same chemistry, material phase, porosity, and pore size, the study implies that the surface features created by adding nanophase compared with conventional titania was a key parameter that enhanced functions of osteoblasts. Such nanophase ceramics (or nanomaterials in general) require further attention as orthopedic tissue engineering materials [83].

PCL is compatible with soft and hard tissue and is used as resorbable suture material, in drug delivery systems, and as bone graft substitutes. It has been incorporated into other synthetic polymers to create an ideal scaffold for the cell growth. Fetal bovine chondrocytes seeded on a nanofibrous PCL scaffold and cultured in serum-free medium maintained their chondrocytic phenotype by expressing cartilage-specific extracellular matrix genes that were evaluated by reverse transcription-polymerase chain reaction (RT-PCR). PCL nanofibers not only promoted the phenotypic differentiation but also promoted chondrocyte proliferation when cultures were maintained in serum-containing medium [131]. Adult bone marrow-derived mesenchymal stem cells cultured in a PCL nanofiber scaffold were induced to form chondrocytes in the presence of transforming growth factor-$\beta 1$ (TGF-$\beta 1$), as evidenced by chondrocyte-specific gene expression and synthesis of cartilage-associated extracellular matrix proteins. The chondrogenic ability of stem cells cultured in a PCL nanofiber scaffold was comparable to that observed for stem cells maintained as cell aggregates or pellets [132]. Yoshimoto et al. (2003) [133] seeded mesenchymal stem cells (MSCs) derived from the bone marrow of neonatal rats on electrospun PCL nanofibers as well and cultured the cells in osteogenic culture medium under dynamic culture conditions. At 4 weeks, the material was covered by multiple cell layers and mineralization occurs together with the existing type I collagen on the PCL nanofiber. Besides supporting the chondrocytes and stem cells, PCL nanofibers have also been observed to support the attachment of cardiomyocytes. Shin et al. (2004) [134] developed a cardiac nanofibrous mesh by culturing cardiomyocytes on electrospun nanofibrous PCL scaffold. The nanofibrous mesh acted as an extracellular matrix for cell growth and was supported by a wire ring that played a role as a passive load for the contracting cardiomyocytes. The cells on the scaffold were observed to start beating after 3 days of culture. In addition, cells cultured *in vitro* for 14 days were found to attach well on the PCL mesh and expressed cardiac-specific proteins such as α-myosin heavy chain, connexin43 and cardiac troponin I. Block copolymers of PLLA and PCL (PLLA-CL) electrospun into non-woven and aligned nanofiber scaffolds have been reported to support the adhesion and proliferation of human coronary artery endothelial and smooth muscle cells, with a future application in blood vessel tissue regeneration [135–137]. In addition, PLGA and PCL nanofibers have been successfully utilized by Li et al. [131, 132, 138] to culture fibroblasts, cartilage, and bone marrow-derived mesenchymal stem cells. These reports showed that biocompatible synthetic polymeric nanofiber scaffolds were capable of enhancing cell attachment and proliferation.

Poly(*p*-dioxane) (PPDO) is a hydrophobic polyester that possesses good biocompatibility, flexibility and tensile strength. It degrades to non-toxic compounds and can be used in the body safely [139]. It has been applied to different applications in the medical field [140–142], such as surgical sutures [143] and drug delivery system [141]. Poly(ethylene glycol) (PEG) is another widely used biocompatible polymer. PEG is a simple linear polyether that exhibits outstanding physiochemical and biological properties, and possesses good hydrophilicity and solubility in water and organic solvents. It lacks toxicity, and does not induce immunological effect, allowing it to be used for many biomedical and biotechnological applications [144, 145]. When PEG is coupled to PPDO, the conjugated product possesses the beneficial properties of both individual polymers. PLLA conjugated to PEG is another block copolymer that possesses properties desirable for tissue engineering applications [146–148]. The synthesis of a novel biocompatible block copolymer consisting of PPDO, PLLA and PEG (PPDO/PLLA-b-PEG) by Bhattarai et al. (2004) [82] takes advantage of the combined properties of these polymers to further enhance biocompatibility. NIH 3T3 fibroblast cells seeded on the PPDO/PLLA-b-PEG nanofibrous structure were shown to maintain their phenotypic shape and guided growth according to the nanofiber orientation. PEO is another biocompatible polymer that has been used in food, cosmetics, personal care products and pharmaceutical. Electrospun poly(ethylene-*co*-vinyl alcohol) nanofiber mat can also support the growth of smooth muscle cells and fibroblast [80].

2.8
Applications of Polymeric Nanofibers in Tissue Engineering

With the advancement in tissue engineering in the past decades, researchers realized that nanofibers have potential as scaffolds in tissue regeneration. These nanofibers can be fabricated from either natural or synthetic polymers for various applications in tissue engineering (Table 2.1). The rational for using nanofibers is based on the principle that cells organize and attach well around fibers with a diameter smaller than that of the cells [74]. In addition, non-woven polymeric nanofibers mimic the nano-scale fibers that are present in the natural extracellular matrix. They provide interconnected pores that facilitate transport of nutrients and growth factors to the cells as well as providing a stable structural support, which are essential for effective tissue engineering. A unique characteristic of the nanofibers is its huge surface area-to-mass ratio, in the range of $10–10^3$ m^2 g^{-1} for a fiber diameter of around 500 nm [98], and this makes them suitable to host various functionalities and be applied to a broad range of applications in life sciences and sensor applications. Therefore, studies have been performed to optimize the properties of the nanofiber scaffolds in the areas of ophthalmology, hepatic biology, nerve, skin, bone and cartilage regeneration, heart and vascular grafts, and stem cell research as well as drug delivery application [149]. Tissues can be engineered at different levels of complexity (Fig. 2.12) [150]. Using this knowledge of nanotechnology, tissue engineered products with highly predictable biological and physical properties can be obtained.

In nerve regeneration, nanofibers have been shown to effectively control the neurite outgrowth rate and orientation that are critical for the repair of the damaged nervous system. Similarly, the ability to control the morphology, proliferation, ad-

Tab. 2.1. Different types of biomaterials in the various biomedical applications.

Applications	Biomaterials
Ophthalmic (Intraocular lenses, contact lenses)	PMMA, PHEMA, PEU, PEO, PGMA, PVA, PLGA, collagen
Liver	PLA, PGA, PGLA, polyorthoesters, polyanhydride, PLGA, PCLEEP
Nerve (Hydrocephalus shunts)	PEU, PLGA, PDLLA, PLLA-CL, PDMS, PTFE, collagen, glycosaminoglycan, PGA
Skin (Facial and hip prostheses, artificial skin)	PEU, PCL, PTFE, PE, Collagen, PGA, PGLA (Vicryl), nylon, collagen-glycosaminoglycan
Orthopedic and cartilage (Bone cement for fracture fixation, sutures, bone repair)	PMMA, polyamides, PP, hydroxyapatite, PGA, PLLA, PGLA+hydroxyapatite fibers, PCL
Heart (Heart valves, artificial heart, ventricular assist devices, pacemaker leads)	PDMS, PEU, PTFE, PE, PSu, PGA
Vascular graft (Blood substitutes)	PTFE, PVP, Polyester (Dacron), polyurethane, expanded PTFE, PGA, PLA, PGLA, PVC, collagen
Stem cells	PCL, PLA
Drug delivery systems	PDMS, PEVA, PLA, PGA, PLGA

PCL: polycaprolactone; PCLEEP: poly(ε-caprolactone-*co*-ethyl ethylene phosphate); PDLLA: poly(D,L-lactic acid); PDMS: polydimethylsiloxane, silicone elastomers; PE: polyethylene; PEO: poly(ethylene oxide); PEU: polyurethanes; PGA: poly(glycolic acid); PGLA, PLAGA, PLGA: copolymers of poly(glycolic acid) and poly(lactic acid); PGMA: poly(glyceryl monomethacrylate); PHEMA: poly(2-hydroxyethyl methacrylate); PLA: poly(L-lactic acid); PLLA-CL: copolymer of PLA and PCL; PMMA: poly(methyl methacrylate); PP: polypropylene; PS: polystyrene; PSu: polysulfone; PTFE: poly(tetrafluoroethylene); PVA: poly(vinyl alcohol); PVC: poly(vinyl chloride); PVP: poly(vinyl pyrrolidone); PEVA: poly(ethylene-*co*-vinyl acetate).

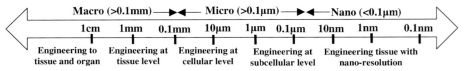

Fig. 2.12. Tissue engineering at different levels.

hesion and alignment of smooth muscles cells and endothelial cells cultured on nanofibrous scaffolds provides a possible avenue for successful tissue engineering of blood vessels. Suitable bioactive nanofibers for proliferation of osteoblasts in guided bone regeneration, and dermal fibroblasts for burns and chronic wound treatment have also been reported. Studies on nanofibers for stem cell proliferation and differentiation are still in their infancy but the potential for biomedical applications is tremendous.

2.8.1
Ophthalmology

Every part of the human eye plays an important role in providing clear vision. The cornea is the primary component of the ocular optical system that refracts light onto the retina for vision and act as a tough protective barrier for the delicate internal eye tissues [151]. It is an avascular and transparent tissue that consists of three main layers: outer stratified epithelium, stroma of cell network within a hydrated collagen–proteoglycan matrix, and inner endothelial layer. Nerve activity is crucial for the maintenance of overall corneal health. Innervation loss can cause "dry eye" [152], a pathological condition that will decrease the sensitivity of the cornea and/ or caused the erosion of the cornea epithelial. When the sensitivity is lost, the cornea can be affected by a wide variety of disorders [153–156] and becomes vulnerable to irreparable injury, ulceration, eventual loss of vision or blindness [157, 158].

Cornea disease is a major cause of vision loss [159] and it affects more than 10 million individuals worldwide [160]. Transplantation of human donor corneal graft is the widely accepted treatment in the hope of visual recovery [151]. However, transplant donors are limited and worldwide demand exceeds supply. This situation will worsen with an aging population and increased used of corrective laser surgery [161]. Patients with conditions such as autoimmune disorders, alkali burns, graft rejection or recurrent graft failures will have a lower success rate in corneal transplantation [162]. Furthermore, there is a risk in the transmission of infectious agent [163]. Therefore, an alternative for these patients is the replacement of the damaged cornea with an artificial substitute. The structural and immunological simplicity of the human cornea, and the importance of nerve innervations for optimal function, make it an ideal tissue for tissue engineering studies. The ideal artificial cornea (keratoprothesis) should consist of materials and structures that support adhesion and proliferation of cornea epithelial cells in order to form an intact continuous epithelial layer [163]. It should be permeable to provide appropriate exchange of nutrient and fluid, light transparent and non-toxicity to cells.

In the early 1950s until the 1990s, a few synthetic polymers were used extensively for the manufacture of keratoprostheses. Poly(methyl methacrylate) (PMMA) [164–168] was used as intraocular lenses into human patients. Poly(2-hydroxyethyl methacrylate) (PHEMA) hydrogels have a long history of proven biocompatibility in the cornea [169–182]. Polyurethanes [183–185], poly(glyceryl monomethacrylate) (PGMA) [186–188], PEO, and poly(vinyl alcohol) (PVA) hydrogels [189] were proposed and/or used episodically as keratoprosthetic materials. Corneal epithelial cells have been preseeded onto PVA and transplanted into rabbit corneas, where they remained adherent and proliferated for 1 to 2 weeks [190–193].

Research in synthetic cornea replacement is essential since most artificial cornea is still plagued with poor adhesion to the host tissues and does not promote rein-nervation [194]. Scientists from Argus Biomedical, Australia, have claimed to successfully develop an artificial cornea, named Alphacor. It is made of a biocompatible, flexible, hydrogel material similar to a soft contact lens. It contains a central clear zone, consisting of a transparent gel PHEMA that provides refractive power and a peripheral skirt or rim made of an opaque, porous, high-water PHEMA that encourages the eye to heal over the device. Alphacor has been under clinical study since 1998 and was FDA-approved in August 2002. The replacement cornea is designed to minimize the risk of complications, and to replace the scarred or diseased cornea of the eye that has a history of multiple graft failures [195]. Using collagen-copolymer implants, Li et al. (2003) [160] reported successful growth of stratified epithelium and stromal fibroblast, and nerve innervations into the implant. In two other studies, collagen-copolymer scaffolds designed to mimic the natural ECM matrix and crosslinked to TERP5 copolymer synthesized from *N*-isopropylacrylamide were shown to allow cell growth on the surface [196] and as well as cell ingrowths [197]. Cell adhesion factors may be incorporated into the biosynthetic matrices to promote cell growth and nerve innervations.

The retina is the other part of the eye that has received extensive research in tissue engineering. It has a complex, multilayered architecture that consists of polarized photoreceptor cells that are closely related to the retinal pigment epithelium (RPE), nerve innervations, and rich supply of blood capillaries [198]. The integrity of the RPE layer is critical to the survival and functions of photoreceptors and hence patterns the vision. Retinal degeneration such as retinitis pigmentosa (RP) and age-related macular degeneration (AMD) affect over a million people in USA alone, where they suffer from irreversible visual disability. Photoreceptor cells will lose their functions when the degeneration happens at the outer retina and this will affect the cells that connect and support the photoreceptors of the inner nuclear layer. Conventional treatment of the disease such as retinal progenitor cell (RPC) grafts is not sufficient to create the complex cytoarchitecture on a large scale, especially when multiple retinal layers have been lost or disrupted [199, 200]. Studies have shown that delivery of RPCs using typical bolus injection of RPCs leads to a large degree of cell death [201]. In addition, the widespread intra-retinal migration frequently observed with these cells may not be desirable in all settings. A tissue engineering approach using a biodegradable polymer scaffold with the appropriate architecture may improve the survival and promote the organized differentiation of grafted RPCs in retinal degeneration and injury.

Recently, RPCs have been isolated from the mature eye [202, 203] to develop retina [204, 205] and these cells may be able to replace photoreceptors [206]. Lavik et al. (2005) [198] (Fig. 2.13) seeded RPCs onto porous degradable PLGA scaffolds fabricated by anisotropic phase inversion [207] and by liquid–liquid phase separa-

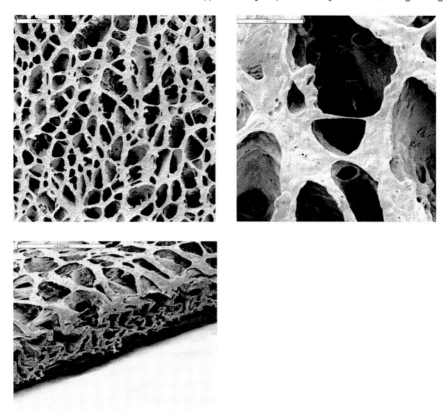

Fig. 2.13. PLGA scaffold prepared using the solid–liquid phase separation technique [198].

tion [208]. The results of this work indicate that degradable polymer scaffolds improve the survival of RPCs in retinal degeneration model, promote differentiation of RPCs, and provide physical guidance to the RPCs resulting in a more normal anatomical organization (Fig. 2.14).

There are very few comprehensive and systematic reports on the functional relationship between nanofibers and applications in ophthalmology. One of the most recent studies is by Zhong et al. (2005) [209] who seeded rabbit conjunctiva fibroblasts (RCF) on electrospun random collagen-glycosaminoglycan (GAG) blended nanofibrous scaffolds with a fiber diameter range 100–600 nm. The results showed that the nano-sized scaffold increased the proliferation of rabbit conjunctiva fibroblast on the scaffolds (Fig. 2.15). Since collagen and GAG are components of the natural ECM, their incorporation into scaffolds has been used extensively for *in vitro* cell-ECM interactions studies and as platform for tissue biosynthesis [210–217]. Traditional collagen gels can self-assemble into a nano-scale fiber-like superstructure [218], but their application is limited due to the poor physical strength. However, aligned nanofibers possess higher mechanical strength than random

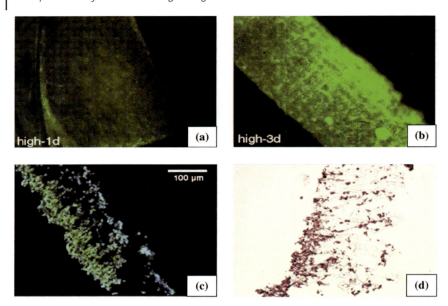

Fig. 2.14. Seeding of RPCs that are stained green on the solid–liquid phase separated scaffolds for (a) 1 and (b) 3 days. (c) Cross section of the scaffold at 3 days post seeding. The green fluorescence RPCs were counterstained blue with 4′,6-diamidino-2-phenylindole (DAPI). (d) Cross section of the same sample stained with hematoxylin and eosin (H&E) [198].

nanofibers [16], and this fiber alignment feature can be used to increase the physical strength of collagen nanofibers. In a study conducted in our laboratory in collaboration with the Singapore Eye Research Institute [219], it was shown that aligned nanofiber scaffold seeded with RCF exhibited lower cell adhesion but higher cell proliferation than the random scaffold. This is likely due to the aligned

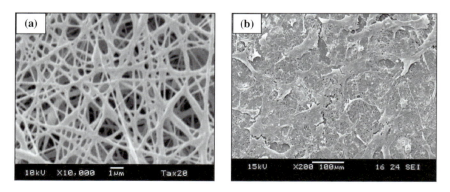

Fig. 2.15. Scanning electron microscope (SEM) micrographs of (a) electrospun collagen-GAG scaffolds, and (b) RCFs on the collagen-GAG scaffolds after 7 days culture [209].

orientation of the fibers, which may control cell orientation and at the same time strengthen the interaction between cells and fibers. This may indirectly increase the overall physical strength of the biological scaffolds. From these results, application of nano-scale collagen or collagen-GAG scaffolds in ophthalmology is significant as the nano-scale dimension of these electrospun nanofibers mimic that of native ECM found in our body and at the same time promote tissue biosynthesis, guide cell growth, and maximize the rate of tissue regeneration [136, 210–221].

2.8.2
Liver

The cells in the liver are highly organized into structural units called liver lobules. These cells or hepatocytes are radially disposed in the lobules, piling up to form one cell thick layers, alike to the bricks of the wall. These cell plates are directed from the periphery of the lobule to its center. In human, these lobules are in close contact with one another, forming sponge-like structures. A unique feature of the liver is its ability to regenerate following injury or resection and tissue engineering methodologies have been applied to assist in liver regeneration. Nano-structured biomaterials have been used to enhance the development of liver replacement devices [222]. In a study by Chua et al. (2005) [18], growth of primary rat hepatocytes on PCLEEP nanofiber scaffolds conjugated with galactose ligands to achieve a biofunctional construct was evaluated. PCLEEP was shown in earlier reports to have good tissue compatibility and low cytotoxicity to the cells [223–225]. Galactose-conjugated substrates have been reported to mediate hepatocyte adhesion, minimize involvement of integrin-mediated signaling pathways, and reduce the loss of hepatocyte phenotype [226]. In this study, hepatocytes cultured on the galactosylated PCLEEP nanofibrous scaffold exhibited comparable functional profiles in terms of cell attachment, ammonia metabolism, albumin secretion, and cytochrome P450 enzymatic activity, when compared with hepatocytes cultured on functional 2D matrix. In addition, hepatocytes cultured on galactosylated nanofiber scaffold formed smaller aggregates of 20–100 µm that engulfed the functional nanofibers and created an integrated spheroid-nanofiber construct, resulting in reduced cell detachment from the nanofibrous scaffold, and form stable immobilized hepatocyte spheroids throughout the period of culture (Figs. 2.16 and 2.17). Such functional nanofiber scaffolds can be incorporated into a bioartificial liver assist device design, and with their textured and porous nature may promote hepatocyte scaffold interaction, improve the stability of the attached cells, maintain their differentiation functions, and, finally, remain stable against the perfusion and shear forces in the bioreactor.

2.8.3
Nerve

Neural tissue engineering (NTE) is a promising and challenging field that involves various disciplines, such as grafting processes, polymer bioprocess and surface

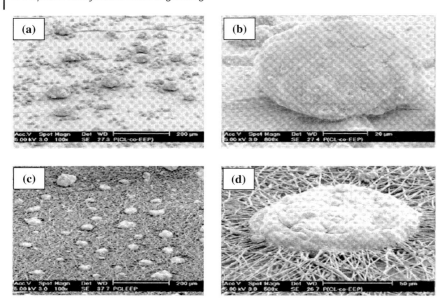

Fig. 2.16. SEM images of hepatocytes after 8 days of culture. Hepatocytes cultured on galactosylated film (Gal-film) formed rounded spheroids that did not integrate with the scaffold (a, b). In contrast, hepatocytes cultured on galactosylated-nanomesh (Gal-nanomesh) showed aggregates that engulfed the functional nanofibers (c, d) [18].

chemistry modifications, as well as involving cell–cell and cell–matrix interactions that occur during development, remodeling and restoration of neural cells [227]. Nerve tissue repair is an important treatment concept in human health care as it directly impacts on the quality of life. However, restoring the nerve function after traumas or diseases has been a great challenge for neurobiologists and neurologists, as the central nervous system (CNS), such as brain and axons, do not regenerate on its own in their native environment. In the peripheral nervous system (PNS), healing of severed peripheral nerves may be impaired by the growth of fibroblastic connective tissues that will disrupt and prevent the proximal neuron from growing towards and reattaching to the distal stump [69]. Attempts to replace lost or dysfunctional neurons using tissue transplantation or peripheral nerve grafting method have been intensely sought for over a century [228]. The purpose is to restore the function in damaged or diseased regions. However, these methods always encountered problems such as donor shortage and immunological problems associated with infectious diseases [229]. To overcome some of these problems, alternative approaches are being investigated that use biomaterials to influence the function and differentiation of cultured or transplanted cells to enhance nerve regeneration [230–233].

The ultimate goal of neural tissue engineering is to achieve suitable biointeractions for a desired cell response that is required to compensate for the loss of tissue function [234] through *in vivo* induction of a neural circuit or *in vitro* fabrication

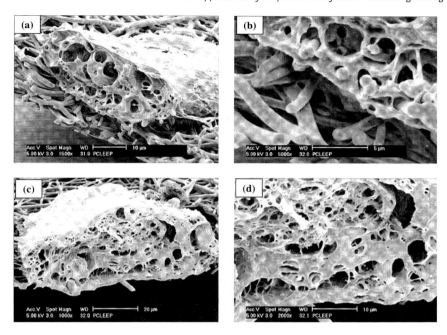

Fig. 2.17. SEM images of freeze-fractured hepatocytes on Gal-nanomesh after 8 days of culture. PCLEEP nanofibers can be found within the hepatocyte aggregates (a, b). However, no fibers were observed in some hepatocyte aggregates, which may be attributed to degradation of the biodegradable PCLEEP nanofibers (c, d) [18].

of a tissue structure. Engineered polymer scaffolds that can serve as extracellular matrix are crucial to support the fundamental cell processes. Recent advances in the NTE provide optimism by creating a permissive environment for nerve regeneration [163, 235–237]. The ECM of the nervous system is crucial in guiding neural outgrowth and may be relevant to the process of regeneration. It consists of the collagen types I, II, III, IV and V, the noncollagenous glycoproteins and the GAGs [238, 239]. A great variety of natural ECM components exist in fibrous form and structure, some examples include collagen, fibronectin and laminin. All of them are characterized by well-organized hierarchical fibrous structures, ranging from nanometer to millimeter scale [240]. Hence, a nano-structured porous scaffold with interconnective pores and large surface area is needed as an alternative to natural ECM for better cell ingrowths in a three-dimensional fashion.

Polymers can be used as scaffold to promote cell adhesion, maintenance of differentiated cell function without interfering proliferation, template to organize and direct the growth of cells and help in the function of extracellular matrix [241]. Several methods have been reported to fabricate polymeric nanofibers or submicron fibers: phase separation [242], electrospinning [243] and self-assembly [244]. Yang et al. (2004) [245] have shown the potential of the PLLA nanofibrous porous scaffold, prepared by phase separation in nerve tissue engineering. The scaffold mimicked the structure of the natural ECM and was shown to support neural stem cells differentiation and outgrowth of the neurites. In another study, Spilker et al. (2001) [246] demonstrated that cells obtained from nerve explants were able to construct and restructure the walls of porous collagen-GAG matrices *in vitro*, suggesting that contractile cells may be capable of restructuring the extracellular component of the nerve wound environment *in vivo*. Electrospinning is able to fabricate fibrous scaffolds with desired properties [136, 247] to create biocompatible thin structures with useful coating design and surface properties that can be deposited on implantable devices to facilitate the integration of these devices into the body. This is exemplified in electrospun silk-like polymer with fibronectin functionality for making biocompatible films, which were used on prosthetic devices aimed to be implanted in the central nervous system [94].

Advanced techniques to produce a complex, degradable guidance channels that precisely mimic a natural repairing process in the human body have been investigated [248–251]. Various biomaterials or polymers have also been investigated for their suitability in nerve tissue engineering applications [101]. These synthetic conduits have been fabricated from materials such as polyurethane, polyorthoester, glycolide trimethylene carbonate, PLGA, PGA, poly(D,L-lactic acid) (PDLLA) and PLLA-CL [63] (Table 2.1). Early nerve regeneration studies used biotextile incorporated with appropriate growth factors to join the two neural stumps. The proximal neuron regenerates and can bridge gaps as wide as 8 mm, so as to re-establish distal neural activity [69]. In another study, the bioresorbable implant (trimethylene carbonate-*co*-ε-caprolactone) seeded with Schwann cells showed a significant increase in growth, ensheathment and peripheral nerve myelination [252]. In a recent study by Bini and colleagues (2004) [101], nerve guidance channel was fabricated by electrospinning PLGA biodegradable polymer nanofibers onto a Teflon mandrel. The biological performance of the conduits was examined in the rat right sciatic nerve model with a 10 mm gap length. There was no inflammatory response after the implantation of the nanofibrous nerve guidance conduit. Five out of eleven rats showed nerve innervation into the conduits after a month of implantation (Fig. 2.18). None of the implanted conduits showed tube breakage. This showed that the nanofibrous nerve guidance conduits have the necessary flexibility to adapt well inside a living system, a porous structure that make it permeable for gaseous and nutrient exchange in the conduit lumen to promote the nerve regeneration, are biocompatible, and showed no inflammatory response, which is clinically desirable in minimizing adhesions of an implanted conduit to surrounding tissues.

Nerve regeneration in the CNS is much more difficult than in the PNS and this is probably due to the glial cells in the CNS, which act as a barrier to the regenerating axons and create an unfavorable environment for axon regeneration after trauma or injury [237, 239]. Another reason would be the dissimilarities in the ECM components between PNS and CNS to assist and promote nervous regeneration [239]. In the PNS, the linear orientation of ECM components at the peripheral stump of a damaged nerve provide terrain well suited for axonal regrowth, whereas such framework is not available in the ECM of the CNS to assist axonal growth

Fig. 2.18. The regenerated nerve cable after a month of implantation [101].

[238, 253]. The nanofibrous scaffolds are well suited for CNS tissue engineering due to two unique properties. Firstly, the morphology and architecture are similar to the natural ECM, which is the foundation of creating reproducible and biocompatible 3D scaffolds for cell attachment, differentiation and proliferation. Secondly, the nanofibrous scaffolds are highly porous structures with a wide variety of pore diameters, allowing fluid transportation while inhibiting glial scars. Moreover, with the electrospinning technique, aligned nanofibrous scaffolds can be fabricated to

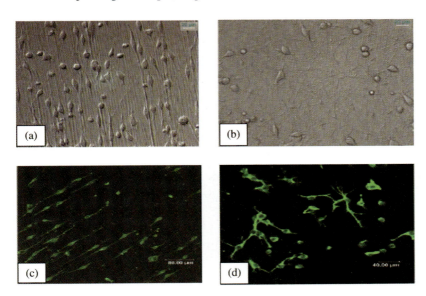

Fig. 2.19. C17.2 cells cultured on aligned (a, c) and random (b, d) PLLA nanofiber scaffolds for 1 day (a, b) and 2 days (c, d). Parts (a) and (b) are phase contrast light microscope images of cells attached on aligned nanofibers and random nanofibers, respectively. Parts (c) and (d) are laser scanning confocal micrographs of cells immunostained for neurofilament 200 kDa when cultured on aligned and random nanofibers, respectively [254].

guide neuronal regeneration in both the central and peripheral nervous systems [254].

Yang and colleagues (2005) [254] designed an electrospun aligned PLLA nanofibrous scaffold to evaluate the efficacy in promoting neuron differentiation and guiding neurite outgrowth of C17.2 cells *in vitro*. The aligned nanofiber scaffold is anticipated to provide better contact guidance effects on the neurite outgrowth. C17.2 is a primordial, multipotent self-renewing cell that can be used as neuron precursors, and is involved in the normal development of cerebellum, embryonic neocortex and other structures upon implantation [251, 255, 256]. The cells are capable of differentiating without interaction with adhesion molecules such as laminin, fibronectin, collagen, poly-L-lysine or Matrigel™, which are generally required as permissive substrates in neurite outgrowth. This offers the convenience of investigating the physical effects of PLLA fibrous scaffolds because the coating of adhesive molecules will affect the surface topography of the scaffold. The *in vitro* results of this study showed that the fiber alignment had a strong effect on the cell phenotype: neural cells on aligned fibers grew parallel to the fiber orientation, and the aligned nanofibers improved neurite outgrowth when compared with random or microfibrous scaffolds (Figs. 2.19 and 2.20). The results suggest that the aligned nanofibrous scaffold may be a suitable nerve guidance channel for both CNS and PNS regeneration.

Researchers have not only explored the use of biomaterials for neural tissue engineering. A silicon-based electronic device for neural network applications has been developed by Fan et al. (2002) [257]. They investigated the adhesion of neural cells from substantia of prenatal rat on silicon wafer with different nanotopographic features of between 20 and 70 nm in dimension produced by etching. The cell adhesion and viability were significantly improved on the nano-featured surface. The results strongly suggested that nanotopography could improve the

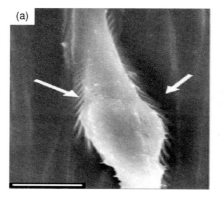 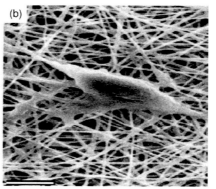

Fig. 2.20. SEM micrographs of neural stem cells seeded on (a) aligned and (b) random nanofiber for 2 days, showing the cell–matrix adhesion between the neural stem cells and PLLA fibers. Bar = 5 μm [254].

response of neural cells and be used in designing neurochips as proposed by researchers such as Weis and Fromherz at the Max-Plank Institute for Biochemistry, Germany, and Pine and colleagues at the California Institute of Technology, Pasadena, California [222].

2.8.4
Skin

Extensive burns or skin damaged by ambustion and injury require comprehensive therapy to repair the conditions, to prevent dehydration and infection. Dressing of wound is aimed to protect, remove exudates, inhibit exogenous microorganism invasion, and improve appearance [81]. Many studies have been done in the past 30 years to develop and enhance the skin regeneration by tissue engineering. In 1962, Winter [258] showed that covering the wound with a polyethylene film significantly improves the epithelialization of the wound in a shorter time. Protection was further accomplished by covering the wound with a dressing such as artificial skin constructed from either natural or synthetic polymers [259]. Wound areas that are kept just damp may heal faster, but accumulation of exudates under the dressing can cause infection [260]. Currently, autograft is a standard treatment in wound dressing. It is done by removing a section of the skin from another part of the body and grafting onto the wound. However, the removal of the dermis and epidermis is a serious operation and the availability of healthy skin can be limited if the burnt areas are widespread [261]. Recent developments in the regeneration of skin by tissue engineering have overcome these difficulties, especially in the use of polymer nanofibers for wound healing.

The objective of tissue engineered wound dressing is to produce an ideal structure, which gives higher porosity and good barrier. The materials must be selected carefully to produce high quality barrier properties and oxygen permeability. Electrospun nanofibrous membranes have potentials as wound dressing. This is because the membrane attained uniform adherence on wet wound surface without any fluid accumulation [82]. It can meet the essential requirements of wound healing such as high gas permeation and protection of wound from infection and dehydration. Khil and colleagues (2003) [81] have shown that the rate of epithelialization was increased and the dermis was well organized in wound covered with electrospun nanofibrous polyurethane membrane, which provided a good support for wound healing. The wound dressing showed controlled evaporative water loss, excellent oxygen permeability and promoted fluid drainage ability due to the porous nature of the nanofibers and the inherent properties of polyurethane. Polyurethane is frequently used in wound dressing because of its good barrier properties and oxygen permeability. Woodley and colleagues (1993) [262] reported that semipermeable dressings, many of which are constructed from polyurethane, enhanced wound healing. The permeability of the wound dressing is crucial to prevent accumulation of fluid between the wound and the dressing. The fluid absorbed in the wound dressing will keep the wound moist and this will inhibit wound desiccation. Researchers have also tried to determine the effect of occlusive dressings on the healing of wound prepared by polyurethane [263–266]. The disadvantage of the semiocclusive dressings is that fluid will accumulate under the dressings after a few days [267] and aspiration of the wound is required to prevent leakage and infection.

Another polymer that is frequently used for wound dressing is PCL, a biodegradable polymer characterized by a resorption time in excess of one year [268] but is known to be susceptible to enzymatic degradation [269]. Various degradative enzymes such as collagenases, matrix metalloproteinases, gelatinase A and stromelysin-1, which are secreted by macrophages, epidermal cells and fibroblasts for cell migration and repair during wound healing have been established to degrade PCL [270]. PCL and collagen nanofiber membranes have been used as wound dressings [107]. This structured membrane improves the mechanical integrity of the matrix, and provides a high level of surface area for cells to attach due to its 3D features and high surface area to volume ratio when compared to polymer film (Fig. 2.21). This PCL/collagen membrane will possess both the mechanical properties and cell binding affinity derived from the unique properties of PCL and collagen, respectively. In this study, fibroblasts were shown to migrate inside the collagen nanofibrous matrices mimicking the structure of the dermal substitute. Collagen synthesized by fibroblasts is a good surface active agent and will en-

PCL nanofibers

Collagen nanofibers

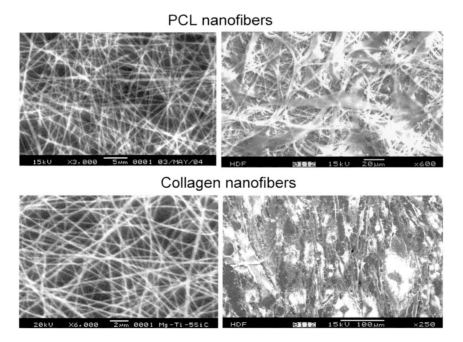

Fig. 2.21. Fibroblast cell attachment and growth on PCL and collagen nanofibers [107].

hance the attachment of keratinocytes to the surface of artificial dermis in serum free medium. It is assumed that the fibroblasts on the artificial dermis can release biologically active substances, such as cytokines, which induce invasion of the wound site by fibroblasts and other immunological cells during the early synthesis of the new skin tissue. The dynamic architecture of the fibers such as pore size and fiber alignment allows the cells to grow into the nanofiber matrices to form a dermal substitute for many types of wound healing. This nanofiber-based fibroblast-cultured dermal substitute maintains the moist environment on the wound surface and thereby promotes wound healing.

The incorporation of collagen in polymer scaffolds has been demonstrated in many studies to promote cell growth and proliferation. Using human osteoblasts, Coombes et al. (2002) [271] demonstrated that PCL composite films consisting of non-crosslinked collagen mats support higher cell growth than unmodified PCL films. In another study, Dai et al. (2004) [272] found that composite films of collagen and PCL are favorable substrates for growth of fibroblasts and keratinocytes and may find utility for skin repair. To improve on this further, Huang and colleagues (2001) [273] developed a convenient, non-toxic, and non-denaturing process to fabricate collagen-containing nanofibers with diameters ranging from 100 to 150 nm, and non-woven fabrics, which they proposed based on the materials excellent properties to have potential applications in wound healing, tissue engineering and as haemostatic agents. Fine fibers of biodegradable polymers have also been developed to be directly sprayed or spun onto the injured location of the skin, forming a nanofibrous mat dressing, which encourages wound healing by the formation of normal skin and elimination of scar tissue that would occur in a conventional treatment [274–277]. Two patents have been filed [278, 279] where researchers produced a skin mask by directly electrospinning fibers that contain pH-adjusting compound onto the skin surface to protect and eventually heal the wound.

The main advantages of non-woven nanofibrous membrane mats for wound dressing are their pore size, usually from 500 nm to 1 μm, which is small enough to protect the wound from bacterial penetration, and the high total nanofiber surface area of 5 to 100 $m^2 g^{-1}$, which is extremely efficient for fluid absorption and dermal delivery. A commonly used wound dressing in hospitals is Tegaderm™ developed by 3M (USA), which is a thin polyurethane membrane that is permeable to both water vapor and oxygen but impermeable to microorganisms. In one study, electrospun nanofibrous membrane showed excellent and immediate adherence to wet wound surface compared with the dermis of wound covered with Tegaderm™ (3M, USA), which was also reported to be inflamed [81]. The rate of epithelialization was increased, the dermis was well organized, and fluid accumulation was not observed in electrospun nanofibrous membrane group [81, 82]. These result indicate that the electrospun nanofibrous membrane provided a good means for wound healing. In addition, histological examination confirmed that epithelialization rate was increased, and the exudates in the dermis were well controlled by covering the wound with the electrospun membrane. Thus, nanofibrous membrane prepared by electrospinning could be properly employed as wound dressings.

2.8.5
Bone and Cartilage

Only in recent years have researchers gained a better understanding of stem cell biology and their potential for bone and cartilage repairs and regeneration, and this has accelerated the development of biomaterial scaffolds for tissue engineering research in this discipline. The surface properties of biomaterial regulate the cell phenotype through their capacity to support and spatially display the crucial adhesion contacts [280, 281]. Nano-scale features have also been used to design the scaffolds for culture of osteoblasts and mesenchymal stem cells [221]. Recent advances in stem cell biology have shown that MSCs can differentiate into cells of mesenchymal tissues such as bone, cartilage, muscle, tendon, ligament and fat, and are expected to play important roles in the repair of skeletal defects [282, 283]. Osteoblasts derived from MSCs of neonatal rats cultured on poly(D,L-lactide-*co*-glycolide) foam were observed to deposit minerals and form 3D bone-like structures in the foam [284]. In another report, a non-woven PCL nanofibrous scaffold was fabricated and its potential evaluated in bone tissue engineering with rat mesenchymal stem cells [133, 285]. It was observed that the polymer constructs were covered with multiple cell layers, mineralized, and impregnated with type I collagen. Besides synthetic polymer, researchers have used natural polymers such as silk to produce nanofibrous scaffolds [6]. Human bone marrow stromal cells were cultured on electrospun silk fibroin fibers with an average diameter of 700 nm. These silk matrices supported bone marrow stromal cell attachment and proliferation with higher cell density than for native silk fibroin matrices.

The 3D structure of these nanofibrous scaffolds are characterized by a wide range of pore diameter, high porosity, high surface area to volume ratio, and morphological architectures similar to the natural structural fibrillar proteins in the native ECM. These physical characteristics promote favorable biological responses of cells by enhancing cell attachment and proliferation as demonstrated by Li and colleagues (2002) [138] when they developed an electrospun PLGA nanofibrous scaffold that has potential as a possible alternative replacement for bone, cartilage and skin tissue engineering. Fibroblast and bone marrow-derived mesenchymal stem cells seeded on this structure tend to maintain phenotypic shape and guided growth according to the nanofiber orientation. Li et al. (2003) [131] further evaluated a novel, 3D non-fibrous PCL scaffold composed of electrospun nanofiber for its ability to maintain chondrocytes in a mature functional state. Nanofibrous cultures maintained in the supplemented serum-free medium produced more sulfated proteoglycan-rich cartilaginous matrix, and supported cellular proliferation than monolayer cultures. In another study, Li et al. (2005) [132] evaluated the effects of growth factors incorporated in the scaffolds on cell growth. When the adult human bone marrow-derived MSCs were cultured on a PCL nanofibrous scaffold in the presence of transforming growth factor-β (TGF-β), the MSCs differentiated to chondrocytic phenotypes, expressing chondrocyte-specific genes such as collagen type II, IX, and XI, aggrecan, and produced cartilage-associated extracellular matrix proteins. The level of chondrogenesis in the nanofibrous scaffold was com-

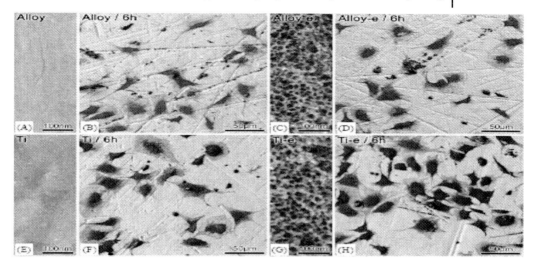

Fig. 2.22. SEM images of osteogenic cells grown on diverse titanium surfaces. Control samples, (A) Ti alloy and (E) cpTi, exhibit a smooth surface, while etched ones, (C) Ti alloy and (G) cpTi, are characterized by a unique nanotopography. At 6 h, no significant differences were detected in cell shape between control (B and F) and nanotextured (D and H) surfaces. The predominant cell shape was polygonal, with cells showing either thin cytoplasmic extensions or large veil-like ones. cp: commercially pure, Ti: titanium [286].

parable to cell aggregates or pellet cultures, which are the established procedures in inducing chondrogenesis. In addition, the mechanical properties of the cell–nanofibrous construct were superior to those developed from pellet cultures. These studies suggested that nanofibrous structure may be a suitable candidate scaffold biomaterial for bone and cartilage tissue engineering applications.

Nanotexturing has also been reported to provide the necessary attachment surfaces on the scaffolds for cell to grow. This was demonstrated by Oliveira et al. (2004) [286] who reported that nanotexturing of titanium-based surfaces upregulates the early expression of bone sialoprotein and osteopontin in osteogenic cell cultures (Fig. 2.22). Badami et al. (2005) [287] determined how chemical and topographical features affect adhesion, morphology, orientation, proliferation, and osteoblastic differentiation of MC3T3-E1 model osteoprogenitor cells, and found that osteoprogenitor cells were able to adhere and proliferate on a random fused fiber topographies and mean fiber diameters ranging from 0.14 nm to 2.1 μm. In the presence of osteogenic factors such as β-glycerophosphate, and L-ascorbate-2-phosphate, cell density on fibers was equal or greater than on smooth surfaces. Cells attached to electrospun substrates exhibited a higher cell aspect ratio than cells on smooth surfaces. This study demonstrated that surface topography introduced by electrospun fibers with an average fiber diameter of 0.14–2.1 μm affects cell morphology and cell proliferation. Carbon nanotubes and nanofibers have several potential properties that make these materials beneficial in the development of novel devices for bone reconstruction [20]. The dimensions of carbon nanofibers or nanotubes mimic that of collagen fibrils (0.1–8 μm in diameter) in the bone. In one study, Elias et al. (2002) [288] investigated the functions of human osteoblasts cultured on carbon nanofibers with diameters of <100 nm. Osteoblasts were observed to increase in proliferation, synthesis of alkaline phosphatase and deposited more extracellular calcium on carbon nanofiber with diameters < 100 nm. The enhanced mineral deposition in this study demonstrated the potential of carbon nanofibers as a material for orthopedic implants. The enhanced function of the osteoblasts is suggested to be due to similarity of the dimension of carbon fibers to that of hydroxyapatite crystals found in physiologic bone. Smaller diameter carbon nanofibers (60–100 nm) have been suggested as suitable for use in orthopedic/dental implant material designs and this may be due to the increase in osteoblast and decrease in osteoblast competitive cells such as fibroblasts, chondrocytes, and smooth muscle cells adhering to the carbon nanofibers [20]. These studies demonstrated that nano-scale carbon fibers could promote osteoblast adhesion, and the application of nanotopography could create an environment favorable for bone cells.

Studies have determined that nanofibrous structure could enhance protein adsorption, including fibronectin and vitronectin, and hence enhance the osteoblast attachment [84, 289]. Woo et al. (2003) [84] hypothesized that nanofibrous PLLA scaffolds, which they developed to mimic the bone extracellular matrix microenvironment, have potential to adsorb serum proteins to enhance cell growth. The 3D porous PLLA scaffolds they developed had interconnected spheroid pores of between 50 and 500 nm with nanofibrous architecture walls. This nanofibrous scaffold adsorbed 4× more serum protein than the non-fibrous scaffold. In another study, Wei and Ma (2004) constructed a nHAP/polymer composite with high porosity from a dioxane–water mixture solvent system using thermally induced phase separation techniques for bone tissue engineering and observed that the protein adsorption was three-fold higher in the fibrous scaffold compared with non-fibrous scaffold [289]. The adsorbed serum proteins may have a role in enhancing cell attachment and proliferation on nano-structured surfaces compared with micro-structured surfaces [290–295]. Incorporation of chemical additives to the nano-fibers has also been reported to enhance cell growth. Recently, Fujihara and colleagues (2005) [296] fabricated a new type of guided bone regeneration membrane from electrospun $PCL/CaCO_3$ composite nanofibers (Fig. 2.23). SEM images showed the presence of $CaCO_3$ nanoparticles on the surface of the PCL nanofibers. *In vitro* studies showed that osteoblast attachment was encouraged on the composite nanofibrous membrane, and good cell proliferation was observed.

In its native environment, self-healing of articular cartilage is rarely possible [297]. Hence, various clinical procedures have been evaluated in articular cartilage repairs, in particular tissue engineering approaches [298, 299]. Successful cartilage tissue engineering depends on the ability of chondrocytes to accumulate into a 3D architecture. Chondrocytes for articular cartilage therapy are isolated from the cartilage of the patients. However, low numbers of chondrocytes can be harvested from each isolation procedure and expansion of the cells is inevitable [131]. This

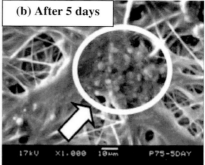

Fig. 2.23. (a) PCL/CaCO$_3$ composite nanofibers, and (b) osteoblast on guided bone regeneration membranes [296].

procedure will also result in the cellular dedifferentiation of the cells [300]. To overcome the dedifferentiation issue, the cells can be redifferentiated upon seeding low passage cells in 3D matrix environments such as alginate, agarose [301, 302] or other scaffolds [303]. Hydrogel formatted biopolymers have limited applications for cartilage tissue engineering and this is due to the lack of long-term mechanical stability and controlled degradation time [303], while synthetic polymers with defined chemical composition provide tailored properties to fit the requirements of engineered tissue scaffolds. The unique architecture of the electrospun 3D synthetic polymer scaffold may provide an ideal environment for both the induction of redifferentiation and the maintenance of differentiated chondrocytes [131]. In cartilage, this synthetic ECM-like structure is crucial to provide mechanical support for the tissue and interacts with the chondrocytes [304, 305].

The above studies implicate the advantages of electrospun nanofibers as scaffolds for bone and cartilage tissue engineering as evidenced by growth of osteoblasts, chondrocytes and mesenchymal stem cells on these fibers. Nanofibrous meshes could provide a better mechanical structure and spatially control cell–scaffold interaction compared with conventional systems. This is based on the observation of osteogenesis and chondrogenesis of the bone marrow-derived stem cells on the nanofibers media.

2.8.6
Heart and Vascular Grafts

Blood vessels play important roles in distributing blood to the body for gaseous and nutrient exchange. One of the major causes of mortality in the world is cardiovascular or vascular disease that results when the heart, arteries or veins are dysfunctional. Due to the high statistics of this disease, much effort in tissue engineering has been geared towards developing a viable vascular substitute as an alternative treatment.

Tissue-engineered blood vessels have a long history, beginning with PMMA tubes as vascular grafts back in the late 1940s [306]. In the early 1950s, textile grafts were first introduced with the development of the first fabric graft by Voorhees (1952) [307]. This subsequently led to the use of PET (Dacron) as a vascular graft substitute by Julian and DeBakey in 1958 [308] who believed that a woven material would be less thrombogenic than a smooth material. However, numerous studies have shown that absorbable vascular grafts would allow the regeneration of new functional arteries [309, 310]. In 1979, Bowald reported a fully bioresorbable graft made of a rolled sheet of Vicryl as vascular substitutes [311, 312]. Bioresorbable woven PGA grafts were also evaluated as vascular substitutes in rabbit models by Greisler (1982) [313] and Greisler et al. (1985) [314]. All these grafts performed only modestly in terms of patency, immunological tolerance, and thrombogenic response.

Currently, a more practical approach for tissue engineering implantation is to use polymer based vascular conduits. However, vascular grafts constructed from synthetic polymers have only been used successfully for large diameter, but not for small diameter, vessels (<6 mm). To date, there are no acceptable synthetic grafts for small diameter blood vessels and autologous grafts have to be harvested from the patients for transplantation. However, this process is time consuming and complications may arise, increasing the time of recovery for patients. In addition, there is a limit on the number of autologous grafts that can be harvested from the patient for transplantation and, at times, the autologous grafts may be diseased and are not suitable for transplantation.

Understanding the basic structure of the blood vessel is important in order to construct a blood vessel scaffold. It consists of three main layers: *tunica intima* (innermost lining), *tunica media* (middle lining) and *tunica adventitia* (outermost lining). The *tunica intima* includes an endothelial lining and an underlying layer of connective tissue of elastic fibers. The *tunica media* contains concentric sheets of smooth muscle tissue in a structure of loose connective tissue, and the *tunica adventitia* contains collagen and elastic fibers [315]. Constructing a blood vessel scaffold is a complex task, and the chances that thrombosis or hyperplasia may arise are high, resulting in vascular graft failure. Rejections can also be another problem when the cellular and humoral immune system detects the graft as a foreign body.

Endothelization is a promising approach to prevent thrombogenesis and intimal hyperplasia. These endothelial cells secrete bioactive substances such as heparin sulfate and nitric oxide, which prevent the smooth muscle cells from overproliferating, as well as cover the scaffold to minimize attachment of platelets to the scaffold to initiate thrombogenesis. Obtaining an endothelized surface on a biodegradable polymer scaffold has become an attractive model for replacing the small diameter blood vessel. Studying the interaction of endothelial cells and polymeric nanofiber scaffold helps researchers optimize the cell growth conditions on the vascular grafts. Surface roughness is one of the parameters that even subtle differences in can affect endothelization. Studies comparing the adhesion of the human vascular artery endothelial cells on a smooth solvent-cast PLLA surface and a rough electrospun PLLA surface showed an inverse relationship between surface roughness and endothelial cells adhesion and proliferation rates [316].

Endothelization is limited when endothelial cells are exposed to blood circulation resulting in cell detachment from the surface. Introduction of biomolecular cell recognition sites on the surface of the vascular graft such as extracellular matrix proteins will enhance cell attachment through receptor–ligand binding with the endothelial cells. In He et al. (2005) [16], human coronary artery endothelial cells (HCAECs) were cultured on collagen-coated PLLA-CL nanofiber mesh (NFM) and its growth compared with cells cultured on PLLA-CL NFM and the tissue culture polystyrene surface (TCPS). The endothelial cells cultured on the collagen-coated NFM had the same typical spreading morphology as the TCPS, in contrast to the rounded morphology of cells cultured on the PLLA-CL NFM (Fig. 2.24). This study clearly demonstrated that the cytocompatibility of PLLA-CL nanofibers was increased when coated with collagen. This is important in achieving an anti-thrombogenic layer for a successful vascular graft. Endothelial cells were cultured on collagen-blended PLLA-CL nanofibers with various PLLA-CL:collagen weight ratios of 4:1, 2:1, 1:1 and 1:2. The cell viability increased with increased collagen content in the blended nanofibers (Fig. 2.25). In addition, there was an enhancement of cell adhesion and spreading on the blended nanofibers when they were grown in culture medium without serum and growth factors [12]. In another

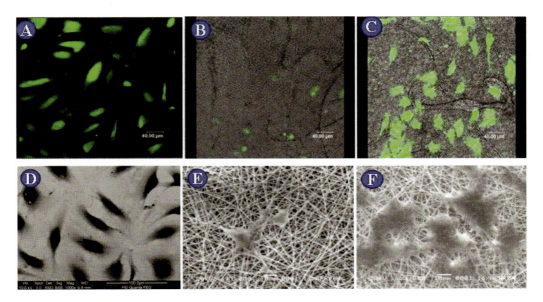

Fig. 2.24. (a)–(c) Laser scanning confocal images of endothelial cells, stained with CMFDA (5-chloromethylfluorescein diacetate), that are cultured on TCPS, PLLA-CL nanofiber mats and collagen-coated PLLA-CL nanofiber mats, respectively. (d)–(f) Scanning electron micrographs of endothelial cells cultured on TCPS, PLLA-CL nanofiber mats and collagen-coated PLLA-CL nanofiber mats, respectively [16].

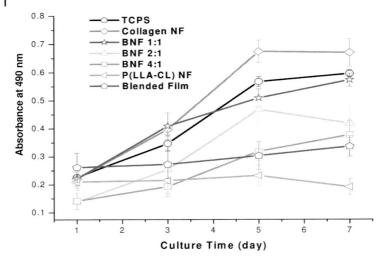

Fig. 2.25. Viability of HCAECs grown on TCPS, collagen nanofibers, collagen-blended PLLA-CL nanofibers (BNF) with different weight ratios of PLLA-CL to collagen (BNF 1:1, BNF 2:1, BNF 4:1), PLLA-CL nanofibers, and collagen-blended PLLA-CL film [12].

study, gelatin grafted PET nanofiber mats were shown to improve endothelial cell proliferation and preserved the cell phenotype. On the gelatin grafted PET nanofiber mats, the cells were observed to exhibit polygonal spread out morphology, in contrast to rounded morphology of cells cultured on the PET nanofiber mats [14]. Alternatively, modifying the physical properties of scaffold surface, such as enhancing hydrophilic property, changing the pore size, and altering the surface roughness, have been reported to increase endothelization on the scaffolds [316].

Another important population of cells in the blood vessels is the smooth muscle cells. Studies have shown that human coronary artery smooth muscle cells (HCASMCs) were able to attach and grow in the direction of the axis of the nanofibers, expressing a spindle-like contractile phenotype when cultured on electrospun aligned PLLA-CL nanofibers (Fig. 2.26), which closely mimics the behavior of cells in native vessels. The adhesion and proliferation of the HCASMCs on the aligned nanofibers was significantly increased compared to that on the solvent-cast polymer film. PLLA-CL aligned nanofiber is seen as a potential scaffold for blood vessel engineering applications.

In the native blood vessel, the most abundant proteins are collagen and elastin. Collagen provides the resistance against rupture and elastin confers elasticity. There are studies showing that pretreatment of nanofibers would enhance cell growth. Buttafoco et al. (2005) [317] have reported that collagen-elastin nanofibrous mesh crosslinked with N-(3-dimethylaminopropyl)-N'-ethylcarbodiimide hydrochloride and N-hydroxysuccinimide enabled smooth muscle cells to grow

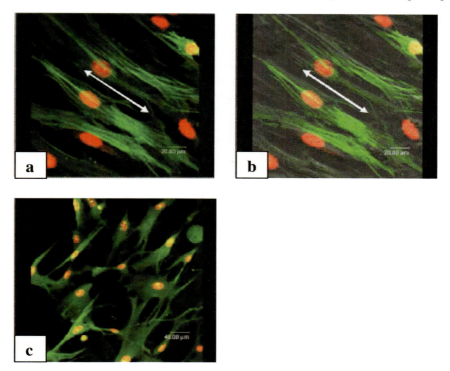

Fig. 2.26. Confocal images of HCASMCs immunostained for α-actin filaments cultured (a) on aligned nanofibrous scaffold, (b) on aligned nanofibrous scaffold, overlay image on the aligned fiber, and (c) on TCPS.

well as a confluent layer. Studies by Boland et al. (2004) [17] showed that pretreatment of small diameter electrospun PGA fibers with concentrated hydrochloric acid significantly increased rat cardiac fibroblast proliferation. To date, constructing a viable blood vessel substitute is a challenging task as many criteria need to be fulfilled. One of the criteria may be the use of nanofiber scaffolds for blood vessel tissue engineering.

2.8.7
Stem Cells

Stem cells are undifferentiated cells that have the capacity for both self-renewal, proliferate and differentiate to one or more types of specialized cells [318]. These unique characteristics of stem cells are crucial in the applications of *in vivo* therapies such as treating leukemia and repairing injured myocardium by the injection of stem cells that will differentiate and proliferate into the desired cell type. However, this approach relies on the pre-existence of extracellular matrix proteins [319]. The two major types of stem cells are embryonic stem cells, derived from the inner cell mass of early aborted fetuses and capable of differentiating into almost any cell type, and adult stem cells of hematopoietic or mesenchymal origin, which are less versatile for tissue engineering applications. However, the use of adult stem cells in therapy incurs less ethical and medical issues since the cells are harvested from the patient, incorporated into a tissue engineered construct and then returned to the individual without the need for immunosuppression [318].

The majority of mammalian cells are anchorage dependent and are mostly derived from mesenchymal stem cells [320]. MSCs are multi-potential cells that can differentiate into bone, cartilage, fat, muscle, marrow stroma, and other cell types when induced by the appropriate biological cues *in vitro* [282, 321]. Their potential for expansion, lineage-specific differentiation, and derivation from autologous sources suggest that human mesenchymal stem cells (hMSCs) can be a possible candidate cells for tissue engineering and regenerative therapies [132]. Research has shown that when MSCs are cultured in the presence of suitable media containing differentiation-promoting agents and growth factors the cells differentiated into osteoblasts [322], adipocytes, and chondrocytes [323]. Human MSCs have been isolated from many adult human tissues [324–331], especially from the bone marrow, which is also a major source and reservoir for hematopoietic stem cells [332]. The capture, expansion, and differentiation of these stem cells are required before incorporation into a tissue engineered construct for successful tissue regeneration.

An ideal biomaterial scaffold for stem cell tissue engineering application should be able to support multi-lineage cell types since most tissues and organs are multiphasic in nature [333]. There are a few successful examples of engineered multiphasic tissues such as the osteochondral construct which consists of bone and cartilage tissues [334–338]. The approach involves the integration of the chondral construct and the osteo-construct after they are separately fabricated from stem cells [336] or differentiated cells [337]. Another approach would be using the cell pellet culture system, which has been widely used to investigate the MSC chondrogenesis [323, 324, 339–341]. In two of those studies, human MSCs exhibited chondrogenic properties when cultured in a cell pellet system [323, 324]. However, constructs derived from the cell pellet culture system are small and uniformly weak in mechanical properties, making them unsuitable for repair of larger cartilage defects.

Alternatively, nanofibrous scaffold is a potential candidate scaffold for a cell-based tissue engineering approach as well as a potential carrier for MSC transplantation. In a study, Li et al. (2005) [132] compared the chondrogenic activities of bone marrow-derived MSCs seeded on PCL nanofibrous scaffolds to cell pellet culture system in the presence of TGF-β1 as chondrogenic growth factor. The results showed that the level of MSC chondrogenesis on nanofibrous scaffolds is enhanced compared to the cell pellet culture system (Fig. 2.27). From this observation, it was proposed that the 3D PCL-based nanofibrous scaffold is able to serve as a bioactive carrier for MSC transplantation for cartilage repair. In a later study, Li et al. (2005) [333] tested a 3D PCL nanofibrous scaffold for its ability to support and maintain multi-lineage differentiation of bone marrow-derived human MSCs.

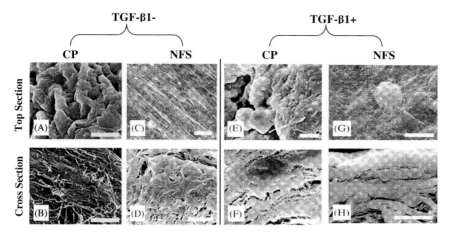

Fig. 2.27. Morphology of day 21 cell pellet and nanofibrous scaffold MSC cultures examined by SEM. (A)–(D) cultures maintained without TGF-β1; (E)–(H) cultures treated with TGF-β1. (A) Top view of cell pellet, revealing a roughened surface; (B) cross-sectional view of cell pellet, showing the presence of native collagen-like fibers; (C) top view of nanofibrous scaffold, showing fibroblast-like cells covering the surface; (D) cross-sectional view of nanofibrous scaffold, showing cells covered with ECM, and integrated with PCL nanofibers; (E) top view of cell pellet, with round chondrocyte-like cells on the surface; (F) cross-sectional view of cell pellet, showing thick ECM; (G) top view of nanofibrous scaffold, showing the presence of round, ECM-embedded chondrocyte-like cells; and (H) cross-sectional view of nanofibrous scaffold, showing a thick, dense ECM-rich layer. Bar = 10 μm [132].

In this *in vitro* study, the cells seeded onto the scaffold were induced to differentiate along adipogenic, chondrogenic or osteogenic lineages by culturing in specific differentiation media. It was reported that multi-lineage differentiation of MSCs is fully supported within the nanofibrous scaffolds. These two studies suggested that the PCL-based nanofibrous scaffold is a promising candidate scaffold for cell-based, multi-phasic tissue engineering. This is further supported by Boudriot and colleagues (2005) [320] who studied the differentiation of human MSCs when cultured onto 3D electrospun PLLA nanofibers for 21 days. The mesenchymal stem cells clearly preferred a guided growth along the nanofibers and revealed no signs of cell death (Fig. 2.28). In addition, osteogenic differentiation of the mesenchymal stem cells was also observed. The results showed that the use of electrospun nanofibers as scaffolds for stem cells is promising in tissue engineering.

2.9
Innovations in Nanofiber Scaffolds

Even though nanofibrous scaffolds are widely used for tissue engineering applications, they are increasingly being used as vectors for controlled delivery systems such as for drugs, and this is due to their high surface area and nano-scale dimen-

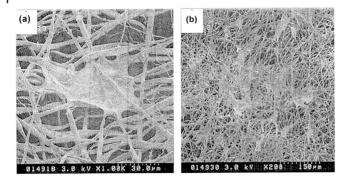

Fig. 2.28. (a) hMSC morphology on nanofiber scaffold; (b) seeded hMSC [320].

sions. Annually, millions of people [342] have been administered with various drugs and proteins such as human growth factors and hormones encapsulated or incorporated with polymers to assist delivery in the treatment of diseases or to improve health [343]. In drug delivery systems, the ideal drug dosage is the minimum amount delivered to the diseased site and effectively absorbed. Uptake of drugs by the body is influenced by the solubility of the compound in aqueous environment, modifications to the compound to assist solubility, and the ease of releasing the drug into the system is facilitated by the degradation of the coating or encapsulating materials of the drugs [35]. Hence, polymeric materials in the form of nano or micro particles, hydrogels and micelle [344] have been developed as vectors for drug delivery. The principle of drug delivery with polymer nanofibers is based on the dissolution rate of a particulate drug which increases with a corresponding increase in the carrier surface area. The nano-dimensions of these polymeric delivery materials will allow the drugs to be more effectively delivered to the target sites and this will improve the therapeutic effects of the drugs and at the same time reduce undesirable side effects associated with the drugs. As such, the usage of polymer nanofiber membranes has become the main focus of research for drug delivery applications [21].

Nanofiber mats prepared from PLA, PEVA and their blend have been incorporated with the drug tetracycline hydrochloride [345, 346]. The PLA nanofibers showed an instantaneous release of the drug while the drug release profiles of PEVA and PLA/PEVA blend gradually increased over 120 h. Zong et al. (2002) [347] have prepared bioabsorbable nanofiber membranes from PDLLA incorporated with the antibiotic Mefoxin and a burst drug release profile was observed within the first 3 h, and a 90% release rate was achieved after 50 h. Even though the release profile of the drug is important, it is essential that the drug remains active upon release. Kim et al. (2004) [344] have evaluated the bioactivity of released Mefoxin incorporated in PLGA nanofibrous membrane to inhibit *Staphylococcus aureus* growth. Incorporation of Mefoxin in the polymer did not diminish the antibiotic activity and the released Mefoxin effectively inhibited bacterial growth.

Drugs incorporated into the nanofibers have a tendency to release in a burst profile when administered directly into the biological system, as shown by the earlier studies [344–347]. This release profile may not be suitable for some drugs that have to be released gradually, such as hypertension drugs. In an effort to better control drug release, Verreck et al. (2003) [348] placed the electrospun hydroxylpropylmethyl cellulose (HPMC) nanofibrous membranes that contained the drug itraconazole into a hard gelatin capsule that is traditionally used to encapsulate drugs. Nanofibers in the capsule gradually released the drug over 20 h compared with the nanofiber membranes without the capsule, which released 100% of the drug within 4 h. This unique approach using pharmaceutical hard gelatin capsule for drug release from nanofiber is able to control the drug release rate by changing the amount of drug incorporated, minimizing the distribution of fiber diameter, and drug distribution in a nanofiber. The diffusion of drugs through the inter-fiber pores in an aqueous environment is another important issue in order to understand the drug release manner from nanofibers. Verrick et al. (2003) [349] used poorly water-soluble drugs, itraconazole and ketanserin, as model compounds while segmented polyurethane was selected as the non-biodegradable polymer. At low drug loading, itraconazole was released from the nanofibers in a linear function. Initial burst release of the drug was not observed. As for ketanserin, a bi-phasic release pattern was observed where two sequential linear release phases were noted. These release phases may be temporally correlated with drug diffusion through the polymer and through formed aqueous pores, where the latter require some time to materialize in the aqueous environment.

Modification to the polymeric nanofibers is an alternative to control drug release. Zeng et al. (2003) [350] have fabricated PLLA nanofibers with cationic, anionic and nonionic surfactants as additives, to evaluate the release profiles of typical drugs such as rifampin (a drug for tuberculosis) and paclitaxel (an anticancer drug). They found that the surfactants can reduce the diameter of electrospun fibers as well as control the size distribution of the nanofiber diameter. The constant rate of drug release is in-line with the degradation rate of the nanofibers. Therefore, the distribution of fiber diameter may play an important role in controlling the drug release rate. Such nanofibers may find clinical applications. Table 2.2 summarizes studies conducted using nanofibers for the drug delivery system. Electrostatic spinning is an efficient and simple technique to prepare the drug-laden nanofiber for potential use in oral, topical, and even *in vivo* delivery of drugs despite its recent application in the pharmaceutical industry. Several recent publications suggest that it may be of high value in the formulation of poorly water-soluble drugs by combining nanotechnology and solid solution/dispersion methodologies [351].

Current therapies to regenerate or replace various tissues (e.g., bone, blood vessels, skin, liver, eye, and nerve) in the body depend on the delivery of growth factors [352]. Growth factors are crucial to achieve successful tissue regeneration as substitutes for biological functions of damaged and injured organs. These growth factors help to create a more favorable environment to enhance tissue regeneration. In addition to the growth factors, this environment can be achieved by providing various biomaterials as cell scaffolds to promote cell proliferation and differen-

Tab. 2.2. Electrospun polymer nanofibers incorporated with different drugs.

Polymer nanofiber	Drug	Function of drug	Ref.
• PLLA • PEVA • PLA/PEVA	Tetracycline hydrochloride	Lyme disease, pneumonia, acne, venereal (sexually transmitted) disease, bladder infections, and ulcers	346
• PDLLA • PLGA • PLGA/PEG-b-PLA [poly(ethylene glycol)-*b*-poly(lactide)]	Mefoxin	A wide ranging drug against Gram-positive and Gram-negative bacteria	344, 347
• Polyurethane • HPMC (Hydroxylpropylmethyl cellulose)	Ketanserin Itraconazole	Treatment for wound healing Treatment of fungal infections, such as aspergillosis, blastomycosis, histoplasmosis, and fungal infection localized to the toenails and fingernails (onychomycosis)	348
• PLLA	Rifampin Paclitaxel	Treats tuberculosis (TB) Treats head and neck cancer, non-small cell lung cancer, small cell lung cancer, and bladder cancer	350

tiation. Growth factors such as fibroblast growth factor (FGF), vascular endothelial growth factor (VEGF), TGF-β, platelet derived growth factor (PDGF), and angiopoietins are often required to promote tissue regeneration. This is because they can induce angiogenesis, promoting a sufficient supply of oxygen and nutrients to effectively maintain the biological functions of cells transplanted for organ substitution. Biodegradable polymers have been developed to provide localized and sustained growth factor release [343, 353–358], as well as to introduce plasmid DNA that encodes the growth factors [359, 360], to desirable targets in human body. However, the success rate of current efforts is limited due to the poor *in vivo* stability of the growth factors, the mode of delivery, and the necessity for numerous signals to complete the regeneration process. Ma (2004) [54] has designed a conceptual model using a bioactive scaffold integrating nanofibrous architecture with 3D biomimetic surface modification to control growth factor release (Fig. 2.29). The design uses the architectural features of collagen to provide a high surface area for cell attachment and new matrix deposition, as well as an open structure that will allow an interactive environment for cell–cell, cell–nutrient, cell–matrix, and

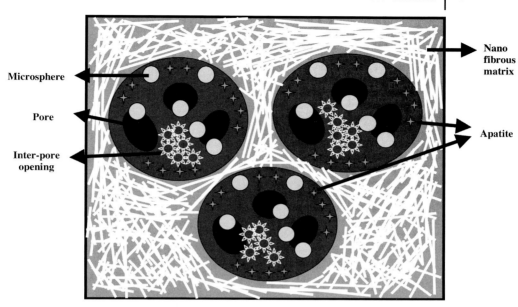

Microsphere

Pore

Inter-pore opening

Nano fibrous matrix

Apatite

Fig. 2.29. Schematic of a "biomimetic nano-scaffold". The scaffold combines the novel nanofibrous architecture of an interconnected pore network with microspheres for controlled release of putative regenerative factors. (Adapted from Figure 10 of Ma [54].)

cell–signal molecule interactions. The challenge of finding an appropriate delivery mechanism for these signaling proteins, and to determine the time and dosage of delivery, could possibly be achieved by capitalizing on the unique characteristics of nano-scale polymeric fibers.

2.10
Conclusion

Nanotechnology is a diverse and interdisciplinary area of research. This research discipline promotes the innovation of new nano-scale materials and devices with improved properties and functionalities. Nanotechnology has the capacity to revolutionize in many areas of applications such as surface microscopy, silicon fabrication, biochemistry, molecular biology, physical chemistry and computational engineering. However, safety concerns of nanomaterials in everyday life and its social, political and regulatory implications have been discussed extensively despite commercialization of some products made of nano-based materials. Nanofiber technology is an important research area in nanotechnology and the understanding of nanofiber processing methodologies is essential for the overall advancement of nanotechnology. Current research in nanomaterials may not only lead to the development of superior functions but also provide a means to deliver these special functions to higher order structures. Nanofibrous scaffolds designed to elicit specific cellular responses through the incorporation of signaling ligands (e.g., growth factors and adhesion peptides) or DNA fragments are particularly promising for advanced tissue repairs and regeneration, while nanoparticles and nanospheres are able to control the release of therapeutic agents, antibiotics, genes and vaccines to the target cells.

The electrospinning technique provides an inexpensive and convenient method to produce nanofibers from various types of polymers with controlled variations in polymer characteristics such as fiber diameter, pore size, high surface area, and lightweight scaffolds. This technology has made it possible to manufacture products such as permeable nanofiber mats and nanofiber tubes for biomedical applications, which have proven difficult to fabricate before and most probably would not have been in existence using fabrication technologies presently available. The availability of new materials will play a critical role in developing advanced applications in disease diagnosis and treatment. Further research and development is required to improve the biocompatibility and potential applications of nanofiber scaffolds. The incorporation of nanofibers into useful devices requires in-depth understanding of the parameters influencing nanofiber processing and the resulting fiber characteristics, cell–nanofiber interactions in the biological system, and the integrative process to existing technologies or applications. These are some of the many challenges that need to be taken into account when using nanofibers for tissue engineering and biomedical applications.

References

1 SKALAK, R., Fox, C. Preface. In: *Tissue Engineering*, Alan R Liss, Inc, New York, **1988**.
2 GOOCH, K.J., BLUNK, T.B., VUNJAK-NOVAKOVIC, G., LANGER, R., FREED, L.E., TENNANT, C.J. *Frontiers in Tissue Engineering*, Pergamon Press, New York, USA, **1998**.
3 LANZA, R.P., LANGER, R., VACANTI, J. *Principles of Tissue Engineering* (2nd Edn.), Academic Press, San Diego, USA, **2000**.
4 DREXLER, K.E. *Engines of Creation: The Coming Era of Nanotechnology*, Anchor Press/Doubleday, New York, **1986**.
5 MIN, B.M., LEE, G., KIM, S.H., NAM, Y.S., LEE, T.S., PARK, W.H. Electrospinning of silk fibroin nanofibers and its effect on the adhesion and spreading of normal human keratinocytes and fibroblasts *in vitro*. *Biomaterials*. **2004**, 25, 1289–1297.
6 JIN, H.J., CHEN, J.S., KARAGEORGIOU, V., ALTMAN, G.H., KAPLAN, D.L. Human bone marrow stromal cell responses on electrospun silk fibroin mats. *Biomaterials*. **2004**, 25, 1039–1047.
7 WNEK, G.E., CARR, M.E., SIMPSON, D.G., BOWLIN, G.L. Electrospinning of nanofiber fibrinogen structures. *Nano Lett*. **2003**, 3, 213–326.
8 MATTHEWS, J.A., WNEK, G.E., SIMPSON, D.G., BOWLIN, G.L. Electrospinning of collagen nanofibers. *Biomacromolecules*. **2001**, 3, 232–238.
9 ZHANG, Y.Z., OUYANG, H.W., LIM, C.T., RAMAKRISHNA, S. Electro-

spinning of gelatin fibers and gelatin/ PCL composite fibrous scaffolds. *J. Biomed. Mater. Res. Part B: Appl. Biomater.* **2004**, 72B, 156–165.

10 GENG, X., KWON, O.H., JANG, J. Electrospinning of chitosan dissolved in concentrated acetic acid solution. *Biomaterials.* **2005**, 26, 5427–5432.

11 MIN, B.M., LEE, S.W., LIM, J.N., YOU, Y., LEE, T.S., KANG, P.H., PARK, W.H. Chitin and chitosan nanofibers: electrospinning of chitin and deacetylation of chitin nanofibers. *Polymer.* **2004**, 45, 7137–7142.

12 HE, W., YONG, T., TEO, W.E., MA, Z. Fabrication and endothelialization of collagen-blended biodegradable polymer nanofibers: Potential vascular graft for the blood vessel tissue engineering. *Tissue Eng.* **2005**, 11, 1575–1589.

13 ZHANG, Y.Z., VENUGOPAL, J., HUANG, Z.M., LIM, C.T., RAMAKRISHNA, S., Characterization of the surface biocompatibility of the electrospun PCL-collagen nanofibers using fibroblasts. *Biomacromolecules.* **2005**, 6, 2583–2589.

14 MA, Z., KOTAKI, M., YONG, T., HE, W., RAMAKRISHNA, S. Surface engineering of electrospun polyethylene terephthalate (PET) nanofibers towards development of a new material for blood vessel tissue engineering. *Biomaterials.* **2005**, 26, 2527–2536.

15 MA, Z.W., GAO, C.Y., JI, J., SHEN, J.C. Protein immobilization on the sur-face of poly-L-lactic acid films for improvement of cellular interactions. *Eur. Polym. J.* **2002**, 38, 2279–2284.

16 HE, W., MA, Z.W., YONG, T., TEO, W.E., RAMAKRISHNA, S. Fabrication of collagen-coated biodegradable copolymer nanofiber and their potential for endothelial cell growth. *Biomaterials.* **2005**, 26, 7606–7615.

17 BOLAND, E.D., TELEMECO, T.A., SIMPSON, D.G., WNEK, G.E., BOWLIN, G.L. Utilizing acid pretreatment and electrospinning to improve bio-compatibility of poly(glycolic acid) for tissue engineering. *J. Biomed Mater Res Part B: Appl. Biomater.* **2004**, 71B, 144–152.

18 CHUA, K.N., LIM, W.S., ZHANG, P., LU, H., WEN, J., RAMAKRISHNA, S., LEONG, K.W., MAO, H.Q. Stable immobilization of rat hepatocyte spheroids on galactosylated nanofiber scaffold. *Biomaterials.* **2005**, 26, 2537–2547.

19 MIN, B.M., YOU, Y., KIM, J.M., LEE, S.J., PARK, W.H. Formation of nanostructured poly(lactic-*co*-glycolic acid)/chitin matrix and its cellular response to normal human keratinocytes and fibroblasts. *Carbohydr. Polym.* **2004**, 57, 285–292.

20 PRICE, R.L., WAID, M.C., HABERSTROH, K.M., WEBSTER, T.J. Selective bone cell adhesion on formulations containing carbon nanofibers. *Biomaterials.* **2003**, 24, 1877–1887.

21 RAMAKRISHNA, S., FUJIHARA, K., TEO, W.E., LIM, T.C., MA, Z. *An Introduction to Electrospinning and Nanofibers*, World Scientific Publishing Co. Pte. Ltd., Singapore, **2005**.

22 ONDARCUHU, T., JOACHIM, C. Drawing a single nanofiber over hundreds of microns. *Europhys. Lett.* **1998**, 42, 215–220.

23 FENG, L., LI, S., LI, H., ZHAI, J., SONG, Y., JIANG, L., ZHU, D. Super-hydrophobic surface of aligned polyacrylonitrile nanofibers. *Angew. Chem. Int. Ed.* **2002**, 41, 1221–1223.

24 ATALA, A., LANZA, R.P. *Methods of Tissue Engineering*, Academic Press, San Diego, USA, **2002**.

25 RAMAKRISHNA, S., JAYARAMAN, K., HUANG, Z.M., ZHANG, Y.Z., MO, X.M. *Advances in Nanoscience & Nanotechnol-ogy*, National Institute of Science Com-munication & Information Resources, New Delhi, **2004**, pp. 113–140.

26 NAM, Y.S., PARK, T.G. Porous biodegradable polymeric scaffolds prepared by thermally induced phase separation. *J. Biomed. Mater. Res.* **1999**, 47, 8–17.

27 CHEN, P. Self-assembly of ionic-complementary peptides: A physicochemical viewpoint. *Colloids Surf. A: Physicochem. Eng. Aspects.* **2005**, 261, 3–24.

28 FORMHALS, A. Process and apparatus

for preparing artificial threads. *US Patent No 1,975,504,* **1934**.

29 Doshi, J., Reneker, D.H. Electrospinning process and applications of electrospun fibers. *J. Electrostatics.* **1995**, 35, 151–160.

30 Boland, E.D., Wnek, G.E., Simpson, D.G., Palowski, K.J., Bowlin, G.L. Tailoring tissue engineering scaffolds using electrostatic processing techniques: A study of poly(glycolic acid) electrospinning. *J. Macromol. Sci. Pure Appl. Chem.* **2001**, A38, 1231–1243.

31 Bornat, A. Production of electrostatically spun products. *US Patent No. 4,689,186,* **1987**.

32 Li, D., Wang, Y., Xia, Y. Electrospinning of polymeric and ceramic nanofibers as uniaxially aligned arrays. *Nano Lett.* **2003**, 3, 1167–1171.

33 Theoron, A., Zussman, E., Yarin, A.L. Electrostatic field-assisted alignment of electrospun nanofibers. *Nanotechnology.* **2001**, 12, 384–390.

34 Dersch, R., Liu, T., Schaper, A.K., Greiner, A., Wendorff, J.H. Electrospun nanofibers: Internal structure and intrinsic orientation. *J. Polym. Sci. Pt. B-Polym. Phys.* **2003**, 41, 545–553.

35 Huang, Z.M., Zhang, Y.Z., Kotaki, M., Ramakrishna, S. A review on polymer nanofibers by electrospinning and their applications in nanocomposites. *Composites Sci. Technol.* **2003**, 63, 2223–2253.

36 Ramakrishna, S. *Textile Scaffolds in Tissue Engineering, Smart Fibers, Fabrics, and Clothing: Fundamentals and Applications,* Woodhead Publishing Limited, Cambridge, UK, **2001**.

37 Lee, K.Y., Bouhadir, K.H., Mooney, D.J. Controlled degradation of hydrogels using multi-functional cross-linking molecules, *Biomaterials.* **2004**, 25, 2461–2466.

38 You, Y., Min, B.M., Lee, S.J., Lee, T.S., Park, W.H. *In vitro* degradation behavior of electrospun polyglycolide, polylactide, and poly(lactide-co-glycolide). *J. Appl. Polym. Sci.* **2005**, 95, 193–200.

39 Shum, A.W.T., Mak, A.F.T. Morphological and biomechanical characterization of poly(glycolic acid) scaffolds after in vitro degradation. *Polym. Degradation Stabil.* **2003**, 81, 141–149.

40 Higgins, S.P., Solan, A.K., Niklason, L.E. Effects of polyglycolic acid on porcine smooth muscle cell growth and proliferation. *J. Biomed. Mater. Res.* **2003**, 67A, 295–302.

41 Zong, X., Bien, H., Chung, C.Y., Yin, L., Fang, D., Hsiao, B.S., Chu, B., Entcheva, E. Electrospun fine-textured scaffolds for heart tissue constructs. *Biomaterials.* **2005**, 26, 5330–5338.

42 Lu, L., Garcia, C.A., Mikos, A.G. In vitro degradation of thin poly(dl-lactic-co-glycolic acid) films. *J. Biomed. Mater. Res.* **1999**, 46, 236–244.

43 Vacanti, C.A., Langer, R., Schloo, B., Vacanti, J.P. Synthetic polymers seeded with chondrocytes provide a template for new cartilage formation. *Plast. Reconstr. Surg.* **1991**, 88, 753–759.

44 Ma, P.X., Schloo, B., Mooney, D., Langer, R. Development of biomechanical properties and morphogenesis of in vitro tissue engineered cartilage. *J. Biomed. Mater. Res.* **1995**, 29, 1587–1595.

45 Freed, L.E., Langer, R., Marvin, I., Pellis, N.R., Vunjak-Novakovic, G. Tissue engineering of cartilage in space. *Proc. Natl. Acad. Sci. U.S.A.* **1997**, 94, 13 885–13 890.

46 Freed, L.E., Vunjak-Novakovic, G., Biron, R.J., Eagles, D.B., Lesnoy, D.C., Barlow, S.K., Langer, R. Biodegradable polymer scaffolds for tissue engineering. *Biotechnology.* **1994**, 12, 689–693.

47 Cao, Y., Vacanti, J.P., Ma, X., Paige, K.T., Upton, J., Chowanski, Z., Schloo, B., Langer, R., Vacanti, C.A. Generation of neo-tendon using synthetic polymers seeded with tenocytes. *Transplant Proc.* **1994**, 26, 3390–3392.

48 Atala, A., Vacanti, J.P., Peters, C.A., Mandell, J., Retik, A.B., Freeman, M.R. Formation of urothelial

structures in vivo from dissociated cells attached to biodegradable polymer scaffolds in vitro. *J. Urol.* **1992**, 148, 658–662.

49 ORGAN, G.M., MOONEY, D.J., HANSEN, L.K., SCHLOO, B., VACANTI, J.P. Enterocyte transplantation using cell-polymer devices to create intestinal epithelial-lined tubes. *Transplant Proc.* **1993**, 25, 998.

50 SHINOKA, T., SHUM-TIM, D., MA, X.P., TANEL, R.E., ISOGAI, N., LANGER, R., VACANTI, J.P., MAYER, J.E. Creation of viable pulmonary artery autografts through tissue engineering. *J. Thorac. Cardiocasc. Surg.* **1998**, 115, 536–546.

51 NIKLASON, L.E., GAO, J., ABBOTT, W.M., HIRSCHI, K.K., HOUSER, S., MARINI, R., LANGER, R. Functional arteries grown in vitro. *Science.* **1999**, 284, 489–493.

52 SHINOKA, T., MA, P.X., SHUM-TIM, D., BREUER, C.K., CUSICK, R.A., ZUND, G., LANGER, R., VACANTI, J.P., MAYER, J.E. JR. Tissue-engineered heart valves. Autologous valve leaflet replacement study in a lamb model. *Circulation.* **1996**, 94, 164–168.

53 HUTMACHER, D.W. Scaffold design and fabrication technologies for engineering tissues – state of the art and future perspectives. *J. Biomater. Sci. Polym. Edn.* **2001**, 12, 107–124.

54 MA, P.X., Scaffolds for tissue engineering. *Materials Today.* **2004**, 7, 30–40.

55 MA, P.X., LANGER, R. *Tissue Engineering Methods and Protocols.* Humana Press, Totowa, NJ, USA, **1999**.

56 LU, L., PETER, S.J., LYMAN, M.D., LAI, H.L., LEITE, S.M., TAMADA, J.A., VACANTI, J.P., LANGER, R., MIKOS, A.G. *In vitro* and *in vivo* degradation of porous poly(dl-lactic-co-glycolic acid) foams. *Biomaterials.* **2000**, 21, 1595–1605.

57 ZELTINGER, J., SHERWOOD, J.K., GRAHAM, D.A., MUELLER, R., GRIFFITH, L.G. Effect of pore size and void fraction on cellular adhesion, proliferation, and matrix deposition. *Tissue Eng.* **2001**, 7, 557–572.

58 YANG, S., LEONG, K.F., DU, Z. and CHUA, C.K. The design of scaffolds for use in tissue engineering. Part II. Rapid prototyping techniques. *Tissue Eng.* **2002**, 8, 1–11.

59 GIORDANO, R.A., WU, B.M., BORLAND, S.W., CIMA, L.G., SACHS, E.M., CIMA, M.J. Mechanical properties of dense polylactic acid structures fabricated by three dimensional printing. *J. Biomater. Sci. Polym.* **1996**, 8, 63–75.

60 PARK, A., WU, B., GRIFFITH, L.G. Integration of surface modification and 3D fabrication techniques to prepare patterned poly(L-lactide) substrates allowing regionally selective cell adhesion. *J. Biomater. Sci., Polym.* **1998**, 9, 89–110.

61 MA, P.X. *Encyclopedia of Polymer Science and Technology.* Vol. 3, John Wiley & Sons, New York, USA, **2004**.

62 MIKOS, A.G., SARAKINOS, G., LEITE, S.M., VACANTI, J.P., LANGER, R. Laminated three-dimensional biodegradable foams for use in tissue engineering. *Biomaterials.* **1993**, 14, 323–330.

63 WIDMER, M.S., GUPTA, P.K., LU, L., MESZLENYI, R.K., EVANS, G.R.D., BRANDT, K., SAVEL, T., GURLEK, A., PATRICK, JR. C.W., MIKOS, A.G. Manufacture of porous biodegradable polymer conduits by an extrusion process for guided tissue regeneration. *Biomaterials.* **1998**, 19, 1945–1955.

64 MCINTYRE, J.E., DENTON, M.J. *Encyclopedia of Polymer Science and Engineering.* Vol. 6, Wiley, New York, **1986**.

65 KOTAKI, M., HUANG, Z.M., RAMAKRISHNA, S. *Handbook of Nanostructured Biomaterials and Their Application in Nanobiotechnology.* Vol. 2, American Scientific Publishers, Los Angeles, USA, **2005**.

66 http://www.fibersource.com/f-tutor/techpag.htm.

67 FRENOT, A., CHRONAKIS, I.S. Polymer nanofibers assembled by electro-spinning. *Curr. Opin. Colloid Interface Sci.* **2003**, 8, 64–75.

68 GUPTA, V.B. *Manufactured Fiber Technology.* Chapman & Hall, London, **1997**.

69 SUMANASINGHE, R.D., KING, M.W. New trends in biotextiles – The

challenge of tissue engineering.
J. Textile Apparel. **2003**, 3, 1–13.

70 GIBSON, P.W., SCHREUDER-GIBSON,
H.L., RIVIN, D. Transport properties
of porous membranes based on
electrospun nanofibers. *Colloids Surf.
A: Physicochem. Eng. Aspects.* **2001**,
187, 469–481.

71 JAYARAMAN, K., KOTAKI, M., ZHANG,
Y.Z., MO, X.M., RAMAKRISHNA, S.
Recent advances in polymer
nanofibers [Review]. *J. Nanosci.
Nanotechnol.* **2004**, 4, 52–65.

72 NALWA, H.S. *Handbook of Nanostructured Materials and Nanotechnology,*
Academic Press, San Diego, **2000**.

73 KO, F.K. Nanofiber technology:
Bridging the gap between nano and
macro world. In: *NATO ASI on
Nanoengineered Nanofibrous Materials.*
Vol. 169, Kluwer Academic Publishers,
Anatalia, Turkey, **2003**.

74 LAURENCIN, C.T., AMBROSIO, A.M.A.,
BORDEN, M.D., COOPER, J.A. Tissue
engineering: orthopedic applications.
Annu. Rev. Biomed. Eng. **1999**, 1,
19–46.

75 KO, F.K., LAURENCIN, C.T., BORDEN,
M.D., RENEKER, D. The Dynamics of
Cell-Fiber Architecture Interaction.
Proceedings, Annual Meeting,
Biomaterials Research Society, San
Diego, USA, April, **1998**.

76 NAIR, L.S., BHATTACHARYYA, S.,
BENDER, J.D., GREISH, Y.E., BROWN,
P.W., ALLCOCK, H.R., LAURENCIN, C.T.
Fabrication and optimization of
methylphenoxy substituted
polyphosphazene nanofibers for
biomedical applications.
Biomacromolecules. **2004**, 5, 2212–2220.

77 ZUSSMAN, E., THERON, A., YARIN, A.L.
Formation of nanofiber crossbars in
electrospinning. *Appl. Phys. Lett.* **2003**,
82, 973–975.

78 MARTIN, G.E., COCKSHOTT, I.D.,
FILDES, F.J.T. Fibrillar product. *US
Patent No 4878908*, **1989**.

79 STENOIN, M.D., DRASLER, W.J., SCOTT,
R.J., JENSON, M.L. *US Patent No
5866217*, **1999**.

80 KENAWY, E.R., LAYMAN, J.M.,
WATKINS, J.R., BOWLIN, G.L.,
MATTHEWS, J.A., SIMPSON, D.G.,

WNEK, G.E. Electrospinning of
poly(ethylene-*co*-vinyl alcohol) fibers.
Biomaterials. **2003**, 24, 907–913.

81 KHIL, M.S., CHA, D.I., KIM, H.Y.,
KIM, I.S., BHATTARAI, N. Electrospun
nanofibrous polyurethane membrane
as wound dressing. *J. Biomed. Mater.
Res. B Appl. Biomater.* **2003**, 67, 675–
679.

82 BHATTARAI, S.R., BHATTARAI, N., YI,
H.K., HWANG, P.H., CHA, D.I., KIM,
H.Y. Novel biodegradable electrospun
membrane: scaffold for tissue
engineering. *Biomaterials.* **2004**, 25,
2595–2602.

83 WEBSTER, T.J., SMITH, T.A. Increased
osteoblast function on PLGA
composites containing nanophase
titania. *J. Biomed. Mater. Res. Part A.*
2005, 74A, 677–686.

84 WOO, K.M., CHEN, V.J., MA, P.X.
Nano-fibrous scaffolding architecture
selectively enhances protein adsorption contributing to cell attachment.
J. Biomed. Mater. Res. A. **2003**, 67,
531–537.

85 BHATTARAI, N., EDMONDSON, D.,
VEISEH, O., MATSEN, F.A., ZHANG, M.
Electrospun chitosan-based nanofibers
and their cellular compatibility.
Biomaterials. **2005**, 26, 6176–6184.

86 GORDON, J.E. *The New Science of
Strong Materials or Why you don't Fall
through the Floor.* Princeton University
Press, Harmondsworth, UK, **1984**.

87 DRESSLHAUS, M.S., DRESSLHAUS, G.,
AVOURIS, P. *Carbon Nanotubes:
Synthesis, Structure, Properties, and
Applications.* Springer-Verlag, Berlin,
2001.

88 NORRIS, I.D., SHAKER, M., KO, F.K.,
MACDIARMID, A.G. Electrostatic
fabrication of ultrafine conducting
fibers: polyaniline/polyethylene oxide
blends. *Synth. Metals.* **2000**, 114,
109–114.

89 EL-AUFY, A., NABET, B., KO, F.K.
Carbon nanotube reinforced
nanocomposites for wearable
electronics. *ACS Polym. Preprints*,
2003, 44.

90 ZHOU, Y., FREITAG, M., HONE, J.,
STAII, C., JOHNSON, A.T., PINTO, N.J.,
MACDIARMID, A.G. Fabrication and

electrical characterization of polyaniline-based nanofibers with diameter below 30 nm. *Appl. Phys. Lett.* **2003**, 83, 3800–3802.

91 LI, G.F., MARTINEZ, C., SEMANCIKA, S., SMITH, J.A., JOSOWICZ, M., JANATA, J. The effect of morphology on the response of polyanilline-based conductometric gas sensors: Nanofibers vs. thin films. *Electrochem. Solid-State Lett.* **2004**, 7, H44–H47.

92 LONG, Y., CHEN, Z., WANG, N., MA, Y., ZHANG, Z., ZHANG, L., WAN, M. Electrical conductivity of a single conducting polyaniline nanotube. *Appl. Phys. Lett.* **2003**, 83, 1863–1865.

93 LONG, Y., ZHANG, L., MA, X., CHEN, Z., WANG, N., ZHANG, Z., WAN, M. Electrical conductivity of an individual polyaniline nanotube synthesized by a self-assembly method. *Macromol. Rapid Commun.* **2003**, 24, 938–946.

94 BUCHKO, C.J., CHEN, L.C., SHEN, Y., MARTIN, D.C. Processing and microstructural characterization of porous biocompatible protein polymer thin films. *Polymer.* **1999**, 40, 7397–7407.

95 FERTALA, A., HAN, W.B., KO, F.K. Mapping critical sites in collagen II for rational design of gene-engineered proteins for cell-supporting materials. *J. Biomed. Mater. Res.* **2001**, 57, 48–58.

96 HUANG, L., MCMILLAN, R.A., APKARIAN, R.P., POURDEYHIMI, B., CONTICELLO, V.P., CHAIKOF, E.L. Generation of synthetic elastin-mimetic small diameter fibers and fiber networks. *Macromolecules.* **2000**, 33, 2989–2997.

97 KING, M.W. Overview of opportunities in medical textiles, Part 1. *Can Textile J.* **2001**, 18, 34–36.

98 RAMAKRISHNA, S., MAYER, J., WINTERMANTEL, E., LEONG, K.W. Biomedical applications of polymer-composite materials: A review. *Composites Sci. Technol.* **2001**, 36, 1189–1224.

99 ANDERSON, J.M. Biocompatibility of tissue engineered implants. In: *Frontiers in Tissue Engineering*, PATRICK, C.W.J., MIKOS, A.G. (Eds.),

Chapter II.8, Pergamon, Oxford, New York, **1998**.

100 MIGLIARESI, C., FAMBRI, L. Processing and degradation of poly(L-lactic acid) fibres. *Macromol. Symp.* **1997**, 123, 155–161.

101 BINI, T.B., GAO, S., TAN, T.C., WANG, S., LIM, A., LIM, B.H., RAMAKRISHNA, S. Electrospun poly(L-lactide-co-glycolide) biodegradable polymer nanofiber tubes for peripheral nerve regeneration, *Nanotechnology.* **2004**, 15, 1459–1464.

102 HADLOCK, T., SUNDBACK, C., HUNTER, D., CHENEY, M., VACANTI, J.P. A polymer foam conduit seeded with Schwann cells promotes guided peripheral nerve regeneration. *Tissue Eng.* **2000**, 6, 119–127.

103 SONDELL, M., LUNDBORG, G., KANJE, M. Vascular endothelial factor stimulates Schwann cell invasion and neovascularization of acellular nerve grafts. *Brain Res.* **1999**, 846, 219–228.

104 MA, Z.W., KOTAKI, M., INAI, R., RAMAKRISHNA, S. Potential of nanofiber matrix as tissue engineering scaffolds. *Tissue Eng.* **2005**, 11, 101–109.

105 KUMAR, M.R. A review of chitin and chitosan applications. *React. Funct. Polym.* **2000**, 46, 1–27.

106 RUSZCZAK, Z. Effect of collagen matrices on dermal wound healing. *Adv. Drug Deliv. Rev.* **2003**, 55, 1595–1611.

107 VENUGOPAL, J., RAMAKRISHNA, S. Biocompatible nanofiber matrices for the engineering of a dermal substitute for skin regeneration. *Tissue Eng.* **2005**, 11, 847–854.

108 MATTHEWS, J.A., BOLAND, E.D., WNEK, G.E., SIMPSON, D.G., BOWLIN, G.L. Electrospinning of collagen type II: A feasibility study. *J. Bioact. Compat. Polym.* **2003**, 18, 125–134.

109 AIEDEH, K., GIANASI, E., ORIENTI, I., ZECCHI, V. Chitosan microcapsules as controlled release systems for insulin. *J. Microencapsul.* **1997**, 14, 567–576.

110 BERGER, J., REIST, M., MAYER, J.M., FELT, O., GURNY, R. Structure and interactions in chitosan hydrogels formed by complexation or

aggregation for biomedical applications. *Eur. J. Pharm. Biopharm.* **2004**, 57, 35–52.

111 PARK, Y.J., LEE, Y.M., PARK, S.N., SHEEN, S.Y., CHUNG, C.P., LEE, S.J. Platelet derived growth factor releasing chitosan sponge for periodontal bone regeneration. *Biomaterials.* **2000**, 21, 153–159.

112 KLOKKEVOLD, P.R., VANDEMARK, L., KENNEY, E.B., BERNARD, G.W. Osteogenesis enhanced by chitosan (poly-N-acetyl glucosaminoglycan) *in vitro. J. Periodontol.* **1996**, 67, 1170–1175.

113 GUTOWSKA, A., JEONG, B., JASIONOW-SKI, M. Injectable gels for tissue engineering. *Anat. Rec.* **2001**, 263, 342–349.

114 YAGI, K., MICHIBAYASHI, N., KURIKAWA, N., NAKASHIMA, Y., MIZOGUCHI, T., HARADA, A., HIGASHIYAMA, S., MURANAKA, H., KAWASE, M. Effectiveness of fructose-modified chitosan as a scaffold for hepatocyte attachment. *Biol. Pharm. Bull.* **1997**, 20, 1290–1294.

115 ZHANG, Y., ZHANG, M.Q. Synthesis and characterization of macroporous chitosan/calcium phosphate composite scaffolds for tissue engineering. *J. Biomed. Mater. Res.* **2001**, 55, 304–312.

116 ZHANG, Y., ZHANG, M.Q. Calcium phosphate/chitosan composite scaffolds for controlled in vitro antibiotic drug release. *J. Biomed. Mater. Res.* **2002**, 62, 378–386.

117 ZHANG, Y., ZHANG, M.Q. Three-dimensional macroporous calcium phosphate bioceramics with nested chitosan sponges for loadbearing bone implants. *J. Biomed. Mater. Res.* **2002**, 61, 1–8.

118 OHKAWA, K., CHA, D.I., KIM, H., NISHIDA, A., YAMAMOTO, H. Electrospinning of chitosan. *Macromol. Rapid Commun.* **2004**, 25, 1600–1605.

119 DUAN, B., DONG, C.H., YUAN, X.Y., YAO, K.D. Electrospinning of chitosan solutions in acetic acid with poly(ethylene oxide). *J. Biomater. Sci-Polym. E.* **2004**, 15, 797–811.

120 NAGAPUDI, K., BRINKMAN, W.T., LEISEN, J.E., HUANG, L., MCMILLAN, R.A., APKARIAN, R.P., CONTICELLO, V.P., CHAIKOF, E.L. Photomediated solid state cross-linking of an elastin-mimetic recombinant protein polymer. *Macromolecules.* **2002**, 35, 1730–1737.

121 WANG, R.Z., CUI, F.Z., LU, H.B., WEN, H.D., LI, H.D. Synthesis of nanophase hydroxyapatite/collagen composite. *J. Mater. Sci. Lett.* **1995**, 14, 490–492.

122 DU, C., CUI, F.Z., ZHU, X.D., DE GROOT, K. Three-dimensional nano-HAp/collagen matrix loading with osteogenic cells in organ culture. *J. Biomed. Mater. Res.* **1999**, 44, 407–415.

123 DU, C., CUI, F.Z., ZHANG, W., FENG, Q.L., ZHU, X.D., DE GROOT, K. Formation of calcium phosphate/collagen composites through mineralization of collagen matrix. *J. Biomed. Mater. Res.* **2000**, 50, 518–527.

124 LIAO, S.S., CUI, F.Z., ZHANG, W., FENG, Q.L. Hierarchically biomimetic bone scaffold materials: Nano-HA/collagen/PLA composite. *J. Biomed. Mater. Res. B.* **2002**, 69B, 158–165.

125 ZHANG, S.M., CUI, F.Z., LIAO, S.S., ZHU, Y., HAN, L. Synthesis and biocompatibility of porous nano-hydroxyapatite/collagen/alginate composite. *J. Mater. Sci. Mater. Med.* **2003**, 14, 1–5.

126 PITT, C.G. Poly-ε-caprolactone and its copolymers. In: *Biodegradable Polymers as Drug Delivery Systems.* CHASIN, M., LANGER, R. (Eds.), New York, Marcel Dekker, **1990**, pp. 71–120.

127 STITZEL, J.D., PAWLOWSKI, K.J., WNEK, G.E., SIMPSON, D.G., BOWLIN, G.L. Arterial smooth muscle cell prolif-eration on a novel biomimicking, biodegradable vascular graft scaffold. *J. Biomater. Appl.* **2001**, 15, 1–12.

128 MA, P.X., ZHANG, R. Porous poly(L-Lactic acid)/apatite composites created by biomimetic process. *J. Biomed. Mater. Res.* **1999**, 45, 285–293.

129 CHEN, V.J., MA, P.X. Nano-fibrous poly(L-lactic acid) scaffolds with interconnected spherical macropores. *Biomaterials.* **2004**, 25, 2065–2073.

130 PATTISON, M.A., WURSTER, S., WEBSTER, T.J., HABERSTROH, K.M. Three-dimensional, nano-structured PLGA scaffolds for bladder tissue replacement applications. *Biomaterials.* **2005**, 26, 2491–500.

131 LI, W.J., DANIELSON, K.G., ALEXANDER, P.G., TUAN, R.S. Biological response of chondrocytes cultured in three-dimensional nanofibrous poly(caprolactone) scaffolds. *J. Biomed. Mater. Res.* **2003**, 67A, 1105–1114.

132 LI, W.J., TULIA, R., OKAFOR, C., DERFOUL, A., DANIELSON, K.G., HALL, D.J., TUAN, R.S. A three-dimensional nanofibrous scaffold for cartilage tissue engineering using human mesenchymal stem cells. *Biomaterials.* **2005**, 26, 599–609.

133 YOSHIMOTO, H., SHIN, Y.M., TERAI, H., VACANTI, J.P. A biodegradable nanofiber scaffold by electrospinning and its potential for bone tissue engineering. *Biomaterials.* **2003**, 24, 2077–2082.

134 SHIN, M., ISHII, O., SUED, T., VACANTI, J.P. Contractile cardiac grafts using a novel nanofibrous mesh. *Biomaterials.* **2004**, 25, 3717–3723.

135 XU, C.Y., INAI, R., KOTAKI, M., RAMAKRISHNA, S. Electrospun nanofiber fabrication as synthetic extracellular matrix and its potential for vascular tissue engineering. *Tissue Eng.* **2004**, 10, 1160–1168.

136 XU, C.Y., INAI, R., KOTAKI, M., RAMAKRISHNA, S. Aligned bio-degradable nanofibrous structure: A potential scaffold for blood vessel engineering. *Biomaterials.* **2004**, 25, 877–886.

137 MO, X.M., XU, C.Y., KOTAKI, M., RAMAKRISHNA, S. Electrospun P(LLA-CL) nanofiber: A biomimetic extracellular matrix for smooth muscle cell and endothelial cell proliferation. *Biomaterials.* **2004**, 25, 1883–1890.

138 LI, W.J., LAURENCIN, C.T., CATERSON, E.J., TUAN, R.S., FRANK, K.K. Electrospun nanofibrous structure: A novel scaffold for tissue engineering. *J. Biomed. Mater. Res.* **2002**, 60, 613–621.

139 DODDI, N., VERSFELT, C.C., WASSER-MAN, D. *U.S. Patent 4,052,988, 20,* **1977**.

140 LIPINSKY, E.S., SINCLAIR, R.G., BROWING, J.D. Degradable polydioxanone-based materials, *US Patent No. 5,767,222,* **1998**.

141 WANG, H., DONG, J.H., QIU, K.U., GU, Z.W. PDON-b-PEG-b-PDON block copolymers and drug delivery system thereof. *J. Appl. Polym. Sci.* **1998**, 68, 2121–2128.

142 SALTZMAN, W.M. Introduction to polymers, in *Tissue Engineering, Engineering Principles for the Design of Replacement Organs and Tissues,* **2004**, Oxford University Press, New York, USA, p. 453.

143 VAINIONPAA, S., ROKKANEN, P., TORMALA, P. Surgical applications of biodegradable polymers in human tissues. *Prog. Polym. Sci.* **1989**, 14, 679–716.

144 HEROLD, D.A., KEIL, K., BRUNS, D.E. Oxidation of polyethylene glycols by alcohol dehydrogenase. *Biochem. Pharmacol.* **1989**, 38, 73–76.

145 HARRIS, J.M. Synthesis of polyethylene glycol derivatives. *J. Macromol. Sci. Rev. Macromol.* **1985**. C25, 325–373.

146 RASHKOV, I., MANOLOVA, N., LI, S.M., ESPARTERO, J.L., VERT, M. Synthesis, characterization, and hydrolytic degradation of PLLA/PEO/PLA triblock copolymers with short poly(L-lactic acid) chains. *Macromolecules.* **1996**, 29, 50–56.

147 DU, Y.J., LEMSTRA, P.J., NIJENHUIS, A.J., VAN AERT, H.A.M., BASTIAANSEN, C. ABA type copolymers of lactide with poly(ethylene glycol) kinetic, mechanistic, and model studies. *Macromolecules.* **1995**, 28, 2124–2132.

148 MOLINA, K.J., LI, S., MARTINEZ, M.B., VERT, M. Protein release from physically crosslinked hydrogels of the PLA/PEO/PLA triblock copolymer-type. *Biomaterials.* **2001**, 22, 363–369.

149 FUKUI, K., SUMPTER, B.G., BARNES, M.D., NOID, D.W. Molecular dynamics studies of the structure and properties of polymer nano-particles. *Comput. Theor. Polym. Sci.* **1999**, 9, 245–254.

150 Wen, X., Shi, D., Zhang, N. Applications of nanotechnology in tissue engineering. In: *Handbook of Nanostructured Biomaterials and their Applications in Nanobiotechnology*, Vol. 1, H.S. Nalwa (Eds.), American Scientific Publishers, Los Angeles, USA, **2005**, pp. 1–23.

151 Chirila, T.V., Hicks, C.R., Dalton, P.D., Vijayasekaran, S., Lou, X., Hong, Y., Clayton, A.B., Ziegelaar, B.W., Fitton, J.H., Platten, S., Crawford, G.F., Constable, I.J. Artificial cornea. *Progr. Polym. Sci.* **1998**, 23, 447–473.

152 Stern, M.E., Beuerman, R.W., Fox, R.I., Gao, J., Mircheff, A.K. and Pflugfelder, S.C. A unified theory of the role of the ocular surface in dry eye. *Adv. Exp. Med. Biol.*, **1998**, 438, 643–651.

153 Hogan, M.J., Zimmerman, L.E. *Ophthalmic Pathology. An Atlas and Textbook.* Saunders, Philadelphia, **1962**, pp. 277–343.

154 Greer, C.H. *Ocular Pathology*, 3rd edn. Blackwell Scientific, Oxford, **1979**, pp. 79–115.

155 Barraquer, J.I., Binder, P.S., Buxton, J.N., Fine, M., Jones, D.B., Laibson, P.R., Nesbum, A.B., Paton, D., Troutman, R.C. *Symposium on Medical and Surgical Diseases of the Cornea.* Mosby Company, St Louis, New Orleans, **1979**.

156 Leibowitz, H.M. *Corneal Disorders, Clinical Diagnosis and Management.* Saunders, Philadelphia, **1984**.

157 Lambiase, A., Rama, P., Bonini, S., Caprioglio, G., Aloe, L. Topical treatment with nerve growth factor for corneal neurotrophic ulcers. *N. Engl. J. Med.* **1998**, 338, 1174–1180.

158 Gilbard, J.P., Rossi, S.R. Tear film and ocular surface changes in a rabbit model of neurotrophic keratitis. *Ophthalmology.* **1990**, 97, 308–312.

159 Whitcher, J.P., Srinivasan, M., Upadhyay, M.P. Corneal blindness: A global perspective. *Bull. World Health Organ.* **2001**, 79, 214–221.

160 Li, F., Carlsson, D., Lohmann, C., Suuronen, E., Vascotto, S., Kobuch, K., Sheardown, H., Munger, R., Nakamura, M., Griffith, M. Cellular and nerve regeneration within a biosynthetic extracellular matrix for corneal transplantation. *Proc. Natl. Acad. Sci. U.S.A.* **2003**, 100, 15 346–15 351.

161 Eye Bank Association of America, *Eye Banking Statistical Report*, Washington DC, USA, **1999**.

162 Chirila, T.V. An overview of the development of artificial corneas with porous skirts and the use of PHEMA for such an application. *Biomaterials.* **2001**, 22, 3311–3317.

163 Langer, R., Vacanti, J.P. Tissue engineering. *Science.* **1993**, 260, 920–926.

164 Ridley, H. Intra-ocular lenses. *Trans. Ophthal. Soc. UK.* **1951**, 71, 617–621.

165 Ridley, H. Intraocular acrylic lenses after cataract extraction. *Lancet.* **1952**, 1, 118–121.

166 Ridley, H. Intra-ocular acrylic lenses; a recent development in the surgery of cataract. *Br. J. Ophthal.* **1952**, 36, 113–122.

167 Chirila, T.V., Constable, I.J., Russo, A.V., Linton, R.G. Ridley intraocular lens revisited: chemical analysis of residuals in the original lens material. *J. Cataract Refr. Stag.* **1989**, 15, 283–288.

168 Apple, D.J., Sims, J. Harold Ridley and the invention of the intraocular lens. *Surv. Ophthal.* **1996**, 40, 279–292.

169 Mester, U., Roth, K., Dardenne, U. Trial with 2-hydroxy-ethyl-methacrylate lenses as keratophakia material. *Ber. Dtsch. Ophthal. Ges.* **1974**, 72, 326–329.

170 Mester, U., Stein, H.J., Meier, J. Permeability of hydrogel plastics (PHEMA) to different substances of corneal metabolism (author's transl). *Gruefes. Arch. Klin. Exp. Ophthal.* **1978**, 205, 207–212.

171 McCarey, B.E., Andrews, D.M. Refractive keratoplasty with intra-stromal hydrogel lenticular implants. *Invest. Ophthal. Vis. Sci.* **1981**, 21, 107–115.

172 McCarey, B.E., Andrews, D.M., Hatchell, D.L., Pederson, H. Hydrogel implants for refractive keratoplasty: Corneal morphology. *Curr. Eye Res.* **1982**, 2, 29–38.

173 BINDER, P.S., DEG, J.K., ZAVALA, E.Y., GROSSMAN, K.R. Hydrogel kerato-phakia in non-human primates. *Curr. Eye Res.* **1982**, 1, 535–542.

174 BINDER, P.S. Hydrogel implants for the correction of myopia. *Curr. Eye Res.* **1983**, 2, 435–441.

175 WERBLIN, T.P., BLAYDES, J.E., FRYCZKOWSKI, A., PEIFFER, R. Re-fractive corneal surgery: The use of implantable alloplastic lens material. *Aust. J. Ophthal.* **1983**, 11, 325–331.

176 SENDELE, D.D., ABELSON, M.B., KENYON, K.R., HANNINEN, L.A. Intracorneal lens implantation. *Arch. Ophthal.* **1983**, 101, 940–944.

177 SAMPLES, J.R., BINDER, P.S., ZAVALA, E.Y., BAUMGARTNER, S.D., DEG, J.K. Morphology of hydrogel implants used for refractive keratoplasty. *Invest. Ophthal. Vis. Sci.* **1984**, 25, 843–850.

178 ZAVALA, E.Y., NAYAK, S., DEG, J.K., BINDER, P.S. Keratocyte attachment to hydrogel materials. *Curr. Eye Res.* **1984**, 3, 1253–1262.

179 BINDER, P.S., BAUMGARTNER, S.D., DEG, J.K. Hydrogel refractive keratoplasty. Lens removal and exchanges. *CLAO J.* **1984**, 10, 105–111.

180 PEIFFER, R.L., WERBLIN, T.P., FRYCZKOWSKI, A.W. Pathology of corneal hydrogel alloplastic implants. *Ophthalmology.* **1985**, 92, 1294–1304.

181 WATSKY, M.A., McCAREY, B.E., BEEKHUIS, W.H. Predicting refractive alterations with hydrogel keratophakia. *Invest. Ophthal. Vis. Sci.* **1985**, 26, 240–243.

182 McCAREY, B.E., VAN RIJ, G., BEEKHUIS, W.H. and WARING, G.O. Hydrogel keratophakia: A freehand pocket dissection in the monkey model. *Br. J. Ophthal.* **1986**, 70, 187–191.

183 SZYCHER, M. Keratoprosthetic polyurethane, *U.S. Patent No. 4,285,073,* **1981**.

184 SZYCHER, M. Process for forming an optically clear polyurethane lens or cornea. *U.S. Patent No. 4,386,039,* **1983**.

185 SZYCHER, M. Keratoprosthetic polyurethane. *U.S. Patent No. 4,424,335,* **1984**.

186 REFOJO, M.F. Glyceryl methacrylate hydrogels. *J. Appl. Polym. Sci.* **1965**, 9, 3161–3170.

187 REFOJO, M.F. Permeation of water through some hydrogels. *J. Appl. Polym. Sci.* **1965**, 9, 3417–3426.

188 DOHLMAN, C.H., REFOJO, M.F., ROSE, J. Synthetic polymers in corneal surgery. I. Glycerylmetacrylate. *Arch. Ophthal.* **1967**, 77, 252–257.

189 HYON, S.H., CHA, W.I., IKADA, Y. Preparation of transparent poly(vinyl alcohol) hydrogel. *Polym. Bull.* **1989**, 22, 119–122.

190 KOBAYASHI, H., IKADA, Y., MORITERA, T., OGURA, Y., HONDA, Y. Collagen-immobilized hydrogel as a material for lamellar keratoplasty. *J. Appl. Biomater.* **1991**, 2, 261–267.

191 TRINKAUS–RANDALL, V., CAPECCHI, J., NEWTON, A., VADASZ, A., LEIBOWITZ, H., FRANZBLAU, C. Development of a biopolymeric keratoprosthetic material. Evaluation in vitro and in vivo. *Invest. Ophthalmol. Vis. Sci.* **1988**, 29, 393–400.

192 THOMPSON, K.P., HANNA, K., WARING, G.O. 3rd, GIPSON, I., LIU, Y., GAILITIS, R.P., JOHNSON-WINT, B., GREEN, K. Current status of synthetic epikeratoplasty. *Refractive Corneal Surg.* **1991**, 7, 240–248.

193 SIPEHIA, R., GARLINKLE, A., JACKSON, W.B., CHANG, T.M.S. Towards an artificial cornea: surface modifications of optically clear, oxygen permeable soft contact lens materials by ammonia plasma modification technique for the enhanced attachment and growth of corneal epithelial cells. *Biomater. Artif. Cells Artif. Organs.* **1990**, 18, 643–655.

194 LEGEAIS, J.M., RENARD, G., PAREL, J.M., SERDAREVIC, O., MUI, M.M., POULIQUEN, Y. Expanded fluorocarbon for keratoprosthesis cellular ingrowth and transparency. *Exp. Eye Res.* **1994**, 58, 41–52.

195 WANG, M., SWARTZ, T., CHU, Y.R., BULIANO, M., ABDELMALAK, H., GULANI, A., KARPECKI, P., PEPOSE,

J.S., YU, K. Peer-reviewed literature: Alphacor. *Cataract Refractive Surg. Today.* **2004**, 23–28.

196 YAMATO, M., UTSUMI, M., KUSHIDA, A.I., KONNO, C., KIKUCHI, A., OKANO, T. Thermo-responsive culture dishes allow the intact harvest of multi-layered keratinocyte sheets without dispase by reducing temperature. *Tissue Eng.* **2001**, 7, 473–480.

197 STILE, R.A., BURGHARDT, W.R., HEALY, K.E. Synthesis and characterization of injectable poly(N-isopropylacrylamide)-based hydrogels that support tissue formation *in vitro*. *Macromolecules.* **1999**, 32, 7370–7379.

198 LAVIK, E.B., KLASSEN, E., WARFVINGE, H., LANGER, R., YOUNG, M.J. Fabrication of degradable polymer scaffolds to direct the integration and differentiation of retinal progenitors *Biomaterials.* **2005**, 26, 3187–3196.

199 MARC, R., JONES, B., WATT, C., STRETTOI, E. Neural remodeling in retinal degeneration. *Prog. Eye Res.* **2003**, 22, 607–655.

200 KLASSEN, H., SAKAGUCHI, D.S., YOUNG, M. Stem cells and retinal repair. *Prog. Eye Res.* **2004**, 23, 149–181.

201 RYAN, S.J. The patho-physiology of proliferative vitreoretinopathy in its management. *Am. J. Ophthalmol.* **1985**, 100, 188–193.

202 REH, T.A., LEVINE, E.M. Multi-potential stem cells and progenitors in the vertebrate retina. *J. Neurobiol.* **1998**, 36, 206–220.

203 TROPEPE, V., COLES, B.L.K., CHIASSON, B.J., HORSFORD, D.J., ELIA, A.J., MCINNES, R.R., VAN DER KOOY, D. Retinal stem cells in the adult mammalian eye. *Science.* **2000**, 287, 2032–2036.

204 SHATOS, M., MIZUMOTO, K., MIZUMOTO, H., KURIMOTO, Y., KLASSEN, H., YOUNG, M. Multipotent stem cells from the brain and retina of green mice. *Regen. Med.* **2001**, 2, 13–15.

205 BHATTACHARYA, S., JACKSON, J.D., DAS, A.V., THORESON, W.B., KUSZYNSKI, C., JAMES, J., JOSHI, S., AHMAD, I. Direct identification and enrichment of retinal stem cells/progenitors by Hoechst dye efflux assay. *Invest. Ophthalmol. Visual Sci.* **2003**, 44, 2764–2773.

206 AKAGI, T., HARUTA, M., AKITA, J., NISHIDA, A., HONDA, Y., TAKAHASHI, M. Different characteristics of rat retinal progenitor cells from different culture periods. *Neurosci. Lett.* **2003**, 341, 213–216.

207 DI LUCCIO, M., NOBREGA, R., BORGES, C.P. Microporous anisotropic phase inversion membranes from bisphenol A polycarbonate: Effect of additives to the polymer solution. *J. Appl. Polym. Sci.* **2002**, 86, 3085–3096.

208 SCHURGENS, C., MAQUET, V., GRANDFILS, Ch., JEROME, R., TEYSSIE, Ph. Polylactide macroporous biodegradable implants for cell transplantation. II. Preparation of polylactide foams by liquid–liquid phase separation. *J. Biomed. Mater. Res.* **1996**, 30, 449–4461.

209 ZHONG, S., TEO, W.E., ZHU, X., BEUERMAN, R., RAMAKRISHNA, S., YUNG, L.Y.L. Formation of collagen-GAG blended nanofibrous scaffolds and their biological properties. *Biomacromolecules.* **2005**, 6, 2998–3004.

210 TAN, W., KRISHNARAJ, R., DESAI, T.A. Evaluation of nanostructured composite collagen-chitosan matrices for tissue engineering. *Tissue Eng.* **2001**, 7, 203–210.

211 BELLAMKONDA, R., RANIERI, J.P., BOUCHE, N., AEBISCHER, P. Hydrogel-based 3-dimensional matrix for neural cells. *J. Biomed. Mater. Res.* **1995**, 29, 663–671.

212 SAITO, N., OKADA, T., HORIUCHI, H., MURAKAMI, N., TAKAHASHI, J., NAWATA, M., OTA, H., NOZAKI, K., TAKAOKA, K. A biodegradable polymer as a cytokine delivery system for inducing bone formation. *Nat. Biotechnol.* **2001**, 19, 332–335.

213 EVANS, G.R.D., BRANDT, K., WIDMER, M.S., LU, L., MESZLENYI, R.K., GUPTA, P.K., MIKOS, A.G., HODGES, J., WILLIAMS, J., GURLEK, A., NABAWI, A., LOHMAN, R., PATRICK, C.W.J. *In vivo* evaluation of poly(L-lactic acid) porous conduits for peripheral nerve regeneration. *Biomaterials.* 1999, 20, 1109–1115.

214 Li, F., Carlsson, D., Lohmann, C., Suuronen, E., Vascotto, S., Kobuch, K., Sheardown, H., Munger, R., Nakamura, M., Griffith, M. Cellular and nerve regeneration within a biosynthetic extracellular matrix for corneal transplantation. *Proc. Natl. Acad. Sci. U.S.A.* 2003, 100, 15 346–15 351.

215 Pek, Y.S., Spector, M., Yannas, M., Gibson, I.V., Ti, L.J. Degradation of a collagen-chondroltin-6-sulfate matrix by collagenase and by chondroitinase. *Biomaterials.* 2004, 25, 473–482.

216 Lee, C.R., Grodzinsky, A.J., Spector, M. The effects of cross-linking of collagen-glycosaminoglycan scaffolds on compressive stiffness, chondrocyte-mediated contraction, proliferation and biosynthesis. *Biomaterials.* 2001, 22, 3145–3154.

217 Doillon, C.J., Watsky, M.A., Hakim, M., Wang, J., Munger, R., Laycock, N., Osborne, R., Griffith, M. A collagen-based scaffold for a tissue engineered human cornea: Physical and physiological properties. *Int. J. Artif. Organs.* 2003, 26, 764–773.

218 Yunoki, S., Nagai, N., Suzuki, T., Munekata, M. Novel biomaterial from reinforced salmon collagen gel prepared by fibril formation and cross-linking. *J. Biosci. Bioeng.* 2004, 98, 40–47.

219 Zhong, S., Teo, W.E., Zhu, X., Beuerman, R., Ramakrishna, S., Yung, L.Y. A study on characterization of aligned nanofibrous collagen scaffold by electrospinning and its effects on *in vitro* culture, unpublished work.

220 Zhu, H.G., Ji, J., Tan, Q., Barbosa, M.A., Shen, J. Surface engineering of poly(DL-lactide) via electrostatic self-assembly of extracellular matrix-like molecules. *Biomacromolecules.* 2003, 4, 378–386.

221 Zhang, R.Y. and Ma, P.X. Synthetic nano-fibrillar extracellular matrices with predesigned macroporous architectures. *J. Biomed. Mater. Res.* 2000, 52, 430–438.

222 Yim, E.K.F., Leong, K.W. Significance of synthetic nanostructures in dictating cellular response. *Nanomed.: Nanotechnol., Biol. Med. 1,* 2005, 10–21.

223 Mao, H.Q., Kadiyala, I., Leong, K.W., Zhao, Z., Dang, W. Biodegradable polymers: poly(phosphoester)s. In: *Encyclopedia of Controlled Drug Delivery,* Mathiowitz, E. (Ed.), Wiley Inc., New York, 1999.

224 Wen, J., Leong, K.W. Synthesis and characterization of poly(ε-caprolactone-co-ethyl ethylene phosphate) for tissue engineering. The 29th International Symposium on Controlled Release of Bioactive Materials, Controlled Release Society, Seoul, Korea, 2002.

225 Wen, J., Kim, G.J., Leong, K.W. Poly(D,L-lactide-co-ethyl ethylene phosphate)s as new drug carriers. *J. Controlled Rel.* 2003, 92, 39–48.

226 Hodgkinson, C.P., Wright, M.C., Paine, A.J. Fibronectin-mediated hepatocyte shape change reprograms cytochrome P450 2C11 gene expression via an integrin-signaled induction of ribonuclease activity. *Mol. Pharmacol.* 2000, 58, 976–981.

227 Woerly, S., Plant, G.W., Harvey, A.R. Neural tissue engineering: From polymer to biohybrid organs. *Biomaterials.* 1996, 17, 301–310.

228 Woerly, S., Marchand, R. 100 ans de neurotransplantation chez les mammifires. 1 Neurochirurgie 1 *Neurochirurgie.* 1990, 36, 71–95.

229 Hudson, T., Evans, G., Schmidt, C. Engineering strategies for peripheral nerve repair. *Orthop. Clin. North. Am.* 2000, 31, 485–498.

230 Ranieri, J.P., Bellamkonda, R., Jacob, J., Vargo, T.G., Gardella, J.A., Aebischer, P. Selective neuronal cell attachment to a covalently patterned monoamine on fluorinated ethylene propylene films. *J. Biomed. Mater. Res.* 1993, 27, 917–925.

231 Ranieri, J.P., Bellamkonda, R., Bekos, E.J., Gardella, J.A. Jr., Mathieu, H.J., Ruiz, L., Aebischer, P. Spatial control of neuronal cell attachment and differentiation on covalently patterned laminin

oligopeptide substrates. *J. Biomed. Mater. Res.* **1994**, 12, 725–735.

232 MATSUDA, T., SUGAWARA, T., INOUE, K. An artificial neural circuit based on surface microphotoprocessing. *ASAIO J.* **1992**, 36, 243–247.

233 WOERLY, S., MAGHAMI, G., DUNCAN, R., SUBR, V., ULBRICH, K. Synthetic polymer derivatives as substrata for neuronal adhesion and growth. *Brain Res. Bull.* **1993**, 30, 423–432.

234 WOERLY, S., LAROCHE, G., MARCHAND, R., PATO, J., SUBR, V., ULBRICH, K. Intracerebral implantation of hydrogel-coupled adhesion peptides: Tissue reaction. *J. Neural Transplant Plast.* **1995**, 5, 245–255.

235 FINE, E.G., VALENTINI, R.E., AEBISCHER, P. Nerve regeneration. In: *Principles of Tissue Engineering*, LANZA, R.P., LANGER, R., VACANTI, J.P. (Eds.), Academic Press, San Diego, **2000**, pp. 785–798.

236 TRESCO, P.A. Tissue engineering strategies for nervous system repair. In: *Progress in Brain Research*, SEIL, F.J. (Ed.), Elsevier Science, New York, **2000**, pp. 349–363.

237 SCHMIDT, C.E., LEACH, J.B. Neural tissue engineering: Strategies for repair and regeneration. *Biomed. Eng.* **2003**, 5, 293–347.

238 RUTKA, J.T., APODACA, G., STERN, R., ROSENBLUM, M. The extracellular matrix of the central and peripheral nervous systems: structure and function. *J. Neurosurg.* **1988**, 69, 155–170.

239 CARBONETTO, S. The extracellular matrix of the nervous systems. *Trends Neurosci.* **1984**, 7, 382–387.

240 DESAI, T.A. Micro- and nanoscale structures for tissue engineering constructs. *Med. Eng. Phys.* **2000**, 2, 595–606.

241 THOMSON, R.C., SHUNG, A.K., YASZEMSKI, M.J., MIKOS, A.G. In: *Principles of Tissue Engineering*, LANZA, R.P., LANGER, R., VACANTI, J.P. (Eds.), Academic Press, San Diego, **2000**, p. 251.

242 MA, P.X., ZHANG, R. Synthetic nano-scale fibrous extracellular matrix. *J. Biomed. Mater. Res.* **1999**, 46, 60–72.

243 BOGNITZKI, M., CZADO, W., FREESE, T., SCHAPER, A., HELLWIG, M., STEINHART, M., GRENIER, A., WENDORFF, J.H. Nanostructured fibers via electrospinning. *Adv. Mater.* **2001**, 13, 70–76.

244 HARTGERINK, D.J., BENIASH, E., STUPP, S.I. Self-assembly and mineralization of peptide-amphiphile nanofibers. *Science.* **2001**, 294, 1684–1687.

245 YANG, F., MURUGAN, R., RAMAK-RISHNA, S., WANG, X., MA, Y.X. Fabrication of nano-structured porous PLLA scaffold intended for nerve tissue engineering. *Biomaterials.* **2004**, 25, 1891–1900.

246 SPILKER, M.H., ASANO, K., YANNAS, I.V., SPECTOR, M. Contraction of collagen-glycosaminoglycan matrices by peripheral nerve cells *in vitro.* *Biomaterials.* **2001**, 22, 1085–1093.

247 XU, C.Y., YANG, F., WANG, S., RAMAKRISHNA, S. *In vitro* study of human vascular endothelial cell function on materials with various surface roughness. *J. Biomed. Mater. Res.* **2004**, 71A, 154–161.

248 HADLOCK, T., ELISSEEFF, J., LANGER, R., VACANTI, J.P., CHENEY, M. A tissue-engineered conduit for peripheral nerve repair. *Arch. Otolaryngol. Head Neck Surg.* **1998**, 124, 1081–1086.

249 CEBALLOS, D., NAVARRO, X., DUBEY, N., WENDELSCHAFER-CRABB, G., KENNEDY, W.R., TRANQUILLO, R.T. Magnetically aligned collagen gel filling a collagen nerve guide improves peripheral nerve regeneration. *Exp. Neurol.* **1999**, 158, 290–300.

250 DUBEY, N., LETOURNEAU, P.C., TRANQUILLO, R.T. Neuronal contact guidance in magnetically aligned fibrin gels: Effect of variation in gel mechano-structural properties. *Biomaterials.* **2001**, 22, 1065–1075.

251 TENG, Y.D., LAVIK, E.B., QU, X., PARK, K.I., OUREDNIK, J., ZURAKOWSKI, D., LANGER, R., SNYDER, E.Y. Functional recovery following traumatic spinal cord injury mediated by a unique polymer scaffold seeded with neural stem cells. *Proc. Natl. Acad. Sci. U.S.A.* **2002**, 99, 3024–3029.

252 KESENCI, K., MOTTA, A., FAMBRI, L., MIGLIARESI, C. Poly(ε-caprolactone-*co*-D,L-lactide)/silk fibroin particles composite materials: Preparation and characterization. *J. Biomater. Sci. Polym. Ed.* **2001**, 12, 337–351.

253 BUNGE, M.B., BUNGE, R.P. Linkage between Schwann cell extracellular matrix production and ensheathment function. *Ann. New York Acad. Sci.* **1986**, 486, 241–247.

254 YANG, F., MURUGAN, R., WANG, S., RAMAKRISHNA, S. Electrospinning of nano/micro scale poly(L-lactic acid) aligned fibers and their potential in neural tissue engineering. *Biomaterials.* **2005**, 26, 2603–2610.

255 SNYDER, E.Y., DEITCHER, D.L., WALSH, C., ARNOLD-ALDEA, S., HARTWIEG, E.A., CEPKO, C.L. Multipotent neural cell lines can engraft and participate in development of mouse cerebellum. *Cell.* **1992**, 68, 33–51.

256 SNYDER, E.Y., YOON, C., FLAX, J.D., MACKLIS, J.D. Multipotent neural precursors can be differentiated toward replacement of neurons undergoing targeted apoptotic degeneration in adult mouse neocortex. *Proc. Natl. Acad. Sci. U.S.A.* **1997**, 94, 11 663–11 668.

257 FAN, Y.W., CUI, F.Z., CHEN, L.N., ZHAI, Y., XU, Q.Y., LEE, I.S. Adhesion of neural cells on silicon wafer with nanotopographic surface. *Appl. Surf. Sci.* **2002**, 187, 313–318.

258 WINTER, G.D. Formation of scab and the rate of epithelialization in superficial wounds of the domestic pig. *Nature.* **1962**, 193, 293–294.

259 YANNAS, I.V., BURKE, J.F. Design of an artificial skin. I. Basic design principles. *J. Biomed. Mater. Res.* **1980**, 14, 65–81.

260 MATSUDA, K., SUZUKI, S., ISSHIKI, N., IKADA, Y. Re-freeze dried bilayer artificial skin. *Biomaterials.* **1993**, 14, 1030–1035.

261 FREYMAN, T.M., YANNAS, I.V., GIBSON, L.J. Cellular materials as porous scaffolds for tissue engineering. *Progr. Mater. Sci.* **2001**, 46, 273–282.

262 WOODLEY, D.T., CHEN, J.D., KIM, J.P., SARRET, Y., IWASAKI, T., KIM, Y.,

O'KEEFE, E.J. Reepithelialization: Human keratinocyte locomotion. *Dermatol. Clin.* **1993**, 11, 641–646.

263 FALANGA, V. Growth factors and wound healing. *Dermatol. Clinics.* **1993**, 11, 667–675.

264 JONES, B.C., BRIGGS, C.D., NORTON, D.A. This new type of I.V. dressing can save you time. *Nursing.* **1982**, 12, 70–73.

265 LEIPZIGER, L.S., GLUSHKO, V., DIBERNADO, B., SHAFAIE, F., NOBLE, J., NICHOLS, J., ALVAREZ, O.M. Dermal wound repair: Role of collagen matrix implants and synthetic polymer dressing. *J. Am. Acad. Dermatol.* **1985**, 12, 409–419.

266 HELFMAN, T.L.O., OVINGTON, L., FALANGA, V. Occlusive dressings and wound healing. *Clin. Dermatol.* **1994**, 12, 121–127.

267 STASO, M.A., RASCHBAUM, M., SLATER, H., GOLDFARB, I.W. Experience with omiderm – A new burn dressing. *J. Burn Care Rehab.* **1991**, 12, 209–210.

268 PITT, C.G., CHASALOW, F.I., HIBIONADA, Y.M., KLIMAS, D.M., SCHINDLER, A. Aliphatic polyesters. I. The degradation of Poly(ε-caprolactone) in vivo. *J. Appl. Polym. Sci.* **1981**, 26, 3779–3787.

269 YANNAS, I.V. Studies on the biological activity of the dermal regeneration template. *Wound Repair. Regen.* **1998**, 6, 518–524.

270 CLARK, R.A.F., SINGER, A.J. *Principles of Tissue Engineering.* LANZA, R.P., LANGER, R., VACANTI, J. (Eds.), Academic Press, San Diego, **2000**, p. 857.

271 COOMBES, A.G.A., VERDERIO, E., SHAW, B., LI, X., GRIFFIN, M., DOWNES, S. Biocomposites of non-crosslinked natural and synthetic polymers. *Biomaterials.* **2002**, 23, 2113–2118.

272 DAI, N.T., WILLIAMSON, M.R., KHAMMO, N., ADAMS, E.F., COOMBES, A.G.A. Composite cell support membranes based on collagen and polycaprolactone for tissue engineering of skin. *Biomaterials.* **2004**, 25, 4263–4271.

273 HUANG, L., NAGAPUNDI, K., CHAIKOF,

E.L. Engineered collagen-PEO nanofibers and fabrics. *J. Biomater. Sci. Polym. Ed.* **2001**, 12, 979–993.

274 COFFEE, R.A. PCT/GB97/01968. **1998**.

275 MARTINDALE, D. Scar no more. *Sci. Am.* **2000**, 34–36.

276 SMITH, D., RENEKER, D.H. *PCT/US00/27737.* **2001**.

277 JIN, H.J., FRIDRIKH, S.V., RUTLEDGE, G.C., KAPLAN, D.L. Electrospinning *Bombyx mori* silk with poly(ethylene oxide). *Biomacromolecules.* **2002**, 3, 1233–1239.

278 KATAPHINAN, W., DABNEY, S., RENEKER, D.H., SMITH, D., AKRON, U.O. Electrospun skin masks and uses therof. *Patent WO0126610.* University of Akron, **2001**.

279 SMITH, D.J., RENEKER, D.H., MCMANUS, A.T., SCHREUDER-GIBSON, H.L., MELLO, C., SENNETT, M.S. Electrospun fibers and an apparatus thereof. *US Patent 6753454.* University of Akron, **2004**.

280 ANSELME, K. Osteoblast adhesion on biomaterials. *Biomaterials.* **2000**, 21, 667–681.

281 BOYAN, B.D., HUMMERT, T.W., DEAN, D.D., SCHWARTS, Z. Role of material surfaces in regulating bone and cartilage cell response. *Biomaterials.* **1996**, 17, 137–146.

282 CAPLAN, A.L. Mesenchymal stem cells. *J. Orthoped. Res.* **1991**, 9, 641–650.

283 PITTENGER, M.F., MACKAY, A.M., BECK, S.C., JAISWAL, R.K., DOUGLAS, R., MARSHAK, D.R. Multilineage potential of adult human mesenchymal stem cells. *Science.* **1999**, 284, 143–147.

284 TERAI, T., HANNOUCHE, D., OCHOA, E., YAMANO, Y., VACANTI, J.P. *In vitro* engineering of bone using a rotational oxygen-permeable bioreactor system. *Mater. Sci. Eng. C.* **2002**, 20, 3–8.

285 SHIN, M., YOSHIMOTO, H., VACANTI, J.P. *In vivo* bone tissue engineering using mesenchymal stem cells on a novel electrospun nanofibrous scaffold. *Tissue Eng.* **2004**, 10, 33–41.

286 DE OLIVEIRA, P.T., NANCI, A. Nanoetexturing of titanium-based surfaces upregulates expression of bone sialoprotein and osteopontin by cultured osteogenic cells. *Biomaterials.* **2004**, 25, 403–413.

287 BADAMI, A.S., KREKE, M.R., THOMPSON, M.S., RIFFLE, J.S., GOLDSTEIN, A.S. Effect of fiber diameter on spreading, proliferation, and differentiation of osteoblastic cells on electrospun poly(lactic acid) substrates. *Biomaterials.* **2006**, 27, 596–606.

288 ELIAS, K.L., PRICE, R.L., WEBSTER, T.J. Enhanced functions of osteoblasts on nanometer diameter carbon fibers. *Biomaterials.* **2002**, 23, 3279–3287.

289 WEI, G., MA, P.X. Structure and properties of nano-hydroxyapatite/polymer composite scaffolds for bone tissue engineering. *Biomaterials.* **2004**, 25, 4749–4757.

290 WEBSTER, T.J., SIEGEL, R.W., BIZIOS, R. Osteoblast adhesion on nanophase ceramics. *Biomaterials.* **1999**, 20, 1221–1227.

291 WEBSTER, T.J., ERGUN, C., DOREMUS, R.H., SIEGEL, R.W., BIZIOS, R. Specific proteins mediate enhanced osteoblast adhesion on nanophase ceramics. *J. Biomed. Mater. Res.* **2000**, 51, 475–483.

292 WEBSTER, T.J., EJIOFOR, J.U. Increased osteoblast adhesion on nanophase metals: Ti, Ti6AL4 V, and CoCrMo. *Biomaterials.* **2004**, 25, 4731–4739.

293 THAPA, A., MILLER, D.C., WEBSTER, T.J., HABERSTROH, K.M. Nanostructured polymers enhance bladder smooth muscle cell function. *Biomaterials.* **2003**, 24, 2915–2926.

294 THAPA, A., WEBSTER, T.J., HABERSTROH, K.M. Polymers with nano-dimensional surface features enhance bladder smooth muscle cell adhesion. *J. Biomed. Mater. Res.* **2003**, 67A, 1374–1383.

295 MILLER, D.C., THAPA, A., HABERSTROH, K.M., WEBSTER, T.J. Endothelial and vascular smooth muscle cell function on poly(lactic-co-glycolic acid) with nano-structured surface features. *Biomaterials.* **2004**, 25, 53–61.

296 FUJIHARA, K., KOTAKI, M., RAMAKRISHNA, S. Guided bone regeneration membrane made of polycaprolactone/calcium carbonate

composite nano-fibers. *Biomaterials.* **2005**, 26, 4139–4147.

297 USHIDA, T., FURUKAWA, K., TOITA, K., TATEISHI, T. Three dimensional seeding of chondrocytes encapsulated in collagen gel into PLLA scaffolds. *Cell Transplantation.* **2002**, 11, 489–494.

298 DUNKELMAN, N.S., ZIMBER, M.P., LEBARON, R.G., PAVELEC, R., KWAN, M., PURCHIO, A.F. Cartilage production by rabbit articular chondrocytes on polyglycolic acid scaffolds in a closed bioreactor system. *Biotechnol. Bioeng.* **1995**, 46, 299–305.

299 FREED, L.E., HOLLANDER, A.P., MARTIN, I., BARRY, J.R., LANGER, R., VUNJAK-NOVAKOVIC, G. Chondrogenesis in a cell polymer-bioreactor system. *Exp. Cell Res.* **1998**, 240, 58–65.

300 ABBOTT, J., HOLTZER, H. The loss of phenotypic traits by differentiated cells. III. The reversible behavior of chondrocytes in primary cultures. *J. Cell Biol.* **1966**, 28, 473–487.

301 VAN OSCH, G.J., VAN DER VEEN, S.W., VERWOERD-VERHOEF, H.L. *In vitro* redifferentiation of culture-expanded rabbit and human auricular chondrocytes for cartilage reconstruction. *Plast. Reconstr. Surg.* **2001**, 107, 433–440.

302 BENYA, P.D., SHAFFER, J.D. Dedifferentiated chondrocytes reexpress the differentiated collagen phenotype when cultured in agarose gels. *Cell.* **1982**, 30, 215–224.

303 MARTIN, I., VUNJAK-NOVAKOVIC, G., YANG, J., LANGER, R., FREED, L.E. Mammalian chondrocytes expanded in the presence of fibroblast growth factor to maintain the ability to differentiate and regenerate three-dimensional cartilaginous tissue. *Exp. Cell Res.* **1999**, 253, 681–688.

304 LEE, K.Y., ALSBERG, E., MOONEY, D.J. Degradable and injectable poly(aldehyde guluronate) hydrogels for bone tissue engineering. *J. Biomed. Mater. Res.* **2001**, 56, 228–233.

305 BUCKWALTER, J.A., MANKIN, H.J. Articular cartilage: Tissue design and chondrocyte–matrix interactions.

J. Bone Joint Surg. Am. **1997**, 79A, 600–611.

306 HUFNAGEL, C.A. Permanent intubation of the thoracic aorta. *Arch. Surgery.* **1947**, 54, 382.

307 VOORHEES, A.B., JR., JARETZKI, A., BLAKEMORE, A.H. The use of tubes constructed from vinyon "N" cloth in bridging arterial defects. *Ann. Surg.* **1952**, 135, 332–336.

308 XUE, L., GREISLER, H.P. Biomaterials in the development and future of vascular grafts. *J. Vasc. Surg.* **2003**, 37, 472–480.

309 WESOLOWSKI, S.A., FRIES, C.C., DOMINGO, R.T., LIEBIG, W.J., SAWYER, P.N. The compound prosthetic vascular graft: A pathologic survey. *Surgery.* **1963**, 53, 19–44.

310 RUDERMAN, R.J., HEGYELI, A.F., HATTLER, B.G., LEONARD, F. A partially biodegradable vascular prosthesis. *Trans. Am. Soc. Artif. Intern. Organs.* **1972**, 18, 30–37.

311 BOWALD, S., BUSCH, C., ERIKSSON, I. Arterial regeneration following polyglactin 910 suture mesh grafting. *Surgery.* **1979**, 86, 722–729.

312 BOWALD, S., BUSCH, C., ERIKSSON, I. Absorbable material in vascular prostheses: A new device. *Acta Chir. Scand.* **1980**, 146, 391–395.

313 GREISLER, H.P. Arterial regeneration over absorbable prostheses. *Arch. Surg.* **1982**, 117, 1425–1431.

314 GREISLER, H.P., KIM, D.U., PRICE, J.B., VOORHEES, A.B. JR. Arterial regenerative activity after prosthetic implantation. *Arch. Surg.* **1985**, 120, 315–323.

315 MARTINI, H.F., OBER, W.C., GARRISON, C.W., WELCH, K., HUTCHINGS, R.T. *Fundamentals of Anatomy and Physiology* (4th Edn.), Prentice Hall, New Jersey, **1998**.

316 CHUNG, T.W., LIU, D.Z., WANG, S.Y., WANG, S.S. Enhancement of the growth of human endothelial cells by surface roughness at nanometer scale. *Biomaterials.* **2003**, 24, 4655–4661.

317 BUTTAFOCO, L., KOLKMAN, N.G., ENGBERS-BUIJTENHUIJS, P., POOT, A.A., DIJKSTRA, P.J., VERMES, I., FEIJEN, J. Electrospinning of collagen

and elastin for tissue engineering applications. *Biomaterials* **2006**, 27, 724–734.

318 BISHOP, A.E., BUTTERY, L.D.K., POLACK, J.M. Embryonic stem cells. *J. Pathol.* **2002**, 197, 424–429.

319 TABATA, Y. The importance of drug delivery systems in tissue engineering. *Pharm. Sci. Technol. Today.* **2000**, 3, 80–89.

320 BOUDRIOT, U., GOETZ, B., DERSCH, R., GREINER, A., WENDORFF, J.H. Role of electrospun nanofibers in stem cell technologies and tissue engineering. *Macromol. Symp.* **2005**, 225, 9–16.

321 CAPLAN, A.I., BRUDER, S.P. Mesenchymal stem cells: Building blocks for molecular medicine in the 21st century. *Trends Mol. Med.* **2001**, 7, 259–264.

322 HAYNESWORTH, S.E., GOSHIMA, J., GOLDBERG, V.M., CAPLAN, A.I. Characterization of cells with osteogenic potential from human marrow. *Bone.* **1992**, 13, 81–88.

323 JOHNSTONE, B., HERING, T.M., CAPLAN, A.I., GOLDBERG, V.M., YOO, J.U. In vitro chondrogenesis of bone marrow-derived mesenchymal progenitor cells. *Exp Cell Res.* **1998**, 238, 265–272.

324 YOO, J.U., BARTHEL, T.S., NISHIMURA, K., SOLCHAGA, L., CAPLAN, A.I., GOLDBERG, V.M., JOHNSTONE, B. The chondrogenic potential of human bone-marrow-derived mesenchymal progenitor cells. *J. Bone Jt. Surg. Am.* **1998**, 80, 1745–1757.

325 ZUK, P.A., ZHU, M., MIZUNO, H., HUANG, J., FUTRELL, J.W., KATZ, A.J., BENHAIM, P., LORENZ, H.P., HEDRICK, M.H. Multilineage cells from human adipose tissue: Implications for cell-based therapies. *Tissue Eng.* **2001**, 7, 211–228.

326 GIMBLE, J., GUILAK, F. Adipose-derived adult stem cells: Isolation, characterization, and differentiation potential. *Cytotherapy.* **2003**, 5, 362–369.

327 NOTH, U., OSYCZKA, A.M., TULI, R., HICKOK, N.J., DANIELSON, K.G., TUAN, R.S. Multilineage mesenchymal differentiation potential of human

trabecular bone-derived cells. *J. Orthop Res.* **2002**, 20, 1060–1069.

328 JANKOWSKI, R.J., DEASY, B.M., HUARD, J. Muscle-derived stem cells. *Gene Ther.* **2002**, 9, 642–647.

329 DE BARI, C., DELL'ACCIO, F., VANDENABEELE, F., VERMEESCH, J.R., RAYMACKERS, J.M., LUYTEN, F.P. Skeletal muscle repair by adult human mesenchymal stem cells from synovial membrane. *J. Cell Biol.* **2003**, 160, 909–918.

330 CATERSON, E.J., NESTI, L.J., LI, W.J., DANIELSON, K.G., ALBERT, T.J., VACCARO, A.R., TUAN, R.S. Three-dimensional cartilage formation by bone marrow-derived cells seeded in polylactide/alginate amalgam. *J. Biomed. Mater. Res.* **2001**, 57, 394–403.

331 TULI, R., TULI, S., NANDI, S., WANG, M.L., ALEXANDER, P.G., HALEEM-SMITH, H., HOZACK, W.J., MANNER, P.A., DANIELSON, K.G., TUAN, R.S. Characterization of multipotential mesenchymal progenitor cells derived from human trabecular bone. *Stem Cells.* **2003**, 21, 681–693.

332 DEVINE, S.M. Mesenchymal stem cells: Will they have a role in the clinic? *J. Cell Biochem.* **2002**, 38, 73–79.

333 LI, W.J., TULI, R., HUANG, X., LAQUERRIERE, P., TUAN, R.S. Multineage differentiation of human mesenchymal stem cells in a three-dimensional nanofibrous scaffold. *Biomaterials.* **2005**, 26, 5158–5166.

334 ANGELE, P., KUJAT, R., NERLICH, M., YOO, J., GOLDBERG, V., JOHNSTONE, B. Engineering of osteochondral tissue with bone marrow mesenchymal progenitor cells in a derivatized hyaluronan-gelatin composite sponge. *Tissue Eng.* **1999**, 5, 545–554.

335 YAYLAOGLU, M.B., YILDIZ, C., KORKUSUZ, F., HASIRCI, V. A novel osteochondral implant. *Biomaterials.* **1999**, 20, 1513–1520.

336 GAO, J., DENNIS, J.E., SOLCHAGA, L.A., AWADALLAH, A.S., GOLDBERG, V.M., CAPLAN, A.I. Tissue-engineered fabrication of an osteochondral composite graft using rat bone

marrow-derived mesenchymal stem
cells. *Tissue Eng.* **2001**, 7, 363–371.

337 SCHAEFER, D., MARTIN, I., SHASTRI,
P., PADERA, R.F., LANGER, R., FREED,
L.E., VUNJAK-NOVAKOVIC, G. *In vitro*
generation of osteochondral composites. *Biomaterials.* **2000**, 21, 2599–
2606.

338 TULI, R., NANDI, S., LI, W.J., TULI, S.,
HUANG, X., MANNER, P.A.,
LAQUERRIERE, P., NOTH, U., HALL,
D.J., TUAN, R.S. Human mesenchymal progenitor cell-based tissue
engineering of a single-unit osteochondral constructs. *Tissue Eng.* **2004**,
10, 1169–1179.

339 MURAGLIA, A., CORSI, A., RIMINUCCI,
M., MASTROGIACOMO, M., CANCEDDA,
R., BIANCO, P., QUARTO, R. Formation
of a chondroosseous rudiment in
micromass cultures of human bone-
marrow stromal cells. *J. Cell Sci.* **2003**,
116, 2949–2955.

340 OSYCZKA, A.M., NOTH, U., O'CONNOR,
J., CATERSON, E.J., YOON, K.,
DANIELSON, K.G., TUAN, R.S.
Multilineage differentiation of adult
human bone marrow progenitor cells
transduced with human papilloma
virus type 16 E6/E7 genes. *Calcif.
Tissue Int.* **2002**, 71, 447–458.

341 MURPHY, J.M., DIXON, K., BECK, S.,
FABIAN, D., FELDMAN, A., BARRY, F.
Reduced chondrogenic and adipogenic
activity of mesenchymal stem cells
from patients with advanced
osteoarthritis. *Arthritis Rheum.* **2002**,
46, 704–713.

342 LANGER, R. Where a pill won't reach.
Sci. Am. **2003**, 288, 50–57.

343 LANGER, R. Drug delivery and
targeting. *Nature.* **1998**, 392, 5–10.

344 KIM, K., LUUC, Y.K., CHANG, C.,
FANG, D., HSIAO, B.S., CHUA, B.,
HADJIARGYROU, M. Incorporation and
controlled release of a hydrophilic
antibiotic using poly(lactide-co-
glycolide)-based electrospun
nanofibrous scaffolds, *J. Controlled
Release.* **2004**, 98, 47–56.

345 KENAWY, E.R., BOWLIN, G.L.,
MANSFIELD, K., LAYMAN, J., SIMPSON,
D.G., SANDERS, E.H., WNEK, G.E.
Release of tetracycline hydrochloride

from electrospun poly(ethylene-*co*-
vinylacetate), poly(lactic-acid), and a
blend. *J. Controlled Release.* **2002**, 81,
57–64.

346 KENAWY, E.R., ABDEL-FATTAH, Y.R.
Antimicrobial properties of modified
and electrospun poly(vinyl phenol).
Macromol. Biosci. **2002**, 2, 261–266.

347 ZONG, X., KIM, K., FANG, D., RAN, S.,
HSIAO, B.S., CHU, B. Structure and
process relationship of electrospun
bioabsorbable nanofiber membranes.
Polymer. **2002**, 43, 4403–4412.

348 VERRECK, G., CHUN, I., PEETERS, J.,
ROSENBLATT, J., BREWSTER, M.E.
Preparation and characterization of
nanofibers containing amorphous
drug dispersions generated by
electrostatic spinning. *Pharm Res.*
2003, 20, 810–817.

349 VERRECK, G., CHUN, I., ROSENBLATT,
J., PEETERS, J., DIJCK, A.V., MENSCH,
J., NOPPE, M., BREWSTER, M.E.
Incorporation of drugs in an
amorphous state into electrospun
nanofibers composed of a water-
insoluble, nonbiodegradable polymer.
J. Controlled Release. **2003**, 92, 349–
360.

350 ZENG, J., XU, X., CHEN, X., LIANG,
Q., BIAN, X., YANG, L., JING, X.
Biodegradable electrospun fibers for
drug delivery. *J. Controlled Release.*
2003, 92, 227–231.

351 BREWSTER, M.E., VERRECK, G., CHUN,
I., ROSENBLATT, J., MENSCH, J., VAN
DIJCK, A., NOPPE, M., ARIEN, A.,
BRUINING, M., PEETERS, J. The use of
polymer-based electrospun nanofibers
containing amorphous drug
dispersions for the delivery of poorly
water-soluble pharmaceuticals.
Pharmazie. **2004**, 59, 387–391.

352 RICHARDSON, T.P., MARTIN, C.P.,
ALESSANDRA, B.E., MOONEY, D.J.
Polymeric system for dual growth
factor delivery. *Nat. Biotechnol.* **2001**,
19, 1029–1034.

353 MURRAY, J., BROWN, L., LANGE, R.
Controlled release of microquantities
of macromolecules. *Cancer Drug Deliv.*
1984, 1, 119–123.

354 EDELMAN, E.R., MATHIOWITZ, E.,
LANGER, R., KLAGSBRUN, M.

Controlled and modulated release of basic fibroblast growth factor. *Biomaterials.* **1991**, 12, 619–626.

355 GOMBOTZ, W.R., PETTIT, D.K. Biodegradable polymers for protein and peptide drug delivery. *Bioconjug. Chem.* **1995**, 6, 332–351.

356 KUO, P.Y.P., SALTZMAN, W.M. Novel systems for controlled delivery of macromolecules. *Crit. Rev. Eukaryot. Gene Expr.* **1996**, 6, 59–73.

357 MAHONEY, M., SALTZMAN, W. Millimeter-scale positioning of a nerve-growth-factor source and biological activity in the brain. *Proc.*

Natl. Acad. Sci. U.S.A. **1999**, 96, 4536–4539.

358 LEE, K.Y., PETERS, M.C., ANDERSON, K.W., MOONEY, D.J. Controlled growth factor release from synthetic extracellular matrices. *Nature.* **2000**, 408, 998–1000.

359 SHEA, L.D., SMILEY, E., BONADIO, J., MOONEY, D.J. DNA delivery from polymer matrices for tissue engineering. *Nat. Biotechnol.* **1999**, 17, 551–554.

360 LU, D., SALTZMAN, W. Synthetic DNA delivery systems. *Nat. Biotechnol.* **2000**, 18, 33–37.

3
Electrospinning Technology for Nanofibrous Scaffolds in Tissue Engineering

Wan-Ju Li, Rabie M. Shanti, and Rocky S. Tuan

3.1
Introduction

Annually, millions of Americans suffer tissue loss or end-stage organ failure at a health care cost exceeding $400 billion, representing nearly one-half of all medical related costs in the United States [1]. Approximately 80 000 of these patients will await solid organ transplantation, and based on the figures from the United Network for Organ Sharing (UNOS) only 27 037 solid organs were transplanted in the United States in 2004. Also, according to the American Association of Tissue Banks (AATB), 1.3 million bone grafts and 94 000 soft tissue (e.g., tendon, meniscus) grafts were performed in the United States in 2003. The current standard-of-care for the replacement of lost or damaged organs and tissues includes transplantation of whole organs from one individual to another (allografts), transplantation of a patient's own tissues from one region of the body to another (autografts), or by the utilization of synthetic materials (alloplasts). Although these therapeutic modalities have had much success in clinical practice, each is associated with significant complications. For instance, allografts are restricted due to limited supply of donor organs and tissues, potential sequelae of chronic immunosuppression, and risk of host immunorejection. Alloplasts or synthetic materials, also have the potential of eliciting an adverse body response, and often fail over time due to wear and fatigue. Therefore, tissue engineering has emerged as a promising alternative for the reconstitution of lost or damaged organs and tissues, circumventing the secondary complications associated with autografts, allografts, and alloplasts.

Three general strategies were devised in the early 1990s for the engineering of new tissue: (1) delivery of cells or cell substitutes, (2) local or systemic delivery of tissue inducing substances, (3) and delivery of scaffolds containing both cells and inductive agents [1]. These approaches typically require some sort of carrier in the form of a biomaterial scaffold to guide the delivery of cells, inductive agents, or a combination of the two to the area of interest. In the same capacity, this biomaterial scaffold can also facilitate the growth and regeneration of new tissue. Figure 3.1 illustrates the general concept of tissue engineering based strategy for therapeutic applications. Briefly, target cells (differentiated/undifferentiated) expanded

Nanotechnologies for the Life Sciences Vol. 9
Tissue, Cell and Organ Engineering. Edited by Challa S. S. R. Kumar
Copyright © 2006 WILEY-VCH Verlag GmbH & Co. KGaA, Weinheim
ISBN: 3-527-31389-3

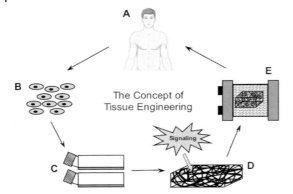

Fig. 3.1. The general concept of tissue engineering. (A) Human tissue or cell donors and recipients. (B) Harvested and isolated cells. (C) Cells expanded in tissue cultures. (D) Cells cultured in tissue engineered scaffolds and induced by biochemical and mechanical signals. (E) Cellular constructs maintained in bioreactors.

in vitro cultures, usually in cell culture flasks, are cultured in three-dimensional (3D), highly porous biomaterial scaffolds (natural/synthetic) under biologically favored conditions. After a period of culture time, the cellular scaffold receiving chemical and physical growth-needed stimuli turns into a natural tissue-like cellular implant. Furthermore, these biologically functional scaffolds play a critical role in the tissue engineering process, for they provide a 3D structure for cellular functions such as attachment, migration, proliferation, and differentiation and/or serve as vehicles for the delivery of cells to the implant site. The success of this process is determined by the biological and functional similarity of the engineered tissue with native tissue. In summary, whether the biomaterial scaffold serves as a 3D matrix for *in vitro* culture or functions as a template to recruit surrounding host cells to conduct the repair process, a principal objective of scaffold design for tissue engineering is to create a structure that can simulate the native extracellular matrix (ECM) until cells seeded within the scaffold and/or derived from the host tissue can synthesize a new, natural matrix.

To achieve this goal, a scaffold material must be carefully selected and the scaffold architecture should be properly designed to ensure biocompatibility with the seeded cells. Ideal characteristics for a scaffold include (a) biocompatibility (no cell toxicity before and after degradation of the scaffold material); (b) promotion of cellular activities such as cell adhesion; (c) biodegradability with a controlled rate of degradation that corresponds with tissue growth within the scaffold; (d) a 3D highly porous structure with an interconnected network of microscopic spaces to allow tissue growth and permeation of nutrient medium; (e) favorable mechanical properties; and (f) a highly reproducible and adaptable fabrication process for different shapes or sizes. These characteristics are determined by the material selection and the method of scaffold fabrication [2].

Both chemical [3] and physical properties [4] of a scaffold affect cell behavior, ultimately determining the fate of a tissue-engineered scaffold. Progress has been made in the improvement of the chemical biocompatibility of scaffolds through the use of natural polymers [5] or synthetic polymers that incorporate bioactive peptides, such as those containing the arginine-glycine-aspartate (RGD) sequence [6]. Conversely, there has been relatively limited work on the effect of physical properties, such as the 3D architecture of a scaffold, on cell behavior. Cells respond differently to geometrically distinct biomaterial substrates [7]. Specifically, previous studies have demonstrated that cells behave differently when cultured on either two-dimensional (2D) or 3D substrates [8, 9]. For instance, chondrocytes dedifferentiate when cultured on a 2D surface [10], but maintain a stable phenotype when cultured in a 3D agarose hydrogel [11]. Fibroblasts cultured in a 3D environment developed a 3D profile of matrix adhesion, leading to a distinct cytoskeletal organization and induction of specific cell signaling pathways, compared with those cultured on 2D substrates, implying the *in vivo* relevance of 3D matrix adhesion. Similarly, the advantage of using 3D structures for cell culture has also been shown to have an effect on the maintenance and differentiation of embryonic stem cells [12]. For instance, biomimetic 3D cultures significantly increase hematopoietic differentiation efficacy of embryonic stem cells over their 2D counterparts. The biological relevance of 3D cultures is most likely a consequence of cell–matrix interactions that proceed in a complicated, bidirectional (outside-in and inside-out) manner [13], and are likely to be critical for effective tissue engineering or regeneration. While much is known about the biology of cell–matrix interactions in 2D cultures, relatively little is known about this process in 3D matrices. To date, two models of 3D environments have been proposed to elucidate the details of this interaction: substrates with a 3D topography, and 3D scaffolds. The former is useful for studying the effects of geometrical variables on cellular activities, but not directly applicable to tissue engineering. Since 3D scaffolds can be fabricated into desired shapes and form interconnected pores to allow tissue ingrowth, they are readily applicable for tissue engineering.

Recent studies have also shown that cells on substrates with varying topography, fabricated using precisely controllable techniques, behave quite differently, suggesting that cells are able to recognize and distinguish geometric properties of substrates, such as shape and/or roughness. For example, Tuan's and Curtis's groups have reported that surface roughness had an effect on osteoblast, endothelial cell, and fibroblast morphology, cytoskeletal properties, and proliferation [14–16]. Other studies have reported that adhesion, proliferation, synthesis of alkaline phosphatase, and deposition of a calcium-containing mineral were all enhanced when osteoblasts were cultured in a nanophase ceramic, compared with micro-grain size ceramics [17, 18]. Increased functions of osteoblasts have also been correlated with a decrease in the diameter of carbon nanofibers [19]. These observations suggest that a scaffold composed of nanometer-scale components is biologically preferred, which is consistent with previous studies reporting that cells favor nanotopographic biomaterial surfaces [18]. As a result, nanometer structural components should be preferred for the fabrication of functional tissue-engineered scaffolds.

This chapter differs from previously published review articles [20–24] that focus on the electrospinning process and the characterization of electrospun nanofibrous scaffolds by covering in much greater detail the breadth of electrospinning technology and its applications in tissue engineering. Notably, by taking advantage of our research expertise in cell and tissue engineering, we particularly focus on the biological mechanisms and activities enhanced by nanofibers while also reviewing the most up-to-date findings for the development of electrospun nanofibrous scaffolds and the engineering of nanofiber-based tissues. In this chapter, we first emphasize the importance of tissue engineering for clinical applications, introduce electrospun nanofibrous scaffolds by discussing the mechanisms and parameters influencing the properties of electrospun nanofibrous scaffolds, and, finally, review the current development of electrospun nanofibrous scaffolds in tissue engineering applications.

3.2
Nanofibrous Scaffolds

Among nanostructures, nanofibers are more suitable for use as the basic component of a scaffold compared with nanoparticles, due to their continuous structure. The advantage of a scaffold composed of ultrafine, continuous fibers are high porosity, variable pore-size distribution, high surface-to-volume ratio, and, most importantly, morphological similarity to natural ECM [25]. The combination of these features makes nanofibrous structures a favorable scaffold for tissue engineering, as shown by several studies described below.

3.2.1
Fabrication Methods for Nanofibrous Scaffolds

Several techniques based on different physical principles have been used to fabricate nanofibrous scaffolds with unique properties.

3.2.1.1   Phase Separation
Phase separation is often utilized as an alternative technique for scaffold fabrication. In most cases, the polymer is dissolved in a solvent at a low temperature, and then the resulting polymer and solvent phases are quenched to create a two-phase solid. The solidified solvent is then removed by sublimation, leaving a porous polymeric scaffold behind. In brief, the polymer is dissolved, gelated, and extracted using various solvents. The advantage of using phase separation to fabricate nanofibrous scaffolds is that the shape and size of the pores can be controlled by the addition of porogens [26]. Conversely, a limitation of the foam-like structures produced by phase separation is their compliance and lack of pore interconnectivity.

3.2.1.2   Self-assembly
By definition, self-assembly is the spontaneous organization of molecules into well-defined and macroscopic aggregates [5, 27]. The principle of molecular self-

assembly is ubiquitous in biological systems, e.g., the aggregation of lipid molecules into micelles in an aqueous environment. In contrast to processes such as electrospinning (see below), which are based on a top-down approach (i.e., materials are broken down into their respective components), self-assembly is based on a bottom-up approach (i.e., materials are assembled molecule by molecule).

On a molecular level, self-assembly is mediated by weak, non-covalent bonds, i.e., ionic bonds, hydrophobic interactions, van der Waals interactions, and water-mediated hydrogen bonds [5]. Aside from the balance of these forces, an additional crucial factor of the self-assembly process is complementary shape among the components [27]. To date, much of the work on self-assembly in tissue engineering has focused on the fabrication of 3D nanofibrous structures from amphiphilic peptides. In brief, when placed within an aqueous solution, the hydrophobic and hydrophilic domains within these peptides interact via the aforementioned forces to form distinct strong and fast-recovering hydrogels, with hydrophobic interactions being the principle force that drives the molecules together [28].

Self-assembled 3D nanofibrous scaffolds also closely mimic the porosity and gross structural scale of the natural ECM. Matrices composed of interwoven nanofibers with a diameter of 5–8 nm, and pores of 50–200 nm diameter [28, 29] have been observed by scanning electron microscopy and atomic force microscopy. The biocompatibility of these scaffolds is demonstrated by their ability to support cell proliferation and differentiation [29, 30]. In extended *in vitro* cultures of bovine chondrocytes within a peptide hydrogel, the chondrocytes encapsulated within the 3D scaffold retained their morphology and developed a cartilaginous extracellular matrix rich in proteoglycans and collagen type II, biochemical markers of a stable chondrocyte phenotype [30].

An advantage of using self-assembly to fabricate nanofibers is that the amino acid residues present may be chemically modified for the addition of bioactive moieties. Furthermore, self-assembly is carried out in aqueous salt solution or physiological media, thus avoiding the use of organic solvents and reducing cytotoxicity.

3.2.1.3 Electrospinning

Electrospinning technology has become popular for the fabrication of tissue engineering scaffolds in recent years (Fig. 3.2). This is a result of the growing interest in nanotechnology and the unique properties and relative ease of fabricating scaffolds using this process. Scaffolds fabricated by electrospinning have a totally different appearance and structure to those made by self-assembly and phase separation. Furthermore, electrospinning technology is exceptionally useful for the fabrication of tissue engineered scaffolds, for not only is the electrospinning equipment economical, but the preparation and fabrication phases are relatively quick compared with phase separation and self-assembly. In addition, the electrospinning system is easy to set up and can be modified to fabricate nanofibrous scaffolds that meet specific requirements for structural or mechanical needs. The advantages of electrospinning technology make it suitable for both small quantity production for laboratory research use and mass production for industrial use. In terms of the properties of scaffolds produced by the different methods, electrospun

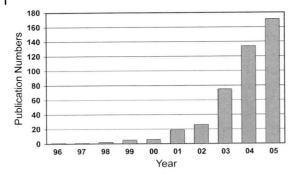

Fig. 3.2. Growth of research using electrospinning technology. The increasing number of scientific publications containing the keyword "electrospinning" indicates the upward trend of electrospinning applications.

nanofibrous scaffolds contain interconnected pores, and possess mechanical and structural stability, that provide superiority over scaffolds fabricated by phase separation and self-assembly. The following sections detail the fabrication technique, properties, and clinical applications of electrospun nanofibrous scaffolds.

3.2.2
The Electrospinning Process

3.2.2.1 History
Electrospinning was developed based on observations of the "electrospray" phenomenon first described by Lord Rayleigh in 1882 [31]. He discovered that a highly charged droplet was unstable and would break down into smaller droplets when passed through a voltage gradient, a property known as the "Rayleigh instability". Rayleigh proposed that disruption of the droplet surface tension was the result of forces generated through Coulombic repulsion. After this initial study, Zeleny further investigated the electrospraying of aqueous solutions [32, 33], and Dole et al. experimented with electrosprays of dilute polymer solutions [34]. In addition, Vonnegut and Neubauer electrosprayed water and other liquids [35], and Drozin found that electrosprayed droplets resembled a highly dispersed aerosol [36]. These seminal studies laid the foundation for what was to become the electrospinning phenomenon.

Electrospinning is a direct extension of the electrospraying phenomenon, as both processes are based on the same physical and electrical mechanisms. The main difference between the two is that electrospraying produces small droplets, whereas electrospinning produces continuous fibers. In 1934, Formhals electrospun fine fibers from a cellulose acetate solution and was granted a series of U.S. patents on this technology [37]. In 1966, Simons found that more viscous solutions favored the formation of longer fibers [38]. Baumgarten designed an apparatus with an infusion pump to electrospin acrylic fibers, and discovered that the diameter of fibers

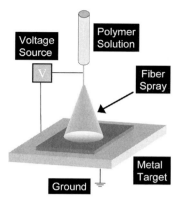

Fig. 3.3. Schema of the electrospinning apparatus (see text for details).

could be controlled by the polymer feed rate from the infusion pump [39]. Finally, Larrondo and Manley electrospun polypropylene and produced polyethylene nanofibers in 1981 [40–42]. The recent, surging interest in nanotechnology has engendered renewed attention to this convenient, economical technology that enables engineers to produce nanofibers for various applications.

3.2.2.2 **Setup**

The electrospinning apparatus is quite simple, with only three major components: a high voltage power supply, a polymer solution reservoir (e.g., a syringe, with a small diameter needle) with or without a flow control pump, and a metal collecting screen (Fig. 3.3). A high voltage power supply with adjustable control should be able to provide up to 50-kV DC output and, depending on the number of electrospinning jets, multiple outputs that function independently are needed. A reservoir is used to store the polymeric solution and is connected to a power supply to form a charged polymer jet. Either a syringe with a metal needle or a capillary with a metal tip in the polymer solution may be used to charge the polymer solution. Polymer flow can be driven by gravity if the syringe is not horizontally placed. However, to eliminate the experimental variables, a syringe pump is usually used to control the precise flow rate. The fiber collecting screen should be conductive and can be either a stationary plate or a rotary platform or substrate. The plate can produce non-woven fibers, while a rotary platform can produce both non-woven and aligned fibers.

Currently, there are two standard electrospinning setups, vertical and horizontal. With increasing interest in this technology, several research groups have developed more sophisticated systems that can fabricate more complex nanofibrous structures in a more controlled and efficient manner [43, 44]. For example, motor-controlled multiple jets and fiber-collecting targets provide a means for fabricating a single nanofibrous scaffold composed of multiple layers, with each layer derived from a different polymer type. Additionally, this technology can be used to fabricate polymer composite scaffolds where the fibers of each layer represent the amalgamation of different polymer types. The multiple jet setup is not only useful for making multi-layer nanofibrous scaffolds, but is also efficient for the production of a large quantity of scaffolding material. Notably, the electrospinning apparatus is usually set up in a chemical fume hood to remove organic vapor. In addition, a closed, non-conductive environment with temperature and humidity control is required to avoid interference from environmental factors, such as air turbulence.

3.2.2.3 Mechanism and Working Parameters

To initiate the electrospinning process, a selected polymer material is dissolved in the appropriate solvent and this solution is loaded into a syringe. A high-voltage electric field is created between the needle and the collecting screen by the use of a power supply and electrodes. When the polymer solution is extruded slowly by a syringe pump and/or gravity, a semispherical polymer solution droplet is formed at the tip of the needle. With increasing voltage, the polymer droplet elongates to form a conical shape known as the Taylor cone [45], causing the surface charge on the polymer droplet to increase with time. Once the surface charge overcomes the surface tension of the polymer droplet, a polymer jet is initiated [46]. The solvent in the polymer jet evaporates during travel to the collecting screen, increasing the surface charge on the jet. This increase in surface charge induces instability in the polymer jet as it passes through the electric field [45]. To compensate for this instability, the polymer jet divides geometrically, first into two jets, and then into many more as the process repeats itself. The formation of nanofibers results from the action of the spinning force provided by the electrostatic force on the continuously splitting polymer droplets. Nanofibers are deposited layer-by-layer on the metal target plate, forming a non-woven nanofibrous mat.

Although the electrospraying/electrospinning technology has been utilized for more than a hundred years, the mechanisms by which nanofibers are formed, much less controlled, have yet to be completely elucidated. Although several studies have been carried out to investigate the mechanism of fiber formation to reproducibly control scaffold design, little theoretical clarity has been achieved. During the electrospinning process, a uniform fibrous structure is created only under optimized operating conditions. Both extrinsic and intrinsic parameters are known to affect the structural morphology of the nanofibers [47]. Specifically, extrinsic parameters, such as environmental humidity and temperature, in addition to intrinsic parameters, including applied voltage, working distance, and conductivity and viscosity of the polymer solution, need to be optimized to produce uniform nanofibers. Generally speaking, the intrinsic parameters are more critical in determining the nanofiber structure. Two major structures are usually found in the nanofibrous mat fabricated by the electrospinning process – a uniform, continuous fibrous structure or a bead-containing fibrous structure. Variation in the relative abundance of these two structures is determined by the relative contributions of the four intrinsic parameters during the electrospinning process.

Polymer Solution Viscosity   Polymer solution viscosity, a parameter directly proportional to the concentration of the polymer solution, is the most critical factor in controlling the structural morphology of the nanofibrous structure. For fiber formation, polymer viscosity should be in a specific range, depending on the type of polymer and solvent used. Fong et al. used polyethylene oxide (PEO) to study nanofiber formation in different PEO viscosities, and found that a range of viscosity between 1 and 20 poise is suitable for electrospinning uniform nanofibers [48]. Below this range, a bead-containing nanofibrous structure was created. With increasing viscosity, spherical beads became elongated into spindle-shaped ones, and the number of beads in the structure was reduced. Similarly, Liu et al. also reported that a different specific range of viscosity was appropriate for the formation of uniform nanofibers composed of cellulose [49]. In addition, recent studies conducted by Deitzel et al. [50] and Demir et al. [51] have shown that a more viscous polymer solution resulted in larger fibers. Taken together, these studies indicate that there exist polymer-specific, optimal viscosity values for electrospinning.

Applied Charge Density   Charge density, as the amount of charge per unit surface area of the polymer droplet, is determined by the applied voltage, the working distance, and the conductivity of the polymer solution. Applied voltage is used to provide the driving force to spin fibers by imparting charge to the polymer droplet. The working distance is defined as the distance between the tip of syringe and the collecting plate and, together with the applied voltage, can influence the structural morphology of nanofibers. Demir et al. suggested that when higher voltages are applied more polymer is ejected to form a larger diameter fiber [51]. Similarly, high voltage conditions also created a rougher fiber structure. Zong et al. proposed an approach to increase charge density on the surface of the droplet by adding salt particles to reduce bead formation [52]. However, their conclusion was that a high charge density produced thinner fibers, a finding not corroborated by Demir et al. [51].

Polymer Solution Conductivity   The conductivity of a polymer solution is mainly determined by the polymer type, solvent used, and the availability of ionizable salts. A more conductive polymer solution carries more electric charge during the electrospinning process, with the as-spun fibers generating a stronger repulsive force, which facilitates the formation of bead-free, uniform fibers. Therefore, the general rule for the production of uniform electrospun nanofibers is to electrospin a highly conductive polymeric solution. Since most synthetic biodegradable polymers such as the commonly used poly(α-hydroxy esters) do not carry a charge, it is preferred to increase the solvent conductivity. As a result, a dipolar aprotic solvent, *N*,*N*-dimethylformamide (DMF), which has a high dielectric constant and dipole moment, is added to enhance the solution conductivity [53]. An additional approach entails the use of salts to increase chargeable functional groups in polymeric solutions. The addition of benzyl triethylammonium chloride (BTEAC) [54] or sodium chloride [55], for instance, enhances polymer solution conductivity, in turn reducing fiber size, while producing bead-free fibrous structures. Natural polymers, though, such as collagens, which contain functional groups, usually carry charges when dissolved in solution; however, the polyelectrolyte characteristic and hydrogen bonding potential of natural polymers limit the choice of solvents. The use of highly polar fluorinated or acid-based solvents can prevent complicated polymer–polymer and polymer–solvent interactions to smooth the electrospinning process.

3.2.3
Properties of Electrospun Nanofibrous Scaffolds

3.2.3.1 Architecture
An electrospun nanofibrous scaffold has a macro-architecture that is defined by the gross structure of the entire scaffold. The macro-architecture can be machined into a desired dimension and shape, depending on the type of tissue being engineered. However, the micro-architecture is defined by the porous, fibrous structure composed of nanofibers, with the more commonly seen micro-architecture being either a non-woven or an aligned fibrous structure.

Non-woven Nanofibrous Scaffolds Upon ejection from the nozzle of a needle, a polymer jet travels spirally at a high speed in the space between the tip of a needle and the fiber-collecting plate. The jet path is a complicated 3D curve and thus nanofibers are deposited on the target platform in a random manner, resulting in a non-woven structure [56]. Nevertheless, under optimal electrospinning conditions, a homogeneous nanofibrous mat will be formed, with fiber direction in the structure being equally oriented in every direction and polygonal, interconnected pores of various sizes formed between nanofibers. Although deposited fibers lay on and barely contact each other, the interlocking between fibers helps to maintain the micro-architecture of the nanofibrous scaffold. Shape and size of pores are altered only when the structure is subjected to loading.

Aligned Nanofibrous Scaffolds While such randomly oriented scaffolds are useful, a significant number of natural tissues exhibit a preferred fiber alignment. This fiber alignment endows tissues with unique functional material properties that depend on the testing direction and location. For example, in tendon and ligaments, tensile properties are 200–500× higher along the fiber direction (the direction over which the force is transmitted) than those perpendicular to the fiber direction [3]. In articular cartilage, tensile properties are greatest in the superficial zone of the tissue and highest along the prevailing collagen (split line) direction [4, 5]. As the goal of tissue engineering is to recapitulate the functional properties of the native tissue, new techniques have been developed for engineering tissue anisotropy.

The strategies in controlling nanofiber alignment generally focus on the as-spun fiber-assembly methods, using either an electrostatic field and/or specially designed collectors such as a rotary target [57] and patterned electrodes [58]. Theron et al. have utilized a rotary disc as a fiber collector to electrically and mechanically align nanofibers [57]. The sharp edge of the disc provides a physically sharp point that serves as an electrode to attract positive-charged nanofibers. The sharp point accumulates charge and generates a strong electrostatic filed. The rotary sharp edge acts like a continuously moving, charged band that forces nanofibers to align on a limited strip that winds around the circumference of the disc. In addition, the rotary movement of the disc provides a tangential force, further aligning and stretching nanofibers at the same time. Fiber alignment in this initial study is relatively satisfactory, but the amount of collected aligned nanofiber is limited due to the narrow sharp edge. To increase aligned nanofiber production for tissue engineering applications, it is suggested that a disc is replaced by a drum or a shaft with a large surface area [59]. Doing so will efficiently produce a larger and thicker nanofibrous mat, but without any precise control of fiber alignment. One of the solutions is to increase the rotational speed. Several studies have shown that the nanofiber alignment in a mat is highly dependent on the rotational speed of the drum or shaft [59]. Nanofiber alignment increases with the rotational speed of the collecting surface, and plateaus after reaching an optimal value.

Another practical approach to align nanofibers utilizes a fiber collector consisting of two rectangular conductive electrodes placed on a highly insulating substrate and separated by an air gap [58]. Nanofibers, due to a preferential electrostatic field, are deposited on the electrodes uniaxially across the air gap. The air humidity, in which the electrospinning is carried out, is the major factor affecting the fiber alignment using this approach. Nanofiber alignment increases with the carried charge of as-spun nanofibers and is also enhanced by the reduction of air humidity, since the higher humidity discharges the nanofibers. In addition, fiber orientation is partially determined by the amount of charge left on the deposited nanofibers. The retained charge on deposited nanofibers improves the orientation through electrostatic repulsions between the fibers. Altering the pattern of the electrode array can change the orientation of uniaxial nanofibers and may be useful to fabricate a scaffold with a complicated aligned nanofibrous pattern.

3.2.3.2 **Porosity**

The internal porous structure of void spaces, or porosity, is a physical component of a biomaterial scaffold that is also dependent on the architectural scale. These pores serve as pathways for mass transport (convection and diffusion), while also providing void space for cells to form new tissues [60]. One technique for measuring the pore diameter distribution, total pore volume, and total pore area of a structure is mercury porosimetry. Utilizing this technology, a 91.63% porosity was reported for electrospun nanofibrous scaffolds, indicating a highly porous structure [25]. This study also reported a total pore volume of 9.69 mL g^{-1}, a total pore area of 23.54 m^2 g^{-1}, and a pore diameter ranging broadly from 2 to 465 μm [25]. While some of these pores are too small to influence the migration of cells and facilitate invasion of blood vessels within the scaffold, nutrients and metabolic wastes are still able to pass through these nano-sized pores, thus enhancing engineering of the tissues. In addition, the high porosity of these scaffolds should present a wider path for mass transport, improving cell survival within the scaffold [60].

Although precise control of the pore size in nanofibrous scaffolds is challenging, efforts have been made to fabricate large-pore nanofibrous scaffolds in a more controllable manner. Lee et al. have combined electrospinning technology and salt leaching/gas foaming methods for the fabrication of an electrospun fibrous scaffold with dual-sized pores [61]. In addition to intrinsic pores formed between nanofibers, micro-sized pores created by salt particles are distributed in the scaffolds using this modified approach. As porous scaffolds fabricated by the conventional salt-leaching method, pore sizes in electrospun nanofibrous scaffolds are determined by the dimension of the salt particles. The dual-pore, electrospun nanofibrous scaffolds represent an enhanced porous structure for more efficient cell migration and nutrition/waste exchange.

3.2.3.3 Mechanical Properties

As previously discussed, tissue engineering aims to utilize natural and synthetic biomaterials to simulate the 3D environment of the natural ECM. Tissue-engineered scaffolds thus also need to match the mechanical properties of the natural environment. The mechanical properties (strength, toughness, modulus, and ductility) of a scaffold are determined by both its structure (macrostructure, microstructure, and nanostructure) and its material properties [60]. Matching the mechanical properties of a scaffold to that of the natural ECM is critically important so that once grafted the progression of tissue healing is not limited by mechanical failure of the scaffold [60]. Furthermore, the mechanical properties of the scaffold can affect cell morphology, proliferation, and differentiation [62].

The process of electrospinning provides ultrafine nanoscale structures that not only geometrically and topologically simulate the natural ECM, but may also mimic the mechanical properties of the natural tissue microenvironment. To accurately investigate the mechanical behavior of nanofibrous scaffolds, it is essential to measure the mechanical properties of individual fibers that make up the scaffold [63]. Such tensile testers must possess the capability of measuring the small load and elongation required for the deformation of these ultrafine fibers, including individual fibers [63, 64]. In a previous study of the mechanical properties of individual electrospun poly(ε-caprolactone) (PCL) nanofibers, it was shown that mechanical properties varied with fiber diameter, i.e., both tensile strength and yield stress decreased with increased fiber diameter, while an increase in fiber diameter resulted in an increase in strain at break [63]. While in this study it was shown that the strain of break increases with an increase in fiber diameter, yield strain was found to decrease with an increase in fiber diameter, and there was also no apparent correlation between Young's modulus (E) and fiber diameter [63]. In summary, the high ductility, and low strength and stiffness of the PCL nanofibers observed in this study depend on the physical properties of the polymer (i.e., low T_g) and fiber diameter.

3.3
Current Development of Electrospun Nanofibrous Scaffolds in Tissue Engineering

3.3.1
Evidence Supporting the Use of Nanofibrous Scaffolds in Tissue Engineering

3.3.1.1 Nanofibrous Scaffolds Enhance Adsorption of Cell Adhesion Molecules

Cell adhesion is one of the most important aspects of cell interaction with a biomaterial. It is the first cellular event to take place after cells are seeded onto a biomaterial surface; cell migration, proliferation, and/or differentiation take place only after cells are securely adhered [65]. In native tissues, cells bind via integrins to specific binding sites in the tissue ECM, consisting of specific peptide sequences recognized by the integrins [66]. However, synthetic polymers do not have natural cell binding sites, and cell adhesion to a polymeric scaffold is therefore necessarily mediated by plasma/serum proteins adsorbed onto the polymer surface. Protein adsorption to the polymer surface is affected by the hydrophilicity [67] and the surface energy of the polymer [68]. Cells attach to many ECM proteins, including fibronectin, vitronectin, collagens, and other matrix proteins [69]. Nikolovski et al. showed that vitronectin, compared with fibronectin, was the predominant matrix protein adsorbed from serum-containing medium onto poly(glycolic acid) (PGA) and poly(lactic acid) (PLA) [70]. Woo et al. recently reported that nanofibrous scaffolds for tissue engineering show enhanced adsorption of cell adhesion ECM molecules, which may therefore enhance cell adhesion [71]. The high surface-area-to-volume ratio of electrospun nanofibrous scaffolds should result in higher adsorption of vitronectin and fibronectin molecules than other scaffolds. In fact, approximately four times as much serum proteins adsorbed to nanofibrous scaffolds compared with scaffolds with solid pore walls. Moreover, fibronectin and vitronectin preferentially adsorbed to the nanofibrous scaffold at a level that was 2–4× higher than those adsorbed to the solid-walled scaffold [71]. Although the mechanisms by which a nanofibrous scaffold acts as a selective substrate are not yet known, it is clear that the enhanced adsorption of cell adhesion matrix molecules enhances cell adhesion.

3.3.1.2 Nanofibrous Scaffolds Induce Favorable Cell–ECM Interaction

In a native environment, cells are in contact with the ECM and neighboring cells, and cell–matrix interaction act as key regulators of cellular activities [72]. Cells synthesize, assemble, organize, and maintain ECM macromolecules, while the ECM functions to provide structural protection to the resident cells and acts as a messenger regulating cellular activities such as proliferation or differentiation [73]. In postnatal life, the ECM continuously undergoes dynamic turnover to respond to environmental changes caused by biochemical stimulation or mechanical loading. In particular, mechanical signals are transmitted from the environment via the ECM network to cells, which respond to these signals by remodeling the ECM to adapt to the environmental change [74]. Extrapolating from these native interactions, a scaffold that serves as a functional, temporary ECM must involve optimal cell–matrix interactions, as well as cell–cell interactions.

Several reports have demonstrated that nanofibers as scaffolds are more favorable than microfibers, suggesting cell activities can be regulated by the size of the fiber [75, 76]. One such study evaluated the influence of the structural properties of biomaterial scaffolds on the biological activities of chondrocytes cultured in microfiber- and nanofiber-based scaffolds [75]. The results show that chondrocytes seeded into microfibrous scaffolds display dedifferentiated, fibroblast-like morphology, whereas chondrocytes seeded onto nanofibrous scaffolds maintain a chondrocyte-like morphology (Fig. 3.4). Large, organized stress fibers, commonly found in cells in monolayer culture also appear in cells seeded onto microfibrous

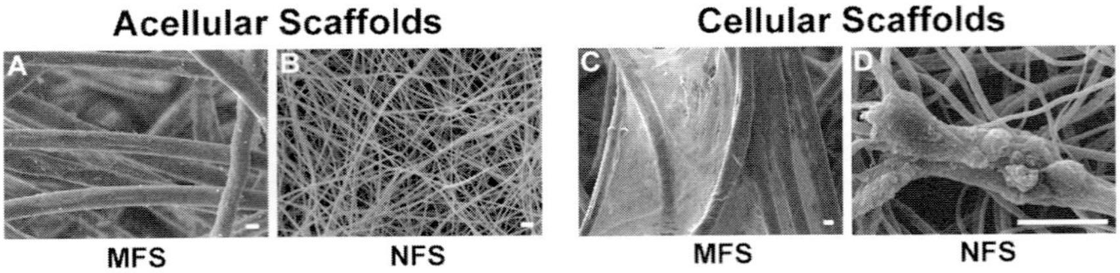

Acellular Scaffolds   ## Cellular Scaffolds

MFS   NFS   MFS   NFS

Fig. 3.4. Ultrastructural morphology of acellular and cellular PLLA microfibrous and nanofibrous scaffolds examined by SEM. (A) A PLLA microfibrous scaffold showing random orientation of microfibers with a diameter ranging from 15 to 20 μm. (B) A PLLA nanofibrous produced by the electrospinning process, showing random orientation of ultrafine fibers with diameters ranging from 500 to 900 nm, defining a matrix with interconnecting pores. (C) Spread cellular sheets composed of fibroblast-like cells spanned between microfibers in the microfibrous culture after 28 days. (D) Cellular aggregates composed of globular, chondrocyte-like cells grew on nanofibers in the nanofibrous culture after 28 days. Bar = 10 μm.

scaffolds. However, cells seeded in nanofibrous scaffolds do not display a similar cytoskeletal structure. Cell activities such as proliferation and the production of a cartilaginous ECM are enhanced in nanofibrous cultures compared with microfibrous cultures. Overall, these results demonstrate the biological effects of different-sized fibrous biomaterial scaffolds, and suggest that a nanofibrous scaffold is a more effective scaffold for cartilage tissue engineering.

Another study also reported a similar conclusion of nanofibers outperforming microfibers [76], in terms of the biological response of neural stem cells cultured on nanofibrous and microfibrous scaffolds. Unlike the comparison of fibers with a two-order difference in Li's study [75], 300 nm nanofibers were compared with 1.5 μm microfibers. The results of this study show that a greater percentage of neural stem cells cultured on nanofibrous scaffolds exhibit a neuron-like morphology when compared with microfibrous scaffolds. In addition, the average neurite length of neural stem cells on a nanofibrous scaffold is significantly longer than that of cells on a microfibrous scaffold, suggesting that nanofibers enhance the neurite outgrowth.

One possible explanation for the enhanced performance of nanofibrous scaffolds is that nano-scale fibers, smaller than a cell by two-orders resembling native ECM, create the 3D environment that has spatial advantage in promoting cell–matrix interaction, whereas a scaffold composed of micro-scale fibers with the same order size as a cell does not have such an advantage.

3.3.1.3 Nanofibrous Scaffolds Maintain Cell Phenotype

Chondrocytes isolated from cartilage and cultured *in vitro* readily undergo dedifferentiation when plated as monolayers, but will redifferentiate and re-express their phenotype maintained in a 3D environment such as collagen gels. Nanofibrous scaffolds have been shown to support the potential of dedifferentiated chondrocytes to redifferentiate *in vitro* [77]. In this study, fetal bovine chondrocytes were seeded onto nanofibrous scaffolds, or as monolayers on standard tissue culture polystyrene (TCPS) as a control substrate. Gene expression analysis shows that chondrocytes seeded on the nanofibrous scaffold and maintained in a serum-free, ITS+ supplemented medium continuously maintain their chondrocytic phenotype by expressing cartilage specific ECM genes, including collagen types II and IX, aggrecan, and cartilage oligomeric matrix protein (COMP). Specifically, expression of the collagen type IIB splice variant transcript, which is indicative of the mature chondrocyte phenotype, is up-regulated. Chondrocytes exhibit either a spindle or round shape on the nanofibrous scaffold, in contrast to the flat, well-spread morphology seen in monolayer TCPS cultures. Organized cytoskeletal actin stress fibers are only observed in the cytoplasm of cells cultured on TCPS. Histologically, nanofibrous cultures maintained in supplemented serum-free medium produce more sulfated proteoglycan-rich, cartilaginous matrix than monolayer cultures. In addition to promoting phenotypic differentiation, the nanofibrous scaffold also supports cellular proliferation when the cultures are maintained in serum-containing medium. These results indicate that the biological activities of chondrocytes are crucially dependent on the architecture of the extracellular scaffolds as well as the composition of the culture medium, and that the nanofibrous scaffold acts as a biologically preferred scaffold/substrate for proliferation and maintenance of the chondrocytic phenotype.

3.3.1.4 Nanofibrous Scaffolds Support Differentiation of Stem Cells

Stem cells, including embryonic and adult stem cells, are promising cell sources for tissue engineering because of their extensive expansion and differentiation capability. Therefore, the biomaterial scaffold used to culture stem cells should be able to support basic stem cell processes, such as proliferation, self-renewal, and differentiation.

Embryonic stem cells (ESCs), different from mesenchymal stem cells (MSCs) isolated from adult tissues, are derived from the inner cell mass of a blastocyst [78]. ESCs have been considered as having the potential of unlimited differentiation, whereas MSCs are more restricted in their differentiation potential. The technique of controlling the fate of ESCs *in vitro* is complicated and far from mature. One of the requirements for the growth of ESCs is the presence of basement membrane matrix or feeder layers of embryonic fibroblasts. Basement membrane matrix or feeder cell layers provide chemical and physical cues to the ESCs, and maintain their ability to self-renew and differentiate [79]. A nanofibrous structure resembling a basement membrane matrix has been used to culture ESCs to maintain their aforementioned "stemness" properties [80]. The results show that ESC-nanofibrous cultures have significantly larger colonies of undifferentiated cells and enhanced proliferation compared with controls cultured on glass coverslips and same-polymer-films, suggesting that the physical cues from the unique geometry of nanofibrous surfaces regulate cellular activities.

Compared with the uncertain future of ESCs in the clinic, owing to ethical and legal issues, adult stem cells provide a promising cell source for stem cell based-tissue engineering. MSCs are undifferentiated, multipotential cells that are capable of giving rise to cells characteristic of several tissues, including the connective tissue lineage [81]. Advantages of using MSCs include the fact that MSCs can be isolated from many different adult tissues, such as bone marrow [81], fat [82], umbilical cord blood [83], muscle [84], and synovial membrane [85]. In addition, nanofibrous scaffolds support multi-differentiation of MSCs [86]. In this study, MSCs from bone marrow maintained in nanofibrous scaffolds differentiated along adipogenic, chondrogenic, or osteogenic lineages when induced with specific differentiation media. Gene expression analysis and immunohistochemical detection of lineage-specific marker molecules confirmed the formation of nanofibrous constructs containing cells differentiated into the specified cell types. These results demonstrate the full support of multi-lineage differentiation of MSCs within nanofibrous scaffolds and the feasibility of tissue-engineering multiphasic constructs using a single cell source, which is of particular relevance to the development of multiphasic tissue constructs.

3.3.1.5 Nanofibrous Scaffolds Promote *in vivo*-like 3D Matrix Adhesion and Activate Cell Signaling Pathway

The observations presented above strongly suggest that nanofibrous scaffolds serve as synthetic ECM networks to provide both chemical and physical stimuli to cells via direct interaction with the cells surface. Unique geometric features of the extracellular matrix, especially ultrafine structure, are generally believed to play a direct and/or indirect role in the regulation of cellular activities. However, studies about cell–nanofiber interactions are new, and the specific mechanisms of how the unique properties of nanofibers enhance cellular activities are largely unknown. From the recently published studies, integrin receptors, cytoskeleton, and signal pathways involving focal adhesion kinase (FAK), Rho, Rac, and Cdc42 GTPases are likely to play a significant role in regulating cells cultured on nanofibrous cultures [87, 88]. Schindler et al. have demonstrated that nanofibrous cultures promote *in vivo*-like cell morphology of both fibroblasts and kidney cells [87]. They also showed that both cells cultured on nanofiber surface have less defined, punctate patterns of vinculin and FAK, molecules mediating cell adhesion, at the edge of lamellipodia. In response to the accumulation of cell adhesion molecules at the edge of cells, a notable increase is observed in the formation of actin cytoskeleton-rich lamellipodia, membrane ruffles, and cortical actin. Such cellular response is also found in 3D cultures on basement membrane matrix, but not on 2D cultures. Instead, in a flat culture on glass, more vinculin and FAK are accumulated in a streaky pattern, promoting the formation of well-defined actin filaments. The decrease of FAK at the adhesion site, characteristic for cells in tissue, is called "3D-matrix adhesion". Different from focal adhesion and fibrillar adhesion commonly formed *in vitro* on flat surfaces, 3D-matrix adhesion takes place in a 3D culture, suggesting that such *in vitro* cultures may bring the cells closer to *in vivo* conditions [13]. Thus, nanofibrous cultures should promote the formation of 3D-matrix adhesion and simulate the *in vivo* microenvironment.

In addition to the formation of 3D-matrix adhesion, nanofibrous cultures also demonstrate different expression profiles of components of cell signaling pathways that mediate cell morphology and cytoskeletal organization. These components in-clude members of the family of Rho GTPases, Rho, Rac, and Cdc42, each control-ling distinct downstream signal pathways [88]. Cells cultured on a nanofibrous sur-face extensively activate Rac, which consequently enhances cell proliferation rate and increases the deposition of fibrillar fibronectin. The activated Rac is also found to accumulate at the lamellopodial edge, intracellular vesicles, and dorsal mem-brane ruffles, often formed in migrating cells and cells cultured in other 3D envi-ronment. The activation of Rho and Cdc42 is not as significant as that of Rac. Both GTPases were mildly elevated in the nanofibrous culture than on glass at the early phase of cell attachment. However, the extent of Cdc42 mediating cell polarity de-creases along the culture. The nanofibrous scaffolds can dramatically induce the preferential activation of Rac GTPase, suggesting that nanofibrous scaffolds can provide physical as well as spatial cues to activate intracellular signaling pathways that are essential to mimic *in vivo*-like tissue growth.

3.3.2
Biomaterials Electrospun into Nanofibrous Scaffolds

Many polymeric biomaterials have been used for tissue engineering applications, including non-biodegradable and biodegradable polymers, with the latter consist-ing of both natural and synthetic polymers. Non-biodegradable polymers can be utilized to engineer tissues requiring substantial mechanical stability, such as liga-ment or muscle. However, their long-lasting nature interferes with tissue turnover and remodeling. Therefore, more attention has been devoted to biodegradable polymers in tissue engineering. Polymer biodegradation, by the combined effect of enzymatic and hydrolytic activities, generates space within the scaffold that facil-itates cellular processes, such as proliferation and the deposition of newly synthe-sized ECM. To date, more than one hundred different biodegradable polymers have been successfully electrospun and over thirty of them have been used for various tissue-engineered applications. Table 3.1 summarizes the polymers used in electrospinning, their potential applications, and the evaluation criteria employed in each study.

3.3.2.1 Natural Polymeric Nanofibrous Scaffolds
Because of their more biologically favorable polymeric chemistry towards cell and tissue growth, natural polymers have recently been fabricated into various 3D, tis-sue engineering scaffolds. Two groups of biocompatible natural polymers electro-spun into nanofibers have been used for tissue engineering applications: protein-based and carbohydrate-based polymers. Protein-based polymers include collagen, gelatin, elastin, silk fibroin, and fibrinogen, and carbohydrate-based polymers in-clude chitin, chitosan, and hyaluronan. Recently, natural polymers have been elec-trospun into nano-/micro-fibers that are structurally similar to ECM protein fibers.

Collagen Collagen, the most abundant protein family in the body, has been exten-sively used for *in vitro* and *in vivo* tissue engineering. Among the at least 19 differ-

Tab. 3.1. Polymers used in electrospinning, their potential applications, and the evaluation criteria employed.

Study	Biomaterial	Applications	Characterization	Ref.
1	PGA	Nonwoven TE scaffold	SEM, TEM, *in vitro* rat cardiac fibroblast culture, *in vivo* rat model	168
2	PGA	Nonwoven, aligned TE scaffold	SEM, mechanical evaluation	129
3	PLGA	Nonwoven TE scaffold	SEM, porosimetry, mechanical evaluation, *in vitro* human mesenchymal stem cell and mouse fibroblast cultures	25
4	PLGA, PLLA, PDLLA	Biomedical applications	SEM, degradation analysis, DSC, WAXD, SAXS	132
5	PLGA	Biomedical applications	SEM, WAXD, SAXS, degradation analysis	147
6	PLGA	Wound healing, drug delivery	SEM	134
7	PLGA	Biomaterials	SEM	135
8	PLGA	Nonwoven TE scaffold, peripheral nerve	SEM, *in vivo* rat model	169
9	PLGA/PEG-PLA	Prevention of postsurgical abdominal adhesions	SEM, *in vivo* rat model	154
10	PCL	Nonwoven TE scaffold	SEM, WAXD, mechanical evaluation	53
11	PCL	Nonwoven TE scaffold bone	SEM, *in vitro* rat mesenchymal stem cell culture	137
12	PCL	Nonwoven TE scaffold bone	SEM, *in vivo* rat model	138
13	PCL	Nonwoven TE scaffold cardiac tissue	SEM, *in vitro* rat cardiomyocyte culture	139
14	PCL	Nonwoven TE scaffold cartilage	SEM, *in vitro* human mesenchymal stem cell culture	140
15	PCL	Nonwoven TE scaffold cartilage	SEM, *in vitro* bovine chondrocyte culture	77

Tab. 3.1 *(continued)*

Study	Biomaterial	Applications	Characterization	Ref.
16	PCL	Nonwoven TE scaffold	SEM, WAXD, mechanical evaluation, *in vitro* MCF-7 mammary carcinoma cell culture	141
17	PCL	Nonwoven TE scaffold	SEM, *in vitro* human coronary smooth muscle cell culture	142
18	PCL	Nonwoven, aligned TE scaffold	SEM, surface contact angle test, *in vitro* human endothelial cell culture	99
19	PCL	Nonwoven TE scaffold	SEM, TEM, *in vitro* human dermal fibroblast culture	143
20	PCL	Nonwoven TE scaffold skin	SEM, *in vitro* human dermal fibroblast culture	144
21	PCL, PLA	Nonwoven TE scaffold blood vessel	SEM, mechanical evaluation, *in vitro* mouse fibroblast and human venous saphenous myofibroblast cultures	145
22	PCL, PCL/ CaCO$_3$	Nonwoven TE scaffold bone	SEM, surface contact angle test, mechanical evaluation, *in vitro* human osteoblast culture	146
23	PCL, PCL/ gelatin, gelatin	Nonwoven TE scaffold	SEM, surface contact angle test, mechanical evaluation, *in vitro* bone marrow stromal cell culture	170
24	PLLA	3D cell substrate	SEM, *in vitro* human chondrocyte culture	171
25	PLLA	Nonwoven TE scaffold	SEM, *in vitro* mouse cerebellum stem cell culture	172
26	PLLA	Surface coating	SEM, AFM, *in vitro* endothelial cell culture	173
27	PLLA	Nonwoven, aligned TE scaffold nerve	SEM, *in vitro* mouse cerebellum stem cell culture	76

Tab. 3.1 *(continued)*

Study	Biomaterial	Applications	Characterization	Ref.
28	PLGA/PEG-PDLLA, PLLA/PLGA, PLLA	Nonwoven, aligned TE scaffold heart	SEM, surface contact angle test, degradation analysis, *in vitro* rat cardiomyocyte culture	174
29	PDLLA-PEG-PDLLA, PDLLA, PLGA	Biomedical applications	SEM, degradation analysis, surface contact angle test, *in vitro* mouse calvaria osteoblast culture	175
30	PLLA, PDLLA, PEG-PLLA, PEG-PDLLA	Nonwoven TE scaffold	SEM, surface contact angle test, *in vitro* mouse calvaria-derived osteoprogenitor culture	176
31	PLLA/MMT	Nonwoven TE scaffold	SEM	177
32	P(LLA-CL)	Aligned TE scaffold blood vessel	SEM, *in vitro* human coronary artery smooth muscle cell culture	160
33	P(LLA-CL)	Nonwoven TE scaffold	SEM, DSC, XRD, *in vitro* human coronary artery endothelial and smooth muscle cell cultures	178
34	P(LLA-CL)	Nonwoven TE scaffold	SEM, mechanical evaluation, *in vitro* human coronary muscle and endothelial cell cultures	179
35	P(LLA-CL)	Nonwoven TE scaffold blood vessel	SEM, flow press analysis, mechanical evaluation	180
36	P(LLA-CL)	Nonwoven TE scaffold blood vessel	SEM, ^{1}H NMR, porosimetry, mechanical evaluation, *in vitro* human umbilical vein endothelial cell culture	181
37	P(LLA-CL)	Nonwoven TE scaffold blood vessel	SEM, porosimetry, mechanical evaluation, *in vitro* human coronary artery endothelial cell culture	157
38	P(CL-EEP)	Nonwoven TE scaffold liver	SEM, *in vitro* rat hepatocyte culture	182
39	PNmPh	Biomedical applications	SEM, NMR, viscosity analysis, DSC, *in vitro* mouse calvarial cell culture	183

Tab. 3.1 *(continued)*

Study	Biomaterial	Applications	Characterization	Ref.
40	PHBV	Nonwoven TE scaffold	SEM, *in vitro* chondrocyte culture	184
41	PHBV	Nonwoven TE scaffold	SEM, viscosity analysis, degradation analysis	54
42	PHBV	Nonwoven TE scaffold	SEM, surface contact angle test, *in vitro* monkey kidney cell culture	185
43	PEU	Nonwoven TE scaffold skeletal muscle	SEM, degradation analysis, mechanical evaluation, *in vitro* murine myoblast, rat myoblast, and human satellite cell cultures	161
44	PEUU	Nonwoven TE scaffold cardiovascular tissues	SEM, mechanical evaluation, *in vitro* rat vascular smooth muscle cell culture	44
45	PEUU/Col I	Nonwoven TE scaffold soft tissue	SEM, circular dichroism spectroscopy, mechanical evaluation, *in vitro* vascular smooth muscle cell culture	186
46	PPDO-PLLA-b-PEG	Nonwoven TE scaffold	SEM, porosimetry, mechanical evaluation, degradation analysis, *in vitro* fibroblast cell culture	187
47	PU	Nonwoven, aligned TE scaffold ligament	SEM, mechanical evaluation, *in vitro* human ligament fibroblast culture	164
48	PU	Nonwoven tissue template wound healing	SEM, *in vivo* guinea pig model	188
49	PU	Nonwoven TE scaffold blood vessel	SEM, flow pressure analysis, mechanical evaluation	189
50	PU	Biomaterial	SEM, porosimetry, mechanical evaluation	190
51	PET	Nonwoven TE scaffold blood vessel	SEM, XPS, surface contact angle test, *in vitro* human coronary artery endothelial cell culture	191

Tab. 3.1 *(continued)*

Study	Biomaterial	Applications	Characterization	Ref.
52	PEVA	Nonwoven TE scaffold	SEM, *in vitro* human aortic smooth muscle cell and dermal fibroblast cultures	148
53	PS	Nonwoven TE scaffold skin	SEM, *in vitro* human fibroblast, keratinocyte, and endothelial single or co-cultures	192
54	PDS	Nonwoven, aligned TE scaffold biomedical applications	SEM, viscosity analysis, mechanical evaluation	193
55	PEG-LMWH/ PLGA, PEG-LMWH/PEO	Biomaterial	SEM, EDX, multiphoton microscopy	194
56	PVA/CA	Biomaterials	SEM, FTIR, WAXD, mechanical evaluation	156
57	Fibrinogen	Nonwoven TE scaffold wound healing	SEM, TEM, mechanical evaluation	115
58	Col I/PEO	Nonwoven TE scaffold wound healing	SEM, TEM, ^{1}H NMR spectroscopy, mechanical evaluation	152
59	Col I, Col III	Nonwoven TE scaffold	SEM, TEM, mechanical evaluation, *in vitro* aortic smooth muscle cell culture	91
60	Col I	Nonwoven TE scaffold wound healing	SEM, degradation, porosimetry, mechanical evaluation, *in vitro* human oral and epithermal keratinocyte cultures, *in vivo* rat model	92
61	Col I/elastin/ PLGA	Nonwoven TE scaffold blood vessel	SEM, mechanical evaluation, *in vitro* bovine endothelial and smooth muscle cell cultures	102
62	Col I/elastin/ PEO, Col I/PEO, elastin/PEO	Nonwoven TE scaffold	SEM, surface tension analysis, viscosity analysis, DSC, amino group detection	103
63	Col I, gelatin, PGA, PGA/PLA, PLA	Nonwoven TE scaffold	SEM, *in vivo* rat model	93

Tab. 3.1 *(continued)*

Study	Biomaterial	Applications	Characterization	Ref.
64	Col I, Col III, elastin	Nonwoven TE scaffold blood vessel	SEM, *in vitro* human umbilical vein endothelial cell, aortic smooth muscle cell, and dermal fibroblast cultures	94
65	Col I, gelatin, PU, PEO	Nonwoven TE scaffold	SEM, laser scanning confocal microscopy, mechanical evaluation	43
66	Col I, gelatin, elastin, tropoelastin	Nonwoven TE scaffold	SEM, AFM, mechanical evaluation, *in vitro* human embryonic palatal mesenchymal cell culture	95
67	Col II	Nonwoven TE scaffold	SEM	96
68	Col II	Nonwoven TE scaffold cartilage	SEM, mechanical evaluation, *in vitro* adult human articular chondrocyte culture	97
69	Elastin-mimetic	Nonwoven TE scaffold	SEM	104
70	Silk fibroin, silk/PEO	Nonwoven TE scaffold	SEM, FTIR, XPS	105
71	Silk, silk/PEO	Nonwoven TE scaffold	SEM, XPS, DSC, mechanical evaluation, *in vitro* human bone marrow stromal cell culture	106
72	Silk	Biomedical applications	SEM, ^{13}C CP/MAS NMR, mechanical evaluation	107
73	Silk	Biomedical applications	SEM, TEM, WAXD	108
74	Silk fibroin	Nonwoven TE scaffold wound healing	SEM, porosimetry, *in vitro* human keratinocyte and fibroblast cultures	109
75	Silk fibroin	Nonwoven, woven TE scaffold wound healing	SEM, ATR-IR, ^{13}C CP/MAS NMR, WAXD, NMR, *in vitro* human keratinocyte culture	110
76	Silk/chitosan	Nonwoven TE scaffold wound dressings	SEM, viscosity analysis, conductivity	111
77	Chitin, chitosan	Wound dressings	SEM, ^{1}H NMR, FTIR, WAXD, DSC, TGA	120

Tab. 3.1 *(continued)*

Study	Biomaterial	Applications	Characterization	Ref.
78	Chitosan	Nonwoven, aligned TE scaffold cartilage	SEM, mechanical evaluation, degradation, *in vitro* canine chondrocyte culture	121
79	Chitosan	Nonwoven biomaterial	SEM	122
80	Chitosan/PEO	Nonwoven TE scaffold, drug delivery, wound healing	SEM, XPS, FTIR, DSC	153
81	Chitosan/PEO	Nonwoven, aligned TE scaffold	SEM, viscosity analysis, *in vitro* human chondrocyte and osteoblast cultures	195
82	Chitin/PLGA	Nonwoven TE scaffold	SEM, *in vitro* human oral keratinocyte and epidermal keratinocyte cultures	158
83	Gelatin	Nonwoven TE scaffold wound healing	SEM, mechanical evaluation	101
84	Oxidized cellulose	Adhesion barriers	SEM, FTIR, WAXD, TGA	196
85	HA	Medical implant	SEM	127

PGA = poly(glycolide), PLGA = poly(lactide-*co*-glycolide), PLLA = poly(ʟ-lactide), PDLLA = poly(ᴅ,ʟ-lactide), PEG = poly(ethylene glycol), PCL = poly(ε-caprolactone), MMT = montmorillonite, P(LLA-CL) = poly(ʟ-lactide-*co*-ε-caprolactone), P(CL-EEP) = poly(ε-caprolactone-*co*-ethyl ethylene phosphate), PNmPh = poly[bis(*p*-methylphenoxy)phosphazene], PHBV = poly(3-hydroxybutyrate-*co*-3-hydroxyvalerate), PEU = polyetherurethane, PEUU = poly(ester urethane)urea, PPDO = poly(*p*-dioxanone), PU = polyurethane, PET = polyethylene terephthalate, PEVA = poly(ethylene-*co*-vinyl alcohol), PS = polystyrene, PDS = polydioxanone, LMWH = low molecular weight heparin, PVA = poly(vinyl alcohol), CA = cellulose acetate, PEO = poly(ethylene oxide), Col I = collagen type I, Col II = collagen type II, Col III = collagen type III, HA = hyaluronic acid, TE = tissue engineering, SEM = scanning electron microscopy, TEM = transmission electron microscopy, DSC = differential scanning calorimetry, WAXD = wide-angle X-ray diffraction, SAXS = small angle X-ray scattering, AFM = atomic force microscopy, XRD = X-ray diffractometry, NMR = nuclear magnetic resonance, XPS = X-ray photoelectron spectroscopy, EDX = energy dispersive X-ray, ATR-IR = attenuated total reflectance infrared spectroscopy, FTIR = Fourier-transform infrared spectroscopy, TGA = thermogravimetric analysis.

ent collagen types, collagen type I is the major collagen in connective tissues and collagen type II is exclusively abundant in hyaline cartilage. Both collagen types I and II have been fabricated into different scaffolds, mostly in a gel-format, to repair tissue defects *in vivo* [89, 90]. Collagen electrospinning provides an *in vitro* method to create a preformed, nanofibrous collagen scaffold that closely mimics the native collagen network [43, 91–97]. For instance, Matthews et al. have demonstrated that, under optimal working parameters, electrospinning can be adapted to fabricate collagen types I and III nanofibers [91]. Electrospun fibers of skin-derived collagen type I and placenta-derived collagen type III were shown to be more uniform than fibers of placenta-derived collagen type I, suggesting that the diameter of electrospun collagen fibers varies with tissue origin and collagen type. Interestingly, electrospun collagen type I nanofibers exhibit the periodic banding pattern typical of native collagen, further demonstrating the potential of electrospinning in the fabrication of natural collagen fibers. Collagen type II has also been successfully electrospun. Being a major fibrous component of the ECM in hyaline cartilage, collagen type II networks are extremely important in cartilage regeneration. Electrospinning collagen type II is a promising approach to reconstruct collagen networks *in vitro*. However, one of drawbacks on using electrospun collagen scaffolds is their rapid degradation in culture. A study by Shields et al., comparing the properties of crosslinked and non-crosslinked scaffolds using glutaraldehyde as the crosslinking agent, found crosslinking to increase the diameter of collagen fibers as well as the thickness of the scaffold [97]. This study also demonstrated that electrospun collagen type II fibers are relatively larger but mechanically weaker than native collagen fibers. Although not fully recapitulating the properties of the native collagen type II network formed in cartilage, electrospun collagen type II scaffolds still hold promise in cartilage tissue engineering due to their ability to closely mimic the natural ECM.

Gelatin Gelatin is the denatured form of collagen type I, a natural protein-based polymer that is extensively used in the biomedical and food industries. Unlike the triple-helical collagen from which is derived, the denatured protein chain of gelatin is readily soluble in water at room temperature. Tissue engineered scaffolds made of gelatin usually require treatment with a crosslinking agent to maintain their structural integrity in a physiological environment, due to their ready dissociation in physiological solution. It is also common for gelatin to be grafted or blended with other polymers in the preparation of tissue engineering scaffolds [98, 99]. Like most protein polymers, gelatin contains amine and carboxylic groups, which are easily ionized in water to carry charge. This polyelectrolyte property of gelatin, coupled with its strong hydrogen bonding makes electrospinning of a gelatin aqueous solution quite challenging [43, 93, 95, 100]. In a study by Huang et al., highly polar fluorinated solvents were used to prepare a gelatin solution with an optimal concentration for electrospinning [101]. This study also showed that the mechanical performance of a gelatin mat is determined by fiber morphology, which is dependent on polymer concentration. For instance, an increase the average diameter of gelatin fibers from 100 nm to 1.9 μm was achieved by increasing polymer concentration, with smaller uniform fibers being mechanically stronger than larger uniform fibers.

Elastin Elastin is a highly water-insoluble protein that functions as a "perfect coil" to provide elasticity to tissues that need to be stretched and recoiled. However, it is rare for elastin to be used alone as a material to fabricate tissue engineering scaffolds [94, 95]. Elastin-containing tissues generally also contain collagen fibers [102, 103]. For instance, tissues such as skin, lung, and blood vessels are composed of collagen fibers for tensile properties and elastin fibers for elastic properties. Nagapudi et al. used a genetic engineering approach to synthesize recombinant elastin-mimetic proteins, and which were electrospun into nanofibers with different properties [104]. Elastin-mimetic proteins dissolved in a highly polar fluorinated solvent were electrospun into smaller fibers, compared with those dissolved in aqueous solution, suggesting that different fibrous morphologies of natural polymers could result from the use of different solvents, such as highly polar fluorinated solvent and water.

Silk Fibroin Besides collagen, silk fibroin is another protein-based, natural polymer most commonly used for electrospinning. With a long history of textile use, beginning with the ancient Chinese, silk has recently been applied for the fabrication of tissue engineering scaffolds [105–111]. One of the most significant properties of silk is its excellent mechanical properties. Natural silk is produced by spiders or silkworms, with different composition and properties among species. A study by Ohgo et al. comparing three electrospun silk fibroin nanofibers suggested that silk fibroins from two different silkworms and genetically engineered silk-like protein each required individual optimal polymer concentration for electrospinning [107]. In addition, the engineered silk-like protein nanofibers were reported to be smaller than the natural silk fibroin nanofibers, and the mechanical strength of silk fibroin nanofibrous mats was also dependent on silk fibroin type.

Fibrinogen Purified from blood plasma, fibrinogen is a globular protein that plays a critical role in wound healing. In the presence of thrombin, fibrinogen gives rise to fibrin which forms fibrous clots that have found use as a clinical fixative, due to its natural role in wound healing. Fibrin has recently attracted scientific attention for use as a scaffold to deliver cells [112, 113] or growth factors [114] for tissue engineering. Due to the high surface area to volume ratio available for clot formation, electrospun nanofibrous fibrinogen mats are highly suitable for wound dressing and hemostatic products. Wnek et al. have fabricated and characterized electrospun fibrinogen scaffolds [115], and found the fibrinogen mats to be composed of uniform, randomly orientated fibrinogen nanofibers. Notably, the electrospun fibrinogen fiber typically exhibited a granular appearance with 22.5 nm banding, characteristic of native fibrinogen. Consistent with the finding of studies using other polymers, the fiber diameter was shown to increase from 80 to 700 nm with increasing in polymer concentration. The electrospun fibrinogen mats were also shown to have substantial structural integrity and good handling, with an elastic modulus comparable to that of PGA. The most important characteristic of electrospun fibrinogen mats was their high surface area-to-weight ratio of $41\,000\ cm^2\ g^{-1}$, making electrospun fibrinogen mats ideal for wound dressing products. However, the risk of immune cross-reactivity may limit the use of fibrinogen for scaffold applications.

Chitin and Chitosan   Chitin and chitosan are biocompatible and biodegradable natural polymers, used in biomedical applications and cosmetics. Chitosan (poly-D-glucosamine) is derived from chitin (poly-*N*-acetyl-D-glucosamine), a polysaccharide formed in shellfish exoskeleton, which has received more attention in biomaterial development than chitin due to its solubility. Therefore, chitosan has been critically considered as a candidate biomaterial for tissue engineering scaffolds [116–118]. Chitosan carries a high cationic charge density and can interact with various anionic polymers, such as chondroitin sulfate, to form a hydrogel scaffold [119]. Different from a hydrated scaffold, an electrospun scaffold is a preformed, nanofiber-based scaffold with a definite structure [120–122]. Min et al. used radiation to depolymerize chitin to increase its solubility for electrospinning [120]. In this study, the electrospun chitin nanofibers started to form at a viscosity at which the required polymer chain entanglement occurs. Fiber diameters vary from 40 to 600 nm, but most fibers are less than 100 nm. The authors also demonstrated that chitosan nanofibrous scaffolds electrospun using deacetylated chitin showed no significant changes in terms of scaffold size and fiber diameter after this transformation. Significantly different from Min's approach, Geng et al. directly electrospun a chitosan solution [122], prepared by dissolving chitosan in concentrated acetic acid, into chitosan nanofibers with an average diameter of 130 nm. The authors suggested that the use of concentrated acetic acid to decrease surface tension of chitosan solution and increase the charge density of the jet, facilitated the formation of uniform chitosan nanofibers. Other factors affecting fiber morphology include chitosan molecular weight, solution concentration, and charge density of electric field. For each parameter, there is a narrow window of optimal working conditions for defect-free nanofiber formation to occur. Generally, a viscous chitosan polymer solution prepared from a higher molecular weight chitosan is preferred, with electrospinning done in a moderate charge density electric field.

Hyaluronan   The most commonly used carbohydrate-based natural polymer in tissue engineering is hyaluronan (hyaluronic acid), a polysaccharide composed of repeating glucuronic acid and *N*-acetylglucosamine. Like chitosan, unmodified hyaluronan is hydrophilic and is commonly produced in a gel-format scaffold. A hyaluronan-based scaffold is inherently unstable and thus has limited application in tissue engineering. Thus, approaches have been developed to increase the resistance of hyaluronan to degradation, including esterification [123] and crosslinking [124]. Modified hyaluronan can be fabricated into scaffolds with pre-formed structures, and several *in vitro* and *in vivo* studies have reported cell proliferation and synthesis of ECM using these modified hyaluronan scaffolds [125, 126]. It is difficult to form uniform size fibers from hyaluronan using electrospinning because of the high viscosity and surface tension of the hyaluronan solution. Um et al. reported on several approaches to prepare a hyaluronan solution that has sufficient molecular entanglement in a rapidly evaporated solvent, while still maintaining low viscosity and surface tension [127]. One of the approaches blended two hyaluronan polymers of low and high molecular weight, resulting in increased molecular entanglement but not viscosity. Another approach consisted of adding ethanol in the polymer solution, which facilitated solvent evaporation as well as reduction in surface tension. However, although the approaches seemed rational, these results, while satisfactory, were limited. Therefore, further approaches entailed adding an air-blow system close to the spinneret, providing both electrical and air-blowing shear forces as well as heat, to facilitate hyaluronan nanofiber formation, and the new setup is named an "electro-blowing" system. In the electro-blowing apparatus, the hyaluronan solution was heated to reduce viscosity and surface tension, and was electrospun by both electrical (voltage) and mechanical (air blow) force. Lastly, air-blowing significantly improved solvent evaporation and fiber stretch, and the electro-blowing system produced hyaluronan nanofibers with a fiber diameter ranging from 49 to 74 nm.

3.3.2.2 Synthetic Polymeric Nanofibrous Scaffolds

Poly(α-hydroxy esters) The most commonly used biopolymers for nanofiber production are of the biodegradable poly(α-hydroxy ester) based polymer family. PLA, PGA, their co-polymers, poly(lactic-*co*-glycolic acid) (PLGA), and poly(ε-caprolactone) (PCL) are biodegradable polymers approved by the U.S. Food and Drug Administration, with a long history in medical applications. *In vivo*, complete degradation of poly(α-hydroxy esters) results in the production of natural metabolites, such as lactic acid, which are subsequently converted into CO_2 and water, and eventually removed by the respiratory and urinary systems, respectively [128]. Thus, poly(α-hydroxy esters) are metabolized into non-toxic end-products that are not accumulated in the body.

PGA is a polymer of glycolic acid and has been extensively used in various biomaterial applications. Due to its high crystallinity, it is insoluble in general organic solvents, with highly fluorinated solvents being the only exception. Consistent with the trend found in other polymers, the PGA fiber diameter, ranging from 110 nm to 1.19 μm, increases with polymer concentrations from 0.05 to 14.3 wt.%. PGA polymer in a higher concentration solution encounters stronger molecular entanglement, resulting in thicker fibers during the electrospinning [129, 130].

PLA, with the addition of a methyl group, is more hydrophobic and more soluble in organic solvents, and degrades more slowly than PGA [131]. The three stereo-isomers of PLA [D, L, and D,L] differ in the position of a methyl group in the lactic acid monomer, and exhibit distinct properties. For example, PLLA has a higher melting temperature than PDLLA because of its higher order crystal structure. However, PDLLA, composed of l and d stereoisomers, degrades faster than PLLA. The properties of various PLA polymers are determined by the stereochemistry of the PLA isomers. PLLA is a semicrystalline polymer, whereas PDLLA is an amorphous polymer. During the preparation process, both PLLA and PDLLA are soluble in most organic solvents, such as chloroform. Commonly, DMF is added to increase the polymer conductivity, enhancing the fiber formation. Electrospun PDLLA and PLLA fibers share similar morphology, fiber density (0.27–0.31 g cm^{-3}), and porosity (75–78%) [132]. However, the average diameter of semicrystalline PLLA fibers is larger than that of amorphous PDLLA fibers when both polymers are used at the same concentration, suggesting that a stronger polymer molecular entanglement exists in PLLA fibers.

PLGA is a widely used biodegradable polymer because of its flexibility of copolymerizing different ratios of PLA and PGA. Depending on the ratio of copolymerization, PLGA has different subtypes with each exhibiting different properties. PLGA has been extensively used in medical products and was one of the first biodegradable polymers electrospun for tissue engineering applications [133]. The PLGA copolymer has an amorphous structure, because the constituent PGA and PLA molecules are unable to pack tightly to one another. Katti et al. and Berkland et al. have both used PLGA for electrospinning to investigate the parameters affecting fiber morphology [134, 135]. Their results showed that, in addition to polymer concentration and charge density, orifice diameter also has an effect on the morphology and diameter of electrospun fibers. Smaller nanofibers tend to be produced from smaller orifices. The area density of electrospun fiber mats increases linearly with electrospinning time. Among the working parameters, polymer concentration is the major parameter affecting fiber morphology. Li et al. have shown that nanofibrous PLGA scaffolds have an interconnected-porous structure with more than 90% porosity and sound mechanical properties, ideal properties for tissue engineered scaffolds [25].

Another member of the poly(α-hydroxy ester) family is poly(ε-caprolactone) (PCL), also a semicrystalline biodegradable polymer. Compared to other polyester family members such as PLA, PGA, and PLGA, PCL has been used less frequently as a material for fabricating biomaterial scaffolds, mainly because of concern over its slower degradation kinetics [136]. However, the improved resistance to hydrolytic attack and lower cost than other biodegradable polymers also make PCL an attractive polymer for the fabrication of electrospun nanofibers [53, 77, 99, 100, 137–146]. The rationale for using a biodegradable polymer with a longer half-life is to provide a structurally stable environment that is able to initiate and promote cell growth for a sufficient period of time. Lee at al. characterized the morphology, crystallinity, and mechanical properties of electrospun PCL meshes produced by electrospinning PCL solutions prepared in different solvents; they found that surface tension, viscosity, and electric conductivity of PCL solutions depended on the addition of DMF [53]. DMF can efficiently enhance the electrospinning process by decreasing fiber diameter by decreasing the surface tension and viscosity of the polymeric solution, and also by increasing polymer conductivity and the dielectric constant. It is also known that the electrospinning process reorganizes PCL polymer chains and, compared with unprocessed PCL, electrospun nanofibers have a reduced crystallinity.

Although poly(α-hydroxy ester) members are classified in the same polymer family, electrospun scaffolds derived from these polymers exhibit distinct degradation and mechanical properties [25, 129, 132, 147]. Studies comparing the degradation of electrospun nanofibers of various poly(α-hydroxy esters) have shown that PLLA and PCL fibers were able to resist hydrolytic attack for the duration of the study. However, all of the remaining polymers underwent severe degradation, including both structural collapse and gross shrinkage. A similar study conducted by Zong et al. demonstrated that PLGA 75:25 and PDLLA degrades at a much faster rate than PLGA 10:90 and PLLA [132]. One of the reasons for this more rapid degradation could be the high surface area of nanofibrous scaffolds, thereby allowing more water contact at the polymer surface, accounting for the vulnerability of small fibers, like nanofibers, to hydrolytic attack. Zong et al. also proposed that the electrospinning process lowers the glass transition temperature of PLGA copolymers to a value very close to their incubation temperature (37 °C); the relaxation of extended amorphous chains near the glass transition temperature thus causes shrinkage under incubation condition. Given the rapid degradation of some nanofiber formulations, many research groups have focused their efforts on electrospinning PLLA or PCL alone, or copolymerized with other biodegradable polymers, to fabricate polymer scaffolds with more desirable, long term stability.

Poly(3-hydroxybutyrate-*co*-3-hydroxyvalerate) (PHBV) PHBV is a member of the polyester family, and represents a new class of biodegradable polymers, recently attracting attention for tissue engineering applications. The co-polymerization with hydroxyvalerate renders PHBV with enhanced mechanical and degradation properties, thus extending the application to scaffold fabrication. Choi et al. investigated the feasibility of PHBV electrospinning and attempt to reduce fiber diameter by increasing the conductivity of the polymeric solution [54]. PHBV fibers electrospun from various concentrations of polymeric solutions range from 1 to 4 μm. However, efforts to manipulate electrospinning parameters such as voltage and concentration were not effective in bringing the fiber diameter below 1 μm. To increase the conductivity of the polymeric solution, favorable for smaller fiber formation, benzyl triethylammonium chloride (BTEAC) was added to the PHBV solution. The addition of BTEAC resulted in a reduction of the size of the electrospun fibers and the needed polymer concentration for producing uniform fibers, suggesting the conductivity of the PHBV solution plays a major role affecting the fiber diameter and morphology of PHBV fibers. In addition, the degradation of the electrospun fibrous mat is relatively slow in the first 12 days, but the disintegration of the structure accelerates after that, and results in a dramatic weight loss.

Poly(ethylene-*co*-vinyl alcohol) (PEVA) PEVA is a semi-crystalline, biocompatible but not biodegradable polymer. Recently, it has been electrospun and evaluated for its potential in tissue engineering scaffolds. PEVA is a hydrophilic polymer that is insoluble in aqueous solution due to the presence of both vinyl alcohol and ethylene groups. Most hydrophobic polymers have the property of slow degradation, whereas most hydrophilic polymers exhibit a rapid degradation rate. The desired property makes PEVA a potential material ideal for tissue engineering scaffolds. PEVA needs to be dissolved in dimethyl sulfoxide or lower alcohols. Kenawy et al. have prepared the PEVA solution by dissolving PEVA in a 70% alcohol solution (rubbing alcohol) at 65 °C for electrospinning [148]. Interestingly, the polymer tends to precipitate when sitting in room temperature for several hours. Therefore, electrospun PEVA needs to be processed before precipitation occurs, and is carried out in a setup designed to maintain the solution temperature. This study also demonstrated a consistent finding of fiber diameter increasing with increase in solution viscosity. The diameter of fibers electrospun from higher concentrations of polymer solutions is higher than the μm level. In these structures, fibers are fused at their contact points, instead of being individually stacked, suggesting that strong molecular entanglement exists in the highly concentrated PEVA polymer solution. Another remarkable observation is that PEVA electrospinning is extremely efficient. The usage of PEVA, disinfected by dissolving in rubbing alcohol, combined with the efficiency of the PEVA electrospinning, could be ideal for wound dressing application.

Poly(ethylene terephthalate) (PET)   PET, referred to also by the trade name "Dacron", has largely been used in biomaterial applications, especially in blood vessel prostheses, since PET is inert and does not interact with blood cells. Ma et al. investigated how electrospinning time affects the thickness and porosity of electrospun PET scaffolds [98]. The electrospun PET nanofibers fabricated using optimal parameters ranged from 200 to 600 nm. The thickness, mass per area, and porosity of the PET fiber mats all increase with increasing electrospinning time. However, there is no linear relation between these factors. A likely explanation is that not all the as-spun nanofibers were collected onto the sampling area, because, in their setup, there was uneven fiber deposition onto the fiber-collecting surface.

Polyurethane (PU)   PU is a non-biodegradable biomaterial with good blood and tissue compatibility, which is primarily used in vascular implants or wound dressing. PU is ideal for applications for tissue engineering products that require stable mechanical properties or structural integrity. Unlike most studies in which the polymer concentration was manipulated to control fiber morphology, Khil et al. have optimized fiber morphology by controlling solvent composition [149]. They electrospun a PU solution, prepared from pure DMF or tetrahydrofuran (THF) or mixed solvents with different DMF/THF ratios, and found that the mixed solvent produced more uniform, small fibers. Electrospun fibers prepared from pure DMF were smaller and irregular, whereas those from pure THF were relatively larger and uniform. Solvents could loosen the polymer coil, thereby affecting the viscosity of a polymer solution. In this study, the different solvent compositions loosened the polymer coil to various extents, resulting in different viscosities. The mixed solvent reduced the surface tension and viscosity of PU solution, and the addition of DMF increased the conductivity of the solution, which together contributed to the formation of uniform fibers.

Poly(ethylene oxide) (PEO)   PEO is a commonly used biomaterial for tissue engineering because of its capability to gel *in situ*. A PEO gel can be directly injected into an irregularly shaped defect site and photopolymerized to help tissue repair [150, 151]. However, the inherent, soft mechanical properties of the PEO hydrogel and the depth limit of the injection site that permits sufficient energy present a major challenge for using PEO hydrogel. Preformed PEO scaffolds with a defined porous structure is another scaffold option. PEO is one of the earliest polymers processed for electrospinning because it is easy to prepare and can be dissolved in both organic and aqueous solvents. Much of the current knowledge on the electrospinning process is in fact based on results using PEO. Because of the ease of forming uniform nanofibers, PEO, is often blended with another polymer in the electrospinning of tissue engineering scaffolds [106, 152, 153].

3.3.2.3   Composite Polymeric Nanofibrous Scaffolds

Current scaffold development aims to incorporate many polymer types for the fabrication of biomaterial scaffolds that are able to respond to the biological activities of cells while meeting specific host tissue site requirements. Regardless of whether a double, triple, even quadruple polymer blend/mix is used, a tissue engineering scaffold made of a polymer blend should still retain the properties of each polymer type. Therefore, it is expected that electrospinning of a polymer blend/mix will create novel composite scaffolds with enhanced performance for tissue engineering. Another practical reason for electrospinning a blend of polymers is that often the polymer of interest cannot be electrospun into uniform fibers without the addition of the "electrospinning-driving" polymer. Regardless of the goal, the challenge of electrospinning a polymer blend is to optimize the standard electrospinning parameters, which are further complicated by the interplay of the properties of multiple polymers. Hence, a sound understanding of polymer and solvent chemistry is important in identifying favorable electrospinning parameters. Three general composite nanofibrous scaffolds, natural–natural, synthetic–synthetic, and natural–synthetic, have been developed and characterized for their properties and potential applications.

Natural–Natural Composites   Electrospinning of multiple natural polymer blends can yield a mixture of natural nanofibers that closely mimic the native ECM. A large percentage of native tissues contains both collagen and elastin fibers that are frequently subjected to tensile and elastic loading, respectively. Electrospun fibrous scaffolds composed of collagen types I and III, and elastin have been fabricated to replicate the native ECM of blood vessels [94]. The fiber diameter found within the native ECM of blood vessels ranges from 270 to 710 nm, which is slightly larger than the 100 to 680 nm diameter range of nanofibers electrospun from blends of collagen types I and III. Aside from protein–protein mixtures, the protein-based silk and carbohydrate-based chitosan blend has been electrospun into nanofibers as well. In the silk fibroin/chitosan blend for electrospinning, the addition of chitosan increases the viscosity and conductivity of the blend solution, thus enhancing the formation of smaller, uniform nanofibers [111]. One possible explanation for this phenomenon is that chitosan carries ionizable, amino groups, and increases the conductivity of the silk fibroin, which is ideal for electrospinning. Nevertheless, the formation of uniform nanofibers discontinues after excessive chitosan is added. Silk fibroin and chitosan in the blend solution may form hydrogen bonds, suggesting that optimizing the electrospinning of a blend natural polymer is more complicated than for pure natural polymers. It is reasonable to expect that the nanofiber mat is composed of individual, blended nanofibers rather than a mixture of different mono-polymer nanofibers.

Synthetic–Synthetic Composites The synthetic–synthetic blend is commonly used to fabricate scaffolds with the combined properties of composite polymers. Polymers, such as PLA and PCL, are biodegradable, biocompatible, and hydrophobic whereas poly(ethylene glycol) (PEG) is hydrophilic, non-immunogenic, and non-biodegradable. In mixing PLA and PEG, the resultant PLA/PEG blend is more hydrophilic than PLA and also more biodegradable than PEG. In addition, the co-existing properties of hydrophilicity and biodegradability can be programmed by manipulating the ratio and the type of composed polymers. For instance, PLGA nanofibrous scaffolds shrink during degradation; thus, by blending of PLGA and PEG-PLA one can effectively resist scaffold shrinkage [154]. In addition to improvements in biodegradation, the polymer blend has been shown to exhibit flexible mechanical properties that can be altered by the ratio of composed polymers. For example, the elastic PEVA nanofibrous mat becomes stiffer after PLA is added for blend electrospinning [155].

Natural–Synthetic Composites The mixing of natural and synthetic polymers can be a major challenge for electrospinning since solvents workable for both polymers are limited. Therefore, the choice of solvent becomes a primary consideration after determining the blend components. Many natural polymers are difficult to electrospin into nanofibers, especially when dissolved in water, since their polyelectrolyte characteristic interferes with fiber formation. One alternative is to add synthetic polymers such as PEO to facilitate nanofiber formation. The natural polymer/ PEO blend in aqueous solution can be electrospun, and the use of organic solvents may be avoided. Duan et al. have systematically characterized the properties of chitosan/PEO blend solutions and their electrospun fibers [153]. They concluded that the chitosan/PEO blend retains conductivity, surface tension, and viscosity, and favors the formation of smaller, uniform nanofibers, compared with pure chitosan or PEO solutions.

Another completely different blending approach is to physically, rather than chemically, mix two polymer nanofibers together using multiple jet electrospinning. In this approach, natural and synthetic polymers are placed in two separate containers and electrospun simultaneously to form a natural/synthetic nanofibrous composite. For example, poly(vinyl alcohol) (PVA) and cellulose acetate (CA) are electrospun onto the same target area [156]. With this method, PVA and CA nanofibers disperse into each other as a direct result of physical blending. When the PVA/CA ratio of the blend is reduced, mechanical properties, such as tensile strength and fiber elongation, are decreased, but the average fiber diameter increases. Further investigations are needed to assess whether the mechanical property changes are due to polymer composition, fiber size, or a combination of the two.

3.3.2.4 Nanofibrous Scaffolds Coated with Bioactive Molecules

A natural ECM scaffold contains bioactive peptides, such as RGD sequences, but is often mechanically too weak or rapidly degraded for tissue engineering applications. Conversely, a synthetic polymeric scaffold has tailorable properties to meet the physical requirements, but lacks the native biocompatibility. There is, therefore, a need to develop methodologies to incorporate and optimize bioactive motifs, such as the RGD peptide, for the synthesis of biocompatible polymeric scaffolds that possess the advantages of both natural and synthetic polymers. Unlike natural–synthetic composite nanofibrous scaffolds fabricated from electrospinning a blend polymer, nanofibrous scaffolds coated with bioactive molecules are produced by coating natural polymers to the surface of synthetic nanofibrous scaffolds. This after-work addition can avoid alteration of the properties of the natural polymer during the electrospinning. The most commonly used approaches for the incorporation of bioactive agents on a polymeric surface include simple coating and covalent linking. The simple coating process is easy and fast but the coated agents detach shortly, depending on the coating surface properties, whereas the covalent linking process is more laborious but the coating should be more stable.

Collagen coated electrospun scaffolds have been developed by simply placing nanofibrous scaffolds in a collagen solution. The simple coating works for most synthetic polymers, but the efficiency and effectiveness of the coating are dependent on polymer surface properties such as charge, chemistry and geometry. Collagen type I has been coated on PCL [142] and P(LLA-CL) [157] electrospun nanofibrous scaffolds, which are used for culturing different cells for various tissue engineering applications. Although there is no direct measurement to evaluate the efficiency of the collagen coating on the scaffolds, cell response, i.e., the promotion of specific biological activities, indirectly demonstrates the presence of collagen coating on electrospun fibrous scaffolds.

Electrospun nanofibrous scaffolds made of either hydrophobic or hydrophilic synthetic polymers are usually inert and lack functional groups for direct covalent linkage of bioactive peptides or domains. To create functional groups for the covalent bonding, surface processing techniques, such as acid/base treatment, chemical activation, and oxidization, are used. For instance, studies have reported on the modification of PET nanofiber surfaces by gelatin grafting [98]. The process starts with the creation of hydroxyl groups on PET nanofibers using formaldehyde, grafts with methacrylic acid, and covalent linkage of gelatin using carbodiimide. The results show that more gelatin is grafted to PET nanofibers than to the film counterpart, and the grafting occurs throughout the entire scaffold surface. In addition, gelatin grafting significantly increases the hydrophilicity of the PET surface, suggesting that covalently grafting bioactive proteins or peptides to a nanofibrous scaffold is an effective approach for the modification of the hydrophobic properties of the scaffold.

3.3.3
Engineered Tissues using Electrospun Nanofibrous Scaffolds

3.3.3.1 **Skin**
Skin is the largest tissue covering the body and provides physical and chemical protection of the body from harmful sources such as heat and microbial organisms. The top layer is the epidermis containing melanocytes and the dermis lies directly underneath the epidermis.

In the body, cells and tissue matrices interact via complex chemical and physical processes. Due to the sophisticated architecture of the skin with its various layers, it is difficult for current synthetic 3D matrices to simulate such processes. Therefore, electrospinning has been used to spin natural and synthetic polymers to fabricate biomaterial scaffolds that more closely mimic the tissue matrix of skin. In a study by Venugopal et al., PCL, collagen type I, and collage type I-coated PCL nanofibers were electrospun to fabricate a substitute for skin regeneration [144]. In this study, fibroblasts from normal human skin attached and spread on all three nanofiber matrices; however, the interaction between cells and nanofiber matrices varied with the scaffold material. For instance, collagen nanofiber matrices showed a significantly higher level of fibroblast proliferation than control monolayer culture. Conversely, PCL fibers showed a significantly lower level of cellular proliferation than control monolayer culture. The results of this study also showed normal morphology and confluence of cells on control monolayer culture and collagen nanofibers. However, while PCL nanofibers were able to support cell growth, they showed no confluence compared with collagen coated PCL nanofibers. As a result, this study proved that PCL fibers are able to partially support the growth of skin fibroblasts, while the presence of collagen on the scaffolds greatly enhances the interactions between cells and nanofibers. The advantage of using a collagen coated PCL composite nanofibrous scaffold is to incorporate the biological and mechanical properties of each polymer, with collagen type I providing a favorable milieu for skin fibroblasts and biodegradable PCL improving the mechanical integrity of the matrix. Therefore, electrospun 3D nanometer sized fibers provide a potential scaffold for skin regeneration. Min et al. electrospun silk fibroin into nanofibers and then coated the fibers with collagen type I, fibronectin, and laminin to evaluate cell response to different ECM coatings on silk fibroin nanofibers [109]. In addition, they also used different cell types, human oral and epithelial keratinocytes and gingival fibroblasts, to measure cell attachment and spreading on the nanofibers. In terms of the effect of various ECM coatings, collagen type I coated nanofibers were shown to more actively promote attachment of keratinocytes, compared with laminin or fibronectin coated nanofibers, and, interestingly, even to a polystyrene surface. Cells adhered to and spread on the surface of silk fibroin nanofibers, and then started to migrate along the fibers, forming a 3D cellular network well integrating with nanofibers. The similar cellular response in a nanofiber scaffold is also demonstrated in the previous study using fibroblasts in PLGA nanofibers [25]. Beside the attempt of using silk fibroin nanofibers as scaffolds for skin regeneration, chitin/PLGA nanofibrous scaffolds cultured with keratinocytes and fibroblasts have also shown great promise in skin tissue engineering since chitin has proved to have good wound healing properties [158]. The chitin/PLGA nanofibrous scaffolds, taking advantages of the desired properties of chitin and PLGA for tissue engineered scaffolds, exhibit improved hydrophilicity and sound mechanical properties. Cells attach better in the chitin/PLGA than the PLGA-only nanofibrous scaffolds. In the preliminary cell study, the same cell types used in the silk fibroin scaffolds show a similar trend of cell attachment and spreading in various ECM coated chitin/PLGA nanofibrous scaffolds. Collagen type I coated chitin/PLGA nanofibrous scaffolds enhances cell attachment and spreading in the scaffolds, which is critical for cellular activities in skin regeneration using the tissue engineering approach.

3.3.3.2 Blood Vessel

Blood vessels are a part of the circulation system, and function as a channel to transport oxygen and nutrient to the body tissues or remove metabolic waste for replenishment. There are three types of blood vessels found in the body: arteries, veins, and capillaries. Each of them has a different structure and various functions. Vascular related disorders are one of the leading fatal diseases around the world, and one of the current treatments is the use of vascular grafts in bypass surgeries. In terms of material types, grafts are either biological (autograft, allograft, and xenograft) or synthetic. Although there is fair success with these conventional grafts, challenges such as the availability of small vascular grafts still remain. Vascular tissue engineering is promising for creating cellular vascular grafts with desired dimensions.

The tissue engineered artery fabricated using electrospun scaffolds has been extensively investigated recently. Boland et al. have fabricated a vascular scaffold by assembling two electrospun nanofibrous tubes composed of different collagen/elastin ratios with cultured dermal fibroblasts, aortic smooth muscle cells, and umbilical vein endothelial cells in the outer, middle, and inner layers of the scaffold, respectively, to simulate the anatomical three layers of an artery [94]. After culturing in a bioreactor, the engineered artery exhibits a tri-cellular layer architecture in cross section. One of the critical requirements for tissue engineered arteries is to have satisfactory mechanical properties to bear compliance strength and burst pressure. Therefore, Stitzel et al. blended PLGA with collagen and elastin to produce an electrospun tri-polymeric fibrous scaffold, thus increasing the strength of the original collagen/elastin scaffold and rendering it mechanically comparable to a native artery [102]. Both endothelial and smooth muscle cells cultured in the scaffolds exhibit favorable cell attachment and growth. Further *in vivo* testing found no signs of toxicity. In addition to studies that focus on the design of a tissue engineered vascular graft, several studies have reported the effects of surface roughness and fiber orientation on the cellular activities of vascular cells [159, 160]. Xu et al. have demonstrated that endothelial cells cultured on electrospun fibrous scaffolds exhibit less-spread cell morphology, compared with those on smooth surfaces [159]. An increase in cell attachment and proliferation was also observed on the smooth surface culture. Another study has shown that smooth muscle cells exhibit a spindle-like contractile phenotype, while cytoskeletal proteins, such as α-actin and myosin filaments, are oriented along the direction of aligned nanofibers [160]. Fiber alignment using electrospinning supports the construction of a tissue engineered vascular graft that contains a middle, smooth muscle layer that mimics the circumferentially oriented nature.

3.3.3.3 Cartilage

Adult cartilage is formed mainly at the articular joint and is a specialized tissue that consists of chondrocytes, ECM macromolecules, and water. Chondrocytes synthesize and maintain the ECM in cartilage while the ECM functions to provide structural protection to residing cells and to act as a regulatory messenger for cell activities. Cartilage tissue engineering aims to replace damaged cartilage and restore biological function. A main research focus in our laboratory is the development of cartilage tissue engineering using the electrospinning technique.

We have recently compared the biological response of chondrocytes seeded onto 3D PCL nanofibrous scaffolds to that of cells seeded as monolayers on standard tissue culture polystyrene (TCPS) [77]. Gene expression analysis (Fig. 3.5) showed that chondrocytes seeded on the nanofibrous scaffold maintained their chondrocytic phenotype by expressing cartilage-specific ECM genes, including collagen types II and IX, aggrecan, and COMP. Specifically, expression of the collagen type IIB spliced variant transcript, indicative of the mature chondrocyte phenotype, was significantly up-regulated. Chondrocytes exhibited a round shape on the nanofibrous scaffolds, in contrast to a flat, well-spread morphology seen in monolayer

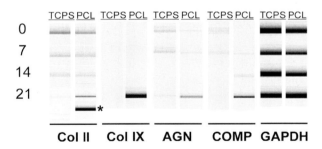

Fig. 3.5. Reverse transcription-polymerase chain reaction (RT-PCR) analysis of expression of cartilage associated genes in fetal bovine chondrocytes seeded onto tissue culture polystyrene (TCPS) or PCL nanofibrous scaffolds. On culture days 0, 7, 14, and 21 total RNA was extracted and RT-PCR performed with gene-specific primer pairs, including collagen type II (Col II), collagen IX (Col IX), aggrecan (AGN), cartilage oligomeric matrix protein (COMP), and glyceraldehyde-3-phosphate dehydrogenase (GAPDH) as a house keeping gene. Cells cultured on nanofibrous scaffolds showed strong induction of mRNA expression of collagen types II and IX, aggrecan, and COMP after 21 days. More importantly, the alternatively spliced mRNA variant of collagen type II, the IIB(*), was expressed only in the nanofibrous scaffold cultures.

H & E **Alcian Blue**

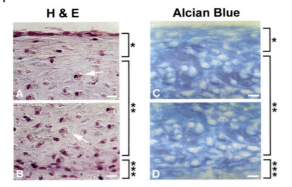

Fig. 3.6. Histological analysis of MSC cultures seeded into a PCL nanofibrous scaffold and maintained in a chondrogenic medium supplemented with TGF-β1 for 21 days. Sections from the upper and lower portions of the 3D constructs were stained with hematoxylin and eosin (H&E) (A, B) and Alcian blue (C, D). H&E staining showed flat fibroblast-like cells on the top zone (*), round chondrocyte-like cells embedded in lacunae (arrows) in the middle zone (**), and small, flat cells at the bottom zone (***). Alcian blue staining showed the presence of sulfated proteoglycan-rich ECM in the construct. Bar = 10 μm.

cultures on TCPS. Histologically, nanofibrous cultures produced more sulfated proteoglycan-rich, cartilaginous matrix than monolayer cultures. These results indicate that the biological activities of chondrocytes are crucially dependent on the dimensionality of the extracellular scaffolds, and that nanofibrous PCL may be a biologically preferred scaffold/substrate for proliferation and phenotype maintenance of chondrocytes.

In another study [140], we specifically compare the chondrogenic activities of MSC–nanofiber cultures to those of high-density cell pellet cultures, a commonly used culture system for studying chondrogenesis of MSCs *in vitro*. MSCs cultured in nanofibrous scaffolds in the presence of TGF-β1 differentiate into a chondrocytic phenotype, as evidenced by chondrocyte-specific gene expression and the synthesis of cartilage-associated ECM proteins. The level of chondrogenesis observed in MSCs seeded within nanofibrous scaffolds is comparable to that observed for MSCs maintained as cell aggregates or pellets. The 3D MSC-seeded constructs display a cartilage-like morphology, containing chondrocyte-like cells surrounded by abundant cartilaginous matrix (Fig. 3.6). Due to the physical nature and improved mechanical properties of nanofibrous scaffolds, particularly in comparison to cell pellets, these findings suggest that the nanofibrous scaffold is a practical carrier for MSC transplantation, and represents a candidate scaffold for cell-based tissue engineering approaches to cartilage repair.

3.3.3.4 Bone

Bone is a hard, solid connective tissue that provides structure and protection to the body. To support external loading and absorb shocks, bone has a unique structure and chemical composition. The bone structure is composed of two layers of different density of bones. The outer layer is compact bone and the inner layer being spongy bone. Bone is always undergoing dynamic remodeling carried out by two different cell types, the osteoblast for building bone and the osteoclast for digesting bone. In addition, the bone marrow contains numerous MSCs that are capable of differentiating into various connective tissue cells to replenish them.

Cells used in bone tissue engineering include osteoblasts, osteoprogenitor cells, and MSCs. Each of these cell types, depending on their nature, requires a different culture environment. MSCs are a highly promising cell source for tissue engineering applications because of their multi-differentiation capabilities and their expandability [81]. Applications of MSCs for bone tissue engineering requires that the cells are seeded within electrospun nanofibrous scaffolds that will permit and/or promote osteogenic differentiation. Yoshimoto et al. and Shin et al. have reported on the osteogenic differentiation of MSCs in PCL nanofibrous scaffolds cultured *in vitro* and *in vivo* [137, 138]. In their study, MSCs from rat bone marrow seeded in PCL nanofibrous scaffolds are cultured in a rotatory oxygen-permeable bioreactor with the osteogenic medium. This environment promotes MSCs to differentiate into osteoblast-like cells producing collagen type I and minerals in the scaffold. In addition, the bioreactor improves oxygen and nutrients diffusion, enhancing cell growth in the nanofibrous scaffold. Initial osteogenic differentiation of MSCs is carried out in the *in vitro* model. Further assessment of bone formation included implanting the MSC loaded nanofibrous scaffold pre-induced and -cultured in a bioreactor in a rat model. The new osteo-matrix is deposited throughout the entire harvested cellular constructs, resulting in a white, smooth appearance and stiffer properties. The preliminary results of new bone formation using MSC–nanofiber constructs are encouraging. To actively induce bone formation, osteoconductive calcium carbonate has been coelectrospun with PCL nanofiber to fabricate a nanofibrous composite scaffold. The addition of calcium carbonate increases the mineralization in the cellular constructs, an early sign of bone formation. The composite scaffold combining the osteogenic promotion of nanofibers and the osteoconductivity of calcium carbonate is also suggested for guided bone regeneration therapy to repair jaw and alveolar bone defects.

3.3.3.5 Muscle

Muscles make up most of the body and account for almost half of body weight. Skeletal muscle covering the skeleton functions to move the body, cardiac muscle making up the heart contracts in a rhythmic movement, and smooth muscle forming internal organs perform the unconscious actions. The three types of muscles are composed of different kinds of cells and ECM protein compositions.

Skeletal muscle tissue is composed of bundles of highly oriented and dense muscle fibers, each representing a multinucleated cell derived from myoblasts. Therefore, the engineering of skeletal muscle tissue necessitates the use of a biomaterial scaffold that will allow cells to align parallel to one another. Electrospinning, as previously illustrated, provides a technique that not only can generate nanoscale non-woven fibers, but this technique also offers the versatility of aligning fibers to yield an ideal scaffold architecture for skeletal muscle tissue engineering. In a study by Riboldi et al., DegraPol®, a degradable block polyesterurethane, was electrospun to fabricate a biomaterial scaffold to engineer skeletal muscle tissue using mouse C2C12 muscle progenitor cells [161]. Preliminary experiments in this study found C2C12 line cells to be compatible with the electrospun scaffold by adhering, proliferating, and differentiation into myotubes on the electrospun scaffolds [161]. More specifically, the authors showed that after 3 days of culture of C2C12 cells in differentiation medium there was formed elongated, multi-nucleated, myosin heavy chain expressing myotubes that preferentially aligned with the direction of the scaffold fibers. C2C12 cells were also viable on the electrospun scaffold up to 7 days of culture. By aligning myofibers to one another, scaffolds may be engineered with the ability to generate sufficient force for contraction of differentiated C2C12 cells. While the tendency of utilizing microfibrous polymeric scaffolds has yielded promising results in driving myofibers development and orientation along the scaffold, the high tensile modulus and low-yield elongation of these microfibrous polymeric scaffolds is inadequate to withstand the mechanical stimulation needed for dynamic culturing methods needed for skeletal muscle tissue engineering [161–163].

A critical factor for successful regeneration of myocardium in a tissue engineered scaffold is that the scaffold structure should be flexible for cardiomyocyte contraction, sustain a reasonable tension for cell morphology maintenance, and have a good integrity for handling. To fulfill this purpose, Shin et al. have cultured primary cardiomyocytes from rat ventricles on PCL nanofiber suspended wire rings [139]. Interestingly, unlike a rigid scaffold, the flexibility of the nanofibrous network applies limited restriction on cell contraction. The cardiomyocytes on the nanofibers start the ubiquitous and synchronized contraction after 3 days of culture. In addition, cells within the scaffold exhibit spindle, multi-angular muscle-cell-like morphology, and also express cardiac-specific markers such as tropomyosin and connexin43. The wire ring applies a tensile force on the suspended nanofibers, which may be transferred to stretch cells, which is critical for the maturation of the cardiomyocyte. In another study, rat cardiac fibroblasts are cultured in acid-treated PGA scaffolds to evaluate the proliferation of cells cultured in PGA scaffolds with different fiber sizes and surface chemistry [130]. Cells proliferate significantly more rapidly in acid treated, smaller nanofibers, suggesting that the acid treated surface exposes more carboxylic and hydroxylic groups that favorably attract cell attachment molecules, such as fibronectins and vitronectins, to enhance cell attachment; thus more cells can grow on a bigger surface area in a scaffold with smaller fibers.

Functional tissue engineered smooth muscle is essential for the development of complex tissue engineered organs. Flexible, anisotropic properties of smooth muscle provide internal organs the capability of constant movement and the expansion for additional content storage. Electrospinning has been applied to microintegrate smooth muscle cells into nanofibrous scaffolds to fabricate a cell–nanofiber composite [44]. The mechanical properties of the composite are in an adequate range to meet the requirements for soft tissue application. After culturing in a perfusion bioreactor, the cellular composite exhibits a high density of smooth muscle cells in a uniform distribution. The smooth muscle is successfully engineered by the microintegration process based on the electrospinning, which can be adopted to tissue engineer other tissues as well.

3.3.3.6 Ligament

The ability of electrospun nanofibrous scaffolds to mimic the architectural scale of the native ECM provides them with great potential for the engineering of ligaments. The ligament is a tough fibrous band of slightly elastic connective tissue that provides stability to skeletal joints. The specialized cells that make up ligaments are called fibroblasts. The matrix of ligaments is made up of collagen fibrils with a diameter of ~60 nm; therefore, nanofibrous scaffolds should provide ligament fibroblasts with a structure that closely mimics the native matrix [164]. A study investigating the effects of fiber alignment on the ECM generation of human ligament fibroblasts (HLF) on PU electrospun nanofibers showed that cells cultured on aligned nanofibers were spindle-shaped and oriented in the nanofiber direction, while cells on non-aligned nanofibers had no directionality [164]. After 7 days of culture, HLF cells on aligned nanofibers showed a significant change in their ECM with increase collagen synthesis compared with non-aligned fibers, although cell proliferation was not affected by fiber directionality [164]. A significant increase in cell attachment and proliferation on nanofiber scaffolds was also observed, as compared with cast PU membranes. Therefore, this study demonstrates that the biomimetic nature of aligned electrospun nanofibers provides an architectural environment similar to that which ligament fibroblasts normally encounter *in vivo*.

3.3.3.7 Nerve

Unlike other parts of the body, the central nervous system (CNS) cannot generate new neurons or regenerate damaged nerves itself. Cell-based tissue engineering strategies using multipotential cells have been investigated as a therapy for specific neurologic disorders (i.e., paralysis of extremities). Due to the complexity of the architectural organization of the CNS, polymeric scaffolds should provide a suitable 3D carrier for cell delivery. Electrospinning presents a sophisticated technique to fabricate an ECM-like matrix. Recent studies have investigated the potential of electrospun 3D scaffolds, both aligned and non-aligned, in neural tissue engineering using a multipotent neural stem cell (NSC) line, C17-2, derived from a neonatal mouse cerebellum [76, 165]. Yang et al. have investigated the morphology and cell adhesion response of NSC to PLLA non-aligned nanofibrous scaffolds [165]. Phase-contrast microscopy revealed that C17-2 cells seeded on the scaffolds adhered to the scaffolds and started to differentiate on the fibrous scaffold 10 h after seeding, and by 24 h around 70% of cells exhibited a spindle-like shape with extended processes. An additional study by Yang et al. compared the response of C17-2 cells to aligned versus non-aligned electrospun PLLA nanofibers [76], and reported that after 2 days of culture significant changes in the phenotype of cells based on directionality had occurred.

Successful nerve regeneration is dependent upon extensive growth of axonal processes. The study by Yang et al. found no significant effect of fiber alignment when comparing aligned (1.5 µm) and non-aligned (1.25 µm) microfibrous scaffolds. However, the directionality did have an effect when comparing aligned (300 nm) and non-aligned (250 nm) nanofibrous scaffolds. The results indicate that the presence or absence of a directionality effect is dependent on architectural scale. The average neurite length of C17-2 cells on aligned nanofibrous scaffolds was calculated as 100 µm and 75–80 µm on aligned and non-aligned microfibrous scaffolds, respectively. These results show a significant relationship of a decrease in fiber diameter increasing neurite outgrowth.

3.4
Current Challenges and Future Directions

To date, many different polymers have been successfully electrospun and characterized for numerous intended applications. However, limited progress has been made on fabricating tissues using nanofibrous scaffolds. Unlike nanofiber applications in composites and electronics, tissue engineering applications of nanofibrous polymeric scaffolds still require further extensive biological analyses in addition to physical characterization to determine if they are suitable for the promotion of tissue growth. Recent studies have focused on the introduction of different new polymers for electrospinning, as well as testing and characterizing the physical properties of electrospun nanofibrous polymeric scaffolds, with less emphasis on biological evaluation. To move to the next developmental phase of nanofibrous scaffold-based tissue engineering, cellular and molecular analyses are necessary to understand the interaction between nanofibers and cells. A better understanding of cell–nanofiber interactions, based on the structural and functional similarity between nanofibers and native ECM, is critical to successful tissue engineering.

Functional nanofibrous scaffolds providing structural and mechanical support for tissue growth as well as actively inducing favorable biological activities should be the future aim of research on electrospun nanofibrous scaffolds. Peptide domains, such as RGD on ECM molecules, bind to cell surface receptors, activating cellular signaling pathways to elicit cellular response. Therefore, to promote cellular activities, efforts should be devoted to developing methodologies to incorporate and optimize bioactive motifs or peptides into the electrospun nanofibrous scaffold, ultimately creating biologically active scaffolds. In addition to matrix-bound signal proteins on ECM, soluble bioactive molecules such as growth factors are important for the regulation of cellular events, including proliferation and differentiation [166]. Biodegradable polymers have long been used as drug delivery systems to deliver therapeutic agents because they can be designed to undergo programmed degradation in a controlled fashion [167]. Incorporation of growth factors, such as transforming growth factor-beta (TGF-β) or insulin-like growth factor (IGF), into a nanofibrous scaffold may be used as an additional strategy for the control of tissue growth. Ideally, it is desirable to fabricate nanofibrous scaffolds that are capable of releasing, in a controlled manner, one or more growth factors to act on specific cellular functions at different times. A nanofibrous scaffold housing multiple growth factors may be produced using a selected combination of biodegradable polymers, each releasing one or more specifically adsorbed growth factors as a function of its own degradation profile. These characteristics, once optimized and taken together with the "tissue-engineering"-friendly nature of the nanofibrous scaffold, should make this a highly promising candidate scaffold system for tissue regeneration applications.

3.5
Conclusion

Electrospinning technology is a useful, economical, and easily set-up means of fabricating of 3D, highly porous, nanofibrous scaffolds that have been shown to support cellular activities and tissue formation. Conversely, tissue engineering is a revolutionary, cell-based therapy with a higher promise for clinical applications. Together, tissue engineering and regeneration with the application of electrospun polymeric nanofibers is an exciting example of the power of the rapidly emerging field of nanotechnology.

References

1 LANGER R, VACANTI JP. Tissue engineering. *Science* **1993**, 260, 920–926.

2 HUTMACHER DW. Scaffold design and fabrication technologies for engineering tissues – state of the art and future perspectives. *J. Biomater. Sci. Polym. Ed.* **2001**, 12, 107–124.

3 LEE JH, JUNG HW, KANG IK, LEE HB. Cell behaviour on polymer surfaces with different functional groups. *Biomaterials* **1994**, 15, 705–711.

4 BHARDWAJ T, PILLIAR RM, GRYNPAS MD, KANDEL RA. Effect of material geometry on cartilagenous tissue formation in vitro. *J. Biomed. Mater. Res.* **2001**, 57, 190–199.

5 ZHANG S. Fabrication of novel biomaterials through molecular self-assembly. *Nat. Biotechnol.* **2003**, 21, 1171–1178.

6 HERSEL U, DAHMEN C, KESSLER H. RGD modified polymers: biomaterials for stimulated cell adhesion and

beyond. *Biomaterials* **2003**, 24, 4385–4415.

7 FLEMMING RG, MURPHY CJ, ABRAMS GA, GOODMAN SL, NEALEY PF. Effects of synthetic micro- and nano-structured surfaces on cell behavior. *Biomaterials* **1999**, 20, 573–588.

8 ELSDALE T, BARD J. Collagen substrata for studies on cell behavior. *J. Cell Biol.* **1972**, 54, 626–637.

9 CUKIERMAN E, PANKOV R, STEVENS DR, YAMADA KM. Taking cell-matrix adhesions to the third dimension. *Science* **2001**, 294, 1708–1712.

10 ABBOTT J, HOLTZER H. The loss of phenotypic traits by differentiated cells. 3. The reversible behavior of chondrocytes in primary cultures. *J. Cell Biol.* **1966**, 28, 473–487.

11 BENYA PD, SHAFFER JD. Dedifferentiated chondrocytes reexpress the differentiated collagen phenotype when cultured in agarose gels. *Cell* **1982**, 30, 215–224.

12 LEVENBERG S, HUANG NF, LAVIK E,

ROGERS AB, ITSKOVITZ-ELDOR J, LANGER R. Differentiation of human embryonic stem cells on three-dimensional polymer scaffolds. *Proc. Natl. Acad. Sci. U.S.A.* **2003**, 100, 12 741–12 746.

13 CUKIERMAN E, PANKOV R, YAMADA KM. Cell interactions with three-dimensional matrices. *Curr. Opin. Cell Biol.* **2002**, 14, 633–639.

14 SINHA RK, MORRIS F, SHAH SA, TUAN RS. Surface composition of orthopaedic implant metals regulates cell attachment, spreading, and cytoskeletal organization of primary human osteoblasts in vitro. *Clin. Orthop.* **1994**, 258–272.

15 DALBY MJ, RIEHLE MO, JOHNSTONE H, AFFROSSMAN S, CURTIS AS. In vitro reaction of endothelial cells to polymer demixed nanotopography. *Biomaterials* **2002**, 23, 2945–2954.

16 DALBY MJ, RIEHLE MO, JOHNSTONE HJ, AFFROSSMAN S, CURTIS AS. Polymer-demixed nanotopography: control of fibroblast spreading and proliferation. *Tissue Eng.* **2002**, 8, 1099–1108.

17 WEBSTER TJ, SIEGEL RW, BIZIOS R. Osteoblast adhesion on nanophase ceramics. *Biomaterials* **1999**, 20, 1221–1227.

18 WEBSTER TJ, ERGUN C, DOREMUS RH, SIEGEL RW, BIZIOS R. Enhanced functions of osteoblasts on nanophase ceramics. *Biomaterials* **2000**, 21, 1803–1810.

19 ELIAS KL, PRICE RL, WEBSTER TJ. Enhanced functions of osteoblasts on nanometer diameter carbon fibers. *Biomaterials* **2002**, 23, 3279–3287.

20 HUANG ZM, ZHANG YZ, KOTAKI M, RAMAKRISHNA S. A review on polymer nanofibers by electrospinning and their applications in nanocomposites. *Composites Sci. Technol.* **2003**, 63, 2223–2253.

21 NAIR LS, BHATTACHARYYA S, LAURENCIN CT. Development of novel tissue engineering scaffolds via electrospinning. *Expert Opin. Biol. Ther.* **2004**, 4, 659–668.

22 LI WJ, MAUCK RL, TUAN RS. Electrospun nanofibrous scaffolds: Production, characterization, and applications for tissue engineering and drug delivery. *J. Biomed. Nanotechnol.* **2005**, 1, 259–275.

23 MA Z, KOTAKI M, INAI R, RAMAKRISHNA S. Potential of nanofiber matrix as tissue-engineering scaffolds. *Tissue Eng.* **2005**, 11, 101–109.

24 VENUGOPAL J, RAMAKRISHNA S. Applications of polymer nanofibers in biomedicine and biotechnology. *Appl. Biochem. Biotechnol.* **2005**, 125, 147–158.

25 LI WJ, LAURENCIN CT, CATERSON EJ, TUAN RS, KO FK. Electrospun nanofibrous structure: A novel scaffold for tissue engineering. *J. Biomed. Mater. Res.* **2002**, 60, 613–621.

26 SMITH LA, MA PX. Nano-fibrous scaffolds for tissue engineering. *Colloids Surf. B Biointerfaces* **2004**, 39, 125–131.

27 WHITESIDES GM, BONCHEVA M. Beyond molecules: self-assembly of mesoscopic and macroscopic components. *Proc. Natl. Acad. Sci. U.S.A.* **2002**, 99, 4769–4774.

28 ZHANG S, MARINI DM, HWANG W, SANTOSO S. Design of nanostructured biological materials through self-assembly of peptides and proteins. *Curr. Opin. Chem. Biol.* **2002**, 6, 865–871.

29 SILVA GA, CZEISLER C, NIECE KL, BENIASH E, HARRINGTON DA, KESSLER JA, STUPP SI. Selective differentiation of neural progenitor cells by high-epitope density nanofibers. *Science* **2004**, 303, 1352–1355.

30 KISIDAY J, JIN M, KURZ B, HUNG H, SEMINO C, ZHANG S, GRODZINSKY AJ. Self-assembling peptide hydrogel fosters chondrocyte extracellular matrix production and cell division: Implications for cartilage tissue repair. *Proc. Natl. Acad. Sci. U.S.A.* **2002**, 99, 9996–10 001.

31 RAYLEIGH JWG. *Lond. Edinburgh Dublin Phil. Mag.* **1882**, 14, 184.

32 ZELENY J. The electrical discharge from liquid points, and a hydrostatic method of measuring the electric intensity at their surface. *Phys. Rev.* **1914**, 3, 69–91.

33 ZELENY J. Instability of electrified liquid surface. *Phys. Rev.* **1917**, 10, 1–6.

34 DOLE M, MACK LL, HINES RL. Molecular beams of macroions. *J. Chem. Phys.* **1968**, 49, 2240–2249.

35 VONNEGUT B, NEUBAUER RL. *J. Colloid Sci.* **1952**, 7, 616.

36 DROZIN VG. The electrical dispersion of liquids as aerosols. *J. Colloid Sci.* **1955**, 7, 616.

37 *Patent 1,975,504.* **1934**.

38 *Patent 3,280,229.* **1966**.

39 BAUMGARTEN PK. Electrostatic spinning of acrylic microfibers. *J. Colloid Interface Sci.* **1971**, 36, 71–79.

40 LARRONDO L, MANLEY RSJ. Electrostatic fiber spinning from polymer melts. 1. Experimental-observations on fiber formation and properties. *J. Polym. Sci. Part B – Polym. Phys.* **1981**, 19, 909–920.

41 LARRONDO L, MANLEY RSJ. Electrostatic fiber spinning from polymer melts. 2. Examination of the flow field in an electrically driven jet. *J. Polym. Sci. Part B – Polym. Phys.* **1981**, 19, 921–932.

42 LARRONDO L, MANLEY RSJ. Electrostatic fiber spinning from polymer melts. 3. Electrostatic deformation of a pendant drop of polymer melt. *J. Polym. Sci. Part B – Polym. Phys.* **1981**, 19, 933–940.

43 KIDOAKI S, KWON IK, MATSUDA T. Mesoscopic spatial designs of nano- and microfiber meshes for tissue-engineering matrix and scaffold based on newly devised multilayering and mixing electrospinning techniques. *Biomaterials* **2005**, 26, 37–46.

44 STANKUS JJ, GUAN J, FUJIMOTO K, WAGNER WR. Microintegrating smooth muscle cells into a biodegradable, elastomeric fiber matrix. *Biomaterials* **2006**, 27, 735–744.

45 TAYLOR GI. Electrically driven jets. *Proc. R. Soc. (London)* **1969**, 453–475.

46 TAYLOR GI. Disintegration of water drops in an electric field. *Proc. R. Soc. (London)* **1964**, 383–397.

47 DOSHI J, RENEKER DH. Electro-spinning process and applications of electrospun fibers. *J. Electrostatics* **1995**, 35, 151–160.

48 FONG H, CHUN I, RENEKER DH. Beaded nanofibers formed during electrospinning. *Polymer* **1999**, 40, 4585–4592.

49 LIU HQ, HSIEH YL. Ultrafine fibrous cellulose membranes from electrospinning of cellulose acetate. *J. Polym. Sci. Part B – Polym. Phys.* **2002**, 40, 2119–2129.

50 DEITZEL JM, KLEINMEYER J, HARRIS D, TAN NCB. The effect of processing variables on the morphology of electrospun nanofibers and textiles. *Polymer* **2001**, 42, 261–272.

51 DEMIR MM, YILGOR I, YILGOR E, ERMAN B. Electrospinning of polyurethane fibers. *Polymer* **2002**, 43, 3303–3309.

52 ZONG XH, KIM K, FANG DF, RAN SF, HSIAO BS, CHU B. Structure and process relationship of electrospun bioabsorbable nanofiber membranes. *Polymer* **2002**, 43, 4403–4412.

53 LEE KH, KIM HY, KHIL MS, RA YM, LEE DR. Characterization of nano-structured poly(ε-caprolactone) nonwoven mats via electrospinning. *Polymer* **2003**, 44, 1287–1294.

54 CHOI JS, LEE SW, JEONG L, BAE SH, MIN BC, YOUK JH, PARK WH. Effect of organosoluble salts on the nanofibrous structure of electrospun poly(3-hydroxybutyrate-co-3-hydroxyvalerate). *Int. J. Biol. Macromol.* **2004**, 34, 249–256.

55 ZONG X, KIM K, FANG D, RAN S, HSIAO BS, CHU B. Structure and process relationship of electrospun bioabsorbable nanofiber membranes. *Polymer* **2002**, 43, 4403–4412.

56 SHIN YM, HOHMAN MM, BRENNER MP, RUTLEDGE GC. Experimental characterization of electrospinning: The electrically forced jet and instabilities. *Polymer* **2001**, 42, 9955–9967.

57 THERON A, ZUSSMAN E, YARIN AL. Electrostatic field-assisted alignment of electrospun nanofibres. *Nanotechnology* **2001**, 12, 384–390.

58 Li D, Ouyang G, McCann JT, Xia Y. Collecting electrospun nanofibers with patterned electrodes. *Nano Lett.* **2005**, 5, 913–916.

59 Li WJ, Mauck R, Cooper JA, Tuan RS. Engineering anisotropy in electrospun biodegradable nano-fibrous scaffolds for musculo-skeletal tissue engineering. In: 5th Combined Meeting of the Orthopaedic Research Societies of Canada, USA, Japan, and Europe, Banff, Alberta, Canada, 2004.

60 Muschler GF, Nakamoto C, Griffith LG. Engineering princi-ples of clinical cell-based tissue engineering. *J. Bone Joint Surg. Am.* **2004**, 86-A, 1541–1558.

61 Lee YH, Lee JH, An IG, Kim C, Lee DS, Lee YK, Nam JD. Electrospun dual-porosity structure and biodegradation morphology of montmorillonite reinforced PLLA nanocomposite scaffolds. *Biomaterials* **2005**, 26, 3165–3172.

62 Hubbell JA. Biomaterials in tissue engineering. *Biotechnology (N Y)* **1995**, 13, 565–576.

63 Tan EP, Ng SY, Lim CT. Tensile testing of a single ultrafine polymeric fiber. *Biomaterials* **2005**, 26, 1453–1456.

64 Sanders JE, Nicholson BS, Mitchell SB, Ledger RE. Polymer microfiber mechanical properties: A system for assessment and investigation of the link with fibrous capsule formation. *J. Biomed. Mater. Res. A* **2003**, 67, 1412–1416.

65 Grinnell F. Cellular adhesiveness and extracellular substrata. *Int. Rev. Cytol.* **1978**, 53, 65–144.

66 Giancotti FG, Ruoslahti E. Integrin signaling. *Science* **1999**, 285, 1028–1032.

67 Smetana K, Jr. Cell biology of hydrogels. *Biomaterials* **1993**, 14, 1046–1050.

68 Boyan BD, Hummert TW, Dean DD, Schwartz Z. Role of material surfaces in regulating bone and cartilage cell response. *Biomaterials* **1996**, 17, 137–146.

69 Loeser RF. Chondrocyte integrin expression and function. *Biorheology* **2000**, 37, 109–116.

70 Nikolovski J, Mooney DJ. Smooth muscle cell adhesion to tissue engineering scaffolds. *Biomaterials* **2000**, 21, 2025–2032.

71 Woo KM, Chen VJ, Ma PX. Nano-fibrous scaffolding architecture selectively enhances protein adsorption contributing to cell attachment. *J. Biomed. Mater. Res.* **2003**, 67A, 531–537.

72 Buckwalter JA, Mankin HJ. Articular cartilage: Tissue design and chondrocyte-matrix interactions. *Instr. Course Lect.* **1998**, 47, 477–486.

73 Scully SP, Lee JW, Ghert PMA, Qi W. The role of the extracellular matrix in articular chondrocyte regulation. *Clin. Orthop.* **2001**, S72–89.

74 Gray ML, Pizzanelli AM, Grodzinsky AJ, Lee RC. Mechanical and physiochemical determinants of the chondrocyte biosynthetic response. *J. Orthop. Res.* **1988**, 6, 777–792.

75 Li WJ, Jiang YJ, Tuan RS. Chon-drocyte phenotype in engineered fibrous matrix is regulated by fiber size. *Tissue Eng.* in the press.

76 Yang F, Murugan R, Wang S, Ramakrishna S. Electrospinning of nano/micro scale poly(L-lactic acid) aligned fibers and their potential in neural tissue engineering. *Biomaterials* **2005**, 26, 2603–2610.

77 Li WJ, Danielson KG, Alexander PG, Tuan RS. Biological response of chondrocytes cultured in three-dimensional nanofibrous poly(epsilon-caprolactone) scaffolds. *J. Biomed. Mater. Res. A* **2003**, 67, 1105–1114.

78 Burdon T, Smith A, Savatier P. Signalling, cell cycle and pluripotency in embryonic stem cells. *Trends Cell Biol.* **2002**, 12, 432–438.

79 Li S, Edgar D, Fassler R, Wadsworth W, Yurchenco PD. The role of laminin in embryonic cell polarization and tissue organization. *Dev. Cell* **2003**, 4, 613–624.

80 Nur-E-Kamal A, Ahmed I, Kamal J, Schindler M, Meiners S. Three-dimensional nanofibrillar surfaces promote self-renewal in mouse embryonic stem cells. *Stem Cells* **2005**, 24, 426–433.

81 PITTENGER MF, MACKAY AM, BECK SC, JAISWAL RK, DOUGLAS R, MOSCA JD, MOORMAN MA, SIMONETTI DW, CRAIG S, MARSHAK DR. Multi-lineage potential of adult human mesenchymal stem cells. *Science* **1999**, 284, 143–147.

82 ZUK PA, ZHU M, MIZUNO H, HUANG J, FUTRELL JW, KATZ AJ, BENHAIM P, LORENZ HP, HEDRICK MH. Multilineage cells from human adipose tissue: Implications for cell-based therapies. *Tissue Eng.* **2001**, 7, 211–228.

83 GOODWIN HS, BICKNESE AR, CHIEN SN, BOGUCKI BD, QUINN CO, WALL DA. Multilineage differentiation activity by cells isolated from umbilical cord blood: Expression of bone, fat, and neural markers. *Biol. Blood Marrow Transplant* **2001**, 7, 581–588.

84 YOUNG HE, STEELE TA, BRAY RA, HUDSON J, FLOYD JA, HAWKINS K, THOMAS K, AUSTIN T, EDWARDS C, CUZZOURT J, DUENZL M, LUCAS PA, BLACK AC, JR. Human reserve pluripotent mesenchymal stem cells are present in the connective tissues of skeletal muscle and dermis derived from fetal, adult, and geriatric donors. *Anat. Rec.* **2001**, 264, 51–62.

85 DE BARI C, DELL'ACCIO F, TYLZANOWSKI P, LUYTEN FP. Multipotent mesenchymal stem cells from adult human synovial membrane. *Arthritis Rheum.* **2001**, 44, 1928–1942.

86 LI WJ, TULI R, HUANG X, LAQUER-RIERE P, TUAN RS. Multilineage differentiation of human mesenchymal stem cells in a three-dimensional nanofibrous scaffold. *Biomaterials* **2005**, 26, 5158–5166.

87 SCHINDLER M, AHMED I, KAMAL J, NUR EKA, GRAFE TH, YOUNG CHUNG H, MEINERS S. A synthetic nanofibrillar matrix promotes in vivo-like organization and morphogenesis for cells in culture. *Biomaterials* **2005**, 26, 5624–5631.

88 NUR EKA, AHMED I, KAMAL J, SCHINDLER M, MEINERS S. Three

dimensional nanofibrillar surfaces induce activation of Rac. *Biochem. Biophys. Res. Commun.* **2005**, 331, 428–434.

89 FRENKEL SR, TOOLAN B, MENCHE D, PITMAN MI, PACHENCE JM. Chondrocyte transplantation using a collagen bilayer matrix for cartilage repair. *J. Bone Joint Surg. Br.* **1997**, 79, 831–836.

90 LEE CR, GRODZINSKY AJ, HSU HP, SPECTOR M. Effects of a cultured autologous chondrocyte-seeded type II collagen scaffold on the healing of a chondral defect in a canine model. *J. Orthop. Res.* **2003**, 21, 272–281.

91 MATTHEWS JA, WNEK GE, SIMPSON DG, BOWLIN GL. Electrospinning of collagen nanofibers. *Biomacromolecules* **2002**, 3, 232–238.

92 RHO KS, JEONG L, LEE G, SEO BM, PARK YJ, HONG SD, ROH S, CHO JJ, PARK WH, MIN BM. Electrospinning of collagen nanofibers: Effects on the behavior of normal human keratinocytes and early-stage wound healing. *Biomaterials* **2006**, 27, 1452–1461.

93 TELEMECO TA, AYRES C, BOWLIN GL, WNEK GE, BOLAND ED, COHEN N, BAUMGARTEN CM, MATHEWS J, SIMPSON DG. Regulation of cellular infiltration into tissue engineering scaffolds composed of sumicron diameter fibrils produced by electrospinning. *Acta Biomater.* **2005**, 1, 377–385.

94 BOLAND ED, MATTHEWS JA, PAWLOWSKI KJ, SIMPSON DG, WNEK GE, BOWLIN GL. Electrospinning collagen and elastin: Preliminary vascular tissue engineering. *Front Biosci.* **2004**, 9, 1422–1432.

95 LI M, MONDRINOS MJ, GANDHI MR, KO FK, WEISS AS, LELKES PI. Electrospun protein fibers as matrices for tissue engineering. *Biomaterials* **2005**, 26, 5999–6008.

96 MATTHEWS JA, BOLAND ED, WNEK GE, SIMPSON DG, BOWLIN GL. Electro-spinning of collagen type II: A feasibility study. *J. Bioactive Compatible Polymers* **2003**, 18, 125–134.

97 SHIELDS KJ, BECKMAN MJ, BOWLIN

GL, WAYNE JS. Mechanical properties and cellular proliferation of electrospun collagen type II. *Tissue Eng.* **2004**, 10, 1510–1517.

98 MA Z, KOTAKI M, YONG T, HE W, RAMAKRISHNA S. Surface engineering of electrospun polyethylene terephthalate (PET) nanofibers towards development of a new material for blood vessel engineering. *Biomaterials* **2005**, 26, 2527–2536.

99 MA Z, HE W, YONG T, RAMAKRISHNA S. Grafting of gelatin on electrospun poly(caprolactone) nanofibers to improve endothelial cell spreading and proliferation and to control cell Orientation. *Tissue Eng.* **2005**, 11, 1149–1158.

100 ZHANG Y, OUYANG H, LIM CT, RAMAKRISHNA S, HUANG ZM. Electrospinning of gelatin fibers and gelatin/PCL composite fibrous scaffolds. *J. Biomed. Mater. Res. B Appl. Biomater.* **2005**, 72, 156–165.

101 HUANG ZM, ZHANG YZ, RAMAKRISHNA S, LIM CT. Electrospinning and mechanical characterization of gelatin nanofibers. *Polymer* **2004**, 45, 5361–5368.

102 STITZEL J, LIU J, LEE SJ, KOMURA M, BERRY J, SOKER S, LIM G, VAN DYKE M, CZERW R, YOO JJ, ATALA A. Controlled fabrication of a biological vascular substitute. *Biomaterials* **2006**, 27, 1088–1094.

103 BUTTAFOCO L, KOLKMAN NG, ENGBERS-BUIJTENHUIJS P, POOT AA, DIJKSTRA PJ, VERMES I, FEIJEN J. Electrospinning of collagen and elastin for tissue engineering applications. *Biomaterials* **2006**, 27, 724–734.

104 NAGAPUDI K, BRINKMAN WT, THOMAS BS, PARK JO, SRINIVASARAO M, WRIGHT E, CONTICELLO VP, CHAIKOF EL. Viscoelastic and mechanical behavior of recombinant protein elastomers. *Biomaterials* **2005**, 26, 4695–4706.

105 JIN HJ, FRIDRIKH SV, RUTLEDGE GC, KAPLAN DL. Electrospinning Bombyx mori silk with poly(ethylene oxide). *Biomacromolecules* **2002**, 3, 1233–1239.

106 JIN HJ, CHEN J, KARAGEORGIOU V, ALTMAN GH, KAPLAN DL. Human bone marrow stromal cell responses on electrospun silk fibroin mats. *Biomaterials* **2004**, 25, 1039–1047.

107 OHGO K, ZHAO C, KOBAYASHI M, ASAKURA T. Preparation of non-woven nanofibers of Bombyx mori silk, Samia cynthia ricini silk and recombinant hybrid silk with electrospinning method. *Polymer* **2003**, 44, 841–846.

108 ZARKOOB S, EBY RK, RENEKER DH, HUDSON SD, ERTLEY D, ADAMS WW. Structure and morphology of electrospun silk nanofibers. *Polymer* **2004**, 45, 3973–3977.

109 MIN BM, LEE G, KIM SH, NAM YS, LEE TS, PARK WH. Electrospinning of silk fibroin nanofibers and its effect on the adhesion and spreading of normal human keratinocytes and fibroblasts in vitro. *Biomaterials* **2004**, 25, 1289–1297.

110 MIN BM, JEONG L, NAM YS, KIM JM, KIM JY, PARK WH. Formation of silk fibroin matrices with different texture and its cellular response to normal human keratinocytes. *Int. J. Biol. Macromol.* **2004**, 34, 281–288.

111 PARK WH, JEONG L, YOO DI, HUDSON S. Effect of chitosan on morphology and conformation of electrospun silk fibroin nanofibers. *Polymer* **2004**, 45, 7151–7157.

112 SILVERMAN RP, PASSARETTI D, HUANG W, RANDOLPH MA, YAREMCHUK MJ. Injectable tissue-engineered cartilage using a fibrin glue polymer. *Plast. Reconstr. Surg.* **1999**, 103, 1809–1818.

113 WORSTER AA, BROWER-TOLAND BD, FORTIER LA, BENT SJ, WILLIAMS J, NIXON AJ. Chondrocytic differentiation of mesenchymal stem cells sequentially exposed to transforming growth factor-beta1 in monolayer and insulin-like growth factor-I in a three-dimensional matrix. *J. Orthop. Res.* **2001**, 19, 738–749.

114 FORTIER LA, MOHAMMED HO, LUST G, NIXON AJ. Insulin-like growth factor-I enhances cell-based repair of articular cartilage. *J. Bone Joint Surg. Br.* **2002**, 84, 276–288.

115 WNEK GE, CARR ME, SIMPSON DG, BOWLIN GL. Electrospinning of nanofiber fibrinogen structures. *Nano Lett.* **2003**, 3, 213–216.

116 SECHRIEST VF, MIAO YJ, NIYIBIZI C, WESTERHAUSEN-LARSON A, MATTHEW HW, EVANS CH, FU FH, SUH JK. GAG-augmented polysaccharide hydrogel: A novel biocompatible and biodegradable material to support chondrogenesis. *J. Biomed. Mater. Res.* **2000**, 49, 534–541.

117 LAHIJI A, SOHRABI A, HUNGERFORD DS, FRONDOZA CG. Chitosan supports the expression of extracellular matrix proteins in human osteoblasts and chondrocytes. *J. Biomed. Mater. Res.* **2000**, 51, 586–595.

118 LU JX, PRUDHOMMEAUX F, MEUNIER A, SEDEL L, GUILLEMIN G. Effects of chitosan on rat knee cartilages. *Biomaterials* **1999**, 20, 1937–1944.

119 DENUZIERE A, FERRIER D, DAMOUR O, DOMARD A. Chitosan-chondroitin sulfate and chitosan-hyaluronate polyelectrolyte complexes: Biological properties. *Biomaterials* **1998**, 19, 1275–1285.

120 MIN BM, LEE SW, LIM JN, YOU Y, LEE TS, KANG PH, PARK WH. Chitin and chitosan nanofibers: Electrospinning of chitin and deacetylation of chitin nanofibers. *Polymer* **2004**, 45, 7137–7142.

121 SUBRAMANIAN A, LIN HY, VU D, LARSEN G. Synthesis and evaluation of scaffolds prepared from chitosan fibers for potential use in cartilage tissue engineering. *Biomed. Sci. Instrum.* **2004**, 40, 117–122.

122 GENG X, KWON OH, JANG J. Electrospinning of chitosan dissolved in concentrated acetic acid solution. *Biomaterials* **2005**, 26, 5427–5432.

123 CAMPOCCIA D, DOHERTY P, RADICE M, BRUN P, ABATANGELO G, WILLIAMS DF. Semisynthetic resorbable materials from hyaluronan esterification. *Biomaterials* **1998**, 19, 2101–2127.

124 VERCRUYSSE KP, MARECAK DM, MARECEK JF, PRESTWICH GD. Synthesis and in vitro degradation of new polyvalent hydrazide cross-linked hydrogels of hyaluronic acid. *Bioconjug. Chem.* **1997**, 8, 686–694.

125 AIGNER J, TEGELER J, HUTZLER P, CAMPOCCIA D, PAVESIO A, HAMMER C, KASTENBAUER E, NAUMANN A. Cartilage tissue engineering with novel nonwoven structured biomaterial based on hyaluronic acid benzyl ester. *J. Biomed. Mater. Res.* **1998**, 42, 172–181.

126 GRIGOLO B, ROSETI L, FIORINI M, FINI M, GIAVARESI G, ALDINI NN, GIARDINO R, FACCHINI A. Transplantation of chondrocytes seeded on a hyaluronan derivative (hyaff-11) into cartilage defects in rabbits. *Biomaterials* **2001**, 22, 2417–2424.

127 UM IC, FANG D, HSIAO BS, OKAMOTO A, CHU B. Electro-spinning and electro-blowing of hyaluronic acid. *Biomacromolecules* **2004**, 5, 1428–1436.

128 HOLLINGER JO. Preliminary report on the osteogenic potential of a biodegradable copolymer of polyactide (PLA) and polyglycolide (PGA). *J. Biomed. Mater. Res.* **1983**, 17, 71–82.

129 BOLAND ED, WNEK GE, SIMPSON DG, PAWLOWSKI KJ, BOWLIN GL. Tailoring tissue engineering scaffolds using electrostatic processing techniques: A study of poly(glycolic acid) electrospinning. *J. Macromol. Sci. – Pure Appl. Chem.* **2001**, A38, 1231–1243.

130 BOLAND ED, TELEMECO TA, SIMPSON DG, WNEK GE, BOWLIN GL. Utilizing acid pretreatment and electrospinning to improve biocompatibility of poly(glycolic acid) for tissue engineering. *J. Biomed. Mater. Res. B Appl. Biomater.* **2004**, 71, 144–152.

131 MIDDLETON JC, TIPTON AJ. Synthetic biodegradable polymers as orthopedic devices. *Biomaterials* **2000**, 21, 2335–2346.

132 ZONG X, RAN S, KIM KS, FANG D, HSIAO BS, CHU B. Structure and morphology changes during in vitro degradation of electrospun poly-(glycolide-co-lactide) nanofiber membrane. *Biomacromolecules* **2003**, 4, 416–423.

133 KO FK, LI WJ, LAURENCIN CT. Electrospun nanofibrous structure for tissue engineering. In: Sixth World Biomaterials Congress, Kamuela, Hawaii, USA, 2000.

134 KATTI DS, ROBINSON KW, KO FK, LAURENCIN CT. Bioresorbable nanofiber-based systems for wound healing and drug delivery: Optimization of fabrication parameters. *J. Biomed. Mater. Res. B* **2004**, 70, 286–296.

135 BERKLAND C, PACK DW, KIM KK. Controlling surface nano-structure using flow-limited field-injection electrostatic spraying (FFESS) of poly(D,L-lactide-co-glycolide). *Biomaterials* **2004**, 25, 5649–5658.

136 PITT CG. Poly-epsilon-caprolactone and its copolymers. In: *Biodegradable Polymers as Drug Delivery Systems.* CHASIN M, LANGER R (Eds.), Marcel Dekker, New York, **1990**, pp. 71–120.

137 YOSHIMOTO H, SHIN YM, TERAI H, VACANTI JP. A biodegradable nanofiber scaffold by electrospinning and its potential for bone tissue engineering. *Biomaterials* **2003**, 24, 2077–2082.

138 SHIN M, YOSHIMOTO H, VACANTI JP. In vivo bone tissue engineering using mesenchymal stem cells on a novel electrospun nanofibrous scaffold. *Tissue Eng.* **2004**, 10, 33–41.

139 SHIN M, ISHII O, SUEDA T, VACANTI JP. Contractile cardiac grafts using a novel nanofibrous mesh. *Biomaterials* **2004**, 25, 3717–3723.

140 LI WJ, TULI R, OKAFOR C, DERFOUL A, DANIELSON KG, HALL DJ, TUAN RS. A three-dimensional nanofibrous scaffold for cartilage tissue engineering using human mesenchymal stem cells. *Biomaterials* **2005**, 26, 599–609.

141 KHIL MS, BHATTARAI SR, KIM HY, KIM SZ, LEE KH. Novel fabricated matrix via electrospinning for tissue engineering. *J. Biomed. Mater. Res. B Appl. Biomater.* **2004**, 72, 117–124.

142 VENUGOPAL J, MA LL, YONG T, RAMAKRISHNA S. In vitro study of smooth muscle cells on polycaprolactone and collagen nanofibrous matrices. *Cell Biol. Int.* **2005**, 29, 861–867.

143 ZHANG YZ, VENUGOPAL J, HUANG ZM, LIM CT, RAMAKRISHNA S. Characterization of the surface biocompatibility of the electrospun PCL-collagen nanofibers using fibroblasts. *Biomacromolecules* **2005**, 6, 2583–2589.

144 VENUGOPAL J, RAMAKRISHNA S. Biocompatible nanofiber matrices for the engineering of a dermal substitute for skin regeneration. *Tissue Eng.* **2005**, 11, 847–854.

145 VAZ CM, TUIJL SV, BOUTEN CVC, BAAIJENS FPT. Design of scaffolds for blood vessel tissue engineering using a multi-layering electrospinning technique. *Acta Biomater.* **2005**, 1, 575–582.

146 FUJIHARA K, KOTAKI M, RAMAKRISHNA S. Guided bone regeneration membrane made of polycaprolactone/ calcium carbonate composite nano-fibers. *Biomaterials* **2005**, 26, 4139–4147.

147 ZONG X, RAN S, FANG D, HSIAO BS, CHU B. Control of structure, morphology and property in electrospun poly(glycolide-co-lactide) non-woven membranes via post-draw treatments. *Polymer* **2003**, 44, 4959–4967.

148 KENAWY EL R, LAYMAN JM, WATKINS JR, BOWLIN GL, MATTHEWS JA, SIMPSON DG, WNEK GE. Electrospinning of poly(ethylene-co-vinyl alcohol) fibers. *Biomaterials* **2003**, 24, 907–913.

149 KHIL MS, CHA DI, KIM HY, KIM IS, BHATTARAI N. Electrospun nanofibrous polyurethane membrane as wound dressing. *J. Biomed. Mater. Res. B Appl. Biomater.* **2003**, 67, 675–679.

150 ELISSEEFF J, ANSETH K, SIMS D, McINTOSH W, RANDOLPH M, LANGER R. Transdermal photopolymerization for minimally invasive implantation. *Proc. Natl. Acad. Sci. U.S.A.* **1999**, 96, 3104–3107.

151 ELISSEEFF J, ANSETH K, SIMS D, McINTOSH W, RANDOLPH M, YAREMCHUK M, LANGER R. Transdermal photopolymerization of poly(ethylene oxide)-based injectable hydrogels for tissue-engineered cartilage. *Plast. Reconstr. Surg.* **1999**, 104, 1014–1022.

152 HUANG L, NAGAPUDI K, APKARIAN RP, CHAIKOF EL. Engineered collagen-

PEO nanofibers and fabrics. *J. Biomater. Sci. Polym. Ed.* **2001**, 12, 979–993.

153 DUAN B, DONG C, YUAN X, YAO K. Electrospinning of chitosan solutions in acetic acid with poly(ethylene oxide). *J. Biomater. Sci. Polym. Ed.* **2004**, 15, 797–811.

154 ZONG X, LI S, CHEN E, GARLICK B, KIM KS, FANG D, CHIU J, ZIMMERMAN T, BRATHWAITE C, HSIAO BS, CHU B. Prevention of postsurgery-induced abdominal adhesions by electrospun bioabsorbable nanofibrous poly(lactide-co-glycolide)-based membranes. *Ann. Surg.* **2004**, 240, 910–915.

155 KENAWY EL R, BOWLIN GL, MANSFIELD K, LAYMAN J, SIMPSON DG, SANDERS EH, WNEK GE. Release of tetracycline hydrochloride from electrospun poly(ethylene-co-vinylacetate), poly(lactic acid), and a blend. *J. Controlled Release* **2002**, 81, 57–64.

156 DING B, KIMURA E, SATO T, FUJITA S, SHIRATORI S. Fabrication of blend biodegradable nanofibrous nonwoven mats via multi-jet electrospinning. *Polymer* **2004**, 45, 1895–1902.

157 HE W, MA Z, YONG T, TEO WE, RAMAKRISHNA S. Fabrication of collagen-coated biodegradable polymer nanofiber mesh and its potential for endothelial cells growth. *Biomaterials* **2005**, 26, 7606–7615.

158 MIN BM, YOU Y, KIM JM, LEE SJ, PARK WH. Formation of nanostructured poly(lactic-co-glycolic acid)/chitin matrix and its cellular response to normal human keratinocytes and fibroblasts. *Carbohydr. Polymers* **2004**, 57, 285–292.

159 XU C, YANG F, WANG S, RAMAKRISHNA S. In vitro study of human vascular endothelial cell function on materials with various surface roughness. *J. Biomed. Mater. Res. A* **2004**, 71, 154–161.

160 XU CY, INAI R, KOTAKI M, RAMAK-RISHNA S. Aligned biodegradable nanofibrous structure: A potential scaffold for blood vessel engineering. *Biomaterials* **2004**, 25, 877–886.

161 RIBOLDI SA, SAMPAOLESI M, NEUENSCHWANDER P, COSSU G, MANTERO S. Electrospun degradable polyesterurethane membranes: Potential scaffolds for skeletal muscle tissue engineering. *Biomaterials* **2005**, 26, 4606–4615.

162 NEUMANN T, HAUSCHKA SD, SANDERS JE. Tissue engineering of skeletal muscle using polymer fiber arrays. *Tissue Eng.* **2003**, 9, 995–1003.

163 CRONIN EM, THURMOND FA, BASSEL-DUBY R, WILLIAMS RS, WRIGHT WE, NELSON KD, GARNER HR. Protein-coated poly(L-lactic acid) fibers provide a substrate for differentiation of human skeletal muscle cells. *J. Biomed. Mater. Res. A* **2004**, 69, 373–381.

164 LEE CH, SHIN HJ, CHO IH, KANG YM, KIM IA, PARK KD, SHIN JW. Nanofiber alignment and direction of mechanical strain affect the ECM production of human ACL fibroblast. *Biomaterials* **2005**, 26, 1261–1270.

165 YANG F, XU CY, KOTAKI M, WANG S, RAMAKRISHNA S. Characterization of neural stem cells on electrospun poly(L-lactic acid) nanofibrous scaffold. *J. Biomater. Sci. Polym. Ed.* **2004**, 15, 1483–1497.

166 TATSUYAMA K, MAEZAWA Y, BABA H, IMAMURA Y, FUKUDA M. Expression of various growth factors for cell proliferation and cytodifferentiation during fracture repair of bone. *Eur. J. Histochem.* **2000**, 44, 269–278.

167 KOST J, LANGER R. Responsive polymeric delivery systems. *Adv. Drug Deliv. Rev.* **2001**, 46, 125–148.

168 BOLAND ED, TELEMECO TA, SIMPSON DG, WNEK GE, BOWLIN GL. Utilizing acid pretreatment and electrospinning to improve biocompatibility of poly(glycolic acid) for tissue engineering. *J. Biomed. Mater. Res. B* **2004**, 71, 144–152.

169 BINI TB, GAO S, XU X, WANG S, RAMAKRISHNA S, LEONG KW. Peripheral nerve regeneration by microbraided poly(L-lactide-co-glycolide) biodegradable polymer fibers. *J. Biomed. Mater. Res. A* **2004**, 68, 286–295.

170 ZHANG Y, OUYANG H, LIM CT, RAMAKRISHNA S, HUANG ZM.

Electrospinning of gelatin fibers and gelatin/PCL composite fibrous scaffolds. *J. Biomed. Mater. Res. B* **2005**, 72, 156–165.

171 FERTALA A, HAN WB, KO FK. Mapping critical sites in collagen II for rational design of gene-engineered proteins for cell-supporting materials. *J. Biomed. Mater. Res.* **2001**, 57, 48–58.

172 YANG F, MURUGAN R, WANG S, RAMAKRISHNA S. Electrospinning of nano/micro scale poly(L-lactic acid) aligned fibers and their potential in neural tissue engineering. *Biomaterials* **2005**, 26, 2603–2610.

173 XU C, YANG F, WANG S, RAMAK-RISHNA S. In vitro study of human vascular endothelial cell function on materials with various surface roughness. *J. Biomed. Mater. Res. A* **2004**, 71, 154–161.

174 ZONG X, BIEN H, CHUNG CY, YIN L, FANG D, HSIAO BS, CHU B, ENTCHEVA E. Electrospun fine-textured scaffolds for heart tissue constructs. *Biomaterials* **2005**, 26, 5330–5338.

175 KIM K, YU M, ZONG X, CHIU J, FANG D, SEO YS, HSIAO BS, CHU B, HADJIARGYROU M. Control of degradation rate and hydrophilicity in electrospun non-woven poly(D,L-lactide) nanofiber scaffolds for biomedical applications. *Biomaterials* **2003**, 24, 4977–4985.

176 BADAMI AS, KREKE MR, THOMPSON MS, RIFFLE JS, GOLDSTEIN AS. Effect of fiber diameter on spreading, proliferation, and differentiation of osteoblastic cells on electrospun poly(lactic acid) substrates. *Biomaterials* **2006**, 27, 596–606.

177 LEE YH, LEE JH, AN IG, KIM C, LEE DS, LEE YK, NAM JD. Electrospun dual-porosity structure and biodegradation morphology of Montmorillonite reinforced PLLA nanocomposite scaffolds. *Biomaterials* **2005**, 26, 3165–3172.

178 MO XM, XU CY, KOTAKI M, RAMAKRISHNA S. Electrospun P (LLA-CL) nanofiber: A biomimetic extracellular matrix for smooth muscle cell and endothelial cell proliferation. *Biomaterials* **2004**, 25, 1883–1890.

179 XU C, INAI R, KOTAKI M, RAMAKRISHNA S. Electrospun nanofiber fabrication as synthetic extracellular matrix and its potential for vascular tissue engineering. *Tissue Eng.* **2004**, 10, 1160–1168.

180 INOGUCHI H, KWON IK, INOUE E, TAKAMIZAWA K, MAEHARA Y, MATSUDA T. Mechanical responses of a compliant electrospun poly(l-lactide-co-epsilon-caprolactone) small-diameter vascular graft. *Biomaterials* **2006**, 27, 1470–1478.

181 KWON IK, KIDOAKI S, MATSUDA T. Electrospun nano- to microfiber fabrics made of biodegradable copolyesters: Structural characteristics, mechanical properties and cell adhesion potential. *Biomaterials* **2005**, 26, 3929–3939.

182 CHUA KN, LIM WS, ZHANG P, LU H, WEN J, RAMAKRISHNA S, LEONG KW, MAO HQ. Stable immobilization of rat hepatocyte spheroids on galactosylated nanofiber scaffold. *Biomaterials* **2005**, 26, 2537–2547.

183 NAIR LS, BHATTACHARYYA S, BENDER JD, GREISH YE, BROWN PW, ALLCOCK HR, LAURENCIN CT. Fabrication and optimization of methylphenoxy substituted polyphosphazene nano-fibers for biomedical applications. *Biomacromolecules* **2004**, 5, 2212–2220.

184 LEE IS, KWON OH, MENG W, KANG IK. Nanofabrication of microbial polyester by electrospinning promotes cell attachment. *Macromol. Res.* **2004**, 12, 374–378.

185 ITO Y, HASUDA H, KAMITAKAHARA M, OHTSUKI C, TANIHARA M, KANG IK, KWON OH. A composite of hydroxyapatite with electrospun biodegradable nanofibers as a tissue engineering material. *J. Biosci. Bioeng.* **2005**, 100, 43–49.

186 STANKUS JJ, GUAN J, WAGNER WR. Fabrication of biodegradable elastomeric scaffolds with sub-micron morphologies. *J. Biomed. Mater. Res. A* **2004**, 70, 603–614.

187 BHATTARAI SR, BHATTARAI N, YI HK, HWANG PH, CHA DI, KIM HY. Novel biodegradable electrospun membrane:

Scaffold for tissue engineering. *Biomaterials* **2004**, 25, 2595–2602.

188 KHIL MS, CHA DI, KIM HY, KIM IS, BHATTARAI N. Electrospun nanofibrous polyurethane membrane as wound dressing. *J. Biomed. Mater. Res. B* **2003**, 67, 675–679.

189 MATSUDA T, IHARA M, INOGUCHI H, KWON IK, TAKAMIZAWA K, KIDOAKI S. Mechano-active scaffold design of small-diameter artificial graft made of electrospun segmented polyurethane fabrics. *J. Biomed. Mater. Res. A* **2005**, 73, 125–131.

190 KIDOAKI S, KWON IK, MATSUDA T. Structural features and mechanical properties of in situ-bonded meshes of segmented polyurethane electrospun from mixed solvents. *J. Biomed. Mater. Res. B Appl. Biomater.* **2006**, 76, 219–229.

191 MA Z, KOTAKI M, YONG T, HE W, RAMAKRISHNA S. Surface engineering of electrospun polyethylene terephthalate (PET) nanofibers towards development of a new material for blood vessel engineer-ing. *Biomaterials* **2005**, 26, 2527–2536.

192 SUN T, MAI S, NORTON D, HAYCOCK JW, RYAN AJ, MacNEIL S. Self-organization of skin cells in three-dimensional electrospun polystyrene scaffolds. *Tissue Eng.* **2005**, 11, 1023–1033.

193 BOLAND ED, COLEMAN BD, BARNES CP, SIMPSON DG, WNEK GE, BOWLIN GL. Electrospinning polydioxanone for biomedical applications. *Acta Biomater.* **2005**, 1, 115–123.

194 CASPER CL, YAMAGUCHI N, KIICK KL, RABOLT JF. Functionalizing electrospun fibers with biologically relevant macromolecules. *Biomacromolecules* **2005**, 6, 1998–2007.

195 BHATTARAI N, EDMONDSON D, VEISEH O, MATSEN FA, ZHANG M. Electrospun chitosan-based nano-fibers and their cellular compatibil-ity. *Biomaterials* **2005**, 26, 6176–6184.

196 SON WK, YOUK JH, PARK WH. Preparation of ultrafine oxidized cellulose mats via electrospinning. *Biomacromolecules* **2004**, 5, 197–201.

4
Nanofibrous Scaffolds and their Biological Effects

Laura A. Smith, Jonathan A. Beck, and Peter X. Ma

4.1
Overview

Natural extracellular matrix (ECM) contains nanofibrous structures. To develop optimal scaffolds (synthetic temporary ECMs) for tissue engineering/regeneration, researchers mimic the natural ECM to recreate fibrous structures at this size scale. These nanofibrous scaffolds may eventually provide a better environment for tissue formation. Three different approaches toward the formation of nanofibrous scaffolds have emerged: self-assembly, electrospinning and phase separation. Each of these approaches is different and has a unique set of characteristics, which lends to its development as a scaffolding system. For instance, self-assembly can generate small diameter nanofibers in the lowest end of the range of natural extracellular matrix collagen, while electrospinning has generated large diameter nanofibers, often on the upper end of the range of natural ECM collagen. Phase separation, however, has generated nanofibers in the same range as natural collagen and allows for the design of macropore structures. Utilizing these three techniques, composite scaffolds have also been formed that contain minerals and biological factors to enable the scaffolds to more effectively mimic the natural ECM's bioactivities. These nanofibrous scaffolds have been shown to promote cellular attachment, proliferation and differentiation when compared to more traditional scaffolds that do not have nanofibrous structures. Nanofibrous scaffolds have been used to engineer various tissues such as cartilage, bone, vascular, cardiac, and neural tissues. This chapter briefly reviews the three fabrication techniques for nanofibrous scaffolds and their applications in tissue engineering.

4.2
Introduction

Tissue engineering is an interdisciplinary field that applies the principles of engineering and the life sciences to the development of biological substitutes that restore, maintain, or improve tissue function [1]. Essentially, there are three possible

Nanotechnologies for the Life Sciences Vol. 9
Tissue, Cell and Organ Engineering. Edited by Challa S. S. R. Kumar
Copyright © 2006 WILEY-VCH Verlag GmbH & Co. KGaA, Weinheim
ISBN: 3-527-31389-3

approaches to tissue engineering [1, 2]: use of isolated cells or cell substitutes to replace the cells that supply a needed function; delivery of tissue-inducing substances such as growth factors to a targeted location; and growing cells in a three-dimensional (3D) scaffold. The first two approaches may be suitable for small, well-contained defects. However, only the third, using a scaffold to direct cell growth, is suitable to engineer larger tissue blocks with predesigned shapes. As such, more tissue engineers are utilizing scaffolds in their studies of neo-tissue formation.

With this approach, biomaterials play a pivotal role [3, 4]. To properly direct cell growth and tissue formation, scaffolds should have certain vital characteristics. The scaffold should provide a suitable surface for the attachment, proliferation, and differentiation of cells and provide an appropriate 3D template to guide the tissue growth to its final shape [5, 6]. To enhance cell adhesion and subsequent tissue formation, a high surface-to-volume ratio is desirable [7, 8]. An open porous network with a suitable pore size should be contained in the scaffold to allow for uniform cell seeding and for mass transport of signaling molecules, nutrients, and removal of metabolic waste. The scaffold should also degrade at the rate of tissue formation, yet have enough mechanical strength to provide temporary support while the tissue is forming. Biocompatibility is important. Neither the scaffold nor its degradation products should be toxic to cells. Finally, because of its importance in cellular behavior with respect to morphology, cytoskeletal structure and functionality, the scaffold should allow cellular interactions similar to those in the natural extracellular matrix (ECM) [9–11]. To meet all these criteria, tissue engineering scaffolds are often designed to mimic the natural ECM until the host cells repopulate and synthesize a new matrix [12, 13].

The most abundant protein in the ECM is collagen. Over 25 distinct collagen chains have been identified, the most abundant of which is type I [14]. Type I collagen molecules are composed of three collagen polypeptide chains wound around each other to form a ropelike superhelix. Collagen molecules assemble further into higher-order polymers called collagen fibrils, which in turn are assembled into collagen fibers. The diameter of collagen fibers typically ranges from 50 to 500 nm.

Type I collagen, as a result of being the primary component of the ECM, has been used in many medical materials, particularly in soft tissue repair [15]. Type I collagen is relatively bioinert because of its helical structure and because the primary sequence of type I collagen is well-conserved across species lines [16]. Despite these advantages, the use of collagen in tissue engineering scaffolds remains questionable for the following reasons: concern of pathogen transmission, difficulties in handling, and limited control of mechanical properties, biodegradability, and batch-to-batch consistency [17]. To combat this, several synthetic tissue engineering scaffolds with nano-scale structures are being developed to mimic the ECM. The success of these scaffolds hinges on their ability to replicate the dimensional scale of living tissue in both the micro- and nanometer scale, which will aid in the maintenance of cell phenotype and provide a structurally, mechanically and biologically compatible cell–material interface [18].

As you progress through this chapter, the methods of forming nanofibrous scaffolds are first addressed along with modification methods to further tailor the scaffolds for biological applications. This is followed by the biological effects of nanofibrous architecture on cultured cells and the neo-tissue formation. The chapter follows the linear development of tissue engineering from scaffold to tissue formation, and provides the most up-to-date comprehensive review of the field.

4.3
Methods of Formation

Three principle methods have been employed in the fabrication of nanofibrous scaffolds for use in tissue engineering. They are self-assembly, electrospinning, and phase separation. Although very different from each other, these approaches are at the cutting edge of technology. Each method is detailed in this section.

4.3.1
Electrospinning

Since it was first patented in the 1930s, electrospinning has been used in polymer-processing [19]. It is a relatively simple process in which an electric field is used to draw a polymer solution (melt) from an orifice to a collector. Application of a high voltage (typically 5–30 kV) to the tip of a needle causing the polymer to form a Taylor cone is required for electrospinning. When the electrical field strength exceeds the surface tension (surpassing a critical voltage), a fiber jet is ejected from the Taylor cone to the collector (Fig. 4.1). The solvent evaporates from the polymer solution as the jet travels through the air. This process leaves behind a randomly oriented fiber matrix on the electrically grounded target [20–23].

Electrospinning typically is used to produce thin two-dimensional (2D) non-woven sheets; however, thicker 3D meshes are possible [24]. To maintain nano-

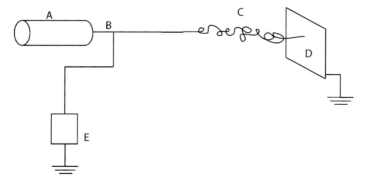

Fig. 4.1. Schematic diagram of a typical electrospinning apparatus: (A) Polymer solution loaded into a syringe; (B) syringe needle; (C) polymer nanofiber jet; (D) grounded target, which can be either stationary or rotating; (E) power supply.

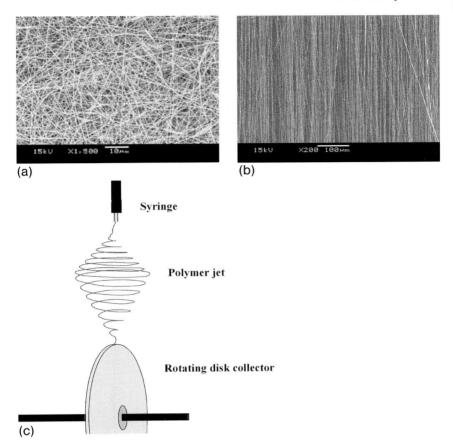

Fig. 4.2. (a) Randomly oriented nanofibers are created when a static collector is used. (b) Aligned nanofibers can be manufactured using a rotating collector. (Reprinted with permission from Ref. [25] © 2005, Elsevier.) (c) Diagram of a rotating collector. (Reprinted with permission from Ref. [27] © 2004, Elsevier.)

scale diameters throughout these meshes, a low extrusion rate must be used, leading to long fabrication times. It has been proposed that 3D meshes could more quickly be produced using multiple orifices simultaneously [24]. In this type of arrangement, the rate of mesh formation will be proportional to the number of orifices. Figure 4.2(a) shows SEM micrograms of a typical electrospun nanofibrous mesh [25]. However, electrospinning has yet to produce designed pore structures. Despite such shortcomings, electrospinning has been widely explored due to its capacity of generating nanofibers and the fabrication simplicity.

Electrospinning has been used to form scaffolds from many different biomaterials, including poly(lactic-*co*-glycolic acid) (PLGA), poly(caprolactone) (PCL), poly(ethylene oxide) (PEO) and poly(vinyl alcohol) (PVA) in water or organic solvent [26]. Nanofibrous mats of collagen, silk protein, elastin-mimetic peptide, fibrinogen, casein and lipase enzyme, and even DNA have been formed using electrospinning [20–23]. The use of electrospinning with regard to naturally occurring biomaterials has been limited in comparison to synthetic polymers because of the difficulty in finding appropriate solvents.

Typically, electrospun fibers produced are randomly oriented on the collection plate. However, random fiber orientation is not always ideal. Many tissues are anisotropic and have a clearly defined directionality. Examples of anisotropic tissues include muscle, bone, blood vessels, and nerves. Several methods have been developed to create mats with oriented fibers for use in these types of tissue engineering projects. The first method involves the use of a rotating drum as the grounded target rather than a static plate. As the drum spins, the fibers naturally align themselves (Fig. 4.2) [27]. A modification on the rotating drum concept that offers more versatility utilizes modified polymer solution system pushed through a microfluidic chamber in close proximity to a rotating collection grid. This system achieves variable fiber alignment from linear to S shaped curves dependent on the rate of collection grid rotation [28].

Besides dynamic collection grids, post-drawing can also be used to create oriented nanofibers [29]. In this method, the fibers are prepared using a static collection grid. Following electrospinning, the mat is drawn by stretching and then annealed under constant strain. Mats created in this manner not only contain uniaxial oriented fibers but also crystal orientation. Uniaxial oriented fibers can also be created with a special collector composed of two pieces of electrically conductive substrates separated by a gap [30]. The electrostatic fields created by the separated conductors cause the fibers to bridge the gap perpendicularly. This method is limited by the electrostatic interactions between the conducting collectors, meaning that only narrow meshes are possible.

The dimensions of electrospun fibers vary, depending on the material used. However, in general, the diameters of electrospun fibers are at least one order of magnitude smaller than those produced by conventional extrusion techniques [24]. Yet, this is still generally at the upper end of the diameters present in the native ECM. Ko and coworkers created nanofibrous PLGA matrices with fibers having diameters between 500 and 800 nm [26], while Vacanti and coworkers fabricated PCL matrices with an average fiber diameter of 400 nm ($\pm$200 nm) [31]. Chaikof and coworkers have also created electrospun nanofibers, with diameters in the 100–150 nm range, by combining collagen and PEO [32]. Pure collagen solutions have also been fabricated using electrospinning techniques [33–35]. Bowlin and coworkers have demonstrated electrospun collagen, with an average diameter of 250 nm, that exhibits the 67-nm banding typical of native collagen fibrils [34].

Although electrospinning is a relatively simple and quick method that can be applied to many different types of biomaterials, there is inherent difficulty in creating scaffolds with well-organized pore architectures with complex geometries. The diameter of most electrospun fibers is near the upper limit for the natural size of collagen fibers (50–500 nm) with some even reaching into the micro-scale. These are the two primary shortcomings of electrospinning.

4.3.2
Self-assembly

Although molecular self-assembly is a rather new laboratory technique for the formation of nano-scale scaffolds, it is prevalent in the natural world. Molecular self-assembly is a "bottom up", rather than a "top down" approach. Self-assembly is described as the reversible process by which preexisting parts are organized into an ordered system. By definition, molecular self-assembly is the spontaneous association of molecules under equilibrium conditions into stable, structurally well-defined aggregates joined by non-covalent bonds [36]. Although non-covalent bonds such as hydrogen bonds, ionic bonds, van der Waals interactions, and hydrophobic interactions are rather weak in isolation, taken in concert they govern the self-assembly of biological macromolecules [37].

As mentioned, self-assembled molecules are ubiquitous in nature. One example of molecules that readily self-assemble into higher order structures is phospholipids. Phospholipids are the principal component of the plasma membrane in cells. In aqueous solutions, phospholipids readily self-assemble to form several structures, including vesicles, micelles, and tubules. One advantage of molecular self-assembly is that fiber diameters can be produced that are much smaller than those produced using electrospinning.

Schnur et al. [38] pioneered the work of molecular self-assembly using lipid tubules. Several molecule types have been utilized in molecular self-assembly. Zhang, Rich, Holmes, et al. [39] have created self-assembled structures using ionic self-complementary oligopeptides. These oligopeptides contain alternating regions of hydrophobic and hydrophilic amino acids, and, in water, they freely organize themselves into stable β-sheet structures. Exposure to monovalent alkaline cations or physiological conditions, the oligopeptides can assemble into hydrogels with interwoven 10–20 nm diameter nanofibers and 50–200 nm diameter pores [39–41].

Peptide-amphiphiles (PA) have also been fabricated and will self-assemble to form nano-structured fibers [42, 43]. Stupp and coworkers [42] demonstrated that a molecule composed of a carbon alkyl tail and several peptide regions will self-assemble to form nanofibers with a diameter of 7.6 ± 1 nm. The nanofibrous scaffolds formed by this method are evident in Fig. 4.3. More recently, Stupp and coworkers [44] have reported preliminary results that indicate that PA self-assembled molecules can be used to encourage angiogenesis.

The PA is synthesized using solid-phase chemistry that ends with the alkylation of the N-terminus of the peptide and contains the following features: a long alkyl tail that conveys hydrophobic characteristics to the molecule; four consecutive cysteine residues that form disulfide bonds to polymerize the structure; a linker region containing three glycine residues to provide the hydrophilic head group the flexibility from the rigid crosslinked regions; a phosphorylated serine residue that interacts strongly with calcium ions and helps to direct mineralization; and Arg-Gly-Asp (RGD), a cell adhesion ligand. The cysteine, phosphorylated serine and RGD sequence are specific characteristics of the peptide portion of the PA [42].

Similar to PAs, diblock copolypeptides containing charged and hydrophobic seg-

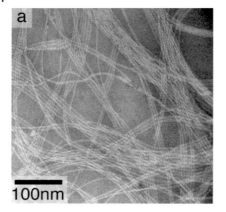

Fig. 4.3. TEM images of PA molecule (a) self-assembled by drying directly onto a TEM grid without adjusted pH, and (b) another PA molecule self-assembled by mixing with CaCl$_2$. Image (a) is negatively stained with phosphotungstic acid whereas (b) is positively stained with uranyl acetate. In both cases the same fibrous morphology is observed, as is seen by pH-induced self-assembly. (Ref. [43] © 2002 by National Academy of Science, USA.)

ment have also been utilized to form rigid hydrogels in water for tissue engineering applications [45–47]. These hydrogels contain 1–30 µm water-containing pores surrounded by a nanofibrous gel matrix [46]. The diblock copolypeptides hydrogels are thermally stable under physiological conditions [45] and reorganize quickly after mechanical deformation due to the low molecular mass of the diblock copolypeptides [47]. However, the pore size is on the lower side and may not be ideal for tissue regeneration.

Molecular self-assembly has also been proven using molecules other than peptides or PAs. For example, Liu et al. reported the self-assembly of polyphenylene dendrimers into µm-long nanofibers [48, 49]. Perutz [50–52] demonstrated the formation of nanotubes from polyglutamines.

As mentioned, tissue engineering scaffolds should ideally have large pore spaces to allow for cell accommodation and mass transport. One of the restrictions of self-assembled scaffolds is the limited ability to form macro size pores. Degradation of self-assembled scaffolds is also an issue that will need to be addressed. In addition, the mechanical properties of these scaffolds will need to be increased before self-assembly can be used in tissue engineering applications that require load bearing. Nevertheless, self-assembled materials are often hydrogels, which are convenient for injection.

4.3.3
Phase Separation

Phase separation processes have been used for some years to create porous polymer membranes for filtration and other applications [53]. However, the use of thermally induced phase separation (TIPS) in the preparation of tissue engineering scaffolds is fairly new [54]. In this process, a temperature change results in the separation of the polymer solution into two phases; one having a low polymer concentration (polymer-lean phase) and one having a high polymer concentration (polymer-rich phase). Conversely, the polymer-lean phase is sometimes referred to as the solvent-rich phase and the polymer-rich phase, in this naming methodology, is referred to as the solvent-lean phase. Once the phase separation has occurred, the solvent is removed by extraction, evaporation, or sublimation. As the polymer-rich phase solidifies, this process leaves behind a network of pores referred to as the polymer foam. The properties of the foam can be tailored to meet specific needs. Thermally induced phase separation does not automatically lead to materials with nanometer scale architecture. For example, our laboratory has used TIPS to develop polymer and polymer/bioceramic composite scaffolds that do not have the nanofibrous structure for tissue engineering [55, 56].

Ma and colleagues have recently developed a novel phase separation process to fabricate nanofibrous materials from polymers [13]. A distinct advantage of the phase separation process is that it can be combined with other manufacturing techniques (such as particulate leaching or three-dimensional printing) to design complex 3D structures with well-controlled pore morphologies [6, 12, 57, 58]. The rest of this section will discuss how such phase separation is used for nanofibrous scaffold fabrication for tissue engineering applications.

An appropriate liquid–liquid phase separation seems critical for nanofibrous structure formation, but does not occur in all solvent systems. The selection of solvent system and phase separation temperature are vital to nanofiber formation. When the system and conditions are right, liquid–liquid phase separation can result in a 3D continuous fibrous network with nano-scaled architecture similar to natural type I collagen [59, 60]. The fibers formed in this manner have diameters ranging from 50 to 500 nm and the scaffolds can have a porosity in excess of 98% [13]. Figure 4.4 is an image of a nanofibrous PLLA scaffold, and as can be seen, this scaffold closely resembles collagen in the ECM.

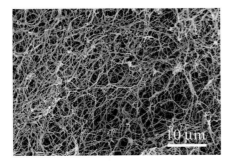

Fig. 4.4. SEM micrograph of a nanofibrous PLLA matrix prepared from 5.0% (wt/v) PLLA/THF solution at a gelation temperature of −18 °C. (Ref. [13] © 1999, John Wiley & Sons.)

The TIPS process for nanofibrous structure formation typically occurs in five steps [13]: polymer dissolution, phase separation and gelation, solvent exchange from the gel with water, freezing, and finally freeze-drying (sublimation) under vacuum. Although each step is necessary, gelation is the step that determines the type of structure that will be formed. At high temperatures, the solution is homogeneous. When the polymer solution is cooled, phase separation into a polymer-rich and polymer-lean phase can occur. Liquid–liquid phase separation usually occurs via one of two routes: nucleation and growth or spinodal decomposition [61].

When a polymer that crystallizes, such as PLLA, is used the phase separation process becomes much more complicated. In addition to phase separation, polymer crystallization can also occur. In these cases, kinetics become important because the solution will experience driving forces for both liquid–liquid phase separation and polymer crystallization. When the liquid–liquid phase separation is faster than the polymer crystallization that leads to the nucleation of crystalline domains in the polymer-rich phase, the morphology of the porous network is largely dependent on the initial liquid–liquid phase separation. The selection of phase-separation temperature should accordingly promote the desired liquid–liquid phase separation [62].

As mentioned earlier, proper selection of solvents is critical in liquid–liquid phase separation. The solvent must have a freezing point (crystallization temperature) lower than the liquid–liquid phase separation temperature of the polymer solution. Otherwise, the solvent may crystallize before the liquid–liquid phase separation occurs. In addition to phase separation temperature and solvent selection, the final morphology of the polymer matrix is dependent on the concentration of polymer solution and the molecular weight of the polymer [59]. For example, a very low concentration of polymer will typically result in a powder-like structure, whereas a higher concentration will result in a foam.

Tissue engineers desire scaffolds not only with fibrous networks that mimic the ECM, but also with designed 3D architecture of interconnected pores. In scaffolds, macroscopic pores (>100 µm) are vital for proper cell seeding, distribution, migration, and neovascularization.

Here we present a few examples to illustrate how to combine liquid–liquid phase separation with other technologies to create scaffolds with the proper 3D shape, pore architecture, and nano-scale fibers. To form a nanofibrous matrix with a macroscopic pore network, TIPS can be combined with particulate leaching techniques [12]. For example, a polymer solution can be dripped over salt or sugar crystals. The system is then cooled, inducing liquid–liquid phase separation. Next the scaffold is immersed in water to simultaneously extract the solvent and leach out the salt/sugar crystals. The scaffold is then frozen and, finally, freeze dried. The resulting scaffolds (Fig. 4.5) have a nano-scale fibrous architecture with macroscopic pores. Although sugar/salt leaching creates well-formed pores, it is difficult to control the amount of interconnectivity. Interconnectivity of pores is important for cell migration, cell signaling, and for mass transport (transport of nutrients to the cells and removal of metabolites from the cells).

To better control the interconnectivity of the macroscopic pores in a scaffold, our

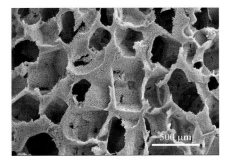

Fig. 4.5. SEM micrograph of a PLLA nanofibrous matrix with particulate macropores prepared from PLLA/THF solution and sugar particles. Particle size 250–500 μm, original magnification ×50. (Ref. [12] © 2000, John Wiley & Sons.)

laboratory has developed a dispersion technique to create paraffin microspheres to be used as porogens. The interconnectivity of the pores can be controlled through thermal bonding of the paraffin microspheres. Paraffin dissolves in THF, and for this reason, THF cannot be used as the polymer solvent, and alternate solvents must be implemented [6]. Following phase separation, hexane is used to leach out the paraffin. The matrix is then frozen and lyophilized. This process results in a scaffold with well-controlled interconnected spherical macropores and nanofibrous pore walls (Fig. 4.6) [6].

Phase separation can also be combined with rapid prototyping techniques or solid free-form (SFF) fabrication. In SFF fabrication a computer-aided design program is used to design a 3D scaffold. To create a nanofibrous scaffold, a negative mold is first generated using SFF fabrication techniques. The polymer solution is then dripped over this mold. Following phase separation, the mold is leached out

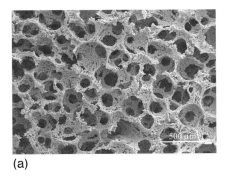

(a)

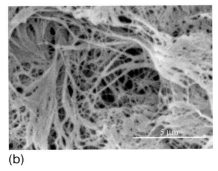

(b)

Fig. 4.6. SEM micrographs of a PLLA nanofibrous matrix with interconnected spherical macropores prepared from a 7.5% (wt/v) PLLA/dioxane/pyridine (dioxane/pyridine = 1:1) solution and paraffin spheres heat treated for 40 min. Sphere size range $d = 250–420$ μm. Original magnification (A) ×50 and (B) ×8000. (Reprinted with permission from Ref. © [6] 2004, Elsevier.)

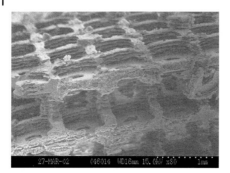

Fig. 4.7. SEM micrograph of the internal pore structure of a scaffold fabricated using the reverse SFF fabrication technique. (Ref. [58] © 2003, John Wiley & Sons.)

and the matrix is processed as described previously. SFF fabrication makes possible the construction of scaffolds with complex 3D structures. Figure 4.7 shows an example of a scaffold formed by combining this reverse SFF fabrication and phase separation [58].

In addition to being combinable with other processing techniques to design macropores, phase separation is also a desirable method for forming tissue engineering scaffolds because it can produce nanofibers in the size range similar to that of natural collagen. Furthermore, phase separation does not require expensive equipment or complicated synthesis schemes and can be easily adapted to a small laboratory setting or large-scale industrial production.

Each technique, molecular self-assembly, electrospinning, and phase separation, provides unique strengths in the area of nano-scale tissue engineering scaffolds. As each area develops further, it is likely that niches will be created where a particular method better suits the needs of certain tissue engineers. As scientists better understand and mimic the ECM, scaffolds will undoubtedly be created that provide the characteristics necessary to move tissue engineering off the laboratory bench and into widespread clinical applications.

4.4
Nanofibrous Composite Scaffolds

Utilizing the above three techniques, composite scaffolds have been formed that contain minerals and biological materials to enable the scaffolds to better mimic the natural ECM's surface chemistry. Although nanofibrous scaffolds provide an improved base architecture, the surface characteristics affect cellular response to a material by influencing the rate and quality of new tissue formation. Specifically, the surface chemistry and topography determine whether protein molecules can adsorb, and the conformation of adsorbed proteins on the surface. This in turn affects cellular attachment, alignment, proliferation and differentiation. Most materials used to fabricate nanofibrous scaffolds lack the biological recognitions of the ECM on their surface. These scaffolds can be altered through the inclusion of molecules and biological recognition sites to more closely mimic natural ECM of the tissues they are trying to emulate.

4.4.1
Inorganic Composites

Most mineralized tissues, like bone, contain both organic and inorganic molecules. In natural mineralized tissues, hydroxyapatite (HAP), a plate or spindle-shaped crystal, is found between collagen fibrils in the ECM [63]. HAP is the main component of the inorganic bone matrix, and the major site of mineral storage in bone and the body. Bone contains 99% of the calcium and 88% of the phosphate in the body. The HAP crystals within the matrix measure up to 200 nm in length and offer a large surface area available for mineral exchange, about 10 m^2 per 1 g of bone [64].

By incorporating HAP into scaffolding, one attempts to increase mineralized tissue formation through mimicking natural mineral content and ECM base architecture. HAP-containing polymer scaffolds possess good mechanical and osteoconductive properties [5]. Nano- and micro-size HAP particles have been incorporated with nanofibrous PLLA scaffolds created by phase separation to yield highly porous scaffolds with improved mechanical strength and increased protein adsorption [65]. Particularly interesting is that the nano-sized HAP provides improved characteristics relative to micro-sized HAP in scaffolds [65].

HAP can also be deposited onto the surface of nanofibrous scaffolds using simulated body fluid. This allows the HAP to have nano-features within the larger deposited masses. This technique has been utilized to deposit HAP on phase separated PLLA scaffolds (Fig. 4.8) [57] and electrospun poly(3-hydroxybutyrate-*co*-3-hydroxyvalerate) nanofibers [66]. Similarly, electrospun PCL and PLGA nanofibers

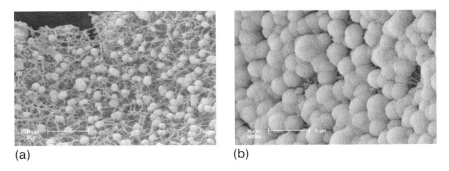

(a)   (b)

Fig. 4.8. SEM micrographs of nanofibrous PLLA/nHAP (90:10) composite scaffolds incubated in 1.5X SBF for varying (a) 4 days and (b) 30 days. (Ref. [57] © 2006, John Wiley & Sons.)

have been combined with calcium carbonate [67], yielding similar advantages for mineralized tissue formation.

4.4.2
Surface Modification

When mineralized tissue is not the goal, yet the base polymer used to create the scaffold does not possess the most desirable surface characteristics, the scaffold surface can be modified to obtain more desirable characteristics to promote cell–scaffold interaction. These modifications can be as simple as plasma treatment or allowing ECM proteins to adsorb to the scaffold [68]. For instance, PCL nanofibers have been air-plasma treated to introduce –COOH groups and then gelatin was allowed to adsorb to the surface [69]. There is a concern over the penetration depth and uniformity of the plasma treatment. More advanced and controlled modifications are desired.

Using electrospinning, a molecule such as gelatin can be blended with polymer and then drawn through the electrospinning process to form nanofibrous scaffolds, as was done by Zhang et al. using PCL and gelatin mixtures [70]. Furthering this concept of multiple species electrospinning, coaxial electrospinning techniques have been develop, which allow for nanofibers to be produced with different core and shell materials. This means that the core and shell material do not necessarily need to be compatible. For instance, Zhang et al. have produced nanofibers with a PCL core and a collagen shell for tissue engineering purposes [71].

Another strategy in surface modification is to use a completely synthetic system. In this approach a biologically active peptide is grafted onto the surface. For instance, a galactose ligand has been covalently conjugated to poly(acrylic acid) spacers, which were UV-grafted onto poly(ε-caprolactone-co-ethyl ethylene phosphate) nanofibers [72]. As such, peptide scaffolds have been made containing the biologically active motifs of laminin I and collagen IV, two important proteins in the basement membrane [73].

Alternatively, peptide amphiphiles containing either biologically active sequences or structures potentially could be assembled on the surface of nanofibrous scaffolds. Fields, Tirrell and coworkers have illustrated the validity of such a coating using PA molecules that consist of a α1 (IV) 1263-1277 collagen sequence Gly-Val-Lys-Gly-Asp-Lys-Gly-Asn-Pro-Gly-Trp-Pro-Gly-Ala-Pro ([IV-H1]) connected to a long-chain mono-or di-alkyl ester lipid [74–78], which assembles into stable triple helices under physiological conditions [78], and have been shown to favorably support cell adhesion when coated on surfaces [76]. This work illustrates that surface chemistry can be mimicked with this type of PAs. However, it has yet to be developed for scaffold modifications.

In our laboratory, a few easier implementing and effective techniques have been developed for 3D surface modification of complex shaped scaffolds. For example, physical entrapment of gelatin along with chemical crosslinking has been used to modify nanofibrous PLLA scaffolds [79]. This is an advancement beyond adsorption because it immobilizes the gelatin on the surface of the nanofibrous scaffold, ensuring that the gelatin does not wash away in a tissue culture environment. After 4 weeks of culture, osteoblasts on nanofibrous scaffolds modified with this method showed increased proliferation and collagen fiber bundle formation over the unmodified control scaffolds [79]. This indicates that this surface modification does not have the depth penetration problems that more simplistic modification like plasma-treated surfaces encounter in 3D applications.

An electrostatic layer-by-layer self-assembly technique has also been used to improve the surface characteristics of nanofibrous PLLA scaffolds with gelatin [80]. Similar to physical entrapment with chemical crosslinking, self-assembly has led to increased osteoblast proliferation over unmodified control scaffolds [80]. Histological sections of nanofibrous scaffolds with the self-assembly modification showed a more even distribution of cells throughout the scaffolds after 2 weeks of culture than unmodified control scaffolds [80].

4.4.3
Factor Delivery Scaffolds

As the drive to mimic the ECM continues, more nanofibrous scaffolds will likely contain biological modifications to more appropriately tailor their characteristics to the desired application. Similarly, investigators are moving beyond simply mimicking the ECM to incorporating the release of factors to promote healing and tissue formation from the scaffolds. To date, the antibiotic Cefoxitin has been released from nanofibrous PLGA scaffolds by several groups [81, 82]. PDLA nanofibers have been used to release Mefoxin, a similar antibiotic to Cefoxitin, by others [83]. Beyond antibiotics, the lipophilic anti-tumor drugs, doxorubicin and paclitaxel, have been released from PLLA nanofibers studied [84]. Ibuprofen, an anti-inflammatory drug, has had its release from PLGA/PEG-g-chitosan nanofibers [85]. Polyurethane nanofibers have been used to study the release of Ketanserin, a selective S2-setotonin antagonist, and Itraconazole, an anti-fungal drug [86].

The release of plasmid DNA from PLGA and PLA-PEG nanofibrous scaffolds has also be studied [2, 87]. Besides DNA, a few biological growth and differentiation factors have been delivered from nanofibers. Human β-nerve growth factor has been released from electrospun PCL/ethyl ethylene phosphate nanofibers [88]. Systems for the delivery of biologically active proteins are often far more difficult to engineer than those for drug or nucleotide release because of the sensitivity of proteins to conformational changes that render them biologically inactive, thus limiting the available methods and necessitating their protection from harsh conditions. Wei et al. have developed a system for release of platelet derived growth factor (PDGF) and/or other biologically active factors from micro-spheres evenly distributed within nanofibrous PLLA scaffolds [89]. Incorporation of the microspheres into the nanofibrous scaffolds decreased the initial burst release of the PDGF and extended the release of biologically active PDGF in a controlled fashion [89].

4.5
Biological Effects of Nanofibers

The ultimate goal of tissue engineering is to develop biological substitutes that restore, maintain or improve tissue function within the body. Developing nano-fibrous scaffolds that better mimic the ECM is one step further down the road to achieving functional neo-tissue formation.

Regardless of the method used to form the nanofibrous scaffolds or further process the scaffolds, these scaffolds seem to possess certain advantages over more conventional scaffolds. They seem advantageous when it comes to cellular adhesion, proliferation, migration and differentiation. Increased cellular adhesion and proliferation, along with more rapid differentiation into mature cell types, have been observed with several cell types [17, 27, 90–93]. However, engineering tissues using nanofibrous scaffolds is still at the early stages. Notably, there is limited data and a systematic study of the biological effects of nanofibers on cell function and tissue development has yet to be completed.

4.5.1
Attachment

Suitable attachment is the first step to successful neo-tissue formation, since most cell types require anchorage to proliferate, migrate and differentiate [94]. Woo et al. [17] found that the cell adhesion proteins fibronectin, vitronectin, and laminin adsorbed to the nanofibrous scaffolds much more than solid-walled control scaffolds. They also found that this effect is not simply due to increased surface area since different proteins adsorbed at different levels, which suggests the selective interaction of the proteins with the nanofibrous matrix [17].

Cells seem to react to this increased adhesion protein presence on the nanofibers by up-regulating integrins, a family of transmembrane proteins involved in adhesion of cells to the ECM and each other. Specifically, normal rat kidney cells upregulate expression of $\beta1$ integrin [91], while chondrocytes upregulate expression of integrin $\alpha2$ on nanofibrous surfaces [92]. Beside up-regulation of $\beta1$ integrin, normal rat kidney cells organize their $\beta1$ integrin in long, slender aggregates on the nanofibers while on glass expression was concentrated in the cell body [91], as seen in Fig. 4.9, on nanofibrous surfaces. Similarly, NIH 3T3 fibroblasts have been shown to express vinculin, a prominent component of focal adhesions that connect the cytoskeleton and plasma membrane with the ECM, strongly in the edge of lamellipodia with more diffuse staining throughout the cell cytoplasm on nanofibrous surfaces [91]. The expression and distribution of these important adhesion molecules are thought to correlate with cellular morphogenesis *in vivo* [91] and the change in cellular interactions between nanofibrous scaffolds and traditional scaffolds, which contain flat walls.

Three days after seeding calvarial cells on both nanofibrous and traditional scaffolds, Woo et al. found that the cells on nanofibrous scaffolds had many long processes that intermingled with the nanofibrous scaffold and connected to neigh-

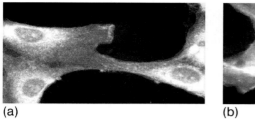

(a)   (b)

Fig. 4.9. Comparison of β1 integrin organization in normal rat kidney cells. Indirect immunofluorescence on (a) glass and (b) nanofibers. (Reprinted with permission from Ref. [91] © 2005, Elsevier.)

boring cells, while cells on traditional scaffolds had only a few processes [95]. Furthermore, cells on the nanofibrous scaffolds were actively synthesizing and secreting molecules [95].

The increased adsorption of adhesion proteins from serum-containing media and the cellular response at least partially explains the increased attachment of numerous cell types on various nanofibrous scaffolds. For instance, neural progenitor cells attach faster to nanofibers than to the control surfaces [90]. Pre-osteoblasts [17], calvarial cells [95], smooth muscle cells [27], fibroblasts [91], normal rat kidney cells [91] and multipotent neural stem cells [90] have also shown preferential adhesion on nanofibers over controls.

4.5.2
Proliferation

After cells attach to a scaffold, they must reproduce by duplicating their contents and dividing in two to develop into new tissue [94]. Several cell types have shown increased proliferative capability on nanofibers over various time periods [27, 91, 92, 96]. Specifically, normal rat kidney cells have shown increased proliferation on nanofibers relative to control at 24 h [91] and high density chondrocyte cultures increased proliferation over 3 weeks on nanofibrous scaffolds relative to control [92]. The authors have also observed this trend with pre-osteoblasts over the course of 12 days. In the context of neo-tissue formation, quicker proliferation will limit the invasion of scar tissue *in vivo* and lead to faster differentiation and functional tissue formation.

4.5.3
Differentiation

Differentiation is the process by which cells undergo a change to an overly specialized cell type from a more immature cell type. This process is different for each cell type. However, nanofibrous materials appear to quicken differentiation of multiple cell types. Fibroblasts have shown more differentiated characteristics on an electrospun nanofibrous substrate [91]. These fibroblasts, along with normal rat kidney cells, express 40–50% less actin on nanofibers than on control substrates at 24 h, which also suggests an increase in differentiated characteristics [91].

Nanofibrous scaffolds have also shown the ability to direct more immature cells to a desired lineage. This has been studied with neural progenitor cells [93], multipotent neural cells [90] and neonatal murine cavarial cells [95]. In these cases, nanofibrous scaffolds have aided in the differentiation of neurons relative to other neuronal support cells. Compared to control scaffolds, nanofibrous scaffolds allow the neonatal murine cavarial cells to differentiate to osteoblasts faster and enhance the expression of phenotypic markers [95]. Specifically, bone sialoprotein was found to be expressed approximately 8–10× higher on nanofibrous scaffolds than on solid-walled scaffolds.

Nanofibrous scaffolds have also shown ability to aid in the redifferentiation of cells that have regressed to a more immature state during cell culture [92]. This particular feature will aid in tissue formation of chondrocytes and hepatocytes, both of which revert to immature cell types during culture.

4.5.4
Migration

To form neo-tissue, cells must be able to infiltrate the scaffold and move throughout it. This movement of cells across the scaffold and pore surfaces is a highly complex process that is dependent on the actin-rich cortex beneath the plasma membrane, which pushes the plasma membrane outward to form a leading edge. Next the actin cytoskeleton must attach to the substrate and provide the traction necessary to move the bulk of the cytoplasm [94]. Many ECM proteins, such as fibronectin, are thought to aid in and direct migrating cells toward their final destination [94]. As discussed in the attachment section, the preferential adsorption of attachment proteins including fibronectin, and the concentration of transmembrane attachment proteins and focal adhesion in localized areas could potentially explain the increased migrant capability of cells on nanofibrous scaffolds. This is particularly evident in a study by Semino et al., in which hippocampal slices where cultured on nanofibrous hydrogel scaffolds [97]. The nanofibrous scaffolds allowed for deeper penetration (∼500 μm relative to the ∼150 μm of the control surface) [97]. Over time the cells migrating into the space were of different lineages. Glia cells were the first to infiltrate the nanofibrous scaffold followed by neural progenitor cells, which then proliferated and differentiated to form a framework for neuron tissue formation within the nanofibrous scaffold [97]. A similar phenomena has been seen by simply placing neural progenitor cells on nanofibrous scaffold [93]. This indicates that cells on nanofibers, independent of tissue type, are capable of increased navigation.

4.6
Tissue Formation

In the human body, tissues and organs are organized into 3D structures. Each tissue or organ has its own specific characteristic architecture depending on its biological function. This architecture provides appropriate channels for mass transport (movement of signaling molecules, nutrients, and metabolic waste) and spatial cellular organization (cell–cell and cell–matrix interactions), both of which are thought to be important for appropriate tissue function [12]. As such, scaffolds for each particular tissue engineering application need to be designed with sensitivity to the cellular requirements of the tissue it is generating.

4.6.1
Connective Tissue

Connective tissue mainly functions to bind and support other tissues. As such, connective tissue contains a sparse cell population scattered throughout an ECM composed of fibrous webs embedded in a uniform foundation, all of which is generally secreted by the cell population [98]. However, there is great diversity in the cellular requirements for various connective tissues. In the context of this discussion we will examine three types, ligaments, cartilage and bone.

4.6.1.1 Ligaments
Ligaments are fibrous connective tissue that connects bones together at joints. To do this, ligaments contain numerous collagenous fibers organized into parallel bundles that maximize non-elastic strength [98]. It is this organization that a scaffold must mimic to successfully generate a new ligament. This has been attempted with scaffolds containing oriented nanofibers seeded with human ligament fibroblasts [96]. Shortly after seeding, the human ligament fibroblasts oriented themselves in the direction of the nanofibers and assumed the appropriate spindle-shaped morphology [96]. After 7 days in culture, oriented bundles were present in a tissue-like formation [96] (Fig. 4.10).

4.6.1.2 Cartilage
Rather than acting as a connector like ligaments, cartilage acts as a cushion between vertebrae and the ends of some bones and reinforces tissue such as the windpipe. As such, the ECM of cartilage is much different from that of ligaments and contains an abundance of collagenous fibers embedded in a rubbery matrix of chondroitin sulfate [98].

Overall, cartilage tissue engineering has been relatively successful even with micro-sized fibers [99, 100]. Nanofibrous materials have been reported to promote chondrocyte proliferation [101, 102], differentiation [92] and the secretion of the suitable ECM [101–103]. Interestingly, in a study by Li et al., chondrocytes on nanofibrous scaffolds were the only ones found to express collagen type IIB, a col-

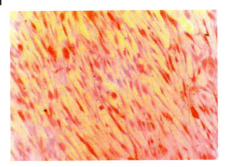

Fig. 4.10. Histological sections of ligament regenerated on nanofibrous scaffolds: human ligament fibroblasts on aligned nanofibers after 7 days of culture. (Reprinted with permission from Ref. [96] © 2005, Elsevier.)

lagen type found only in mature chondrocytes [92]. While in another study, Li et al. found collagen X down-regulated on nanofibrous scaffolds, and yet the chondrocytes had morphology similar to native cartilage and still produced the correct sulfated matrix [103]. This seems to suggest that, in this study, the scaffold potentially interacted with the cells in a way similar to that of collagen X, triggering the decrease in its expression.

4.6.1.3 Bone

Bone is a mineralized connective tissue that supports the body and a major skeletal component that is constantly changing and remodeling itself based on the current needs of the organism. This remodeling is done to help bone fulfill one or more of its primary functions within the organism. These functions include supporting and protecting soft tissues, providing the rigidity necessary for locomotion, regulating calcium and phosphate, and providing a site for hematopoiesis [104]. Although bone can remodel and repair itself, it cannot repair large defects caused by trauma or degenerative disease on its own, which is why tissue engineers are interested in *de novo* bone generation.

After 1 week or more of *in vitro* culture on a nanofibrous scaffold, collagen type 1 has been observed throughout the scaffold [31, 95, 105]. After three or more weeks of *in vitro* culture, up-regulation of bone markers such as alkaline phosphatase, bone sialoprotein and osteocalcin on nanofibrous scaffolds relative to control have been observed [103, 106]. The notable exception to this is collagen type I, a major component of the bone ECM [103]. As with cartilage, it is possible that the nanofibrous scaffolds interact with the cells in a similar manner to native collagen, limiting the cells' need to express it. This has been supported by Woo et al., who administered 3,5-dehydroproline to inhibit collagen fibril formation and found alpha-2 integrin and osteogenic markers expressed on the nanofibrous scaffold but not the control scaffolds [95, 106].

In a 10-day study of osteoblastic phenotype development in neonatal murine cavarial cells, α1 collagen and Runx2 expression were shown to be initially up-regulated on the nanofibrous scaffolds compared to solid-walled scaffolds. However, as the experiment continued the expression levels for both markers approached the expression levels on the nanofibrous scaffolds, while both bone sialoprotein and osteocalcin once observed were substantially higher on the nanofibrous scaffolds compared with solid-walled scaffolds [95]. The authors have also observed the up-regulation of bone sialoprotein and osteocalcin and the down-regulation of collagen type I over the course of 6 weeks with pre-osteoblasts on nanofibrous scaffolds compared with solid-walled scaffolds.

Small globular mineral deposits have been observed within nanofibrous scaffolds in as early as 2 weeks using primary calvarial cells [95]. In this study, large mineral deposits existed along the nanofibers, indicating mineral development along the nanofibers. This is supported by the authors' own observations of mineralization emanating from within areas of the synthetic nanofibrous scaffold. This architecture allows for cellular penetration within the synthetic matrix [107].

In terms of successful bone generation, Vacanti and coworkers found that after 4 weeks *in vitro* and 4 weeks *in vivo* vascularization, mineralization and embedded osteocyte-like cells can be seen in electrospun nanofibrous scaffolds [105].

4.6.2
Neural Tissue

Neural tissue is composed of neurons and support cells. Neurons transmit signals to direct other tissues' function via nerve impulses throughout the body while the support cells provide structure for the tissue and protect, insulate and generally support the neurons. This tissue has a limited capacity to regenerate, but is vital to normal body function.

Nerve impulses are passed from one neuron to the next at synapses where a projection from the first neuron approaches the body of a second neuron [98]. To achieve this sort of order within a scaffold, neurons grow neurite projections with growth cones to interact with other cells. Several nanofibrous scaffolds have showed increased neurite outgrowth relative to control [25, 90, 93, 108]. At times, this increased outgrowth has been as much as 100 μm [25]. It has been generally observed that the nanofibrous scaffold acts as a guide for neurite outgrowth [25, 108]. When the nanofibers are aligned this growth tends to be in the direction parallel to the nanofibers and the cells tend to have a more elongated shape similar to *in vivo* [25].

Since neuronal progenitor cells are difficult to isolate, much of neuronal tissue engineering relies on the migration of progenitor cells from tissue. This has been documented with nanofibrous scaffolds [93, 97]. Once these progenitor cells migrate, they must differentiate to provide the necessary neurons and support cells for proper function of the neuronal tissue. Nanofibrous scaffolds promote increased differentiation compared to control scaffolds [90, 93, 97]. Although complete *de novo* neural tissue has yet to be formed on nanofibrous scaffolds, great potential and interest exist to pursue *de novo* tissue formation of this tissue type.

4.6.3
Cardiovascular Tissue

The cardiovascular system is responsible for the transport of nutrients and wastes to and from cells. The heart acts as the pump driving the system and the vascular tissue acts as tubes directing the path of collection and exchange of nutrients and waste. As such, disruption in the function of this system is life threatening and must be repaired quickly. This section concentrates on cardiac muscle and blood vessels.

4.6.3.1 Cardiac Muscle

The regenerative capability of cardiac muscle is particularly limited, making it the most challenging tissue to regenerate in the cardiovascular system [109]. To combat this, nanofibrous self-assembling RAD12-II peptide scaffolds have been developed by Lee and coworkers that can be injected into the heart to promote cardiac muscle regeneration. When injected into the left ventricle wall of adult mice, endothelial cells invaded, organized and matured into the appropriate structure to support smooth muscle cell migration, which allowed for the formation of arterioles [110]. Red blood cells located within the arterioles indicate that they were functionally connected to the vascular system. This led to the migration of cardiac progenitor cells that differentiated into myocytes [110]. Inclusion of myocytes in the scaffold greatly increases myocyte density within the scaffold over a much shorter period of time. When implanted with embryonic stem cells this same self-assembling RAD12-II peptide scaffold promoted the controlled differentiation of the cells to cardiomyocytes [110]. During *in vitro* cultures, the self-assembling RAD12-II peptide scaffold promoted endothelial and myocardial cell survival and synchronized contraction of the generated muscle [111].

4.6.3.2 Blood Vessel

Blood vessels are composed of smooth muscle and endothelial cells. Similar to previously discussed tissue types, alignment of nanofibers aids in blood vessel function. The aligned nanofibrous scaffolds have been shown to prompt the alignment of smooth muscle cells parallel to the fiber direction, which increases vessel strength [27]. This alignment has also been shown to increase initial adhesion and proliferation on the scaffold, which leads to quicker functional blood vessel formation [27].

The second cellular component of blood vessels, endothelial cells, also benefit from culture on nanofibrous scaffolds. Human aortic endothelial cells cultured on nanofibrous scaffolds form confluent layers similar to those formed on collagen type I gel or Matrigel [73]. The scaffolds also promote nitric oxide release which is a marker of endothelial cell phenotype [73].

4.6.4
Liver Tissue

Liver is a multipurpose organ that aids in the digestion of fats, the uptake and release of glucose, synthesis of plasma proteins and the detoxification of many

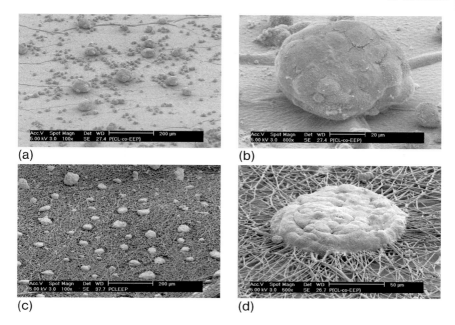

Fig. 4.11. SEM images of hepatocytes after 8 days of culture: (a, b) Hepatocytes cultured on a Gal-film formed spheroids that did not integrate with the substrate; (c, d) in contrast, hepatocytes cultured on Gal-nanomesh afforded aggregates that engulfed the functional nanofibers. (Reprinted with permission from [72] © 2005, Elsevier.)

poisons and metabolic waste [98]. Regeneration of liver tissue in a functional form has proven incredibly difficult. However, nanofibrous scaffolds have brought renewed efforts. Nanofibrous scaffolds appear to increase hepatocyte function in comparison to control scaffolds [72, 112]. Hepatocytes on nanofibrous scaffolds also exhibit a more uniform spheroid-like distribution than on control scaffolds [72]. Figure 4.11 illustrates these differences in interaction between hepatocytes and nanofibrous and non-fibrous materials.

4.7
Conclusion

To date the most successful neo-tissue generation has involved connective tissues [100, 101, 113]. As evident by the developments in cardiac muscle formation, headway is being made with more metabolically demanding and complex tissues. However, the holy grail of tissue engineering still remains the development of complete patient-specific organs available on demand [2, 114]. To achieve this, scaffolds must not only provide structural support but also biological cues [115]. Increased knowledge about cellular interactions with the native ECM and nanostructured scaffolds will facilitate more functional scaffold development.

Hopefully, this chapter has provided some insight into the crucial role tissue engineering scaffolding plays in this drive for 3D neo-tissue formation [4, 17, 116] through its focus on nanofibrous scaffolding fabrication and the biological advantages of nanofibrous scaffolding structures in mimicking the ECM.

References

1 LANGER, R., J. VACANTI, Tissue engineering. *Science*, **1993**, 260, 920–926.

2 MA, P., Scaffolds for tissue fabrication. *Mater. Today*, **2004**, 7, 30–40.

3 MA, P., R. ZHANG, *J. Biomed. Mater. Res.*, **1999**, 46, 60–72.

4 HUBBELL, J., *Biotechnology*, **1995**, 13(6), 565–575.

5 LIU, X., P. MA, Polymeric scaffolds for bone tissue engineering. *Ann. Biomed. Eng.*, **2004**, 32, 477–486.

6 CHEN, V.J., P.X. MA, Nano-fibrous poly(L-lactic acid) scaffolds with interconnected spherical macropores. *Biomaterials*, **2004**, 25(11), 2065–2073.

7 ELSDALE, T., J. BARD, *J. Cell Biol.*, **1972**, 54, 626–637.

8 MA, P., R. LANGER, Fabrication of biodegradable polymer foams for cell transplantation and tissue engineering, in *Tissue Engineering Methods and Protocols*, M. YARMUSH, J. MORGAN (Eds.), **1999**, Humana Press Inc., Totowa, NJ, pp. 47–56.

9 ABBOTT, A., Cell culture: Biology's new dimension. *Nature*, **2003**, 424, 870–872.

10 CUKIERMAN, E., R. PANKOV, D. STEVENS, K. YAMADA, Taking Cell-matrix adhesions to the third dimensions. *Science*, **2001**, 294, 1708–1712.

11 SCHMEICHEL, K.L., M.J. BISSELL, Modeling tissue-specific signaling and organ function in three dimensions. *J. Cell Sci.*, **2003**, 116(Pt 12), 2377–2388.

12 ZHANG, R., P.X. MA, Synthetic nano-fibrillar extracellular matrices with predesigned macroporous architectures. *J. Biomed. Mater. Res.*, **2000**, 52(2), 430–438.

13 MA, P.X., R. ZHANG, Synthetic nano-scale fibrous extracellular matrix. *J. Biomed. Mater. Res.*, **1999**, 46(1), 60–72.

14 ALBERTS, B., *Molecular Biology of the Cell*, 4th edn. **2002**, Garland Science, New York, xxxiv, 1463.

15 PACHENCE, J.M., Collagen-based devices for soft tissue repair. *J. Biomed. Mater. Res.*, **1996**, 33(1), 35–40.

16 ANSELME, K., C. BACQUES, G. CHARRIERE, D.J. HARTMANN, D. HERBAGE, R. GARRONE, Tissue reaction to subcutaneous implantation of a collagen sponge. A histological, ultrastructural, and immunological study. *J. Biomed. Mater. Res.*, **1990**, 24(6), 689–703.

17 WOO, K.M., V.J. CHEN, P.X. MA, Nano-fibrous scaffolding architecture selectively enhances protein adsorption contributing to cell attachment. *J. Biomed. Mater. Res. A*, **2003**, 67(2), 531–537.

18 DESAI, T., Micro- and nanoscale structures for tissue engineering constructs. *Med. Eng. Phys.*, **2000**, 22, 595–606.

19 FORMHALS, A., *Process and Apparatus for Preparing Artificial Threads*, U.S.P. Office, Editor. **1934**, Schreiber-Gastell, Richard and Formhals, Anton: United States. pp. 1–7.

20 DEITZEL, J.M., J.D. KLEINMEYER, J.K. HIRVONEN, N.C. BECK TAN, Controlled deposition of electrospun poly(ethylene oxide) fibers. *Polymer*, **2001**, 42(19), 8163–8170.

21 JAEGER, R., M.M. BERGSHOEF, C. MARTÍN I BATLLE, H. SCHÖNHERR, G.J. VANCSO. Electrospinning of ultrathin polymer fibres. In: *Macromolecular Symposia*. **1998**, 127, 141–150.

22 RENEKER, D.H., I. CHUN, Nanometre

diameter fibres of polymer, produced by electrospinning. *Nanotechnology*, **1996**, 7(3), 216–223.

23 KENAWY EL, R., J.M. LAYMAN, J.R. WATKINS, G.L. BOWLIN, J.A. MATTHEWS, D.G. SIMPSON, G.E. WNEK, Electrospinning of poly(ethylene-co-vinyl alcohol) fibers. *Biomaterials*, **2003**, 24(6), 907–913.

24 MA, Z., M. KOTAKI, R. INAI, S. RAMAKRISHNA, Potential of nanofiber matrix as tissue-engineering scaffolds. *Tissue Eng.*, **2005**, 11(1–2), 101–109.

25 YANG, F., R. MURUGAN, S. WANG, S. RAMAKRISHNA, Electrospinning of nano/micro scale poly(L-lactic acid) aligned fibers and their potential in neural tissue engineering. *Biomaterials*, **2005**, 26, 2603–2610.

26 LI, W.J., C.T. LAURENCIN, E.J. CATERSON, R.S. TUAN, F.K. KO, Electrospun nanofibrous structure: A novel scaffold for tissue engineering. *J. Biomed. Mater. Res.*, **2002**, 60(4), 613–621.

27 XU, C., R. INAI, M. KOTAKI, S. RAMAKRISHNA, Aligned biodegradable nanofibrous structure: A potential scaffold for blood vessel engineering. *Biomaterials*, **2004**, 25, 877–886.

28 KAMEOKA, J., H.G. CRAIGHEAD, Fabrication of oriented polymeric nanofibers on planar surfaces by electrospinning. *Appl. Phys. Lett.*, **2003**, 83(2), 371–373.

29 ZONG, X., S. RAN, D. FANG, B.S. HSIAO, B. CHU, Control of structure, morphology and property in electrospun poly(glycolide-co-lactide) non-woven membranes via post-draw treatments. *Polymer*, **2003**, 44(17), 4959–4967.

30 LI, D., Y.L. WANG, Y.N. XIA, Electrospinning of polymeric and ceramic nanofibers as uniaxially aligned arrays. *Nano Lett.*, **2003**, 3(8), 1167–1171.

31 YOSHIMOTO, H., Y.M. SHIN, H. TERAI, J.P. VACANTI, A biodegradable nanofiber scaffold by electrospinning and its potential for bone tissue engineering. *Biomaterials*, **2003**, 24(12), 2077–2082.

32 HUANG, L., K. NAGAPUDI, R.P.

APKARIAN, E.L. CHAIKOF, Engineered collagen-PEO nanofibers and fabrics. *J. Biomater. Sci. Polym. Ed.*, **2001**, 12(9), 979–993.

33 LI, M., M. MONDRINOS, M. GANDHI, F. KO, A. WEISS, P. LELKES, Electrospun protein fibers as matrices for tissue engineering, *Biomaterials*, **2005**, 26, 5999–6008.

34 MATTHEWS, J.A., G.E. WNEK, D.G. SIMPSON, G.L. BOWLIN, Electrospinning of collagen nanofibers. *Biomacromolecules*, **2002**, 3(2), 232–238.

35 BOLAND, E., J. MATTHEWS, K. PAWLOWSKI, D. SIMPSON, G. WNEK, G. BOWLIN, Electrospinning collagen and elastin: Preliminary vascular tissue engineering. *Front Biosci.*, **2004**, 9, 1422–1432.

36 ROEMER, D., H.H. BUESCHER, R.C. HILL, J. PLESS, W. BAUER, F. CARDINAUX, A. CLOSSE, D. HAUSER, R. HUGUENIN, A synthetic enkephalin analogue with prolonged parenteral and oral analgesic activity. *Nature*, **1977**, 268(5620), 547–549.

37 ZHANG, S., Fabrication of novel biomaterials through molecular self-assembly. *Nat. Biotechnol.*, **2003**, 21(10), 1171–1178.

38 SCHNUR, J.M., R. PRICE, P. SCHOEN, P. YAGER, J.M. CALVERT, J. GEORGER, A. SINGH, Lipid-based tubule microstructures. *Thin Solid Films*, **1987**, 152(1–2), 181–206.

39 HOLMES, T.C., S. DE LACALLE, X. SU, G. LIU, A. RICH, S. ZHANG, Extensive neurite outgrowth and active synapse formation on self-assembling peptide scaffolds. *Proc. Natl. Acad. Sci. U.S.A.*, **2000**, 97(12), 6728–6733.

40 ZANG, S., T. HOLMES, C. LOCKSHIN, A. RICH, Spontaneous assembly of a self-complementary oligopeptide to form a stable macroscopic membrane. *Proc. Natl. Acad. Sci. U.S.A.*, **1993**, 90, 3334–3338.

41 ZANG, S., T. HOLMES, C. DIPERSIO, R. HYNES, X. SU, A. RICH, Self-complementing olgopeptide matrices support mammalian-cell attachment. *Biomaterials*, **1995**, 16, 1385–1395.

42 HARTGERINK, J.D., E. BENIASH, S.I. STUPP, Self-assembly and mineraliza-

tion of peptide-amphiphile nano-fibers. *Science*, **2001**, 294(5547), 1684–1688.

43 HARTGERINK, J., E. BENIASH, S. STUPP, Peptide-amphiphile nanofibers: A versatile scaffold for the preparation of self-assembing materials. *Proc. Natl. Acad. Sci. U.S.A.*, **2002**, 99, 5133–5138.

44 SERVICE, R.F., Nanofibers seed blood vessels. *Science*, **2005**, 308(5718), 44b–45.

45 PAKSTIS, L., B. OZBAS, K. HALES, A. NOWAK, T. BEMING, D. POCHAN, Effect of chemistry and morphology on the biofunctionality of self-assembling diblock copolypeptide hydrogels. *Biomacromolecules*, **2004**, 5, 312–318.

46 POCHAN, D., L. LPAKSTIS, B. OZBAS, SANS and Cryo-TEM study of Self-assembled diblock copolypeptide hydrogels with rich nano-through micoscale morphology. *Macromolecules*, **2002**, 35, 5358–5360.

47 NOWAK, A., V. BREEDVELD, L. PAKSTIS, B. OZBAS, D. PINE, D. POCHAN, T. DEMING, Rapidly recovering hydrogel scaffolds from self-assembling diblock copolypeptide amphiphiles. *Nature*, **2002**, 417, 424–428.

48 LIU, D.J., S. DE FEYTER, M. COTLET, U.M. WIESLER, T. WEIL, A. HERR-MANN, K. MULLEN, F.C. DE SCHRYVER, Fluorescent self-assembled poly-phenylene dendrimer nanofibers. *Macromolecules*, **2003**, 36(22), 8489–8498.

49 LIU, D.J., H. ZHANG, P.C.M. GRIM, S. DE FEYTER, U.M. WIESLER, A.J. BERRESHEIM, K. MULLEN, F.C. DE SCHRYVER, Self-assembly of polyphenylene dendrimers into micrometer long nanofibers: An atomic force microscopy study. *Langmuir*, **2002**, 18(6), 2385–2391.

50 PERUTZ, M.F., A.H. WINDLE, Cause of neural death in neurodegenerative diseases attributable to expansion of glutamine repeats. *Nature*, **2001**, 412(6843), 143–144.

51 PERUTZ, M.F., Glutamine repeats and neurodegenerative diseases. *Brain Res. Bull.*, **1999**, 50(5–6), 467.

52 PERUTZ, M.F., B.J. POPE, D. OWEN, E.E. WANKER, E. SCHERZINGER, Aggregation of proteins with expanded glutamine and alanine repeats of the glutamine-rich and asparagine-rich domains of Sup35 and of the amyloid beta-peptide of amyloid plaques. *Proc. Natl. Acad. Sci. U.S.A.*, **2002**, 99(8), 5596–5600.

53 VAN DE WITTE, P., P.J. DIJKSTRA, J.W.A. VAN DEN BERG, J. FEIJEN, Phase separation processes in polymer solutions in relation to membrane formation. *J. Membr. Sci.*, **1996**, 117(1–2), 1–31.

54 ATALA, A., R.P. LANZA, *Methods of Tissue Engineering*. **2001**, Academic Press, San Diego, xli, 1285 pp.

55 ZHANG, R., P.X. MA, Poly(alpha-hydroxyl acids)/hydroxyapatite porous composites for bone tissue engineering. I. Preparation and morphology. *J. Biomed. Mater. Res.*, **1999**, 44(4), 446–455.

56 MA, P.X., R. ZHANG, G. XIAO, R. FRANCESCHI, Engineering new bone tissue in vitro on highly porous poly(alpha-hydroxyl acids)/hydroxyapatite composite scaffolds. *J. Biomed. Mater. Res.*, **2001**, 54(2), 284–293.

57 WEI, G., P.X. MA, Macro-porous and nano-fibrous polymer scaffolds and polymer/bone-like apatite composite scaffolds generated by sugar spheres. *J. Biomed. Mater. Res.*, **2006**, in the press.

58 MA, P.X., Tissue engineering, in *Encyclopedia of Polymer Science and Technology*, J.I. KROSCHWITZ (Ed.), **2003**, John Wiley & Sons, Inc., Hoboken, NJ. p. www.mrw.interscience.wiley.com/epst.

59 ZHANG, R., P.X. MA, Processing of polymer scaffolds: Phase separation, in *Methods of Tissue Engineering*, A. ATALA, R. LANZA (Eds.), **2001**, Academic Press, San Diego, pp. 715–724.

60 CHEN, V.J., P.X. MA, Phase separation for polymer scaffolds, in *Scaffolding in Tissue Engineering*, P.X. MA, J. ELISSEEFF (Eds.), CRC Press, Boca Raton, FL, **2005**, pp. 121–133.

61 LARSON, R.G., *The Structure and Rheology of Complex Fluids*. Oxford University Press, New York, **1999**, xxi, 663 pp.

62 CHEN, V., G. WEI, P. MA, Nanostructured scaffolds for tissue engineering and regeneration, in *Handbook of Nanostructured Biomaterials and their Applications in Nanobiotechnology*, H. NALWA (Ed.), American Scientific Publishers, Los Angeles, **2005**, pp. 415–435.

63 ROSSERT, J., B. DE CROMBUGGHE, Type I collagen: Structure, synthesis, and regulation, in *Principles of Bone Biology*, J. BILEZIKIAN (Ed.), Academic Press, San Diego, **2002**, pp. 189–210.

64 SALO, J., Bone resorbing osteoclasts reveal two basal plasma membrane domains and transcytosis of degraded matrix material, in *Department of Anatomy and Cell Biology*, **2002**, University of Oulu, Oulu, Finland, p. 59.

65 WEI, G., P. MA, Structure and properties of nano-hydroxyapatite/polymer composite scaffolds for bone tissue engineering. *Biomaterials*, **2004**, 25, 4749–4759.

66 ITO, Y., H. HASUDA, M. KAMITAKAHARA, C. OHTSUKI, M. TANIHARA, I. KANG, O. KOWON, A composite of hydroxyapatite with electrospun biodegradable nanofibers as a tissue engineering material. *J. Biosci. Bioeng.*, **2005**, 100, 43–49.

67 FUJIHARA, K., M. KOTAKI, S. RAMAKRISHNA, Guided bone reneration membrane made of polycaprolactone/calcium carbonate composite nano-fibers. *Biomaterials*, **2005**, 26, 4139–4147.

68 MIN, B., G. LEE, S. KIM, Y. NAM, T. LEE, W. PARK, Electrospinning of silk fibroin nanofibers and its effect on the adhesion and spreading of normal human keratinocytes and fibroblasts in vitro. *Biomaterials*, **2004**, 25, 1289–1297.

69 MA, Z., W. HE, T. YONG, S. RAMAKRISHNA, Grafting of gelatin on electronspun poly(caprolactone) nanofibers to improve endothelial cell spreading and proliferation and to control cell orientation. *Tissue Eng.*, **2005**, 11(7/8), 1149–1158.

70 ZHANG, Y., H. OUYANG, C. LIM, S. RAMAKRISHNA, Z. HUANG, Electro-spinning of gelatin fibers and gelatin/PCL composite fibrous scaffolds. *J. Biomed. Mater. Res., B* **2005**, 72, 156–165.

71 ZHANG, Y., J. VENUGOPAL, Z. HUANG, C. LIM, S. RAMAKRISHNA, Characterization of the surface biocompatibility of the electrospun PCL-collagen nanofibers using fibroblasts. *Biomacromolecules*, **2005**, 6, 2583–2589.

72 CHUA, K., W. LIM, P. ZHANG, H. LU, J. WEN, S. RAMAKRISHNA, K. LEONG, H. MAO, Stable immobilization of rat hepatocyte spheroids on galactosylated nanofibers. *Biomaterials*, **2005**, 26, 2537–2547.

73 GENOVE, E., C. SHEN, S. ZHANG, C. SEMINO, The effect of functionalized self-assembling peptide scaffolds on human aortic endothelial cell function. *Biomaterials*, **2005**, 26, 3341–3351.

74 MIKOS, A.G., A.J. THORSEN, L.A. CZERWONKA, Y. BAO, R. LANGER, D.N. WINSLOW, J.P. VACANTI, Preparation and characterization of poly(-lactic acid) foams. *Polymer*, **1994**, 35(5), 1068–1077.

75 BERNDT, P., G. FIELDS, M. TIRRELL, *J. Am. Chem. Soc.*, **1995**, 117, 9515–9522.

76 MALKAR, N., J. LAUER-FIELDS, D. JUSKA, G. FIELDS, *Biomacromolecules*, **2003**, 4(3), 518–528.

77 HAVERSTICK, K., T. PAKALNS, Y. YU, J. MCCARTHY, G. FIELDS, M. TIRRELL, *Polym. Mater. Sci. Eng.*, **1997**, 77, 584–585.

78 YU, Y., V. ROONTGA, V. DARAGAN, K. MAYO, M. TIRRELL, G. FIELDS, *Biochemistry*, **1999**, 38, 1659–1668.

79 LIU, X., Y. WON, P. MA, Surface modification of interconnected porous scaffolds. *J. Biomed. Mater. Res., A* **2005**, 74, 84–91.

80 LIU, X., L. SMITH, G. WEI, Y. WON, P. MA, Surface engineering of nanofirous poly(L-lactic acid) scaffolds via self-assembly technique for bone tissue engineering. *J. Biomed. Nanotechnol.*, **2005**, 1(1), 54–60.

81 KIM, K., Y. LUU, C. CHANG, D. FANG, B. HSIAO, B. CHU, M. HADJIARGYROU, Incorporation and controlled release of a hydrophilic antibiotic using poly(lactide-co-glycolide)-based electrospun nanofibrous scaffolds. *J. Controlled Release*, **2004**, 98, 47–56.

82 KATTI, D., K. ROBINSON, F. KO, C. LAURENCIN, Bioresorbable nanofiber-based systems for wound healing and drug delivery: Optimization of fabrication parameters. *J. Biomed. Mater. Res. Part B: Appl Biomater.*, **2004**, 70, 286–296.

83 ZONG, X., K. KIM, D. FANG, S. RAN, B. HSIAO, B. CHU, Structure and process relationship of electrospun bioabsorbable nanofiber membranes. *Polymer*, **2002**, 43, 4403–4412.

84 ZENG, J., L. YANG, Q. LIANG, X. ZHANG, H. GUAN, X. XU, X. CHEN, X. JING, Influence of the drug compatibility with polymer solution on the release kinetics of electrospun fiber formulation. *J. Controlled Release*, **2005**, 20, 43–51.

85 JIANG, H., D. FANG, B. HSIAO, B. CHU, W. CHEN, Preparation and characterization of ibuprofen-loaded poly(lactid-co-glycolide)/poly(ethylene glycol)-g-chitosan electrospun membranes. *J. Biomater. Sci. Polym. Edn.*, **2004**, 15, 279–296.

86 VERRECK, G., I. CHUN, J. ROSENBLATT, J. PEETERS, A. DIJCK, J. MENSCH, M. NOPPE, M. BREWSTER, Incorporation of drugs in an amorphous state into electrospun nanofibers composed of a water-insoluble, nonbiodgradable polymer. *J. Controlled Release*, **2003**, 92, 349–360.

87 LUU, Y., K. KIM, B. HSIAO, B. CHU, M. HADJIARYROU, Development of a nanostructured DNA delivery scaffold via electrospinning of PLGA and PLA-PEG block copolymers. *J. Controlled Release*, **2003**, 89, 341–353.

88 CHEW, S., J. WEN, E. YIM, K. LEONG, Sustained release of proteins from electrospun biodegradable fibers. *Biomacromolecules*, **2005**, 6, 2017–2024.

89 WEI, G., Q. JIN, W. GIANNOBILE, P.

MA, Nano-fibrous scaffolds for controlled delivery of recombinana human PDGF-BB. *J. Controlled Release*, **2006**, 112(1), 103–110.

90 YANG, F., C. XU, M. KOTAKI, S. WANG, S. RAMAKRISHNA, Characterization of neural stem cells on electrospun poly(L-lactic acid) nanofibrous scaffolds. *J. Biomater. Sci. Polym. Edn.*, **2004**, 15(12), 1483–1497.

91 SCHINDLER, M., I. AHMED, J. KAMAL, A. NUR-E-KAMAL, T. GRAFE, H. CHUNG, S. MEINERS, A synthetic nanofibrillar matrix promotes in vivo-like organization and morphogenesis for cells in culture. *Biomaterials*, **2005**, 26, 5624–5631.

92 LI, W., K. DANIELSON, P. ALEXANDER, R. TUAN, Biological response of chondrocytes cultured in three-dimensional nanofibrous poly(E-caprolactone) scaffolds. *J. Biomed. Mater. Res. A*, **2003**, 67, 1105–1114.

93 SILVA, G., C. CZEISLER, K. NIECE, E. BENIASH, D. HARRINGTON, J. KESSLER, S. STUPP, Selective differentiation of neural progenitor cells by high-epitope density nanofibers. *Science*, **2004**, 303, 1352–1355.

94 ALBERTS, B., A. JOHNSON, J. LEWIS, M. RAFF, K. ROBERTS, P. WALTER, *Molecular Biology of the Cell.* 4th edn., Graland Science, New York, **2002**.

95 WOO, K., J. JUN, V. CHEN, J. SEO, J. BAEK, H. RYOO, G. KIM, M. SOMERMAN, P. MA, Nano-fibrous scaffolding promotes osteoblast differentiation and biomineralization, *Biomaterials*, **2006**, in press.

96 LEE, C., H. SHIN, I. CHO, Y. KANG, I. KIM, K. PARK, J. SHIN, Nanofiber alignment and direction of mechanical strain affect the ECM productions of human ACL fibroblast. *Biomaterials*, **2005**, 26, 1261–1270.

97 SEMINO, C., J. KASAHARA, Y. HAYASHI, S. ZHANG, Entrapment of migrating hippocampal neural cells in three-dimensional peptide nanofiber scaffolds. *Tissue Eng.*, **2004**, 10(3/4), 643–655.

98 CAMPBELL, N., J. REECE, L. MITCHELL, *Biology.* 5 edn., BenjaminCummings, Menlo Park, **1999**.

99 Ma, P.X., B. Schloo, D. Mooney, R. Langer, Development of biomechanical properties and morphogenesis of in vitro tissue engineered cartilage. *J. Biomed. Mater. Res.*, **1995**, 29(12), 1587–1595.

100 Ma, P., R. Langer, Morphology and mechanical function of long-term in vitro engineered cartilage. *J. Biomed. Mater. Res.*, **1999**, 44, 217–221.

101 Li, W., R. Tuli, C. Okafor, A. Derfoul, K. Danielson, D. Hall, R. Tuan, A three-dimensional nanofibrous scaffold for cartilage tissue engineering using human mesenchymal stem cells. *Biomaterials*, **2005**, 26, 599–609.

102 Kisiday, J., M. Jin, B. Kurz, H. Hung, C. Semino, S. Zhang, A. Grodzinsky, Self-assembling peptide hydrogel fosters chondrocyte extracellular matrix production and cell dividsion: Implications for catrilage tissue repair. *Proc. Natl. Acad. Sci. U.S.A.*, **2002**, 99, 9996–10 001.

103 Li, W., R. Tuli, X. Huang, P. Laquerriere, R. Tuan, Multilineage differentiation of human mesnchymal stem cells in a three-dimensional nanofibrous scaffold. *Biomaterials*, **2005**, 26, 5158–5166.

104 Marks, S., P. Odgren, Structure and development of the skeleton, in *Principles of Bone Biology*, G. Rodan (Ed.), Academic Press, San Diego. **2002**, pp. 3–16.

105 Shin, M., H. Yoshimoto, J. Vacanti, In vivo bone tissue engineering using mesenchymal stem cells on a novel electrospun nanofibrous scaffold. *Tissue Eng.*, **2004**, 10(1/2), 33–41.

106 Woo, K., J. Jun, V. Chen, J. Baek, S. Seo, G. Kim, S. Ko, P. Ma, Mineralization in the mouse calvarial cell-seeded nano-fibrous scaffolds. *J. Biomed. Mater. Res.*, **2004**, 19, S212.

107 Chen, V.J., L.A. Smith, P.X. Ma, Bone regeneration on computer designed nano-fibrous scaffolds, *Biomaterials*, **2006**, 27(21), 3973–3979.

108 Yang, F., R. Murugan, S. Ramakrishna, X. Wang, Y. Ma, S. Wang, Fabrication of nano-structured porous PLLA scaffold intended for nerve tissue engineering. *Biomaterials*, **2004**, 25, 1891–1900.

109 Leor, J., Y. Amsalem, S. Cohen, Cells, scaffolds and molecules for myocardial tissue engineering. *Pharmacol. Therapeut.*, **2005**, 105, 151–161.

110 Davis, M., B. Motion, D. Narmoneva, T. Takahashi, D. Hakuno, R. Kamm, S. Zhang, R. Lee, Injectable self-assembling peptide nanofibers create intramyocardial microenvironments for endothelial cells. *Circulation*, **2005**, 111, 442–450.

111 Narmoneva, D., R. Vukmirovic, M. Davis, R. Kamm, R. Lee, Endothelial cells promote cardiac myocyte survival and spatial reoganization. *Circulation*, **2004**, 110, 962–968.

112 Semino, C., J. Merok, G. Crane, G. Panagiotakos, S. Zhang, Functional differentiation of hepatocyte-like spheroid structures from putative liver progenitor cells in three-deminional peptide scaffolds. *Differentiation*, **2003**, 71, 262–270.

113 Li, W., R. Tuan, Polymeric scaffolds for cartilage tissue engineering. *Macromol. Symp.*, **2005**, 227, 65–75.

114 Mooney, D., A. Mikos, Growing new organs. *Sci. Am.*, **1999**, 280, 60–67.

115 Langer, R., D. Tirrell, Designing materials for biology and medicine. *Nature*, **2004**, 428, 487–492.

116 Ma, P., J. Choi, Biodegradable polymer scaffolds with well-defined interconnected spherical pore network. *Tissue Eng.*, **2001**, 7, 23–33.

5

Nanophase Biomaterials for Tissue Engineering

Ramalingam Murugan and Seeram Ramakrishna

5.1
Introduction: Problems with Current Therapies

Tissue or organ failure, resulting from traumatic or non-traumatic destruction, gives rise to a major health problem that directly affects quality and length of patient's life. These circumstances often call for surgical treatments to repair, replace, maintain or augment the functions of affected tissue or organ using some additional functional components to facilitate quality life to the patient. They have been traditionally treated with the help of tissue or organ procured from the donor site. Depending on the location of reimplantation of the procured tissue or organ (also called graft), they are termed as autograft, allograft, or xenograft (Fig. 5.1) [1]. If the graft is implanted in the same patient, it is termed as autograft and if it is placed in another individual of the same species it is termed as allograft. If the graft is placed in another species, then it is termed as xenograft. Among them, autograft is considered as a gold standard and has long been used with good clinical results, but the supply of autograft is limited. Conversely, allograft and xenograft are not much preferred due to the possibility of pathogen transfer and graft rejection. Furthermore, procurement of living tissue/organ is complex, expensive, and requires an additional surgery. Notably, over 8 million surgical procedures are performed annually to treat tissue and organ failures in the United States of America itself, as per medical statistics [2–4]. Over 80 000 patients are still awaiting tissue and organ transplantations. In this regard, a potential solution is needed to overcome the limitations of traditional tissue transplantation methods and, at the same time, to increase the accessibility and long-time survivability of the tissue implants.

These limitations have instigated the search for alternative approaches and have resulted in the emergence of the concept of tissue engineering, whereby tissue or organ failure is addressed by using synthetic materials cultured with appropriate cells harvested from patient or donor and then implanted in the patient's body where tissue regeneration is required. The synthetic materials used for this purpose are generally called biomaterials. Biomaterial by definition is a substance or a combination of substances, other than drugs, derived either from natural or syn-

Nanotechnologies for the Life Sciences Vol. 9
Tissue, Cell and Organ Engineering. Edited by Challa S. S. R. Kumar
Copyright © 2006 WILEY-VCH Verlag GmbH & Co. KGaA, Weinheim
ISBN: 3-527-31389-3

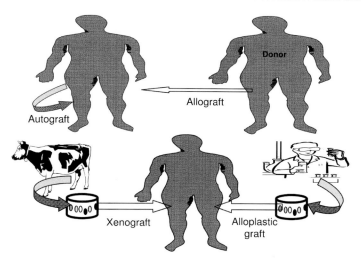

Fig. 5.1. A schematic of tissue transplantation. (Adopted and modified from Ref. [1].)

thetic origin that can be used for any period of time as a whole or as a part of the system that treats, augments, or replaces any tissue, organ, or functions of body [5]. Biomaterials have the advantage of eliminating the surgical process in procuring tissue or organ encountered in the traditional graft procedures, and eliminating the chance of graft rejection or transmission of infectious diseases. In addition, the availability, reproducibility, and cost-effectiveness make the biomaterials more suitable for tissue engineering applications. Figure 5.2 gives some of the clinical uses of biomaterials. Although biomaterials have been used in human health care systems for more than 50 years, the medical community still faces many problematic cases due to the biomaterials' failure before fulfilling their intended functions. There are many reasons for such failure, but one of the prime factors is related to lack of sufficient tissue regeneration around the biomaterials immediately after implantation, which occurs mainly due to the poor surface interaction of biomaterials with the host tissue. In this concern, it is essential to design biomaterials with superior surface properties (e.g., surface area and surface roughness) to facilitate favorable host tissue interactions for their long-term survivability, and to enhance tissue regeneration. High surface area is one of the key properties of biomaterials, which influences cell adhesion and cell density. Surface roughness is another key determinant factor for the occurrence of sufficient host tissue interactions. It has been hypothesized that cell–matrix interactions are governed by surface properties and their imperative interactions found to occur within 1 nm of the implant surface [6]. Apart from surface chemistry, it is also worth mentioning that the cells live in a complex mixture of pores, ridges, and fibers of extra cellular matrix (ECM) in a nano-featured environment. This is because cells attach, organize, and grow well around the fibers with diameters smaller than those of the cells [7]. These information eventually lead to the concept of nanophase biomaterials

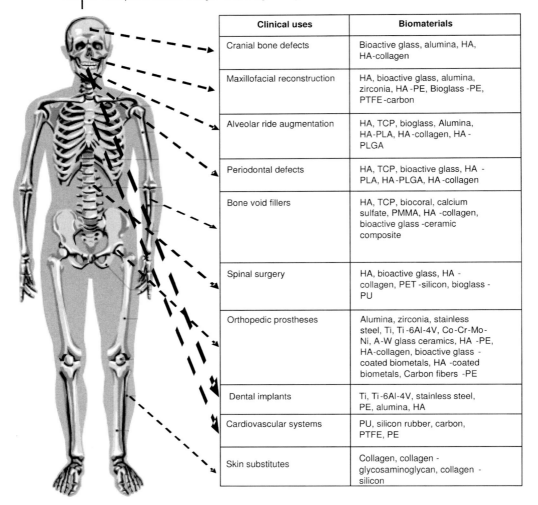

Clinical uses	Biomaterials
Cranial bone defects	Bioactive glass, alumina, HA, HA-collagen
Maxillofacial reconstruction	HA, bioactive glass, alumina, zirconia, HA-PE, Bioglass-PE, PTFE-carbon
Alveolar ride augmentation	HA, TCP, bioglass, Alumina, HA-PLA, HA-collagen, HA-PLGA
Periodontal defects	HA, TCP, bioactive glass, HA-PLA, HA-PLGA, HA-collagen
Bone void fillers	HA, TCP, biocoral, calcium sulfate, PMMA, HA-collagen, bioactive glass-ceramic composite
Spinal surgery	HA, bioactive glass, HA-collagen, PET-silicon, bioglass-PU
Orthopedic prostheses	Alumina, zirconia, stainless steel, Ti, Ti-6Al-4V, Co-Cr-Mo-Ni, A-W glass ceramics, HA-PE, HA-collagen, bioactive glass-coated biometals, HA-coated biometals, Carbon fibers-PE
Dental implants	Ti, Ti-6Al-4V, stainless steel, PE, alumina, HA
Cardiovascular systems	PU, silicon rubber, carbon, PTFE, PE
Skin substitutes	Collagen, collagen-glycosaminoglycan, collagen-silicon

Fig. 5.2. Clinical uses of biomaterials.

(herein called nanobiomaterials) for tissue repair and regeneration. The nanobiomaterials are therefore perceived to be more beneficial in tissue engineering than their microscale counterparts.

This chapter introduces the role of the nanobiomaterials in the success of tissue engineering. Numerous articles describe the potential of biomaterials in tissue engineering applications, but only a very few studies emphasis the impact of nanobiomaterials in constructing tissues or organs. This chapter focuses mainly on nanobiomaterials, in particular ceramic and polymeric nanobiomaterials, and examines prospects for their applications in tissue engineering. The following sections discuss the basic aspects of tissue engineering and the key factors for their success, fabrication methodologies and properties of widely used nanobiomaterials, and current scenarios of nanofibrous scaffold processing techniques. The chapter culminates with a discussion on how these materials influence cell growth functions with illustrated experimental examples.

5.2
Tissue Engineering: A Potential Solution

Tissue engineering is emerging as a significant solution for the traditional tissue and organ transplantations. Tissue engineering is a multidisciplinary theme that applies the principles and methods of engineering and life sciences towards the development of viable substitutes capable of repairing or regenerating the functions of damaged tissue that fails to heal spontaneously. Tissue repair is a process, which involves the usual inflammatory cell cascade, followed by matrix deposition and then a remodeling process, which attempts to regenerate the damaged tissue into healthy tissue. In contrast, tissue regeneration involves gradual replacement of damaged tissue with identical healthy tissue. The major advantage of the tissue engineering approach is that tissues can be perfectly reconstructed in such a way that they match exactly the patient's requirements and can be transplanted into the patient's body with a minimal surgical intervention, which eventually conquers several limitations encountered in the conventional tissue transplantations.

The prime concept of tissue engineering is to harvest a small biopsy of specific cells from the donor site, seed them on a scaffold to culture a specific tissue, and transplant the cultured tissue into the defective site of the patient's body that needs tissue regeneration. The ability to satisfy these criteria is largely dependent on three key factors: (a) the cells that create tissue, (b) the scaffold that gives structural support to cells, and (c) cell–matrix (scaffold) interactions that direct the tissue growth (Fig. 5.3). Among them, cells are the prime determinant factor for the success of tissue engineering because they are the basic units of all life. There are different types of cells that could be used for tissue reconstruction, which are commonly categorized as mature cells (non-stem) and stem cells. Although the use of mature cells dates far back, they are generally differentiated cells, which give rise to low rate of proliferation. As an alternative, stem cells have gained much recognition recently, owing to their ability to overcome many of the limitations of mature cells. The potential of stem cells are briefly discussed in Section 5.3, with experimental examples. The scaffold is another important determinant factor for the success of tissue engineering, which basically provides a structural support for the cells to grow in the three-dimensional (3D) space into a specific tissue. The success of the scaffold typically depends on the quality of starting material and the fabrication methodology (much discussed in Sections 5.4 and 5.5, respectively). Cell–matrix (scaffold) interactions is another key determinant factor for the success of tissue engineering because it greatly influences tissue growth. As aforementioned, cell–matrix interactions are governed by surface properties and their imperative interactions found to occur within 1 nm of the implant surface. Furthermore, the

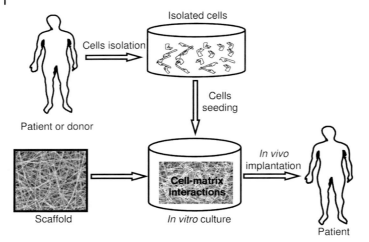

Fig. 5.3. Key factors contributing to the success of tissue engineering.

growth of new tissue depends on the way the seeded cells interact with the scaffold. In this regard, cell seeding density must be optimized to enhance the cell–matrix interactions. However, several growth factors and biological molecules are also involved in the cell–matrix interactions (Section 5.6).

5.3
Stem Cells: The Essentials

A key factor in tissue engineering approach for repair or regeneration of defective tissue is the use of appropriate cell source. Currently, there are various cells available corresponding to specific tissues. Although the use of mature cells has a successful history and remains an important source for tissue engineering, rapid advances are being made in identifying new cell types that has the capacity to transform into specific cell type that may be required to produce appropriate tissue. Stem cells have recently been recognized as a promising alternative to mature cells owing to their enormous potential in generating a spectrum of tissues with adequate functions. With the advances of cell biology, several well-characterized tissue-specific stem cells are currently available for tissue engineering.

Stem cells, by definition, are immature or undifferentiated cells that are able to renew by themselves and differentiate into more specialized type of cells/tissues in response to appropriate signals and cell plasticity. A diverse range tissues can now be engineered using stem cells, from epithelial surfaces to skeletal tissues. This ability allows them to act as a good repair system for the defective tissue or organ of our body. Stem cells are generally classified into three types: totipotent (e.g., zygote), pluripotent (e.g., embryonic stem cells), and multipotent (e.g., mesenchymal stem cells). The totipotent stem cells are capable of producing any type of cells/tissues. The pluripotent stem cells are capable of producing most of the cells/tissues, whilst multipotent stem cells produce a limited number of cells/tissues with specific functions. Stem cells are also categorized into embryonic or adult cells according to their source of origination. Embryonic stem cells, as their name suggests, are derived from the early stage of an embryo. Although embryonic stem cells research is thought to have much greater potential than adult stem cells, several ethical and legal controversies still exist concerning their use for humans. Besides, some of the important clinical and safety issues also hinder their use to a certain extent. Conversely, adult stem cells have considerable biological and clinical interests. Basically these are undifferentiated cells found in differentiated adult tissues. They are also called somatic stem cells. The potential advantage of using adult stem cells is that the patient's own cells could be easily isolated, expanded in culture with appropriate functions, and then transplanted into the patient's body where tissue regeneration is required. By this approach, there is no chance of immune rejection and certain ethical and legal issues can also be conquered. These cells are typically harvested from bone marrow. The bone marrow chiefly constitutes two distinct stem cells: (a) hematopoietic stem cells (HSCs), which are responsible for the formation of blood cells, and (b) mesenchymal stem cells (MSCs), which are responsible for the formation of bone cells.

Mesenchymal stem cell transplantation is considered as a promising method to improve various tissue functions of mesenchymal origin. MSCs are one of the well-characterized populations of stem cells residing in various adult mesenchymal tissues; thereby they are well adopted for tissue engineering with considerable interest and, subsequently, for clinical practice. These cells were first recognized by Friedenstein and group in the 1960s, wherein they noticed that an adherent, fibroblast-like cell population could have the ability to regenerate the essentials of normal bone [8]. Notably, the early investigations using these cells are mainly focused on bone tissue engineering. Nowadays, these cells are being applied to various tissue engineering, including cartilage, myocardium, liver, nerve, spinal cord, and dermal tissues. The advantages of using MSCs are that they can be easily harvested from the donor site and cultured into a spectrum of tissue-specific phenotypes of mesenchymal origin, such as bone, cartilage, muscle, marrow stroma, and tendon (Fig. 5.4) [9–11]. These features attracted the attention of the biomedical community towards stem cell-based therapy, in particular MSCs. The list of tissues with the potential to be engineered has recently grown progressively due to the advances in stem cell biology and sophisticated characterization techniques, which indicates the prospects for future MSCs-based tissue engineering.

Notably, once cells are isolated from their natural environment, they tend to lose their phenotype and differentiated functions. It is really a great challenge to refurbish the cell phenotypic expression and differentiated functions without hindering proliferation under laboratory conditions. A few studies have reported the direct delivery of cell suspension without using scaffolds [12, 13], but these processes encountered difficulties in having poor control over the localization of transplanted cells; thereby, notably, the cell expression may not be achieved without a suitable scaffold. Numerous investigations have reported the effective delivery of MSCs us-

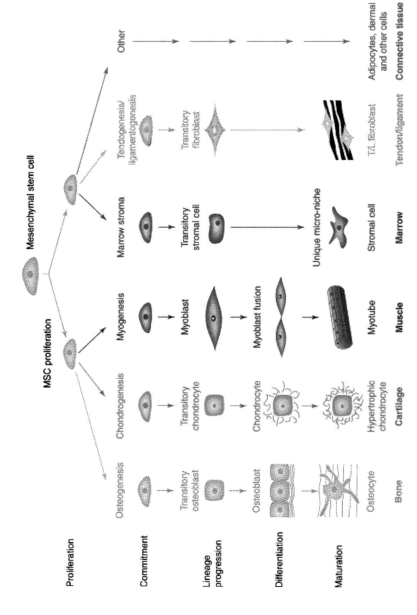

Fig. 5.4. Schematic of mesengenic process, showing stepwise cellular transitions from the putative mesenchymal stem cell to highly differentiated phenotypes. (Adopted with permission from Ref. [11].)

ing scaffolds made of polymers, ceramics, and their composites [14–16]. Recently, Li et al. [17] have reported the ability of MSCs to differentiate into chondrocytic phenotype, a cartilage-specific tissue. First, they fabricated the nanoporous biodegradable polymer scaffolds using poly(ε-caprolactone) (PCL) by the electrospinning method followed by *in vitro* assessment of chondrogenesis, in the presence of transforming growth factor-β (TGF-β). The results indicated that the MSCs are capable of giving rise to chondrocytes and also stimulate the synthesis of cartilage-associated ECM proteins throughout the scaffold. Therefore, they suggested that the electrospun PCL scaffold is a practical carrier system for MSCs transplantation in tissue engineering-based cartilage repair. Another interesting study, reported by Yoshimoto et al. [18], demonstrated the ability of MSCs to differentiate into osteoblastic phenotype. To facilitate this, the authors manufactured the nano-featured PCL scaffolds by electrospinning, and their ability of MSCs differentiation was studied under *in vitro* cell culture conditions. They found that the scaffolds provided a suitable environment to stimulate the MSCs to differentiate into osteoblastic phenotype, a bone-specific tissue. Furthermore, the scaffolds supported the synthesis of ECM of type I collagen throughout the matrix, which is a good sign of mineralized tissue formation; thereby, the authors suggested that the PCL scaffold may be a potential candidate for bone tissue engineering. This experimental information suggests that most of the cells are anchorage-dependent and will not survive if delivered without a suitable scaffold. Processing a perfect scaffold with all the qualities similar to natural ECM is, therefore, of great importance for tissue engineering.

5.4
Nanobiomaterials: A New Generation Scaffolding Material

Recent progress in nanoscience and nanotechnology has drastically increased the success rate of nanobiomaterials as a scaffolding material in tissue engineering. Scaffold, by definition, is a temporary supporting structure for growing cells and tissues. It is also called synthetic ECM, which plays a significant role in supporting the cells to accommodate. These cells then undergo proliferation, migration, and differentiation in 3D and eventually lead to the formation of a specific tissue with appropriate functions. Numerous studies have reported the prospects of nanobiomaterials as tissue scaffolds. Nanobiomaterials generally refers to biomaterials with the basic structural unit less than 100 nm (nanostructured), crystalline solids with a grain size less than 100 nm (nanocrystals), ultrafine powders with an average particle size less than 100 nm (nanopowders), and extremely small fibers with a diameter less than 100 nm (nanofibers). A nanometer is a billionth of a meter (10^{-9} m). For ease of understanding, Fig. 5.5 illustrates the scale of some of natural and manmade things, highlighting nanoscale items [19]. Notably, all hard and soft tissues of our body contain plenty of cells living in ECM at the nanoscale hierarchical structure elegantly designed by the Mother Nature. For example, bone tissue can be considered as an assembly of various levels of hierarchical structural

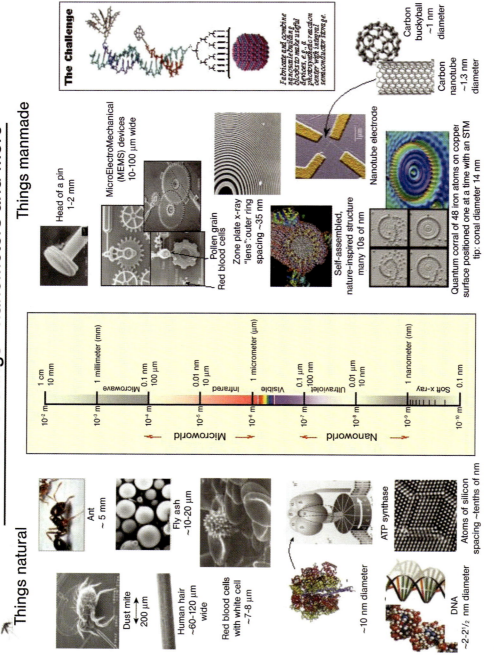

Fig. 5.5. The scale of various natural and manmade things. (Adopted by courtesy from Ref. [19].)

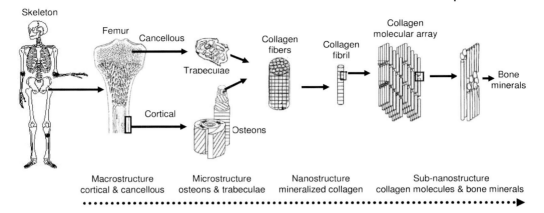

Fig. 5.6. Hierarchical structure of bone, from macro- to nanoassembly. (Adapted with permission from Ref. [20].)

units designed on many length scales ranging from macro to nano, using essential
organic and inorganic components, to facilitate multiple functions required for tissue formation (Fig. 5.6) [20]. The nano- or micro-featured environment of the ECM
is critical for the proper functions of cells and tissues. The cells can, reportedly, attach and organize better around the nanobiomaterials than around their microscale counterparts because of their typical surface properties [21]. The fact is that
nanobiomaterials have more atoms and crystal grains at the surfaces, and possess
higher surface area to volume ratio, than conventional microscale biomaterials,
making the surface of nanobiomaterials more reactive to cultured cells (during
in vitro) and to host tissue (during *in vivo*), and thus greatly enhancing the cell–
matrix interactions, leading to faster tissue regeneration. In this regard, the rate
of tissue regeneration will be greater for nanobiomaterials than for conventional
biomaterials. Nanobiomaterials are, therefore, perceived to be beneficial for tissue
engineering applications as a new generation scaffolding material. To facilitate
these measures, the scaffold should possess a few essential characteristics, which
are discussed below.

5.4.1
Characteristics of Scaffold

The structural and behavioral characteristics of the tissue scaffold are critical to ensure normal cell activities and performance of the cultured tissues. An ideal scaffold for tissue engineering should possess all the qualities of a natural ECM and
should function in the same way as that of ECM under physiological conditions.
However, there is no clear guideline as to which characteristic defines the so-called
ideal scaffold. Notably, the characteristics of a scaffold vary according to the tissue
types where the scaffold is to be applied. For example, scaffold for engineering

Tab. 5.1. Characteristics of tissue scaffold.

Characteristics	General remarks
Biocompatible	Biologically compatible to host tissue, i.e., should not provoke any rejection, inflammation, and immune responses
Biodegradable	The rate of degradation must perfectly match the rate of tissue regeneration and the degraded product(s) should not harm the living tissues
Vascular supportive	Should provide channels for blood supply for fast and healthy tissue regeneration
Non-toxic	Should not evoke toxicity to tissues
Non-immunogenic	Should not evoke immunogenic response to tissues
Non-corrosive	Should not corrode at physiological pH and at body temperature
Porosity with interconnected pores	To maximize the space for cellular adhesion, growth, ECM secretion, revascularization, adequate nutrition and oxygen supply without compromising mechanical strength
3D structure	To assist cellular ingrowth and transport of nutrition and oxygen
High surface area to volume ratio	To accommodate high density cells
Surface modifiable	To functionalize chemical or biomolecular groups to improve tissue adhesion
Adequate mechanical strength	To withstand hemodynamic and other biological forces *in vivo*
Sterilizable	To avoid toxic contamination

bone tissues requires an osteoconductive feature but it is not required for engineering nerve or dermal tissues. Table 5.1 gives some of the basic characteristics of a perfect scaffold for generalized tissue engineering applications. Any scaffold, irrespective of applications, should be biocompatible, meaning that it should not provoke any rejection, inflammation, and immune responses. The scaffold should provide a 3D template with interconnected porous architecture for the maximum loading of cells, tissue in-growth, and transportation of nutrients and oxygen. It should be able to facilitate several biochemical and biological process, in synthesizing specific proteins, required for a healthy tissue growth in the bodily environment. Notably, most tissue engineering applications require scaffolds that are biodegradable, but the rate of degradation must match the rate of tissue regeneration. The scaffold should be mechanically strong to withstand hemodynamic and other biological forces *in vivo*, otherwise it may hinder tissue formation. Besides, the scaffold should have high surface area to facilitate better cell adhesion and interaction with each other, and should also have a surface roughness similar to that of natural tissues for the occurrence of enhanced host tissue interactions. The scaffold should also be vascular supportive because cells will not survive without an adequate blood supply. Finally, the scaffold should be stable during storage, and must be sterilizable, to avoid toxic contaminations, without compromising any structural and other related properties. The pragmatic success of tissue scaffold is not only dependent on the above measures but also on appropriate communication between the cells, tissues, and the host system as a whole.

5.4.2
Types of Scaffolding Materials

The materials used to manufacture scaffold play a key role in supporting the growth of cells, which are cultured on the scaffold. In this regard, selection of materials is of great importance for the success of tissue engineering. Based on the facts discussed in Section 5.4, nanobiomaterials are perceived to be more beneficial than their microscale counterparts owing to their superior surface characteristics. Nanobiomaterials currently investigated as scaffolding materials for tissue engineering applications can be classified into four types: metals, ceramics, polymers, and composites thereof (Table 5.2). Each type of material has its own distinct properties that can be advantageous for specific tissue engineering. The choice of materials depends on the type of tissue to be reconstructed. For example, metals and ceramics are widely used for manufacturing scaffolds for hard tissue applications, whilst polymers are used for soft tissue applications, and composites are considerably used for both the applications due to their mechanical properties (Table 5.3) [22–24]. This information helps clinicians to choose a right material for a specific application. In the following sections we focus on the widely used nanobiomaterials, in particular ceramic and polymeric nanobiomaterials, for tissue engineering applications.

5.4.2.1 **Ceramic Nanobiomaterials**
Ceramics represent a class of nanobiomaterials that is widely used as a scaffolding material in hard tissue repair and regeneration (e.g., bone). The ceramics used in tissue engineering can be classified into three major groups corresponding to their ability to interact with host tissue: (a) bioinert (e.g., alumina and zirconia), (b) bioactive (e.g., HA and bioglass), and (c) bioresorbable [e.g., tricalcium phosphate (TCP)]. It is desirable, for bone tissue engineering, that the scaffolding material mimic the natural bone in chemical composition and phase structure, to facilitate better osteointegration and other related functions. In this regard, HA [hydroxyapatite, $Ca_{10}(PO_4)_6(OH)_2$] may be considered a good choice of scaffolding material. We, therefore, focus on the properties and the various processing methodologies of

Tab. 5.2. A broad classification of nanobiomaterials.

Nanobiomaterials	Advantages[a]	Disadvantages	Applications	Examples
Metals & alloys	Strong, tough, ductile	Dense, may corrode, difficult to make	Load-bearing bone implants, dental restoration, etc.	Nanostructured titanium and Ti-6Al-4V alloys
Ceramics	Bioinert Bioactive Bioresorbable High resistance to wear, corrosion resistance	Brittle, low toughness, not resilient	Low weight-bearing bone implants, dental restoration, tissue scaffolds, bone drug delivery, etc.	Nano alumina Nano HA Nano TCP
Polymers	Flexible, low density, resilient, surface modifiable, chemical functional groups	Low stiffness, may degrade	Tissue scaffolds, drug delivery, breast implant, sutures, skin augmentation, blood vessels, heart valves, etc.	Collagen and PLLA nanofibers
Composites	Strong, design flexibility, enhanced mechanical reliability than monolithics	Properties might be varied with respect to fabrication methodology	Tissue scaffolds, drug delivery, dental restoration, spinal surgery, etc.	Nano HA-collagen, nano HA-PLLA

[a] Common characteristics of nanobiomaterials, such as biocompatibility, high surface area, etc. are not highlighted.

HA in detail. HA possesses most of the qualities required for bone tissue engineering, in particular biocompatibility, bioactivity, and osteoconductivity. Table 5.4 gives a compilation of some of the physical, mechanical, chemical, and biological properties of HA that make it an appropriate scaffolding material [1, 23–29]. As per the literature survey, several methods for processing HA, either from natural sources (e.g., coral exo-skeleton and animal bone) or from synthetic sources (e.g., inorganic chemical synthesis), have been reported [30–36], but these studies concentrated on microscale HA.

Recently, nanoscale HA has received much attention owing to its superior surface functional properties over its microscale counterpart, particularly surface area and surface roughness, which are the most imperative properties of nanobiomate-

Tab. 5.3. Mechanical properties of biological tissues and biomaterials. (Compiled from Refs. [22–24].)

Materials	Young's modulus (GPa)	Tensile strength (MPa)
Soft tissues		
Articular cartilage	10.5	27.5
Fibrocartilage	159.1	10.4
Ligament	303.0	29.5
Tendon	401.5	46.5
Skin	0.1–0.2	7.6
Hard tissues		
Cortical bone	7–30	50–150
Cancellous bone	1–14	7.4
Dentine	11–17	21–53
Enamel	84–131	10
Metals		
Ti	110	300–740
Stainless steel	190	500–950
Ti-6Al-4V alloy	120	860–1140
Co-Cr alloy	210	665–1200
Ceramics		
Alumina	380	310
Zirconia, partially stabilized	200	420
HA	30–100	50–190
Polymers		
Biodegradable		
Poly(L-lactic acid)	2.7	50
Poly(D,L-lactic acid)	1.9	29
Poly(caprolactone)	0.4	16
Non-biodegradable		
Polyethylene	0.88	35
Polyurethane	0.02	35
Poly(methyl methacrylate)	2.55	59
Poly(ethylene terephthalate)	2.85	61
Composites		
HA/PE (40/60)	4.29	20.67
Bioglass/PE (40/60)	2.54	10.15
Glass-ceramic/PE (40/60)	2.84	14.87

Tab. 5.4. Physical, mechanical, chemical, and biological properties of HA. (Compiled from Refs. [1, 23–29].)

Properties	Experimental data
Chemical composition	$Ca_{10}(PO_4)_6(OH)_2$
Ca/P (molar)	1.67
Color	White
Crystal system	Hexagonal
Space group	$P6_3/m$
Cell dimensions (Å)	$a = b = 9.42, c = 6.88$
Young's modulus (GPa)	80–110
Elastic modulus (GPa)	114
Compressive strength (MPa)	400–900
Bending strength (MPa)	115–200
Density (g cm^{-3})	3.16
Relative density (%)	95–99.5
Fracture toughness (MPa m$^{-1/2}$)	0.7–1.2
Hardness (HV)	600
Decomposition temp. (°C)	>1614
Melting point (°C)	1000
Dielectric constant	7.40–10.47
Thermal conductivity (W cm^{-1} K^{-1})	0.013
Biocompatibility	High
Bioactivity	High
Biodegradation	Low
Cellular compatibility	High
Osteoinduction	Nil
Osteoconduction	High

rials because of enhanced cell–matrix interactions. For the past few years, significant research effort has been devoted to the production of nanosize HA [37–39], to obtain high surface reactive materials with enhanced physicochemical and biological properties compared with their microscale counterparts and, at the same time, quite similar to natural bone mineral. It has also been proved that the nano HA, compared with conventional HA, promotes osteoblast adhesion, differentiation and proliferation, which leads to enhanced formation of new bone tissue within a short period [21, 40]. With reference to this information, nano HA may be considered as a unique class of ceramic scaffolding material for bone tissue engineering. The nano HA can be synthesized by many different methods, including solid state [41], wet chemical [42, 43], hydrothermal [44, 45], mechanochemical [46], pH shock wave [47], and microwave processing [48]. Table 5.5 depicts some of the general remarks on these methods, and the detailed processing conditions of nano HA are described in the following sections.

Tab. 5.5. Methods involved in the synthesis of nano HA.

Methods	Grain sizes (nm)	General remarks	Ref.
Solid state	500	Inhomogeneous, large grain sizes, irregular shapes, reaction temp.: 900–1300 °C	41
Wet chemical	20–200	Nano size grains, low crystallinity, homogeneous, reaction conditions: room temp. to 100 °C at elevated pH	42, 43
Hydrothermal	10–25	Homogeneous, ultrafine particles, reaction condition: high temp. and high pressure atmosphere	44, 45
Mechanochemical	<20	Easy production, semi-crystallinity, ultrafine crystals, room temp. process	46
pH shock wave	300	High-energy dispersing, nonporous, monocrystalline particles	47
Microwave	100–300	Uniformity, nanosize particles, time and energy saving	48

Solid-state Reaction Solid-state reaction has generally been used for the processing of HA at high temperature. HA powders synthesized by this method usually have irregular shapes with larger grain size, and they quite often exhibit heterogeneity in the phase composition, owing to chemical reactions resulting from small diffusion coefficients of ions within the solid. A general formation of HA using the solid-state method is based on Reaction 1:

$$4CaCO_3 + CaHPO_4 \cdot 2H_2O \rightarrow Ca_{10}(PO_4)_6(OH)_2 + 4CO_2$$
$$+ 14H_2O \text{ at } 1000\text{–}1300 \text{ °C} \qquad (1)$$

The key ingredients of this reaction are basically in the solid phase, in which $CaCO_3$ acts as a calcium precursor and $CaHPO_4 \cdot 2H_2O$ acts as a phosphorous precursor in the formation of HA. The solid-state reaction typically takes place at a very high temperature ($\sim$1000 °C). A method for the preparation of HA fibers was introduced using a solid-phase reaction [41]. With this method, the HA fibers were produced by heating a compact consisting of calcium metaphosphate fibers with calcium hydroxide particles at 1000 °C in air, and treated subsequently with dilute aqueous HCl solution to remove unwanted secondary phase substitution such as CaO. The obtained fibers were characterized by various analytical methods, and the results confirmed the formation of HA with nanostructural features. Therefore, the solid-state reaction can be considered as one of the methods for the production of nano HA.

Wet Chemical Method   This method is one of the promising methods, widely used in synthesizing nano HA at low temperature, in contrast to the solid-state reaction. HA powders synthesized by using this method have a homogeneous phase composition, but are poorly crystallized owing to the low temperature process. One common wet chemical method of synthesizing HA is based on Reaction (2), which uses calcium hydroxide and orthophosphoric acid as calcium and phosphorous precursors, respectively.

$$10Ca(OH)_2 + 6H_3PO_4 \rightarrow Ca_{10}(PO_4)_6(OH)_2 + 13H_2O \tag{2}$$

Recently, rod-like nano HA was synthesized by this method [42]. In brief, a H_3PO_4 solution (0.3 M) was added drop-wise to a 0.5 M $Ca(OH)_2$ solution under continuous stirring at room temperature, while the pH was kept above 10.5 by the addition of ammonia solution. Stirring was maintained for a further 16 h after complete addition of the reactants. The precipitate thus obtained was further aged for a week and was examined for phase purity and chemical composition using various analytical methods. The results indicated the possibility of producing phase pure, nanosize, and rod-like HA by a wet chemical method. Our group, Murugan et al. [38, 43], have also synthesized HA nanoparticles by using this method, and have extensively studied their physicochemical and physiological properties at body pH. The results clearly showed the enhanced resorbable characteristics of nano HA compared with their microscale counterpart, owing to their larger surface area to volume ratio, which ultimately makes the nano HA highly surface reactive. Based on these experimental data, the wet chemical method is one of the most promising methods for the production of nano HA.

Hydrothermal Process   This hydrothermal process involves the reaction of an aqueous solution of calcium and phosphorous precursors at ambient temperature and pressure for the production of nano HA. It enables the synthesis of crystallized HA with homogenous phase composition. The HA so-prepared is easily sinterable, owing to the effects of high temperature and high pressure aqueous solutions. Zhang and Consalves [44] have synthesized nano HA by precipitating the precursors under hydrothermal conditions, and they studied its thermal stability in detail. By using this approach, they obtained rod-like, nano-sized HA with an average crystal size of 10 to 24 nm, according to the composition of the precursors. They prepared various HA powders with a Ca/P ratio ranging from 1.66 to 1.73. In another interesting study [45], the efficacy of nano HA as filler in fabricating tissue scaffold has been reported. The authors produced nano HA by a hydrothermal method. In brief, 2 g of ammonium phosphate dissolved in 37.5 mL of de-ionized water was added to 5.9 g of calcium nitrate dissolved in 22.5 mL of de-ionized water under stirring, and the pH of solution was adjusted to 12 by adding ammonia. The solution was then autoclaved under hydrothermal conditions (170 °C and 19.5 MPa) and subsequently aged for 5 h. The precipitates thus obtained were cooled to room temperature, repeatedly washed with distilled water to remove traces of ammonia and ammonium nitrate, dried at 80 °C overnight, and then ground to a fine powder. Characterization of the powders indicated the formation and purity of HA. The HA particles were ellipsoidal-shaped with needle-like morphologies. The particles were, on average, approximately 25 nm wide and 150 nm long with aspect ratios ranging from 6 to 8, indicating nanostructure features. The authors further investigated the efficacy of nano HA as filler in polymer matrix in fabricating porous 3D tissue scaffolds. The results showed that the scaffold consisting of nano HA and polymer should be suitable for non-load sharing tissue engineering applications as compared with scaffold made of polymer alone, owing to their enhanced compression modulus.

Mechanochemical Route   Production of nano HA by the mechanochemical route has gained interest in recent years, owing to its simplicity. The processing conditions involved in this method slightly differ from the conventional wet chemical method. Notably, the purity of material synthesized by conventional wet chemical method depends on the reaction pH, whereas reaction through the mechanochemical route need not be under precise pH control. Nakamura et al. [46] have reported the synthesis of nano HA by using this method. The HA was prepared directly by milling a mixture of $Ca(OH)_2$, H_3PO_4 and a dispersant, an ammonium salt of polyacrylic acid. The reaction was carried out at room temperature. The prepared HA powders have relatively crystallized particles in the nanometer regime. The average crystallite size was 20 nm. The authors suggested some of the merits of this method, which include (a) assisting deprotonation from brushite to form HA and (b) keeping high dispersibility and low viscosity of reaction sol. This method is of particular interest in synthesizing nano HA with ultrafine crystallite sizes.

pH Shock Wave Method   Another interesting method for the production of nano HA is pH shock wave. Koumoulidis et al. [47] have reported the synthesis of HA in the form of lath-like monocrystalline particles using the precursors of $Ca(H_2PO_4)_2 \cdot H_2O$ and $CaCl_2$. They employed high-energy dispersing equipment to synthesize nano HA. The prepared HA was characterized by various analytical methods to determine the phase composition, structure, and texture. The results showed that the material has a different Ca/P molar ratio (1.43 to 1.66) with respect to processing conditions. The crystal grains of HA were 140–1300 nm long, 20–100 nm wide, and 10–40 nm thick. With reference to experimental results, this method seems to be quite interesting because it creates appropriate hydrodynamic conditions for lath-like particle growth in the [001] direction, which could not be obtained by using conventional mechanical stirring equipment.

Microwave Processing   Microwave-assisted synthesis of ceramic nanobiomaterials is a relatively new method, which has recently been of interest in synthesizing nano HA. Sarig and Kahana [48] reported the processing conditions for the rapid formation of nanocrystalline HA. Briefly, the recipe for the synthesis of nano HA is: the HA powder was precipitated from $CaCl_2$ 10 mM aqueous solution (A) and NaH_2PO_4 6 mM aqueous solution (B). The solution A was used without admixture, whereas the solution B was made with a concentration of 25 ppm with respect to L-aspartic acid and 150 ppm with respect to $NaHCO_3$. Both solutions were kept at pH 7.4 by admixture of Trizma. This pH was selected to make the HA appropriate for medical applications. Equal volumes of solutions A and B (250 mL each) were introduced simultaneously into a 1000 mL beaker, which was put into a 2450 MHz microwave oven at maximum power for 5 min. The microwave-irradiated mixture was then quenched in ice for 30 min. The resultant precipitate was filtered off, washed with water, and dried overnight at 55 °C. The powder thus obtained was characterized and found to be bouncy and free-flowing. It was composed of spherulites of about 2–4 μm in diameter, and the spherulite was composed of ultrafine platelets of about 300 nm. Notably, the powder exhibited a quite peculiar behavior on storage, in contrast with commercial apatites. Commercial powders usually tend to aggregate and form large solid lumps, whereas the powder produced by this method remained free-flowing after 3 years of storage in non-hermetically closed containers. Yang et al. [49] further reported the sintering effect of HA by using microwaves. They synthesized HA by precipitating the calcium and phosphorous precursors at 95 °C with a pH of 9 to 11.5. The HA green samples were sintered by microwave heating and the results were compared with other HA green samples sintered by a conventional heating method. The results clearly indicated that the HA prepared by microwave processing was denser and had a finer grain size than those prepared by the conventional heating method, which is the merit of microwave processing. Overall, microwave-assisted synthesis offers the advantages of uniform heating throughout the volume with very efficient transformation energy within a short period; thereby, it may be considered as a suitable method for the production of HA nanocrystals.

5.4.2.2 Polymeric Nanobiomaterials

Polymers, the largest class of biomaterials, are made up of repeated small and simple chemical units called monomers. The term "polymer" was derived from two Greek words, *polys* meaning "many" and *meros* meaning "units". Polymers in the form of nanofibers are considered as a unique class of scaffolding material and are in demand for various tissue engineering applications compared with other types of nanobiomaterials, owing to their functional properties, design flexibility, and surface modifiability. The polymer nanofibers used for this purpose are commonly called polymeric nanobiomaterials. A nanofiber generally refers to a fiber having a diameter size in the range of 1 to 100 nm. Although the polymer nanofibers have many desirable characteristics, they tend to possess low mechanical strength (e.g., low stiffness) compared with nanophase metals and ceramics; thereby, they are often used in soft tissue reconstructions, but are also used in hard tissue reconstructions in conjunction with nanophase ceramics. The following sections focus on the major classification of polymers and emphasize widely used biodegradable polymers in the process of nanofibrous scaffolds for tissue engineering applications.

Polymers used in tissue engineering can be grouped into (a) naturally-derived and (b) synthetic polymers. Naturally-derived polymers are biodegradable. Collagen, gelatin, and chitosan are a few notable examples. Synthetic polymers can be either biodegradable or non-biodegradable. Poly(lactic acid) (PLA), poly(glycolic acid) (PGA), and poly(lactic-*co*-glycolic acid) (PLGA) are notable examples of biodegradable polymers, and poly(ethylene) (PE), poly(ethylene terephthalate) (PET), and poly(tetrafluoroethylene) (PTFE) are notable examples of non-biodegradable polymers. In most circumstances, biodegradable polymers, either from natural or from synthetic origin, are considered as good choice for tissue engineering rather than non-biodegradable polymers. However, degraded products from the biodegradable polymers must be non-toxic and should not elicit any foreign body reaction. Based on their astonishing qualities, we further focus our attention on some of the widely used biodegradable polymers, both naturally-derived and synthetic, in tissue engineering applications.

Naturally-derived Biodegradable Polymers Collagen is the most widely used naturally-derived polymer in designing scaffolds for tissue engineering. It is a primary structural protein of the natural ECM and hence natural collagen has various functional characteristics favorable for cells and tissue growth. Although many types of collagen exist in a living organism, the most abundant forms in native tissue are types I and III. Type I collagen is composed of two α_1 (I) chains and one α_2 (I) chain in the triple helix pattern with a fiber diameter of about 50 nm. Type III collagen is composed of three α_1 (III) chains with fibers ranging from 30 to 130 nm in diameter. Structurally, collagen is composed of three polypeptides (α-chains) that are each coiled into a left-handed helical pattern, and then these three chains are wrapped around each other into a right-handed helical pattern, resulting in well-organized rope-like fibers of great structural strength. The triple-helical domain has a characteristic primary structure, where glycine in every third amino acid generates repeating (Gly X–Y) *n* units, where X is alanine or proline, and Y is hydroxyproline. In general, collagen extracted from natural tissues is capable of eliciting an immunogenic response upon implantation; thereby direct use of this type of collagen is limited. Nowadays, a purified form of collagen known as reconstituted collagen, which has relatively lesser immunogenic response, is produced by various biochemical methods and is commercially available for tissue engineering applications.

Gelatin is a denatured form of collagen, obtained by acidic and alkaline process, widely used in tissue engineering as a scaffolding material. There are two types of gelatin, namely A and B, which are identified by the method of processing from the native collagen. Gelatin A can be obtained by the extraction of collagen by acidic treatment, and gelatin B can be obtained by the extraction of collagen by alkaline treatment. Gelatin B has a higher content of carboxylic groups than gelatin A owing to their biochemical processing. Although gelatin is a denatured form of collagen, it has its own functional properties. Gelatin is a highly biocompatible, bioresorbable, non-toxic, and non-immunogenic natural polymer. It does not elicit any noticeable antigenic response and has a low coagulation activity towards platelets. Based on these characteristics, gelatin has some individual importance as a scaffolding material for various tissue engineering.

Chitosan is also a naturally-derived polymer (polysaccharide), which is obtained by alkaline deacetylation of chitin. The chitin is extracted from the exoskeleton of shellfish. Basically, this process consists of two steps: (a) deproteination of the shells with a dilute sodium hydroxide solution and (b) decalcification with a dilute hydrochloric acid solution. The chitin thus obtained is subjected to N-deacetylation by treatment with a 40–45% of sodium hydroxide solution followed by purification procedures, resulting in chitosan. Chemically, it is a copolymer of $\beta(1-4)$ linked 2-acetamido-2-deoxy-D-glucopyranose and 2-amino-2-deoxy-D-glucopyranose. It is a weak base and thus it is insoluble in water and in a few organic solvents, but it is soluble in dilute aqueous acidic solution with pH < 6.5. It is a biocompatible, biodegradable, and non-toxic polymer. It shows good antimicrobial and antifungal activities. Due to its natural abundance and specific physicochemical and biological properties, it is used in tissue engineering applications. Further, its plasticity and adhesiveness make chitosan a suitable binder for various biomedical applications.

Synthetic Biodegradable Polymers Synthetic biodegradable polymers play an increasingly pivotal role in tissue engineering. These polymers offer many advantages over naturally-derived polymers in that they can be easily tailored to provide a wide range of functional properties; in particular, freedom from concerns of immunogenicity. Further, they are easily processable, surface modifiable, and sterilizable. Notably, not all the polymers are widely used in tissue engineering, but there are a few polymers (e.g., PLA, PGA, and PLGA) that have been approved for clinical applications by the US Food and Drug Authority, which can be considered for tissue engineering applications.

PLA is one of the most frequently used synthetic polymers in designing tissue scaffolds, owing to its functional characteristics that favor cell and tissue growth. It has two enantiomeric forms: (a) the left-handed (L-lactide), and (b) the right-handed (D,L-lactide). These two enantiomers have different degradation rates. Hydrolysis is the principal mode of degradation and it is greatly affected by the hydrophilicity and crystallinity of the polymer. The degradation products of PLA are non-toxic. The L-lactide is widely used for tissue engineering applications owing to its superior biocompatibility and prolonged biodegradation compared with D,L-lactide. Furthermore, it possesses high strength and modulus. The D,L-lactide possesses, in contrast, low strength and modulus, and undergoes rapid biodegradation.

PGA is the simplest form of linear aliphatic polyester, being used in various tissue engineering applications. It is also a biocompatible and a biodegradable polymer. It can be synthesized by ring-opening polymerization of glycolic acid, but does not form any enantiomers like PLA. It is highly crystalline and is a hydrophilic polymer. The percent crystallinity ranges from 45 to 55. Owing to its hydrophilic nature, it is completely absorbed within 4 to 6 months upon implantation. It exhibits a high tensile strength, modulus, and is also stiff, making it suitable for biomedical applications.

PLGA is a copolymer that consists of PLLA and PGA macromolecular units. By changing the composition of copolymeric units, the rate of degradation and mechanical strength of PLGA can be manipulated. This kind of copolymer has less crystallinity and degrades faster than homopolymers. For example, a copolymer of 50% PGA and 50% PLLA degrades much faster than each homopolymer. The composition of copolymer also significantly affects its percent crystallinity. For example, a copolymer of 90% PGA and 10% PLLA has high crystallinity, whereas a PGA-to-PLLA ratio from 25/75 to 75/25 is relatively amorphous. Owing to its biocompatibility and tunable biodegradability, PLGA is widely used in various tissue engineering applications.

5.5
Nanofibrous Scaffold Processing: Current Scenarios

Processing nanofibrous scaffolds with structural features similar to natural ECM is essential for tissue engineering. Numerous studies have been conducted in processing tissue scaffolds quite similar to natural ECM in the authors' laboratory [50–52]. These scaffolds possess high surface area, high aspect ratio, high porosity, small pore size, and low density. These features are essential for the improvement of cell adhesion, which is a significant issue in the initial stage of tissue engineering because cell migration, proliferation and differentiation functions typically depend on the cell adhesion. Although tissue scaffolds can be manufactured by multiple methods, very few methods have the ability to produce scaffolds with nanofibrous structure, which include self-assembly [53, 54], phase separation [55, 56], and electrospinning [57, 58]. Table 5.6 depicts the comparative merits and demerits of these methods. Some of the key aspects of these methods, along with illustrated examples, are briefly described in the following sections.

5.5.1
Self-assembly

The past few years has seen great interest in self-assembling processes for manufacturing scaffolds suitable for tissue engineering. Self-assembly is simply a process in which individual, pre-existing components organize themselves into an ordered structure required for specific functions without human intervention. The key challenge in self-assembly is to design molecular building blocks that can undergo spontaneous organization into a well-defined pattern that mimics complex biological systems. Self-assembly takes place by non-covalent bonding, which typically includes hydrogen bonding, ionic bonding, hydrophilic and van der Waals

Tab. 5.6. Comparison of various nanofibrous scaffold processing methods.

Scaffold processing methods	General descriptions	Level of processability	Level of productivity	Merits	Demerits
Self-assembly	A process in which atoms, molecules, and supramolecular aggregates organize and arrange themselves into an ordered structure through weak and non-covalent bonds; typically involves a bottom-up approach	Difficult	Lab scale	Mimic the biological process in certain circumstances	Complex process, limited to a few polymers, unable to produce long and continuous fibers with control over fiber orientation
Phase separation	A process that involves various steps, typically raw material dissolution, gelation, solvent extraction, freezing and drying, leading to the formation of nanofibrous foam-like structure	Easy	Lab scale	Simple process, tailorable mechanical properties	Limited to a few polymers, longer processing time, unable to produce long and continuous fibers with control over fiber orientation
Electro-spinning	A process that essentially employs electrostatic forces for the production of polymer nanofibrous scaffolds; typically involves a top-down approach	Relatively easy	Lab and industrial scales	Simple and cost effective process, capable of producing long and continuous fibers with control over fiber orientation; versatile to many polymers	Use of high voltage apparatus

interactions. Although each of these forces is rather weak, their collective interactions could produce a very stable structure that could match the structural features of biological systems.

Recently, peptide-amphiphile (PA) nanofibers were developed by a self-assembling process for cell transplantation [53] and to direct biomineralization of HA [54]. Beniash et al. [53] have demonstrated the feasibility of PA molecules to self-assemble into 3D nanofibrous scaffold under physiological conditions, and their efficacy in cell entrapment was studied. The results indicated that cells entrapped in the nanofibrous network can survive and proliferate well, which suggests that the self-assembled PA system could be used for cell transplantation or other related tissue engineering applications. Hartgerink et al. [54] have further revealed the potential of self-assembled PA nanofibrous scaffold to direct self-biomineralization. They reported a PA molecule that self-assembles into a nano-structural gel and mimics some of the key features of the natural ECM of bone tissue. They processed the PA nanofibers with negatively charged surfaces because it is believed that this promotes biomineralization by establishing local ion supersaturation. The experimental results confirmed that the PA fibers are capable of nucleating HA on their surfaces. It was also noticed that the c-axes of HA crystals are co-aligned with the long axes of PA fiber that nucleates the structural orientation between collagen and HA observed in the natural bone tissues. Therefore, this self-assembled system could be promising for mineralized tissue repair.

5.5.2
Phase Separation

Phase separation is an interesting method used for manufacturing tissue scaffolds, which involves various processing steps, typically raw material dissolution, gelation, solvent extraction, freezing, and drying. Of course, these steps to convert the polymer into a nanofibrous structure would take a prolonged time. The unique advantage of this method lies in the possibility of designing 3D nanostructures without using any special equipment. Furthermore, the porosity and mechanical properties of scaffold can be controlled to some extent, but it is very difficult and perhaps not feasible to maintain the fiber orientation. Tissue scaffolds using poly(L-lactic acid) (PLLA) were recently fabricated by the authors' group using this phase separation method [55]. They processed scaffolds with highly porous and fibrous structures close to natural ECM. The fiber diameters ranged from 50 to 300 nm with an average of 150 to 250 nm, according with the polymer concentration used in the experimental process. The scaffolds were further evaluated with neural stem cells under *in vitro* cell culture conditions. The results indicated that the cells differentiated well on the nanofibrous scaffolds and the scaffolds served as a positive cue in supporting neurite outgrowth. This study, therefore, suggests that PLLA nanofibrous scaffolds would be promising for neural tissue engineering. Another study, reported by Chen and Ma [56], manufactured nanofibrous PLLA scaffolds with interconnected spherical macropores by a phase separation method, with the aim of mimicking the fibrous architecture of type I collagen protein. By using this method, they were able to control the structure of the scaffolds, in particular the macroscopic shape, the spherical pore size, interfiber distance, and the fiber diameter at the nanometer size; thereby, the phase separation method has gained considerable interest in processing scaffolds for tissue engineering.

5.5.3
Electrospinning – A New Approach

Although the above methods have been employed to produce nanofibrous scaffolds, they often encountered some difficulties, especially in controlling the fiber orientation of the scaffolds. For the success of tissue engineering, it is desirable to control not only the fiber diameter but also the fiber orientation of scaffolds. For various tissue engineering applications, it is necessary to produce scaffolds with control over fiber orientation because alignment of the fibers predominantly influences the cell growth. Therefore, importance is given to a method that facilitates a well-defined fiber orientation and control over fiber thickness. Recently, considerable efforts have been made world-wide to process nanofibrous scaffolds with fiber orientation by so-called electrospinning.

Electrospinning is a straightforward, cost-effective, and versatile technique that essentially employs electrostatic forces to produce polymer fibers, ranging in diameter from a few microns down to tens of nanometers. This technique was first introduced by Zeleny in 1914 [59]. Later, Formhals contributed much to the development of electrospinning and, prolifically, obtained several patents in the 1930s and 1940s [60–64]. In those days it was called electrostatic spray or electrostatic spinning, and subsequently renamed as electrospinning in the 1990s. However, this technique was not adopted for tissue engineering as a tool for manufacturing tissue scaffolds at the time of invention, but it was used in other industrial applications. Recently, it has been revitalized and has been successfully applied in processing nanofibrous scaffolds suitable for tissue engineering. The precise merit of this technique is that it could produce scaffolds with most of the structural features required for tissue engineering. It also offers many advantages over conventional scaffold methodologies, e.g., it can produce ultrafine fibers with spatial orientation, high aspect ratio, high surface area, and having control over pore geometry. These are some of the favorable characteristics to be considered for better cell growth because they directly influence the cell adhesion, cell expression, and transportation of oxygen and nutrients. In this light, electrospun scaffold can serve as a good tissue scaffold and could provide spatial environment for the growth of new tissue with appropriate functions.

5.5.3.1 Experimental System
Figure 5.7 illustrates schematically the basic configuration of electrospinning [24], which consists of three major components: (a) a spinneret, (b) a fiber collector, and (c) a high-voltage power system. As illustrated in the figure, the spinneret is directly connected to a syringe, which acts as a reservoir for the polymer solution to be electrospun. This polymer solution can be fed through the spinneret with the

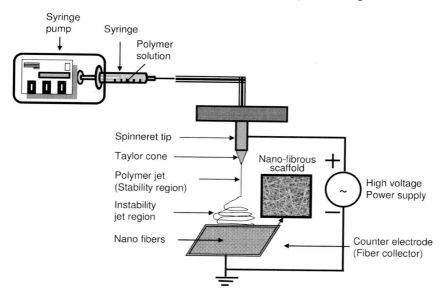

Fig. 5.7. Schematic of the electrospinning system. (Adopted and modified from Ref. [24].)

help of a syringe pump at a steady and controllable feed rate. The feeding rate can be controlled, corresponding to the concentration of polymer solution. Notably, the material to be spun by this technique must be viscous, but the viscosity may vary, depending on the type of materials used. The fiber collecting device is positioned immediately below the spinneret, with an appropriate gap. A high-voltage/low-current power system is required for the conversion of polymer solution into a charged polymer jet. The electric power (usually up to 30 kV) is applied across the spinneret and the grounded metallic counter electrode (fiber collector) to facilitate ejection of the charged jet from the spinneret tip towards the surface of the fiber collector.

5.5.3.2 Spinning Mechanism
Although the experimental design and functional components of electrospinning seems to be extremely simple, mechanism involved in the spinning of polymer nanofibers is rather complicated. Upon applying an optimized electric potential to the spinneret, a pendent droplet of the polymer solution at the tip of the spinneret gets electrified, thereby inducing charge accumulation on the surface of the droplet that subsequently allows the droplet to deform into a cone, known as Taylor's cone [65]. This deformation is commonly caused by two electrostatic forces: (a) electrostatic repulsion between the surface charges of the droplet and (b) Columbic force exerted by the strong external electric field applied [66]. Once the applied electric field surpasses a critical value (threshold), the electrostatic force tends to exceed the viscoelastic force and the surface tension of the polymer droplet; thereby, a

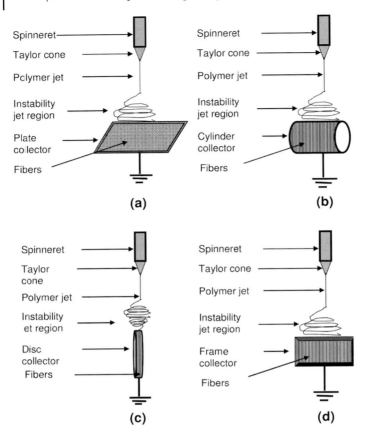

Fig. 5.8. Types of fiber collector.

fine charged polymer jet is forced to eject from the tip of the Taylor cone. This polymer jet then moves toward the counter electrode, accompanied by rapid evaporation of the solvent molecules and, while in transit, the different polymer strands in the jet are separated out due to the mutual repulsion, a phenomenon called "splaying" [67], which gives rise to a series of ultrafine dry fibers. These fibers can be collected on a grounded metallic target (fiber collector) and are typically in the range of a few micrometers to nanometers.

Notably, different types of fiber collectors are available, corresponding to spatial orientation (Fig. 5.8). For example, rotating-type collectors can be used to obtain aligned nanofibers, whilst static-type collectors can be used to obtain random nanofibers. The alignment of the fibers is rather complicated and numerous parameters influence the fiber orientation, including the rotational speed and shape of counter electrode, the concentration of polymer solution, the dielectric constant of the solvent, the strength and shape of the electric field, and the environmental conditions such as relative humidity and temperature.

Fig. 5.9. SEM micrographs of electrospun nanofibrous scaffolds made of (a) collagen, (b) poly(L-lactide), showing scattered nanopores, (c) poly(L-lactide), showing randomly-oriented fibers, and (d) poly(L-lactide), showing well-aligned fibers. (Adopted with permission from Refs.[57, 68, 58, and 58], respectively.)

5.5.3.3 Electrospun Nanofibrous Scaffolds

This section explicates how electrospinning can be adopted to produce nanofibrous scaffolds using a few biodegradable polymers suitable for tissue engineering. Matthews et al. [57] have processed collagen nanofibrous scaffolds; Fig. 5.9(a) shows a representative scanning electron microscopy (SEM) image of such a scaffold. The results indicate that the scaffold is composed of randomly-oriented nanofibers with an average diameter of 100 ± 40 nm. The morphology of collagen structure is quite analogous to the natural ECM; therefore, they suggested that it may be used as a good scaffolding material in tissue engineering. Bognitzki et al. [68] have revealed the possibility of spinning PLLA nanofibrous scaffolds with nanopores. Figure 5.9(b) is a representative SEM micrograph of such a scaffold, showing that each fiber has a beautiful porous structure with well-defined nanopores. The average pore was 100 nm wide and 250 nm long. The size and density of the pores can be controlled by manipulating the spinning parameters. Elongation of the pores along the fiber axis, which is obvious from the SEM micrograph of Fig. 5.9(b), is the result of a uniaxial extension of the polymer jet in the high electric potential.

Recently, our group, Yang et al. [58], also reported the feasibility of spinning PLLA nanofibrous scaffolds suitable for tissue engineering. A representative SEM micrograph of the scaffold is shown in Fig. 5.9(c), which provides a complete detail on size, shape, and pores of the scaffold architecture. The average fiber diameter was 700 nm and the fibers are randomly orientated. They revealed that spatial orientation of the nanofibers mainly depends on the concentration of polymer solution and rotational speed of the fiber collector. Yang et al. [58] further reported the feasibility of spinning PLLA nanofibrous scaffolds with well-defined fiber orientation perfectly suitable for tissue engineering. Briefly, the spinning conditions to obtain such scaffolds are: first, PLLA was dissolved in dichloromethane and N,N-dimethylformamide (70:30) at a concentration of 2%. The polymer solution was extruded at a constant flow rate (1.0 mL h^{-1}) by keeping the distance between the spinneret tip and the fiber collector at 10 cm. A high voltage power supply (12 kV) was applied to the tip of the spinneret using a regulated DC power system. Once the applied voltage reaches the threshold, continuous long and ultrafine fibers eject from the spinneret tip and deposit on the surface of the fiber collector. Note that a sharp-edge rotating disc was used as a fiber collector in this study. A representative SEM micrograph of such a nanofibrous scaffold is shown in Fig. 5.9(d). The results show that a high order of alignment on the fiber orientation was obtained, with fiber diameters ranging from 100 to 400 nm (average is 300 nm). The nanofibrous scaffolds with fiber orientation could be potentially ideal for tissue engineering since the fiber orientation of the scaffolds very much influences cell orientation and phenotypic expression [69]. Therefore, manufacturing nanofibrous scaffolds with fiber orientation is of great importance for the success of tissue engineering, which is now practically possible by the so-called electrospinning method.

5.6
Cell–Matrix (Scaffold) Interactions

5.6.1
Cell–Ceramic Scaffold Interactions

As mentioned above, ceramic nanobiomaterials, particularly nano HA, have considerable interest in tissue engineering, owing to their excellent cell compatibility and bio-activity. Several studies have reported *in vitro* and *in vivo* assessments of the biological responses to nano HA. Table 5.7 lists a few notable examples of tissue scaffolds made from HA-based nanobiomaterials [70–76]. Recently, Huang et al. [70] have reported the *in vitro* assessment of the biological response to nano HA. They prepared the nano HA by a precipitation reaction, and the nano HA suspension was subsequently electrosprayed onto glass substrates using a novel processing route to maintain nanocrystals of HA. The *in vitro* cell-compatibility of nano HA was studied with human osteoblast-like cells. The obtained results are shown in Fig. 5.10. Figure 5.10(a) shows that the prepared HA particles are in the nanometer size range (50–80 nm long), and there are no secondary phase observed as evidenced by X-ray diffraction analysis (Fig. 5.10b), which confirms the phase

Tab. 5.7. Examples of tissue scaffolds made of HA-based nanobiomaterials.

Tissue scaffolds	Cells/growth factors studied	Ref.
Nano HA	Osteoblast-like cells	70
Nano HA	Mesenchymal stem cells	71
Nano HA	Osteoblast cells	72
NanoHA/collagen	Mesenchymal cells	73
NanoHA/collagen	Osteoblast cells/rhBMP-2	74
NanoHA/collagen/PLA	rhBMP-2	75
NanoHA/collagen/PLA	Osteoblast cells	76

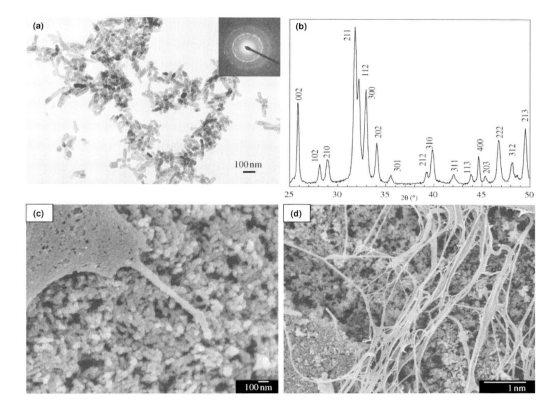

Fig. 5.10. Characteristic results of nano HA produced by wet chemical method: (a) transmission electron micrograph and selected area of diffraction pattern (inset) of nano HA, (b) X-ray diffraction pattern of nano HA, and scanning electron micrographs showing (c) the attachment of human osteoblast-like cells onto nano HA and (d) ECM produced by the cells on nano HA-sprayed glass substrate. (Adopted with permission from Ref. [70].)

purity of nano HA. The results of *in vitro* cell culture studies confirmed the ability of nano HA in supporting cell adhesion and growth. The cells were able to attach to the nano HA surfaces and maintain their osteoblastic morphology with visible filapodia attached to nano HA particles (Fig. 5.10c). There were large areas of confluent cells after seven days of culture and fiber-like ECM was produced (Fig. 5.10d), which provides evidence for the better cell–matrix interactions.

The *in vitro* response of sol–gel derived nano HA with human MSCs was investigated to confirm their biocompatibility and bioactivity [71]. The *in vitro* results showed that the human MSCs adhered to nano HA and exhibited better proliferation signals on their surface. Evidence of proliferation of MSCs was observed with a tendency to spread onto the surface as if forming groups, thus supporting not only biocompatibility but also enhanced bioactivity of the material. This behavior is considered important, since it simulates, on an artificial surface, several steps of the cell-cycle occurring in a tissue environment, with no need for an interfacial protein scaffold. The authors concluded that these surfaces interact with the MSCs in a manner that might be useful for tissue engineers to design scaffolds as an alternative to biological ECM. Porous biomaterials are of great interest in tissue engineering as cells and tissues invade pores for their extensive growth, and they also help the transportation of nutrient and oxygen supply to the cells. Besides, the pores provide a mechanical interlock, thereby leading to a firmer fixation of the biomaterials upon implantation. As bone tissue grows well into the pores, it is expected to increase the strength of the implanted biomaterial. *In vitro* studies on the enhanced functions of osteoblasts (bone-forming cells) on ceramic nanobiomaterials have been reported [72]. The authors extensively studied select functions of osteoblasts on nano HA (with grain size of 67 nm) and conventional HA (with grain size of 179 nm) using *in vitro* cellular models. Osteoblasts were seeded onto the HA substrates and cultured under standard cell culture conditions. The results of cell culture studies provided evidence of increased osteoblast proliferation onto nano HA when compared with conventional HA and with to reference substrate (borosilicate glass). Although the osteoblast proliferation trend was similar on nanophase and conventional HA after 1 day of culture, it was significantly greater on nano HA after 3 and 5 days of culture, indicating that the nanophase provides highly adhesive substrate to facilitate better cell adhesion and long-term cell growth. The results further provided evidence of enhanced long-term functions, in particular synthesis of alkaline phosphatase and concentration of calcium in the ECM, of osteoblasts cultured on nano HA than on conventional HA.

Notably, though the nano HA serves as an excellent bone tissue scaffold, it behaves as a typical brittle material; thereby it is used only in certain bone tissue repair applications. To improve reliability, it is necessary to introduce some biocompatible reinforcement agents or matrix materials. However, introduction of foreign materials may decrease the reliability of HA; thereby, choosing the reinforcement agents or matrix materials is of great importance. A composite of nano HA with collagen is perceived to be beneficial in improving the reliability of nano HA, and, at the same time, both are the key components of natural bone tissue (Fig. 5.6). Osteogenic cell-engineered HA/collagen nanocomposite scaffold structures were developed using culture techniques as well as conventional methods, and their *in vitro* cellular functions were studied [73]. It was noticed that the scaffolds supported well the cellular growth and related functions, leading to new bone formation. Later, a 3D bone-resembling nanocomposite matrix using nanoHA/collagen/osteoblasts was developed in conjunction with PLA [76]. This system supported cellular adhesion, proliferation, and migration. Interestingly, the cells penetrated deep into the matrix to about 200–400 μm within a short period (12 days), probably owing to its composition and structural similarity to natural bone tissue, thereby providing a promising scaffold for bone tissue engineering. In this light, ceramic nanobiomaterials clearly represent a unique and promising class of scaffolding material for tissue engineering applications, but they are limited to hard tissue repair and regeneration.

5.6.2
Cell–Polymer Scaffold Interactions

This section briefly describes how the polymer nanofibrous scaffolds influence cell growth and related functions. Matthews et al. [57] have processed collagen nanofibrous scaffolds and the *in vitro* cellular compatibility was studied using smooth muscle cells. The results suggest that the scaffold has excellent biocompatibility, cellular compatibility, and capability of promoting cells in-growth and related functions (Fig. 5.11a). This study thus demonstrated the feasibility of using collagen as a good tissue scaffold. Yang et al. [58] processed the PLLA nanofibrous scaffolds, with randomly-oriented fibers, and the *in vitro* cellular compatibility was studied using neural stem cells. The results suggest that the scaffold has excellent biocompatibility, cellular compatibility, and is highly supportive to cell differentiation and related functions. The results further indicated that most of the cells have a bipolar shape with extended neurites and the neurites outgrowth was randomly spread over the scaffold (Fig. 5.11b), a good sign for cell–matrix interactions under *in vitro* cell culture conditions.

Yang et al. [58] also reported the feasibility of processing PLLA nanofibrous scaffolds with well-aligned fibers and the *in vitro* cellular compatibility was studied using neural stem cells. They observed the influence of fiber orientation on the cell growth. Figure 5.11(c) shows a representative phase contrast light microscopy image of cell cultured scaffold. The results indicated that cells were well attached to the scaffold and showed an extensive neurite-like outgrowth, which is a good sign for their differentiation. Interestingly, the results revealed that the cells elongated and their neurites outgrew along the direction of the fiber orientation, which shows the significance of spatial orientation of the fibers on cellular growth behavior. To observe expression of the cytoskeleton proteins of neural stem cells as well as the relationship between the cytoskeleton proteins and fiber orientation of the scaffolds, laser scanning confocal microscopy (LSCM) was performed; Fig. 5.11(d) shows a representative micrograph. The results show that the cells exhibited a classical contact guidance by growing parallel to the fibers. In addition, most of the differentiated cells show a bipolar shape with two extended neurites, which

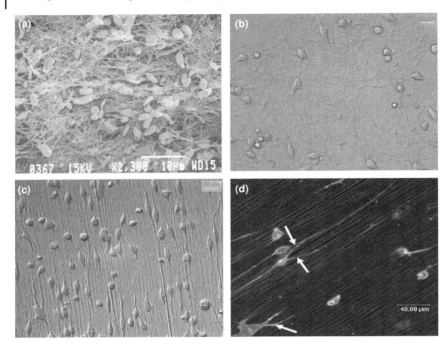

Fig. 5.11. SEM micrographs showing cell–matrix interactions within the electrospun nanofibrous scaffolds made of (a) collagen, (b) randomly-oriented poly(L-lactide) nanofibers. Phase contrast light microscopy (c) and LSCM (d) images of well-aligned poly(L-lactide) nanofibers, evaluated in the authors' laboratory. [Adopted with permission from Ref. [57] for part (a) and Ref. [58] for parts (b)–(d).]

emerged from the regions of the somas, parallel to aligned fibers and were symmetrically distributed around the soma. However, some exceptional cases were also noticed (indicated by arrows). Although neurites outgrowth was not affected by the fiber orientation during the initial period, they all turned through large angles to grow in the direction of fiber orientation, which implies the significance of fiber orientation in directing cell growth. The information thus obtained from the above experimental examples provides evidence for the enhanced performance of polymeric nanobiomaterials for tissue engineering applications.

5.7
Concluding Remarks

Being the first inventor, nature sets high standards for researchers who design implants for tissue repair and regeneration. Tissue engineering is emerging as a potential alternative to traditional tissue transplantations, whereby tissue or organ failure is addressed by implanting tissue-engineered nanobiomaterials that holds the ability to enhance the tissue growth with appropriate functions. Although pre-liminary investigations seem to support the impact of nanobiomaterials in tissue engineering, significant advancements are necessary to realize their full potential in clinical use. On the other hand, the nanobiomaterials are relatively less available than their microscale counterparts. Presently, there is a trend towards using nano-biomaterials due to their sophisticated surface functional properties, which are fa-vorable for cell growth and host tissue interactions. This is an exciting time to be involved in nanobiomaterials in order to formulate them as a clinically ideal scaf-folding system for tissue engineering, with great challenges and also great expect-ations ahead.

Acknowledgments

Financial support of the National University of Singapore and the Singapore Mil-lennium Foundation is gratefully acknowledged.

Abbreviations

μm	Micrometer
3D	Three-dimensional
Ca/P	Calcium to phosphorous ratio
ECM	Extracellular matrix
HA	Hydroxyapatite
HSCs	Hematopoietic stem cells
LSCM	Laser scanning confocal microscopy
MSCs	Mesenchymal stem cells
nm	Nanometer
PA	Peptide-amphiphile
PCL	Poly(caprolactone)
PE	Polyethylene
PET	Poly(ethylene terephthalate)
PGA	Poly(glycolic acid)
PLA	Poly(lactic acid)
PLLA	Poly(L-lactide)
PLGA	Poly(lactide-*co*-glycolide)
PMMA	Poly(methyl methacrylate)
PTEF	Poly(tetrafluoroethylene)
PU	Polyurethane
rhBMP	Recombinant human bone morphogenetic protein
SEM	Scanning electron microscopy
TCP	Tricalcium phosphate
TGF-β	Transforming growth factor-β
Ti	Titanium

Glossary

Adult stem cell: An undifferentiated cell found in a differentiated tissue that is able to self-renew and to differentiate into tissue-specific cell types from which it originated.

Allograft: Tissue or organ transplanted from one individual to another of the same species.

Alloplast graft: Any synthetic material substituted to repair or replace defective parts of the body.

Artificial organ: A medical device or implant intended to replace the body organs.

Autograft: Tissue or organ transplanted within the same body.

Bioactivity: Ability of the implant, to play a vital role in the metabolic processes of living body.

Bioceramics: Inorganic and nonmetallic materials that are compatible with biological tissues.

Biocompatibility: Ability of the implant to perform with an appropriate host response in a specific application.

Biodegradability: Susceptibility of implant to be decomposed by a living organism.

Bioinert: No host response to the material.

Biomaterial: Any synthetic material that is biocompatible with the tissues and the body upon implantation. It can be metal, ceramic, polymer, and a composite of each.

Biomineralization: A conversion process mediated by an organic matrix in which inorganic derivatives are produced by living organisms, especially microorganisms.

Biopsy: Removal of a small portion of tissue, usually for the purpose of making a diagnosis.

Bone: A rigid, yet dynamic connective tissue consisting of calcium phosphate based minerals embedded with collagen fibers in conjunction with osteogenic cellular elements.

Bone marrow stromal cell: A kind of stem cell found in bone marrow that has the ability to generate bone, cartilage, and other connective tissues.

Cell: Fundamental, structural, and functional unit of all living beings that is composed of an outer membrane enclosing protoplasm and nucleus.

Collagen: A fibrous structural protein that function to hold tissues together.

Composite: A heterogeneous combination of two or more materials.

Corrosion: A chemical or electrochemical degradation of metals due to surrounding environmental factors.

Differentiation: A process whereby an unspecialized cell acquires the features of a specialized cell.

Electrospinning: A fiber processing technique, which utilizes electrostatic forces to produce extremely fine fibers ranging from a few micrometers to nanometers in diameter.

Embryonic stem cell: An undifferentiated cell found in an embryo that is able to produce a spectrum of specialized cell/tissue types.

Extracellular matrix: The surrounding material of a cell, including ground substances and fibers.

Fiber: The individual strands of material that form a nonwoven construct.

Graft: A transplant.

Growth factors: A heterogeneous group of substances capable of enhancing tissue growth.

Hard tissue: The general term for calcified structures of the body (e.g., bone and tooth).

Hematopoietic stem cells: A stem cell from which all blood cells develop.

Hydroxyapatite: A calcium phosphate-based material, with chemical composition $Ca_{10}(PO_4)_6(OH)_2$, that is abundant in bone minerals.

Immunogenic: Capable of stimulating an immune response.

Implant: Any medical device or prosthesis inserted or grafted in the human body.

Implantation: A surgical procedure by which medical device or prosthesis is placed in human body either temporarily or permanently.

In vitro: A biological study performed in the laboratory. In other words, outside the living body.

In vivo: A biological study performed inside the living body.

Mesenchymal stem cell: A cell derived from the immature embryonic connective tissue.

Metabolism: A general term used to designate all biochemical changes that occur to substances within the body by either anabolism or catabolism.

Micro: A unit prefix meaning one millionth (1/1 000 000).

Modulus: One of several measures of strain versus applied stress (e.g., Young's modulus).

Monolithic: Made from a single material.

Multipotent: Ability of a single stem cell to develop into a small number of cell types of the body that have a specific function.

Nano: A unit prefix meaning one billionth (1/1 000 000 000).

Nanobiomaterials: Biomaterials composed of particles or grains having nanometric tolerances; a nanometer is equal to 10^{-9} m.

Non-trauma: Any injury or wound caused by disease.

Organ: A differentiated part of an organism adapted for a definite function.

Osteoconduction: An action associated with in-growth of capillaries and migration of bone-forming cells from the host into 3D matrix.

Pathogen: Any organism that is capable of producing disease.

Phenotype: Observable characteristics of an organism, resulting from the interaction of its genotype with the environment.

Plasticity: Ability of stem cells from one adult tissue to generate the differentiated cell type of another tissue.

Pluripotent: Ability of a single stem cell to develop into many different cell types of the body.

Polymer: Long-chain high molecular weight material consisting of repeated monomer units.

Porosity: A ratio of void volume to total volume expressed in terms of percentage.

Proliferation: Expansion of cell population by continuous division of a single cell into two identical daughter cells.

Prosthesis: A medical device that is capable of replacing the organs or tissues.

Protein: A large biomolecule composed of one or more chains of amino acids in a specific sequence.

Resorption: Dissolution of a substance.

Scaffold: A temporary structural construct or matrix used to support the cells for accommodation during tissue fabrication.

Stem cell: An undifferentiated cell that is capable of producing specialized types of cells.

Stem cell-therapy: A treatment in which stem cells are induced to differentiate into specific types of cells/tissues for the purpose of tissue repair.

Surface area: The total area of exposed surface of an object.

Tissue: A collection of similar cells and their surrounding intercellular substances.

Tissue engineering: Development of human tissues or organs in the laboratory from cells removed from the patient or other sources.

Totipotent: Ability of a single stem cell to develop into all types of cells/tissues of the body.

Toughness: The amount of energy absorbed by a material before breakage.

Transplantation: Surgical transfer of tissue or organ from one place to another.

Trauma: Any injury or wound caused by an external force.

Vascular: A medical term pertaining to blood vessel.

Xenograft: Tissue or organ transplanted from one species onto a different species.

References

1 Murugan, R., Ramakrishna, S. Bioactive nanomaterials in bone grafting and tissue engineering. In: Nalwa, H.S. (Ed.), *Handbook of Nanostructured Biomaterials and their Applications in Nanobiotechnology*, American Scientific Publishers, Stevenson Ranch, CA, **2005**, pp. 141–168.

2 Niklason, L.E., Langer, R. Prospects for organ and tissue replacement., *J. Am. Med. Assoc.* **2001**, 285, 573–576.

3 MOTTEP® Facts and Figures, http://www.nationalmottep.org/statistics.shtml.

4 Stock, U.A., Vacanti, J.P. Tissue engineering: current state and prospects., *Ann. Rev. Med.* **2001**, 52, 443–451.

5 WILLIAMS, D.F. Definitions in biomaterials. In: *Proceedings of a Consensus Conference of the European Society for Biomaterials*, Elsevier, Amsterdam, **1987**.

6 KESEMO, B., LAUSMAA, J. Surface science aspects on inorganic biomaterials., *CRC Crit. Rev. Biocomp.* **1986**, 2, 335–380.

7 LAURENCIN, C.T., AMBROSIO, A.M., BORDEN, M.D., COOPER, JR. Tissue engineering: Orthopedic applications., *Annu. Rev. Biomed. Eng.* **1999**, 1, 19–46.

8 FRIEDENSTEIN, A.J., PETRAKOVA, K.V., KUROLESOVA, A.I. Heterotopic transplants of bone marrow. Analysis of precursor cells for osteogenic and haemopoietic tissues., *Transplantation* **1968**, 6, 230–247.

9 PITTENGER, M.F., MACKAY, A.M., BECK, S.C., JAISWAL, R.K., DOUGLAS, R., MOSCA, J.D., MOORMAN, M.A., SIMONETTI, D.W., CRAIG, S., MARSHAK, D.R. Multilineage potential of adult human mesenchymal stem cells., *Science* **1999**, 284, 143–147.

10 SESHI, B., KUMAR, S., SELLERS, D. Human bone marrow stromal cell: Co-expression of markers specific for multiple mesenchymal cell lineages., *Blood Cells Mol. Dis.* **2000**, 26, 234–246.

11 CAPLAN, A.I., BRUDER, S.P. Mesenchymal stem cells: Building blocks for molecular medicine in the 21st century., *Trends Mol. Med.* **2001**, 7, 259–264.

12 BRITTBERG, M., LINDAHL, A., NILSSON, A., OHLSSON, C., ISAKSSON, O., PETERSON, L. Treatment of deep cartilage defects in the knee with autologous chondrocyte transplanta-tion., *New Engl. J. Med.* **1994**, 331, 889–895.

13 PONDER, K.P., GUPTA, S., LELAND, F., DARLINGTON, G., FINEGOLD, M., DEMAYO, J., LEDLEY, F.D., CHOWDHURY, J.R., WOO, S.L. Mouse hepatocytes migrate to liver parenchyma and function indefinitely after intrasplenic transplantation. *Proc. Natl. Acad. Sci. U.S.A.* **1991**, 88, 1217–1221.

14 BEHRAVESH, E., MIKOS, A.G. Three-dimensional culture of differentiating marrow stromal osteoblasts in biomimetic poly(propylene fumarate-co-ethylene glycol)-based macroporous hydrogels., *J. Biomed. Mater. Res. A* **2003**, 66, 698–706.

15 CINOTTI, G., PATTI, A.M., VULCANO, A., DELLA ROCCA, C., POLVERONI, G., GIANNICOLA, G., POSTACCHINI, F. Experimental posterolateral spinal fusion with porous ceramics and mesenchymal stem cells., *J. Bone Joint Surg. Br.* **2004**, 86, 135–142.

16 PARK, D.J., CHOI, B.H., ZHU, S.J., HUH, J.Y., KIM, B.Y., LEE, S.H. Injectable bone using chitosan-alginate gel/mesenchymal stem cells/BMP-2 composites., *J. Cranio-Maxillofacial Surg.* **2005**, 33, 50–54.

17 LI, W.J., TULI, R., OKAFOR, R., DERFOUL, A., DANIELSON, K.G., HALL, D.J., TUAN, R.S. A three-dimensional nanofibrous scaffold for cartilage tissue engineering using human mesenchymal stem cells., *Biomaterials* **2005**, 26, 599–609.

18 YOSHIMOTO, H., SHIN, Y.M., TERAI, H., VACANTI, J.P. A biodegradable nanofiber scaffold by electrospinning and its potential for bone tissue engineering., *Biomaterials* **2003**, 24, 2077–2082.

19 PATRICIA M. DEHMER. *The Scale of Things*, Office of Basic Energy Sciences, Office of Science, U.S. Department of Energy. http://www.sc.doe.gov/production/bes/scale_of_things.html.

20 MURUGAN, R., RAMAKRISHNA, S. Development of nanocomposites for bone grafting., *Comp. Sci. Technol.* **2005**, 65, 2385–2406.

21 PARK, G.E., WEBSTER, T.J. A review of nanotechnology for the development of better orthopedic implants., *J. Biomed. Nanotechnol.* **2005**, 1, 18–29.

22 THOMPSON, I., HENCH, L.L. Medical applications of composites. In: *Comprehensive Composite Materials*, Elsevier Science, Amsterdam, **2000**, pp. 727–753.

23 BLACK, J., HASTINGS, G.W. *Handbook*

of Biomaterials Properties, Chapman & Hall, London, **1998**.

24 MURUGAN, R., RAMAKRISHNA, S. Nanostructured biomaterials. In: *Encyclopedia of Nanoscience and Nanotechnology*, NALWA, H.S. (Ed.), American Scientific Publishers, Stevenson Ranch, CA, **2004**, pp. 595–613.

25 DE GROOT, K. In: *Chemistry of Calcium Phosphates*, YAMMAMURO, T., HENCH, L.L. (Eds.), CRC Press, Boca Raton, FL, **1990**.

26 HENCH, L.L. Bioceramics., *J. Am. Ceram. Soc.* **1998**, 81, 1705–1728.

27 LEGEROS, R.Z., LEGEROS, J.P. In: *An Introduction to Bioceramics*, HENCH, L.L., WILSON, J. (Eds.), World Scientific, Singapore, **1993**, pp. 139–180.

28 KAY, M.I., YOUNG, R.A., POSNER, A.S. Crystal structure of hydroxyapatite., *Nature* **1964**, 204, 1050–1052.

29 MURUGAN, R., RAMAKRISHNA, S. Development of nanocomposites for bone grafting., *Composite Sci. Technol.* **2005**, 65, 2385–2406.

30 ROY, D.M., LINNEHAN, S.K. Hydroxyapatite formed from coral skeletal carbonate by hydrothermal exchange., *Nature* **1974**, 247, 220–222.

31 MURUGAN, R., RAO, K.P., KUMAR, T.S.S. Microwave synthesis of bioresorbable carbonated hydroxyapatite using goniopora., *Bioceramics* **2002**, 15, 51–54.

32 MURUGAN, R., RAMAKRISHNA, S. Coupling of therapeutic molecules onto surface modified coralline hydroxyapatite., *Biomaterials* **2004**, 25, 3073–3080.

33 MURUGAN, R., RAMAKRISHNA, S. Porous bovine hydroxyapatite for drug delivery., *J. Appl. Biomater. Biomech.* **2005**, 3, 93–97.

34 MURUGAN, R., KUMAR, T.S.S., RAO, K.P. Fluorinated bovine hydroxyapatite: Preparation and characterization., *Mater. Lett.* **2002**, 57, 429–433.

35 MURUGAN, R., RAMAKRISHNA, S. Crystallographic study of hydroxyapatite bioceramics derived from various sources., *Crystal Growth Design* **2005**, 5, 111–112.

36 SUCHANEK, W., YOSHIMURA, M. Processing and properties of hydroxyapatite-based biomaterials for use as hard tissue replacement implants., *J. Mater. Res.* **1998**, 13, 94–117.

37 MURUGAN, R., RAMAKRISHNA, S. Aqueous mediated synthesis of bioresorbable nanocrystalline hydroxyapatite., *J. Crystal Growth* **2005**, 274, 209–213.

38 MURUGAN, R., RAMAKRISHNA, S. Bioresorbable composite bone paste using polysaccharide based nano hydroxyapatite., *Biomaterials* **2004**, 25, 3829–3835.

39 WANG, F., LI, M.S., LU, Y.P., QI, Y.X., LIU, Y.X. Synthesis and microstructure of hydroxyapatite nanofibers synthesized at 37 °C., *Mater. Chem. Phys.* **2006**, 95, 145–149.

40 WEBSTER, T.J., ERGAN, C., DOREMUS, R.H., SIEGEL, R.W., BIZIOS, R. Enhanced functions of osteoblasts on nanophase ceramics., *Biomaterials* **2000**, 21, 1803–1810.

41 OTA, Y., IWASHITA, T. Novel preparation method of hydroxyapatite fibers., *J. Am. Ceram. Soc.* **1998**, 81, 1665–1668.

42 HUANG, J., BEST, S.M., BONFIELD, W., BROOKS, R.A., RUSHTON, N., JAYASINGHE, S.N., EDIRISINGHE, M.J. In vitro assessment of the biological response to nano-sized hydroxyapatite., *J. Mater. Sci. Mater. Med.* **2004**, 15, 441–445.

43 MURUGAN, R., RAMAKRISHNA, S. Aqueous mediated synthesis of bioresorbable nanocrystalline hydroxyapatite., *J. Crystal Growth* **2005**, 274, 209–213.

44 ZHANG, S., CONSALVES, K.E. Preparation and characterization of thermally stable nanohydroxyapatite., *J. Mater. Sci. Mater. Med.* **1997**, 8, 25–28.

45 KOTHAPALLI, C.R., SHAW, M.T., WEI, M. Biodegradable HA-PLA 3-D porous scaffolds: Effect of nano-sized filler content on scaffold properties., *Acta Biomater.* **2005**, 1, 653–662.

46 NAKAMURA, S., TSOBE, T., SENNA, M. Hydroxyapatite nano sol prepared via

a mechanochemical route., *J. Nanopart. Res.* **2001**, 3, 57–61.

47 KOUMOULIDIS, G.C., VAIMAKIS, T.C., SDOUKOS, A.T., BOUKOS, N.K., TRAPALIS, C.C. Preparation of hydroxyapatite lathlike particles using high-speed dispersing equipment., *J. Am. Ceram. Soc.* **2001**, 84, 1203–1208.

48 SARIG, S., KAHANA, F. Rapid formation of nanocrystalline apatite., *J. Crystal Growth* **2002**, 237–239, 55–59.

49 YANG, Y., ONG, J.L. Rapid sintering of hydroxyapatite by microwave processing., *J. Mater. Sci. Lett.* **2002**, 21, 67–69.

50 YANG, F., MURUGAN, R., RAMAK-RISHNA, S., WANG, X., MA, Y.-X., WANG, S. Fabrication of nano-structured porous PLLA scaffold intended for nerve tissue engineering., *Biomaterials* **2004**, 25, 1891–1900.

51 YANG, F., MURUGAN, R., WANG, S., RAMAKRISHNA, S. Electrospinning of nano/micro scale poly(L-lactic acid) aligned fibers and their potential in neural tissue engineering., *Biomaterials* **2005**, 26, 2603–2610.

52 MURUGAN, R., RAMAKRISHNA, S. Nano-featured scaffolds for tissue engineering: Spinning methodologies., *Tissue Eng.* **2006**, 12, 3, 435–447.

53 BENIASH, E., HARTGERINK, J.D., STORRIE, H., STENDAHL, J.C., STUPP, S.I. Self-assembling peptide amphiphile nanofiber matrices for cell entrapment., *Acta Biomater.* **2005**, 1, 387–397.

54 HARTGERINK, J.D., BENIASH, E., STUPP, S.I. Self-assembly and mineralization of peptide-amphiphile nanofibers., *Science* **2001**, 294, 1684–1688.

55 YANG, F., MURUGAN, R., RAMAK-RISHNA, S., WANG, X., MA, Y.-X., WANG, S. Fabrication of nano-structured porous PLLA scaffold intended for nerve tissue engineering., *Biomaterials* **2004**, 25, 1891–1900.

56 CHEN, V.J., MA, P.X. Nano-fibrous Poly(L-lactic acid) scaffolds with interconnected spherical macropores., *Biomaterials* **2004**, 25, 2065–2073.

57 MATTHEWS, J.A., WNEK, G.E., SIMPSON, D.G., BOWLIN, G.L. Electrospinning of collagen nanofibers., *Biomacromolecules* **2002**, 3, 232–238.

58 YANG, F., MURUGAN, R., WANG, S., RAMAKRISHNA, S. Electrospinning of nano/micro scale poly(L-lactic acid) aligned fibers and their potential in neural tissue engineering., *Biomaterials* **2005**, 26, 2603–2610.

59 ZELENY, J. The electrical discharge from liquid points and a hydrostatic method of measuring the electric intensity at their surfaces. *Phy. Rev.* **1914**, 3, 69–91.

60 FORMHALS, A. Electrical spinning of fibers from solutions., *US Patent No. 2123992*, **1934**.

61 FORMHALS, A. Apparatus for production and electric treatment of artificial fibers., *US Patent No. 2109333*, **1938**.

62 FORMHALS, A. Artificial threads., *US Patent No. 2187306*, **1940**.

63 FORMHALS, A. Apparatus for producing artificial fibers from fiber-forming liquids by an electrical spinning method., *US Patent No. 2323025*, **1943**.

64 FORMHALS, A. Spinner for synthetic fibers., *US Patent No. 2349950*, **1944**.

65 TAYLOR, G.I. Electrically driven jets., *Proc. R. Soc. London* **1969**, A313, 453–475.

66 RAMAKRISHNA, S., FUJIHARA, K., TEO, W.E., LIM, T.C., MA, Z. *An Introduction to Electrospinning and Nanofibers*, World Scientific, Singapore, **2005**.

67 RENEKER, D.H., CHUN, I. Nanometer diameter fibers of polymer produced by electrospinning., *Nanotechnology* **1996**, 7, 261–223.

68 BOGNITZKI, M., CZADO, W., FRESE, T., SCHAPER, A.S., HELLWIG, M., STEINHART, M., GREINER, A., WENDROFF, J.H. Nanostructured fibers via electrospinning., *Adv. Mater.* **2001**, 13, 70–72.

69 FRENOT, A., CHRONAKIS, I.S. Polymer nanofibers assembled by electro-spinning., *Curr. Opin. Colloid Interface Sci.* **2003**, 8, 64–75.

70 HUANG, J., BEST, S.M., BONFIELD, W., BROOKS, R.A., RUSHTON, N., JAYASINGHE, S.N., EDIRISINGHE, M.J. In vitro assessment of the biological response to nano-sized hydroxyapatite., *J. Mater. Sci. Mater. Med.* **2004**, 15, 441–445.

71 MANSO, M., OGUETA, S., FERNANDEZ, P.H., VAZQUEZL, L., LANGLET, M., RUIZ, J.P.G. Biological evaluation of aerosol–gel-derived hydroxyapatite coatings with human mesenchymal stem cells., *Biomaterials* **2002**, 23, 3985–3990.

72 WEBSTER, T.J., ERGAN, C., DOREMUS, R.H., SIEGEL, R.W., BIZIOS, R. Enhanced functions of osteoblasts on nanophase ceramics., *Biomaterials* **2000**, 21, 1803–1810.

73 DU, C., CUI, F.Z., ZHU, X.D., DE GROOT, K. Three-dimensional nano-HAp/collagen matrix loading with osteogenic cells in organ culture., *J. Biomed. Mater. Res.* **1999**, 44, 407–415.

74 LIAO, S.S., CUI, F.Z., ZHANG, W., FENG, Q.L. Hierarchically biomimetic bone scaffold materials: Nano-HA/collagen/PLA composite., *J. Biomed. Mater. Res. Appl. Biomater.* **2004**, 69, 158–165.

75 LIAO, S.S., GUAN, K., CUI, F.Z., SHI, S.S., SUN, T.S. Lumbar spinal fusion with a mineralized collagen matrix and rhBMP-2 in a rabbit model., *Spine* **2003**, 28, 1954–1960.

76 LIAO, S.S., CUI, F.Z., ZHU, X.D. Osteoblasts adherence and migration through three-dimensional porous mineralized collagen based composite: nHAC/PLA., *J. Bioactive Compat. Polym.* **2004**, 19, 117–130.

6
Orthopedic Tissue Engineering Using Nanomaterials

Michiko Sato and Thomas J. Webster

6.1
Preface

Over the past nine decades of administering bioimplants to humans, most synthetic prostheses consist of material particles and/or grain sizes with conventional dimensions (on the order of 1 to 10^4 μm). But the lack of sufficient bonding of synthetic implants to surrounding body tissues has, in recent years, led to the investigation of novel material formulations. One such classification of materials, nanomaterials (or materials with constituent components less than 100 nm in at least one direction), can be used to synthesize implants with similar surface roughness to that of natural tissues. Natural tissues have numerous nanometer features available for cellular interactions since they are composed of many nanostructures (specifically, proteins).

Several nanophase biomedical implants are currently being investigated, and are likely to gain approvals for clinical use in the near future. The critical factor for this drive is the increasingly documented special, biologically-improved material properties of nanophase implants compared with conventional grain size formulations of the same material chemistry. In this manner, the present chapter highlights a novel property of nanophase materials that makes them attractive for use as implants: enhanced cyto-compatibility leading to increased tissue regeneration. Active works are focused in the domains of orthopedic, dental, bladder, neurological, vascular, cartilage, and cardiovascular applications. However, only orthopedic applications, which are the closest to clinical applications, will be emphasized here. In fact, the field of incorporating nanotechnology into orthopedics has matured enough to present exciting *in vivo* data. This entry will thus articulate the seeming revolutionary changes and the potential gains nanostructured materials can make for bone implant technology. This chapter will first cover problems with current orthopedic implants and then how nanotechnology is providing solutions as pertaining to the use of ceramics, metals, polymers, and composites thereof. Lastly, critical hurdles that need to be addressed before nanophase materials can be used in orthopedics are discussed. This chapter is different from other chapters on the use of nanotechnology towards orthopedic tissue engineering as it covers, for the first

Nanotechnologies for the Life Sciences Vol. 9
Tissue, Cell and Organ Engineering. Edited by Challa S. S. R. Kumar
Copyright © 2006 WILEY-VCH Verlag GmbH & Co. KGaA, Weinheim
ISBN: 3-527-31389-3

time, *in vivo* studies that highlight increased bone growth on nanostructured compared with currently used materials.

6.2
Introduction: Problems with Current Implants

Since 1990, the total number of hip replacements, which is the replacement of the femoral and hip bones, has been steadily increasing [1–7]. In fact, the 152 000 total hip replacements in 2000 is a 33% increase from the number performed in 1990 and a little over half of the projected number of total hip replacements (272 000) by the year 2030 in the United States alone [1–4]. However, in 1997, 12.8% of the total hip arthroplasties were simply due to revision surgeries of previously implanted failed hip replacements [1–4].

The fact that such a high percentage of hip replacements performed every year are revision surgeries is not surprising when you consider the life expectancy of the implant versus that of the patient receiving the implant. Consistently, over 30% of those requiring total hip replacements have been below the age of 65 and even those at 65 have a life expectancy of 17.9 years [5–7]. Females, which make up the majority of those receiving total hip replacements, have a life expectancy of 19.2 years at the age of 65. Since the longevity of implants ranges only from about 12–15 years, even most of those that receive bone implants at the age of 65 will require at least one revision surgery before the end of their lives [5–7].

For the dental community the story is not any better. Since dental implants may be necessary for the young and old alike, it is imperative that they are able to last for the duration of the patient's life. Recent studies have found that dental implants that have been used in over 300 000 cases in the United States have up to a 96% success rate (meaning that the implant was not mobile and was non-inflamed) after five years, 80% after ten, and less than 75% after 15 years [1–7].

The above strongly suggests that the longevity of the prostheses is a reoccurring problem for the orthopedic/dental community that has to be dealt with since current approaches clearly fail. Orthopedic implant failure can be due to numerous reasons, including poor initial bonding of the implant to juxtaposed bone, generation of wear debris that lodges between the implant and surrounding bone to cause bone cell death, and/or stress and strain imbalances between the implant and surrounding bone causing implant loosening and eventual failure [8]. Although there are many reasons why implants fail, a central one is the lack of sufficient bone regeneration around the implant immediately after insertion [8]. Shockingly, about a quarter of dental implant failures are attributed to incomplete healing of the implant to juxtaposed bone (for those that failed between three and six months) [1–7]. Importantly, this leads to eventual implant loosening and regions for possible wear debris to situate between the implant and surrounding bone further complicating bone loss [8–11].

To improve this performance and hence extend the lifetime of bone implants, it is essential to design biomaterial surface characteristics that interface optimally with select proteins and subsequently with pertinent bone cell types. That is, im-

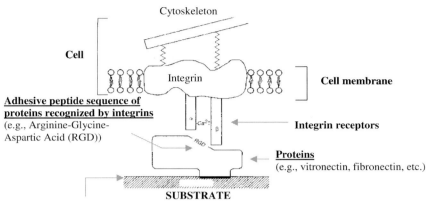

Cytoskeleton

Cell

Integrin

Cell membrane

Adhesive peptide sequence of proteins recognized by integrins
(e.g., Arginine-Glycine-
Aspartic Acid (RGD))

Integrin receptors

Proteins
(e.g., vitronectin, fibronectin, etc.)

SUBSTRATE

Surface properties affecting protein interactions: Chemistry, Wettability; Topography; etc.

Fig. 6.1. Cell recognition of biomaterial surfaces controlled by initial protein interactions. Initial protein interactions can influence cell adhesion and, thus, the degree of bone tissue formation on biomaterials.

Changing the material properties will alter the protein interactions and influence subsequent cell function. (Adapted and redrawn from Ref. [13], with permission.)

mediately after implantation, proteins will adsorb from plasma to biomaterial surfaces to control cell attachment and eventual tissue regeneration (Fig. 6.1) [12, 13]. Initial protein interactions that mediate cell function depend on many biomaterial properties, including chemistry, charge, wettability, and topography [12, 13]. Of significant influence for protein interactions is surface roughness [14–18], and this represents the promise of nanophase materials in bone implant applications.

6.3
A Potential Solution: Nanotechnology

Nanotechnology embraces a system whose core of materials is in the range of nanometers $(10^{-9}$ m) [19–29]. The application of nanomaterials for medical diagnosis, treatment of failing organ systems or prevention and cure of human diseases can generally be referred to as nanomedicine [28, 29]. The branch of nanomedicine devoted to the development of biodegradable or non-biodegradable prostheses falls within the purview of nano-biomedical science and engineering [28, 29]. Although various definitions are attached to the word "nanomaterial" by different experts, the commonly accepted concept refers nanomaterials as that material with the basic structural unit in the range 1–100 nm (nanostructured), crystalline solids with grain sizes 1–100 nm (nanocrystals), individual layer or multilayer surface coatings in the range 1–100 nm (nanocoatings), extremely fine powders with an average particle size in the range 1–100 nm (nanopowders) and, fibers with a diameter in the range 1–100 nm (nanofibers) [19, 20].

Since nature itself exists in the nanometer regime, especially tissues in the hu-

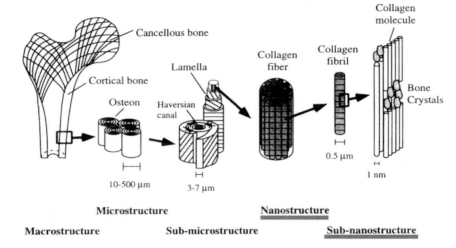

Fig. 6.2. Nanocomponents of bone provide a high degree of nanostructured surface roughness for bone cells. (Adapted and redrawn from Ref. [31], with permission.)

man body [30], clearly, nanotechnology can play an integral role in the biological milieu. Specifically, bone is composed of numerous nanostructures – like collagen and hydroxyapatite that, most importantly, provide a unique nanostructure for protein and bone cell interactions in the body (Fig. 6.2) [31]. Although mimicking constituent components of bone is novel in itself, there are additional reasons to consider nanomaterials for orthopedic applications: their special surface properties compared with conventional (or micron constituent component structured) materials [24–27]. For example, a nanomaterial has increased numbers of atoms at the surface, grain boundaries or material defects at the surface, surface area, and altered electron distributions compared with conventional materials (Fig. 6.3) [27]; in summary, nanophase material surfaces are more reactive than their conventional counterparts. In this light, clearly, proteins that influence cell interactions that lead to tissue regeneration will be quite different on a nanophase compared with conventional implant surfaces (Fig. 6.1).

Despite this, the evolution of orthopedic implants has centered on the use of materials with non-biologically inspired micron surface features [32, 33], mostly changing in chemistry or micron roughness but not degree of nanometer roughness (Fig. 6.4). In this manner, it should not be surprising why the optimal material to regenerate bone has not been found.

6.3.1
Current Research Efforts to Improve Implant Performance Targeted at the Nanoscale

Nanoscale materials currently being investigated for bone tissue engineering applications can be placed in the following categories: ceramics, metals, polymers, and

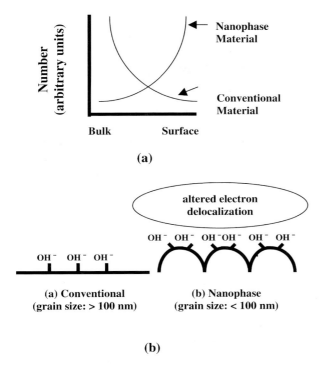

(a)

(b)

Fig. 6.3. Special surface properties of nanophase materials. (a) Higher number of atoms at the surface for nanophase compared with conventional materials. (b) Nanophase materials have higher surface areas, possess greater numbers of material defects at the surface, and altered electron delocalization. Such special properties will influence protein interactions for controlling cell functions. (Adapted and redrawn from Ref. [25], with permission.)

composites thereof. Each type of material has distinct properties that can be advantageous for specific bone regrowth applications. For example, hydroxyapatite, a ceramic mineral present in bone (Fig. 6.2), can also be made synthetically. Ceramics, however, are not mechanically tough enough to be used in bulk for large-scale bone fractures; nonetheless, they have found applications for a long time as bioactive coatings due to their ionic bonding mechanisms, which are favorable for osteoblast (or bone-forming cells) function [34].

Unlike ceramic materials, metals are not found in the body. However, due to their mechanical strength and relative inactivity with biological substances, metals (specifically, Ti, Ti6Al4V, and CoCrMo) have been the materials of choice for large bone fractures [32, 33]. Polymers exhibit unique properties (such as viscosity, malleability, moldability) and possess mechanical strength that is comparable to many soft (not hard) tissues in the body [35]. To date, because of their excellent friction properties, polymers (like ultrahigh molecular weight polyethylene) have been primarily used as articulating components of orthopedic joint replacements [35].

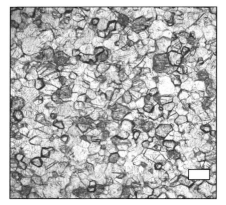
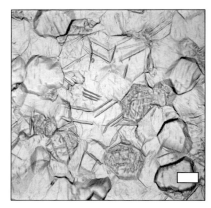
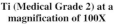

**Ti (Medical Grade 2) at a
magnification of 100X**

**Ti (Medical Grade 2) at a
magnification of 400X**

Fig. 6.4. Conventional grain size of currently used orthopedic
implants. Bar = 10 and 2.5 μm for the left and right
micrograph, respectively.

Additionally, some polymers (particularly the polyester family) can be resorbed or
degraded in the body, which opens the window for controllable repair of damaged
bone that is actively being investigated in tissue engineering circles. Lastly, composites of any or all of the above can be synthesized to provide a wide range of material properties to increase bone implant performance [36]; such ability to tailor
composite properties to specific orthopedic applications makes them attractive.

Due to the numerous materials currently being used and investigated in orthopedics, this review will cover selected efforts to create nanoscale surfaces in all of
these categories: ceramics, metals, polymers, and composites. Several current and
potential materials that have shown promise in nanotechnology for bone biomedical applications, as well as needed future directions, will be emphasized.

6.3.1.1 Ceramic Nanomaterials

Increased Osteoblast Functions   The first report correlating increased bone cell
functions with decreased material grain or particulate size into the nanometer
regime dates back to 1998 and involves ceramics [37]. Such reports described that
in vitro osteoblast (bone-forming cell) adhesion, proliferation, differentiation (as
measured by intracellular and extracellular matrix protein synthesis such as alkaline phosphatase), and calcium deposition was enhanced on ceramics with particulate or grain sizes less than 100 nm [37–46]. Specifically, this was first demonstrated for a wide range of ceramic chemistries, including titania (Fig. 6.5),
alumina, and hydroxyapatite [39]. For example, four, three, and two times the
amount of calcium-mineral deposition was observed when osteoblasts were cultured for up to 28 days on nanophase compared with conventional alumina, tita-

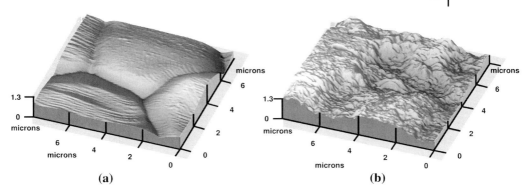

Fig. 6.5. (a) Conventional and (b) nanophase titania. One of the first studies correlating increased osteoblast function with decreasing ceramic grain size was carried out on titania as pictured here. (From Ref. [39], with permission.)

nia, and hydroxyapatite, respectively [41]. Notably, for each respective nanophase and conventional ceramic mentioned in these first reports, similar chemistry and material phase were studied [37–46]. That is to say, only the degree of nanometer surface features was altered between respective nanophase and conventional alumina, titania, and hydroxyapatite. This is important since as previously discussed, it is well known that alterations in surface chemistry will influence bone cell function [8, 32–36], but this was the first time changes in the degree of nanometer roughness alone were reported to enhance bone cell responses [37].

Although these studies provided preliminary evidence that osteoblast functions can be promoted on nanostructured compared with conventional materials regardless of ceramic chemistry, Elias et al. further described a study where the topography of compacted carbon nanometer fibers were transferred to poly(lactic-*co*-glycolic acid) (PLGA) using well-established silastic mold techniques [47]. The same was done for compacts composed of conventional carbon fibers. Figure 6.6 illustrates the successful transfer of nanometer compared with micron surface features from the carbon nanometer compared with conventional fiber compacts, respectively [47]. Importantly, osteoblast adhesion increased on PLGA molds made from nanometer compared with conventional carbon fibers [47]. Increased osteoblast functions were also observed on the starting materials of nanometer compared with conventional carbon fiber compacts [47]. In this manner, this study provided further evidence of the importance of nanometer surface features (and not chemistry) in promoting functions of bone-forming cells.

Equally as interesting, a step-function increase in osteoblast performance has been reported at distinct ceramic grain sizes: specifically, at alumina and titania spherical grain sizes below 60 nm [39]. This is intriguing since when creating alumina or titania ceramics with average grain sizes below 60 nm, a drastic increase in osteoblast function was observed compared with respective ceramics with grain sizes just 10 nm higher (i.e., those with average grain sizes of 70 nm) [39]. This

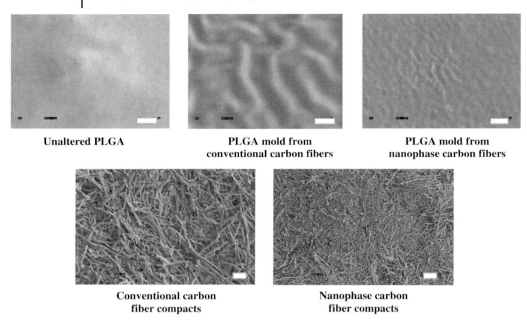

| Unaltered PLGA | PLGA mold from conventional carbon fibers | PLGA mold from nanophase carbon fibers |

| Conventional carbon fiber compacts | Nanophase carbon fiber compacts |

Fig. 6.6. Poly(lactic-*co*-glycolic acid) (PLGA) molds of conventional and nanophase carbon fiber compacts. To highlight the importance of nanometer surface roughness regardless of substrate chemistry, studies have shown increased functions of osteoblasts on PLGA molds of nanophase compared with conventional carbon compacts. In addition, increased functions of osteoblasts have been found on compacts composed of nanometer compared with conventional carbon fibers. Bar = 1 μm. (From Ref. [47], with permission.)

critical grain size for improving osteoblast function is also of paramount importance since numerous other special properties (such as mechanical, electrical, catalytic, etc.) of materials have been reported when grain size is specifically reduced to below 100 nm [19–27]. With this information, evidence has been provided to show for the first time that the ability of nanophase ceramics to promote bone cell function is indeed limited to grain sizes (or subsequent surface features) below 100 nm, specifically those below 60 nm [39]. Thus, another novel size-dependent property of nanostructured ceramics has been elucidated by these pioneering studies.

Although an exact explanation as to why greater bone regeneration is observed on smaller grain size ceramics into the nanometer regime is not known, the importance of this specific grain size in improving osteoblast function is connected with interactions of vitronectin (a protein known to mediate osteoblast adhesion with linear dimensions remarkably similar to the critical grain size of 60 nm mentioned above) [40, 44]. Moreover, as previously mentioned, several studies have indicated that vitronectin and other proteins important for osteoblast adhesion are more well-spread and, thus, expose amino acid sequences to a greater extent when interacting with nanometer compared with conventional ceramics [40, 44]. Intriguingly, numerous investigators have confirmed that the minimum distance between protein ligands (such as arginine-glycine-aspartic acid or RGD) necessary for cell attachment and spreading is in the nanometer regime (specifically, from 10 to 440 nm, depending on whether the study was completed with full proteins, protein fragments, or single RGD units) [48–53]. Therefore, an underlying substrate surface that mediates protein spreading (as opposed to protein folding) to expose such ligands coupled with a nanometer surface roughness, to further project such ligands to the cell, may promote cell adhesion due to this optimal ligand spacing.

Increased Osteoclast Functions   In addition to studies highlighting enhanced osteoblast functions on nanophase ceramics, increased functions of osteoclasts (bone-resorbing cells) have been reported on nanospherical compared with larger grain size alumina, titania, and hydroxyapatite (HA) [45]. Specifically, osteoclast synthesis of tartrate-resistant acid phosphatase (TRAP) and subsequent formation of resorption pits was up to two-times greater on nanophase than on conventional ceramics such as hydroxyapatite. Coordinated functions of osteoblasts and osteoclasts are imperative for the formation and maintenance of healthy new bone juxtaposed to an orthopedic implant [8]. Frequently, newly formed bone juxtaposed to implants is not remodeled by osteoclasts and thus becomes unhealthy or necrotic [34]. At this time, the exact mechanism of greater functions of osteoclasts on nanophase ceramics is not known, but it may be tied to the well-documented increased solubility properties of nanophase compared with conventional materials [25]. In other words, due to larger numbers of grain boundaries at the surface of smaller grain size materials, increased diffusion of chemicals (such as TRAP) may be occurring to subsequently result in the formation of more resorption pits.

Collectively, results of promoted functions of osteoblasts coupled with greater functions of osteoclasts imply increased formation and maintenance of healthy bone juxtaposed to an implant surface composed of nanophase ceramics. In fact, although not compared with conventional grain size apatite coated metals, some studies have indeed demonstrated increased new bone formation on metals coated with nanophase apatite [54]. As shown in Fig. 6.7, bone formation can be clearly seen on the surface of metals coated with nano-apatite, whereas there is no indication of new bone formation on the underlying metal without the coating [54]. Incidentally, coating metals with nanophase HA has been problematic [55]. For example, due to their small grain size, techniques that use high temperatures (like plasma spray deposition) are not an option since they will result in HA grain growth into the micron regime [55]. To circumvent such difficulties, some investigators have allowed nanophase HA to precipitate on metal surfaces; this can be time consuming and not very controllable [54]. In contrast, others have developed novel techniques that use high-pressure based processes that do not significantly create elevated temperatures to coat nanophase ceramics on metals so as to retain their bioactive properties (Fig. 6.8) [56].

Decreased Competitive Cell Functions   Importantly, it has also been shown that competitive cells do not respond in the same manner to nanophase materials as osteoblasts and osteoclasts do [40, 47, 57]. In fact, decreased functions of fibro-

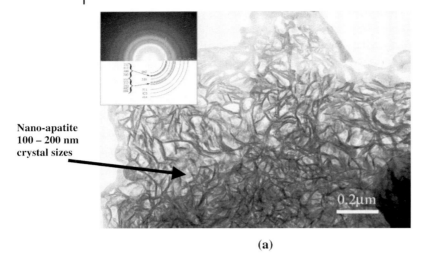

**Nano-apatite
100 – 200 nm
crystal sizes**

0.2μm

(a)

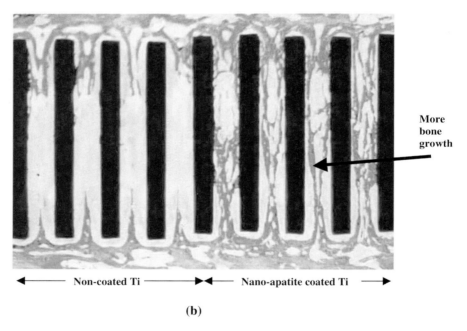

**More
bone
growth**

◄─────── Non-coated Ti ───────►◄─── Nano-apatite coated Ti ───►

(b)

Fig. 6.7. Increased *in vivo* bone regeneration on titanium coated with nanophase apatite. Scanning electron micrograph of nanometer dimensioned apatite (100–200 nm) is depicted in (a). Increased bone regeneration in titanium cages when coated with nano-apatite is depicted in (b). (From Ref. [54], with permission.)

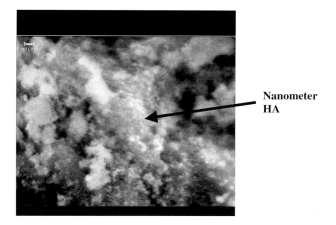

microns

**Nanometer
HA**

Fig. 6.8. Nanophase hydroxyapatite (HA) coated on titanium. Due to elevated temperatures, traditional coating techniques, such as plasma spray deposition, cannot be used to coat metals with nanophase ceramics. This process, developed by Spire Biomedical (Bedford, MA), uses high pressure at low temperatures so as to not allow for grain growth. Bar = 1 μm (upper left).

blasts (cells that contribute to fibrous encapsulation and callus formation events that may lead to implant loosening and failure [8]) and of endothelial cells (cells that line the vasculature of the body) have been observed on nanophase compared with conventional ceramics [40]. Indeed, the ratio of osteoblast to fibroblast adhesion increased from 1:1 on conventional alumina to 3:1 on nanophase alumina [40].

Previously, such selectively in bone cell function on materials has only been observed through delicate surface chemistry (such as through the immobilization of peptide sequences such as Lys-Arg-Ser-Arg or KRSR) [58]. It has been argued that immobilized delicate surface chemistries may be compromised once implanted due to macromolecular interactions that render such epitopes non-functional *in vivo*. For these reasons, notably, studies demonstrating selective enhanced osteoblast and osteoclast functions with decreased functions of competitive cells on nanophase compared with conventional materials have been conducted on surfaces that have not been chemically modified by the immobilization of proteins, amino acids, peptides, or other entities [40, 47, 57]. Rather it is the unmodified, raw material surface that is specifically promoting bone cell functions.

Fibroblast function was also investigated in the same study that was previously mentioned in which Elias et al. transferred the topography of compacted carbon nanometer compared with conventional fibers to PLGA using well-established silastic mold techniques (again please refer Fig. 6.6) [47]. Similar to the observed greater osteoblast adhesion already noted, decreased fibroblast adhesion was measured on PLGA molds synthesized from carbon nanometer compared with conventional fibers [47]. Again, this was the same trend observed on the starting material of carbon nanometer compared with conventional fiber compacts [47]. Thus, this

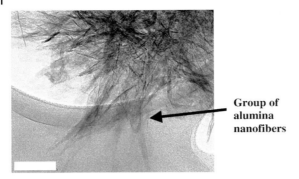

Fig. 6.9. Transmission electron microscope image of alumina nanofibers. Compared with spherical conventional alumina, increased functions of osteoblasts have been reported on nanophase fibrous alumina. Scale bar = 10 nm. (From Ref. [57], with permission.)

study demonstrated the importance of a nanometer surface roughness (and not chemical composition of the material) in decreasing functions of fibroblasts that may lead to undesirable fibrous encapsulation and callus formation events inhibiting osseointegration of orthopedic implants with surrounding bone.

Increased Osteoblast Functions on Nanofibrous Materials Recently, researchers have further modified nanophase ceramics to simulate not only the nanometer dimension but also the aspect ratio of proteins and hydroxyapatite crystals found in the extracellular matrix of bone [57]. For example, consolidated substrates formulated from nano-fibrous alumina (diameter: 2 nm, length > 50 nm; Fig. 6.9) increased osteoblast functions in comparison with similar alumina substrates formulated from the aforementioned nanospherical particles [57]. Specifically, Price et al. determined a two-fold increase in osteoblast cell adhesion density on nanofiber versus conventional nanospherical alumina substrates, following only a 2 h culture [57]. Greater subsequent functions leading to new bone synthesis has also been reported on nanofibrous compared with nano and conventional spherical alumina [57]. Thus, perhaps not only is the nanometer grain size of components of bone important to mimic in materials, but the aspect ratio may also be key to simulate in synthetic materials to optimize bone cell response.

Another classification of novel biologically-inspired nanofiber materials that has been investigated for orthopedic applications is self-assembled helical rosette nanotubes [59]. These organic compounds are composed of guanine and cytosine DNA pairs that self-assembled when added to water to form unique nanostructures (Fig. 6.10). These nanotubes have been reported to be 1.1 nm wide and up to several millimeters long [59]. Compared with currently used titanium, recent studies have indicated that osteoblast function is increased on titanium coated with helical rosette nanotubes (Fig. 6.10) [59]. Although in these studies it has not been possi-

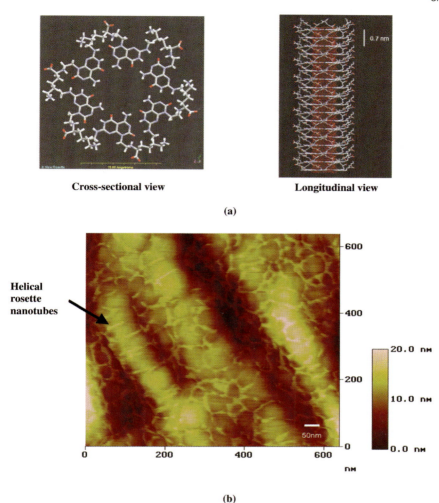

Cross-sectional view   **Longitudinal view**

(a)

Helical
rosette
nanotubes

(b)

Fig. 6.10. Helical rosette nanotubes. (a) Drawings of the cross-sectional and longitudinal views of self-assembled helical rosette nanotubes. (b) Helical rosette nanotubes coated on titanium. Note the nanophase dimension of these organic tubes. Increased osteoblast function has been observed on helical rosette nanotubes coated on Ti. (From Ref. [59], with permission.)

ble to separate the influence of nanometer dimensions from the effects of nanotube chemistry on cell functions, these nanotubes are clearly another category of novel nanostructured materials that can be used to promote bone formation. It is also intriguing to consider what role self-assembled nanofibers may play in orthopedics since bone itself is a self-assembled collection of nanofibers.

In this context, notably, only nanophase materials can mimic the unique aspect ratio of hydroxyapatite and proteins found in the extracellular matrix of bone; it is not possible for micron-sized materials to simulate the unique nanometer constituent components of bone. As previously mentioned, results concerning the importance of nanofibrous materials in promoting functions of osteoblasts have been reported for carbon and polymer molds of carbon nanofibers (Fig. 6.6) [47]. These findings consistently testify to the unprecedented ability to create nanomaterials to mimic the dimensions of components of physiological bone to promote new bone formation.

6.3.1.2 Metal Nanomaterials

Although much more work has been conducted on nanophase ceramics to date, several recent studies have focused on the analysis of bone regeneration on nanophase metals. Metals investigated to date include titanium, Ti_6Al_4V, and CoCrMo [60]. While many have attempted to create nanostructured surface features using chemical etchants (such as HNO_3) on titanium, results concerning increased bone synthesis have been mixed [35]. Moreover, through the use of chemical etchants it is unclear to what the cells may be responding – changes in chemistry or changes in topography. For this reason, as was done for the ceramics in this review, it is important to focus on studies that have attempted to minimize large differences in material chemistry and focus only on creating surfaces that alter in their degree of nanometer roughness.

One such study by Ejiofor et al. utilized traditional powder metallurgy techniques without the use of heat to avoid changes in chemistry to fabricate different particle size groups of Ti, Ti6Al4V, and CoCrMo (Fig. 6.11) [60]. Increased osteoblast adhesion, proliferation, synthesis of extracellular matrix proteins (like alkaline phosphatase and collagen), and deposition of calcium containing mineral was observed on respective nanophase compared conventional metals [60]. This was the first study to demonstrate that the novel enhancements in bone regeneration previously seen in ceramics by decreasing grain size can be achieved in metals.

Interestingly, when Ejiofor et al. examined spatial attachment of osteoblasts on the surfaces of nanophase metals, they observed directed osteoblast attachment at metal grain boundaries (Fig. 6.12) [60]. Because of this, the authors speculated that the increased osteoblast adhesion may be due to more grain boundaries at the surface of nanophase compared with conventional metals. As was the case with nanophase ceramics [40, 44], it is plausible that protein adsorption and conformation at nanophase metal grain boundaries may be greatly altered compared with nongrain boundary areas and/or conventional grain boundaries; in this manner, protein interactions at grain boundaries may be key for osteoblast adhesion.

6.3.1.3 Polymeric Nanomaterials

For ceramics and metals, most studies conducted to date have created desirable nanometer surface features by decreasing the size of constituent components of the material, e.g., a grain, particle, or fiber. However, due to the versatility of poly-

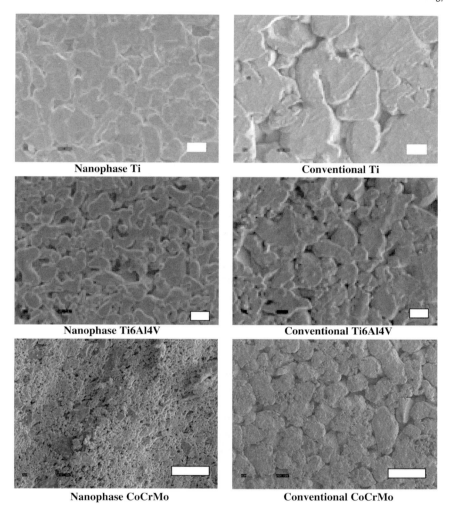

Fig. 6.11. Scanning electron micrographs of nanophase metals. Increased functions of osteoblasts have been observed on nanophase compared with conventional c.p. Ti, Ti6Al4V, and CoCrMo. Scale bar = 1 μm for nanophase Ti/Ti6Al4V and 10 μm for conventional Ti/Ti6Al4V. Scale bar = 10 μm for nanophase and conventional CoCrMo. (From Ref. [60], with permission.)

mers, many additional techniques exist to create nanometer surface roughness values. In addition, polymers contribute even further to rehabilitating damaged tissue by, possibly, providing a degradable scaffold that dissolves within a controllable time while the native tissue reforms. Techniques utilized to fabricate nanometer features on polymers include e-beam lithography, polymer demixing, chemical

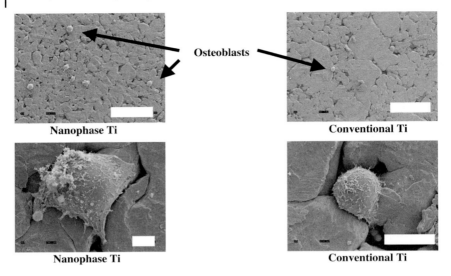

Fig. 6.12. Scanning electron micrographs of adherent osteoblasts on nanophase c.p. Ti. Directed osteoblast adhesion on nanophase metal grain boundaries has been reported. Scale bar = 100 μm for top and 10 μm for bottom; adhesion time = 30 min. (From Ref. [60], with permission.)

etching, cast-mold techniques, and the use of spin-casting [60–68]. For those that have been applied to orthopedics, chemical etching followed by mold casting and polymer demixing techniques have received the most attention [61, 62].

For chemical etching techniques, polymers investigated to date include poly(lactic-*co*-glycolic acid) (PLGA; Fig. 6.13), polyurethane, and polycaprolactone [62, 64–66]. The idea proposed by Kay et al. has been to treat acidic polymers with basic solutions (i.e., NaOH) and basic polymers with acidic solutions (i.e., HNO$_3$) to create nanosurface features [62]. While only on two-dimensional (2D) films, Kay et al. observed greater osteoblast adhesion on PLGA treated with increasing concentrations and exposure times of NaOH. As expected, data was also provided indicating larger degrees of nanometer surface roughness with increased concentrations and exposure time of NaOH on PLGA. Park et al. took this one step further and fabricated three-dimensional (3D) tissue engineering scaffolds by NaOH treatment of PLGA [64]. When comparing osteoblast functions on such scaffolds, even though similar porosity properties existed between non-treated and NaOH treated PLGA (since similar amounts and sizes of NaCl crystals were used to create the pores through salt-leaching techniques), greater numbers of osteoblasts were counted on and in NaOH treated PLGA [64]. Unfortunately, due to these fabrication techniques, it is unclear whether the altered PLGA chemistry or nano-etched surface promoted osteoblast adhesion; however, in light of the previous studies mentioned in this review, the authors of that study suggested the

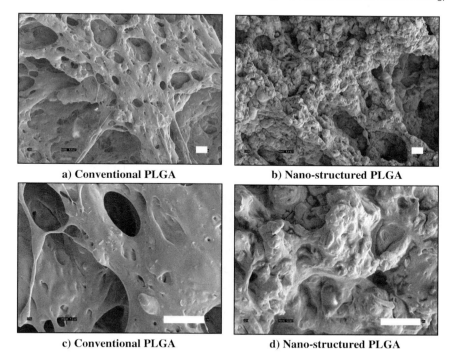

a) Conventional PLGA   **b) Nano-structured PLGA**

c) Conventional PLGA   **d) Nano-structured PLGA**

Fig. 6.13.   Scanning electron micrographs of conventional and nanophase PLGA scaffolds. Increased osteoblast functions have been demonstrated on nanophase PLGA scaffolds. Scale bar = 10 μm. (From Ref. [64], with permission.)

nanometer surface roughness of the NaOH-treated PLGA played an important role [64].

Studies have also been conducted on cell responses to polymers with changes in nanometer surface roughness without changes in chemistry. Specifically, Li et al. utilized polymer demixing techniques to create well-controlled nanometer islands of polystyrene and poly(bromo-styrene) [67]. Although osteoblast functions have not been tested on these constructs to date, fibroblast morphology was significantly influenced by incremental nanometer changes in polymer island dimensions (Fig. 6.14). Again, this study points to the unprecedented control that can be gained over cell functions by synthesizing materials to have nanometer surface features.

Although not related to orthopedic applications, vascular and bladder cell responses have also been promoted by altering the topography of polymeric materials in the nanometer regime [62, 64, 66, 68]. In these studies, chondrocytes [62], bladder [66], and vascular smooth muscle cell [65] adhesion and proliferation were greater on 2D nanometer surfaces of biodegradable polymers such as PLGA, polyurethane, and polycaprolactone; similar trends have recently been reported on 3D PLGA scaffolds [68].

**Polymer
Demixed
Nanoislands**

**Fibroblast
Filopodia**

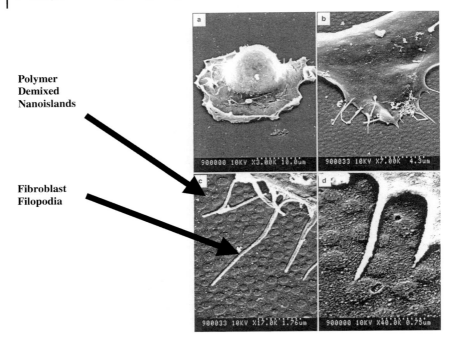

Fig. 6.14. Polymer nanoislands created by demixing
polystyrene and poly(bromo-styrene). Altered cell functions
have been observed on polymer nanoislands compared with
conventional polymer topographies; (a)–(d) represent
increased magnification. (From Ref. [67], with permission.)

6.3.1.4 Composite Nanomaterials

Due to the previous information of increased osteoblast function on ceramics and
polymers, investigators have also determined bone cell function on nanophase ceramic polymer composites (Fig. 6.15). Specifically, studies conducted to date show
promoted osteoblast responses on composites of PLGA combined separately with
nanophase alumina, titania, and hydroxyapatite (30:70 wt.% PLGA:ceramic) [62,
69]. For example, up to three times more osteoblasts adhered to PLGA when it contained nanophase compared with conventional titania particles [62]. Since similar
porosity (both % and diameters) existed between PLGA with conventional compared with nanophase titania, another novel property of nanophase ceramic composites was elucidated in this study: increased osteoblast functions. This is in
addition to numerous reports in the literature highlighting greater toughness of
nanophase compared with conventional ceramic:polymer composites [21–23].

Moreover, promoted responses of osteoblasts have also been reported when
carbon nanofibers were incorporated into polymer composites; specifically, three

Conventional TiO₂

PLGA: Conventional TiO₂
(70:30 wt.%)

Nanophase TiO₂

PLGA: Nanophase TiO₂
(70:30 wt.%)

Fig. 6.15. Scanning electron micrographs of poly(lactic-*co*-glycolic acid) (PLGA):titania composites. Increased osteoblast function has been observed on polymer composites containing nanophase compared with conventional ceramics. Scale bar = 10 μm. (From Ref. [69], with permission.)

times the number of osteoblasts adhered on polyurethane (PU) with increasing weight percentages of nanometer not conventional dimension carbon fibers (Fig. 6.16) [70]. As mentioned, reports in the literature have demonstrated higher osteoblast adhesion on carbon nanofibers in comparison with conventional carbon fibers (or titanium (ASTM F-67, Grade 2) [70]), but this study demonstrated greater osteoblast adhesion with only a 2 wt.% increase of carbon nanofibers (CN) in the PU matrix. Up to three and four times the number of osteoblasts that adhered on the 100:0 PU:CN wt.% adhered on the 90:10 and the 75:25 PU:CN wt.% composites, respectively [70]. This exemplifies the unprecedented ability of nanophase materials to increase functions of bone cells whether used alone or in polymer composite form.

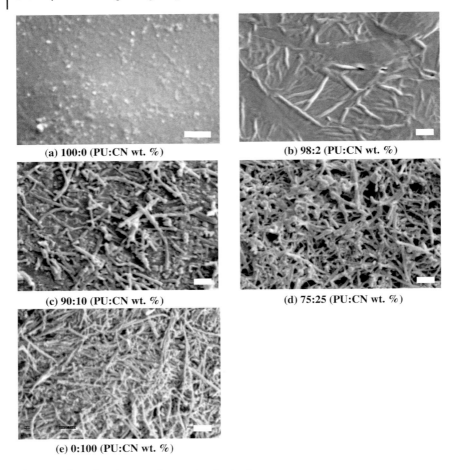

(a) 100:0 (PU:CN wt. %)

(b) 98:2 (PU:CN wt. %)

(c) 90:10 (PU:CN wt. %)

(d) 75:25 (PU:CN wt. %)

(e) 0:100 (PU:CN wt. %)

Fig. 6.16. Scanning electron micrographs of poly-ether-urethane (PU):carbon nanofibers (CN) (wt.%) composites. Increased functions of osteoblasts have been observed on polymer composites containing carbon nanofibers. Scale bar = 1 μm. (From Ref. [70], with permission.)

6.3.2
In Vivo Compared with In Vitro Studies

Of course, in any new field of biomaterials, it is completely natural to conduct extensive *in vitro* analysis before conducting *in vivo* studies. The *in vitro* studies emphasized in the previous sections provided key preliminary promise for the use of these materials in orthopedics. Moreover, they provided an important mechanism as to why osteoblasts prefer nanophase over conventional materials. To date, though, there have been few *in vivo* studies that specifically address increased new

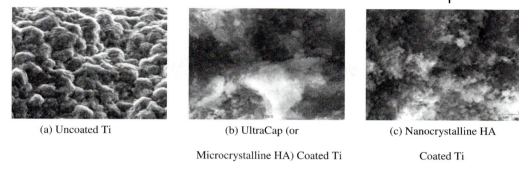

(a) Uncoated Ti   (b) UltraCap (or   (c) Nanocrystalline HA

Microcrystalline HA) Coated Ti   Coated Ti

Fig. 6.17. SEM pictures of uncoated and coated porous titanium. Coating process maintained the nanoscale features of hydroxyapatite (HA) after coating on titanium.

bone growth on materials of the same chemistry but altering only in grain or particle size. That is, while some *in vivo* studies were presented in the early sections, they did not specifically evaluate the influence of grain size on *in vivo* new bone growth. However, one study exists that utilized a novel coating process developed at Spire Biomedical (call IonTite™) to deposit nanocrystalline particles of hydroxyapatite (HA) onto porous titanium scaffolds (Fig. 6.17). Importantly, when implanted into the calvaria of rats for 6 weeks, increased bone infiltration was seen only on titanium scaffolds coated with nanocrystalline HA (Fig. 6.18). That is, little to no bone ingrowth was measured on either uncoated titanium or titanium coated with conventional (or microcrystalline HA) after the same time period. This exciting *in vivo* data confirms the promising *in vitro* data of increased responses from osteoblasts on nano compared with micron HA. Furthermore, while much work is still needed (as outlined in the next section), this *in vivo* study may begin to pave the way for the widespread use of nanophase materials either in bulk or as coatings on traditionally used bone implant materials.

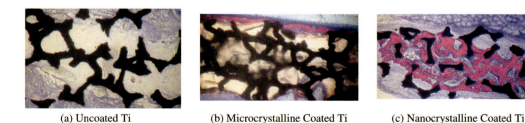

(a) Uncoated Ti   (b) Microcrystalline Coated Ti   (c) Nanocrystalline Coated Ti

Fig. 6.18. Increased bone ingrowth for Ti coated with nanocrystalline HA by IonTite™. Implantation time = 6 weeks. Stain = Sanford (blue indicates bone and red indicates collagen).

6.4
Considerations and Future Directions

Although preliminary attempts to incorporate nanophase materials into orthopedic implants seem promising, numerous urgent questions still remain about this new field. First and foremost, the safety of nanoparticles once in the human body remains largely unanswered both from a manufacturing point of view and when used in full or as a component of an implantable device. Since such particles are smaller than many pores of biological tissues, this information will, clearly, have to be obtained before further consideration of implantable nanomaterials is undertaken. Such nanoparticles can easily become dislodged from implants during surgical implantation or from fragmentation of articulating components of a joint prosthetic composed of nanophase materials. Although preliminary *in vitro* studies highlight a less adverse influence of nanometer compared with micron particulate wear debris on bone cell viability [71, 72], many more experiments are needed especially *in vivo* to evaluate their efficacy.

In addition, once exact optimal nanometer surface features are elucidated for increasing bone regeneration, inexpensive tools that can be used in industry will be required. In this context, if the only nanofabrication devices that can be used to synthesize desirable nanometer surface features for bone regeneration are e-beam lithography or other equally expensive techniques, industry may not participate in this boom of nanotechnology at the intersection of tissue engineering. Inexpensive, but effective, nanometer synthesis techniques must continually be a focus of many investigators.

Still, the direction of the nanotechnology should be and is geared towards dealing with these issues. For example, according to the U.S. government's research agenda, the current and future broad interests in nano biomedical activity can be categorized in three broad related fronts [28, 29]:

1. Development of pharmaceuticals for inside-the-body applications – such as drugs for anticancer and gene therapy.
2. Development of diagnostic sensors and lab-on-a-chip techniques for outside-the-body applications – such as biosensors to identify bacteriological infections in biowarfare.
3. Development of prostheses and implants for inside-the-body uses.

Whereas the European governments emphasize commercial applications in all three fronts above, according to Malsch [29], the U.S. government tends to gear towards fundamental research on biomedical implants and biodefense, leaving commercial applications to industry. Both classifications identify nanophase biomedical implants as of potential interest. The biological and biomimetic nanostructures to be used as an orthopedic implant involve some sort of an assembly in which smaller materials later on assume the shape of a body part, such as hipbone. These final biomimetic, bulk nanostructures can start with a predefined nanochemical (like an array of large reactive molecules attached to a surface) or nanophysical (like a small crystal) structure. It is believed that by using these fundamental nano-structured building blocks as seed molecules or crystals, a larger bulk material will self-assemble or keep growing by itself.

In summary, significant evidence exists that highlights changes in cellular functions on materials with nanoscale surface features, particularly in the bone arena. Clearly, nanomaterials as mentioned here are at their infancy and require much more testing before their full potential can be realized. However, even if nanophase materials never make it to the marketplace due to safety concerns, we have already learned much about how cells interact with nanostructured surfaces through their application in the orthopedic environment.

Acknowledgments

The authors (M.S. and T.W.) would like to thank the NSF and NIH for funding part of the research summarized here through the Bio-nanotechnology National Initiative.

References

1 http://www.aaos.org/wordhtml/press/arthropl.htm

2 http://www.azcentral.com/health/0617newhips17/html

3 http://www.aaos.org/wordhtml/press/joinrepl.htm

4 http://www.aaos.org/wordhtml/press/hip_knee.htm

5 MININO, A.M. and SMITH, B.L., Vital statistics, *National Vital Statistics Rep.*, **2001**, 49, 12.

6 http://www.cdc.gov/nchs/fastats/lifexpec.htm

7 DOWSON, D., New joints for the millennium: wear control in total hip replacement hip joints, *Proc. Institution Mechanical Eng. Part H – J. Eng. Med.*, **2001**, 215(4), 335.

8 KAPLAN, F.S., HAYES, W.C., KEAVENY, T.M., BOSKEY, A., EINHORN, T.A., IANOTTI, J.P., Bone in *Orthopedic Basic Science*, SIMON, S.P. (Ed.), American Academy of Orthopedic Surgeons, Columbus, OH, **1994**, pp. 127–185.

9 OPARAUGO, P.C., CLARKE, I.C., MALCHAU, H., HERBERTS, P., Correlation of wear debris-induced osteolysis and revision with volumetric wear-rate of polyethylene: A survey of

8 reports in literature, *Acta Orthopaed. Scand.*, **2001**, 72, 22.

10 URBAN, R.M., JACOBS, J.J., TOMILSON, M.J., GAVRILOVIC, J., BLACK, J., PEOC'H M., Dissemination of wear particles in the liver, spleen, and abdominal lymph nodes of patients with hip or knee replacement, *Am. J. Bone Joint Surg.*, **2000**, 82, 457.

11 LEROUGE, S., HUK, O., YAHIA, L.H., SEDEL, L., Characterization of in vitro wear debris from ceramic-ceramic total hip arthroplastics, *J. Biomed. Mater. Res.*, **1996**, 32, 627.

12 HORBETT, T.A., Proteins: Structure, properties, and adsorption to surfaces, in *Biomaterials Science: An Introduction to Materials in Medicine*, RATNER, B.D., HOFFMAN, A.S., SCHOEN, F.S., LEMMONS, J.E. (Eds.), Academic Press, New York, **1996**, p. 133.

13 SCHAKENRAAD, J.M., Cells: Their surfaces and interactions with materials, in *Biomaterials Science: An Introduction to Materials in Medicine*, RATNER, B.D., HOFFMAN, A.S., SCHOEN, F.S., LEMMONS, J.E. (Eds.), Academic Press, New York, **1996**, pp. 133–140.

14 BRUNETTE, D.M., The effects of implant surface topography on the behavior of cells, *Int. J. Oral Maxillofac Implants*, **1988**, 3, 231.

15 MARTIN, J.Y., SCHWARTZ, Z.Z., HUMMERT, T.W., SCHRAUB, D.M., SIMPSON, J., LANKFORD, J.J., DEAN, D.D., COCHRAN, D.L., BOYAN, B.D., Effect of titanium surface roughness on proliferation, differentiation, and protein synthesis of human osteoblast-like cells (MG63), *J. Biomed. Mater. Res.*, **1995**, 29, 389.

16 WEN, H.B., CUI, F.Z., ZHU, X.D., Microstructural features of non-union of human humeral shaft fracture, *J. Mater. Sci.: Mater. Med.*, **1998**, 9(3), 121.

17 LARSSON, C., THOMSEN, P., ARONSSON, B.O., RODAL, M., LAUSMAA, J., KASEMO, B., ERICSON, L.E., Bone response to surface modified titanium implants: Studies on electropolished implants with different oxide thickness and morphology, *Biomaterials*, **1994**, 15(13), 1325.

18 BORDJI, K., JOUZEAU, J.Y., MAINARD, D., PAYAN, E., WETTER, P., RIE, K.T., STUCKY, T., HAGE-ALI, M., Cytocompatibility of Ti6Al4V ad Ti5Al2.5Fe alloys according to three surface treatments using human fibroblasts and osteoblasts, *Biomaterials*, **1996**, 17(9), 929.

19 SIEGEL, R.W., FOUGERE, G.E., Mechanical properties of nanophase metals, in *Nanophase Materials: Synthesis-Properties Applications*, HADJIPANAYIS, G.C., SIEGEL, R.W. (Eds.), Kulwer, Dordrecht, **1994**, p. 233.

20 ROCO, M.S., WILLIAMS, R.S, ALIVISATOS, P., Nano-technology research directions, *Nano-Technology Research Directions: IWGN Workshop Report*, **1999**.

21 SIEGEL, R.W., FOUGERE, G.E., Mechanical properties of nanophase metals, *Nanostruct. Mater.*, **1995**, 6, 205.

22 SIEGEL, R.W., Creating nanophase materials, *Sci. Am.*, **1996**, 275, 42.

23 SIEGEL, R.W., HU, E., ROCO, M.C., *Nano-Structure Science and Technology*, Kluwer Academic Press, Boston, **1999**.

24 BARATON, M.I., CHEN, X., GONSALVES, K.E., FTIR study of nanostructured alumina nitride powder surface: determination of the acidic/basic sites by CO, CO_2, and acetic acid adsorptions, *Nanostruct. Mater.*, **1999**, 8, 435.

25 KLABUNDE, K.J., STRAK, J., KOPER, O., MOHS, C., PART, D., DECKER, S., JIANG, Y., LAGADIC, I., ZHANG, D., Nanocrystals as stiochiometric reagents with unique surface chemistry, *J. Phys. Chem.*, **1996**, 100, 12141.

26 WU, S.J., DEJONG, L.C., RAHAMAN, M.N., Sintering of nanophase γ-Al_2O_3 powder, *J. Am. Ceram. Soc.*, **1996**, 79, 2207.

27 MARTYANOV, I.N., KLABUNDE, K.J., Photocatalytic oxidation of gaseous 2-chloroethyl ethyl sulfide over TiO_2, *Environ. Sci. Technol.*, **2003**, 37, 3448.

28 MALSCH, I., Biomedical applications of nanotechnology, *The Industrial Physicist*, June/July, 15, **2002**.

29 MALSCH, I., The nano-bosy:sense and non-sense on biomedical applications of nanotechnology, Lecture for COST and NanoSTAG Conference, Leuven, Oct. 29, 2001.

30 AYAD, S., BOOT-HANDFORD, R., HUMPHRIES, M.J., KADLER, K.E., SHUTTLEWORTH, A., *The Extracellular Matrix Factsbook*, Academic Press, San Diego, **1994**, pp. 29–149.

31 COWIN, R., *Handbook of Bioengineering*, McGraw Hill, New York, **1987**.

32 BRUNETTE, D., The effect of surface topography on cell migration and adhesion, in *Titanium in Medicine*, BRUNETTE, D.M., TENGVALL, P., TEXTOR, M., THOMSEN, P. (Eds.), Springer-Verlag, New York, **2001**, pp. 486–512.

33 BUSER, D., NYDEGGER, T., OXLAND, T., COCHRAN, D.L., SCHENK, R.K., HIRT, H.P., SNETIVY, D., NOLTE, L.P., Interface shear strength of titanium implants with a sandblasted and acid-etched surface: A biomechanical study in the maxilla of miniature pigs, *J. Biomed. Mater. Res.*, **1999**, 45(2), 75.

34 HENCH, L.L., ETHRIDGE, E.C.,

Histochemical responses at a biomaterial's interface, *J. Biomed. Mat. Res.*, **1974**, 8, 49.

35 LITSY, A.S., SPECTOR, M., Biomaterials, in *Orthopedic Basic Science*, SIMON, S.P. (Ed.), American Academy of Orthopedic Surgeons, Columbus, **1994**, p. 482.

36 NIKOLOVSKI, J., MOONEY, D.J., Smooth muscle cell adhesion to tissue engineering scaffolds, *Biomaterials*, **2000**, 21, 2025.

37 WEBSTER, T.J., SIEGEL, R.W., BIZIOS, R., An in vitro evaluation of nanophase alumina for orthopedic applications, in *Bioceramics 11: Proceedings of the 11th International Symposium on Ceramics in Medicine*, LEGEROS, R.Z., LEGEROS, J.P. (Eds.), World Scientific, New York, **1998**, pp. 273–276.

38 WEBSTER, T.J., SIEGEL, R.W., BIZIOS, R., Design and evaluation of nanophase alumina for orthopaedic/ dental applications, *Nanostruct. Mater.*, **1999**, 12, 983.

39 WEBSTER, T.J., SIEGEL, R.W., BIZIOS, R., Osteoblast adhesion on nanophase ceramics, *Biomaterials*, **1999**, 20, 1221.

40 WEBSTER, T.J., ERGUN, C., DOREMUS, R.H., SIEGEL, R.W., BIZIOS, R., Specific proteins mediate osteoblast adhesion on nanophase ceramics, *J. Biomed. Mater. Res.*, **2000**, 51(3), 475.

41 WEBSTER, T.J., SIEGEL, R.W., BIZIOS, R., Enhanced functions of osteoblasts on nanophase ceramics, *Biomaterials*, **2000**, 21, 1803.

42 WEBSTER, T.J., SIEGEL, R.W., BIZIOS, R., Enhanced surface and mechanical properties of nanophase ceramics for increased orthopaedic/dental implant efficacy, in *Bioceramics 13: Proceedings of the 13th International Symposium on Ceramics in Medicine*, GIANNINI S. and MORONI, A. (Eds.), World Scientific, New York, **2000**, p. 321.

43 WEBSTER, T.J., The future orthopedic and dental implant materials, in *Advances in Chemical Engineering Vol. 27*, YING J.Y. (Ed.), Academic Press, New York, **2001**, pp. 125–166.

44 WEBSTER, T.J., SCHADLER, L.S., SIEGEL, R.W., BIZIOS, R., Mechanisms of enhanced osteoblast adhesion on nanophase alumina involve vitronectin, *Tissue Eng.*, **2001**, 7(3), 291–301.

45 WEBSTER, T.J., ERGUN, C., DOREMUS, R.H., SIEGEL, R.W., BIZIOS, R., Enhanced functions of osteoclast-like cells on nanophase ceramics, *Biomaterials*, **2001**, 22(11), 1327–1333.

46 WEBSTER, T.J., SIEGEL, R.W., BIZIOS, R., Nanoceramics surface roughness enhances osteoblast and osteoclast functions for improved orthopaedic/ dental implant efficacy, *Script. Mater.*, **2001**, 44, 1639–1642.

47 ELIAS, K.L., PRICE, R.L., WEBSTER, T.J., Enhanced functions of osteoblasts on nanometer diameter carbon fibers, *Biomaterials*, **2002**, 23, 3279–3287.

48 DANILOV, Y.N. and JULIANA, R.L., (Arg-Gly-Asp)n-albumin conjugates as a model substratum for integrin-mediate cell adhesion, *Exp. Cell Res.*, **1989**, 182, 186.

49 HUGHES, R.C., PENA, S.D.J., CLARK, J., DOURMASHKIN, R.R., Molecular requirements for adhesion and spreading of hamster fibroblasts, *Exp. Cell Res.*, **1979**, 121, 307.

50 HUMPHRIES, M.J., AKIYAMA, S.K., KOMORIYA, A., OLDEN, K., YAMADA, K.M., Identification of an alternatively spliced site in human plasma fibronectin that mediates cell type-specific adhesion, *J. Cell Biol.* **1986**, 103, 2637.

51 SINGER, I.I., KAWKA, D.W., SCOTT, S., MUMFORD, R.A., LARK, M.W., Cell surface distribution of fibronectin and vitronectin receptors depends on substrate composition and extra-cellular matrix accumulation, *J. Cell Biol.* **1987**, 104, 573.

52 UNDERWOOD, P.A. and BENNETT, F.A., A comparison of the biological activites of the cell-adhesive proteins: Vitronectin and fibronectin, *J. Cell Sci.*, **1989**, 93, 641.

53 MASSIA, S.P., HUBBELL, J.A., Human endothelial cell interactions with surface-coupled adhesion peptides on a nonadhesive glass substrate and two polymeric biomaterials, *J. Cell Biol.* **1991**, 14(5), 1089.

54 Li, L., Biomimetic nano-apatite coating capable of promoting bone ingrowth, *J. Biomed. Mater. Res.* **2003**, 66, 79–85.

55 Thull, R., Grant, D., Titanium surface modification, in *Titanium in Medicine*, Brunette, D.M., Tengvall, P., Textor, M., Thomsen P. (Eds.), Springer-Verlag, New York, **2001**, pp. 284–302.

56 Sato, M., Slamovich, E.B., Webster, T.J., Novel nanophase hydroxyapatite coatings on titanium, 29th International Conference on Advanced Ceramics and Composites, Cocoa Beach, FL, 2005.

57 Price, R.L., Gutwein, L.G., Kaledin, L., Tepper, F., Webster, T.J., Osteoblast functions on nanophase alumina: Influence of chemistry, phase, and topography, *J. Biomed. Mater. Res.*, **2003**, 67(4), 1284–1293.

58 Dee, K.C., Andersen, T.T., Rueger, D.C., Bizios, R., Conditions with promote mineralization at the bone-implant interface: A model in vitro study, *Biomaterials*, **1996**, 17, 209.

59 Chun, A., Moralez, J., Fenirri, H., Webster, T.J., Helical rosette nanotubes: A more effective orthopaedic implant materials, *Nanotechnology*, **2004**, 15, S234–S239.

60 Ejiofor, J.U. and Webster, T.J., Increased osteoblast functions on nanostructured metals, *ASM Conference*, Las Vegas, NV, **2004**.

61 Dalby, M.J., Riehle, M.D., Johnstone, H., Afrossman, S., Curtis, A.S.G., In vitro reaction of endothelial cells to polymer demixed nanotopography, *Biomaterials*, **2002**, 23, 2945–2954.

62 Kay, S., Thapa, A., Haberstroh, K.M., Webster, T.J., Nanostructured polymer:nanophase ceramics composites enhance osteoblast and chondrocyte adhesion, *Tissue Eng.*, **2002**, 8, 753–761.

63 Zhang, R. and Ma, P.X., Porous poly(L-lactic acid)/apatite composites created by biomimetic process, *J. Biomed. Mater. Res.*, **1999**, 45(4), 285–293.

64 Park, G.E., Park, K., Webster, T.J., Accelerated chondrocyte functions on NaOH-treated PLGA scaffolds, *Biomaterials*, **2005**, 26, 3075.

65 Miller, D.C., Thapa, A., Haberstroh, K.M., Webster, T.J., Endothelial and vascular smooth muscle cell functions on poly(lactic-co-glycolic acid) with nanostructured surface features, *Biomaterials*, **2004**, 25, 53–61.

66 Thapa, A., Webster, T.J., Haberstroh, K.M., Nano-structured polymers enhance bladder smooth muscle cell function, *J. Biomed. Mater. Res.*, **2003**, 67, 1374–1383.

67 Li, W.-J., Laurencin, C., Caterson, E.J., Tuan, R.S., Ko, F.K., Electrospun nanofibrous structure: a novel scaffold for tissue engineering, *J. Biomed. Mater. Res.*, **2002**, 60, 613–621.

68 Pattison, M., Webster, T.J., Haberstroh, K.M., Three-dimensional nanostructured PLGA scaffolds for bladder tissue replacement applications, *Biomaterials*, **2005**, 26, 3075–3082.

69 Smith, T.A., Webster, T.J., Improved osteoblast functions on polymer:nanophase ceramics composites, *J. Biomed. Mater. Res.*, **2005**, 74, 677–686.

70 Price, R.L., Waid, M.C., Haberstroh, K.M., Webster, T.J., Select bone cell adhesion on formulations containing carbon nanofibers, *Biomaterials*, **2003**, 24(11), 1877.

71 Gutwein, L.G., Webster, T.J., Osteoblast and chondrocyte proliferation in the presence of alumina and titania nanoparticles, *J. Nanoparticle Res.*, **2002**, 4, 231–238.

72 Price, R.L. and Webster, T.J., Increased osteoblast viability in the presence of smaller nano-dimensional carbon fibers, *Nanotechnology*, **2004**, 15(8), 892–900.

7

Hydroxyapatite Nanocrystals as Bone Tissue Substitute

Norberto Roveri and Barbara Palazzo

7.1
Overview

Nanoscience, where the properties of materials are exploited to innovative amazing applications, is involved in the size-dependent chemical and biological activity of bone substitute materials, which is indeed a fascinating field. The present chapter reviews the synthesis and chemical–physical characteristics of hydroxyapatite (HA) nano-crystals, which have excellent properties to represent an elective material covering a wide range of applications for bone substitution. We start from an examination of biogenic bone and tooth hydroxyapatite nanocrystals morphological and chemical–physical characteristics. The highlighted concepts have been used to review up-to date main new ideas on the preparation of synthetic apatitic bone substitutes mimicking the above biogenic properties, among which the nano-size is the basis of their self-assembly, self-mineralization and bone regeneration ability.

Hydroxyapatite (HA) nanocrystals with high bioreabsorbability containing foreign ions and mimicking bone HA chemical–physical and physiological behaviour have been described. We have pointed to their possible use in preparing scaffolds with a porosity simulating that of spongy bone and upon which cells can be seeded, thereby developing *"in vitro* autologous bone"*. Biologically inspired HA nanocrystals/collagen composites have been reviewed, focusing on the role of a self-assembling strategy in conditioning the bone repairing activity of this biomaterial. Furthermore, considering that calcium phosphate/collagen composites are not limited to loading application, the possibility of preparing bio-inspired coating on the surface of metallic implants could be an advantageous approach.

Finally, surface functionalization of HA nano-crystals with bioactive molecules makes them able to transfer information and to act selectively on the biological environment and can be considered one of the main future challenges for innovative bone substitute materials.

Nanotechnologies for the Life Sciences Vol. 9
Tissue, Cell and Organ Engineering. Edited by Challa S. S. R. Kumar
Copyright © 2006 WILEY-VCH Verlag GmbH & Co. KGaA, Weinheim
ISBN: 3-527-31389-3

7.2
Introduction

Biomaterials represent one of the most interesting and interdisciplinary areas of science, where chemical, biological, engineering scientists are contributing to human health care and improving the quality of life. Skeletal deficiencies resulting from trauma, tumors or abnormal development are common, and are usually treated by surgical intervention. As a consequence, there is an increasing demand for materials that can potentially replace, repair or even regenerate injured or diseased bone tissue. Currently, autografts and allografts, or, alternatively, surface-treated metals, inorganic, polymeric and polymeric/inorganic hybrid materials, are applied for the reconstruction of skeletal defects [1]. Due to the limited supply of autografts, and the potential for pathogen transfer and possible stress shielding around the surrounding bones, none of the currently used biomaterials can provide completely the required properties. The enormous need for bone grafts and recent progress in biomaterial science have resulted in the newly evolving approach of "bone tissue engineering" [2–4]. This may be performed by using cell transplantation and culturing on biodegradable scaffolds for the development of hybrid constructs aiming at the regeneration of bone tissue [5].

Calcium phosphates are an interesting subject of research and development in the preparation of biomaterials for bone substitution. Calcium phosphates used as bone grafts can be classified either bioinert or bioactive. Bioinert phosphates have no influence in the surrounding living tissue; by contrast, bioactive materials exhibit the ability to bond with bone tissues. Initially, bioinert materials used in clinical applications were chosen because of their lack of chemical reactions with the environment, which could yield an undesirable outcome. The bioactivity of a bone substitute material can be evaluated through its chemical reactivity with the environment. This reactivity can be evaluated *in vitro* by means of an artificial solution simulating body fluids, and *in vivo* when the material is in contact with physiological fluids, towards the production of newly formed bone. The bioactivity of some reabsorbable calcium phosphate bioceramics is so high as to induce, in repairing a skeletal section, not only replacement of the damaged part but to substitute it, regenerating new bone.

This review focuses on nanosized hydroxyapatite, which is an elective biomaterial for bone substitution. Hydroxyapatite nano-crystals are similar to the mineral component of bone for composition and attempt to mimic the features, structure and morphology of natural bone crystals by their nanosize, blade-like shape, low degree of crystallinity and high surface reactivity.

7.3
Biogenic Hydroxyapatite: Bone and Teeth

Bone and teeth of vertebrates are natural composite materials constituted of calcium phosphate in the form of hydroxyapatite (HA) and organic matter prevalently

Fig. 7.1. Deproteinized bone hydroxyapatite nanocrystals observed by transmission electron microscopy. Scale bar = 100 nm. (Reproduced by permission from Ref. [10].)

formed of proteins and polysaccharides. Biological HA can only be represented ideally with the formula $Ca_{10}(PO_4)_6(OH)_2$. It is not stoichiometric as many cations replace Ca^{2+}, such as Mg^{2+}, Sr^{2+}, Na^+, K^+, and some anions replace PO_4^{3-} and OH^-, such as CO_3^{2-}, HPO_4^{2-}, $P_2O_7^{4-}$, SiO_4^{4-}, and CO_3^{2-}, F^-, Cl^-, Br^-. The carbonate content of bone hydroxyapatite is about 4–8 wt.% [6] and increases with the age of the individual [7, 8] with increasing CO_3^{2-} groups substituting OH^- sites (type A carbonate apatite). Conversely, CO_3^{2-} groups substituting PO_4^{3-} sites (type B carbonate apatite) is the prevalent type in young humans [9]. The nanosized carbonate hydroxyapatite bone crystals have a blade shape approximately 25 nm wide, 2–5 nm thick and about 60 nm long, as can be appreciated in the Transmission Electron Microscopy (TEM) image shown in Fig. 7.1 [10].

Bone tissues can be defined as mineral–organic composite materials. The organic–mineral phase ratio can vary, depending on species, tissue location, age, diet and pathologies [11], [12].

In mammal long bone (cortical bone), the organic matrix represents 22 wt.%, mineral 69 wt.% and the rest (9 wt.%) is the water associated with the organic matrix or with the mineral. The organic matrix is mainly composed of type I collagen (90 wt.%) that would act as template upon which the first mineral crystals were formed [13]. The not yet completely elucidated mechanism of cell-mediated collagen mineralization may be considered a sequence of events requiring the interaction of many different promoting or inhibiting factors [14]. It is widely accepted that matrix vesicles are formed by release of budding from osteoblasts surfaces [15–18], inside which the amount of calcium and phosphate raise to a saturation level, favorable for deposition of amorphous calcium phosphate, octacalcium phosphate and/or brushite with later transformation into HA [19]. The apatite nuclei from matrix vesicles act as templates for new crystal proliferation, which spread into the adjacent collagenous matrix [18]. Type I collagen is produced by osteoblasts and then processed outside the cell, ending in the surrounding extracellular space, where the fibrils self-assemble into mature collagen fibrils that undergo bio-mineralization. For this reason, bone is the typical example of an "organic matrix-mediated" mineralization process [10]. Each collagen I polypeptidic chain contains a helical domain with 338 contiguous repetitions of the sequence (Glycine-X-Y) where X and Y are often proline and hydroxyproline. Each individual chain coils in a triple-stranded helix, stabilized by interchain crosslinking. The triple-stranded helical filaments of uniform size (280 nm long, 1.5 nm wide) and a molecular mass of about 285 kDa present N and C non-helicoidal terminal domains [20]. Figure 7.2(a) reports the collagen fibril structure as a revised quarter-stagger model

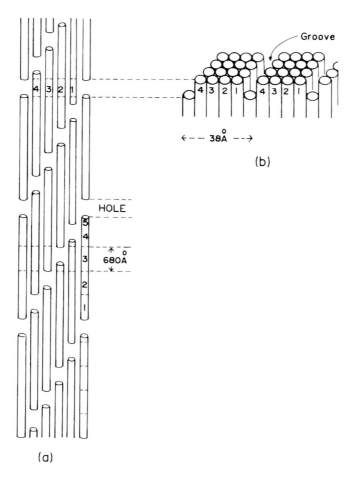

Fig. 7.2. Quarter stagger model of collagen fibers: (a) alignment of collagen gap regions to form groves. In them HA crystals grow with crystallographic *c*- and *a*-axes preferentially aligned parallel to the collagen fiber axis and groove directions, respectively (b). (Reproduced by permission from Ref. [10].)

where molecules are lined up head-to-tail in rows that are staggered of 64 nm along their long axis. Each molecule is transposed by 64 nm with respect to the previous one, and is divided into five sectors, the first four are 64 nm long and the fifth is 25 nm long. This arrangement, stabilized by strong intermolecular crosslinks, produces a regular array of small gaps, 40 nm long and about 5 nm wide, referred to as "hole zones" which are considered to be the *loci* of the nucleation and growth of hydroxyapatite nanocrystals. Adjacent hole zones overlap to form grooves structured in parallel rows along the main axis of the three-dimensional (3D) fibril structure, allowing hydroxyapatite to grow into plate-shaped nano-crystals, oriented such that the crystallographic *c*-axis is preferentially aligned along the collagen fibril axis (Fig. 7.2b) [21]. Utilizing the coincidence, in orientation, between HA crystallographic *c*-axis and collagen fibrils main axis, X-ray diffraction patterns recorded using conventional and synchrotron radiation sources at both wide and low angles from single osteons and osteonic lamellae have allowed to determination of the orientation of hydroxyapatite crystallites and, consequently, collagen fibrils in bone [22–25]. The bone of vertebrates can be considered as a "living biomaterial" since inside it there are cells under permanent activity. The osteoblasts remain trapped inside the mineral phase, evolving towards osteocites, which continuously maintain the bone formation activity.

In teeth, dentine, which resides within the central regions of the tooth, is similar in structure and composition to bone. Enamel on the outside of tooth, however, has a high inorganic content, close to 95 wt.%, which is constituted mainly of long thin ribbon-like crystals of carbonated hydroxyapatite. The "spaghetti-shaped" nano-crystals are arranged in bundles oriented in three different directions, associated with a matrix of proteins excluding collagen (Fig. 7.3). Amelogenins, present in relatively large amount in the early stages of enamel formation, are degraded and removed by up to 5 wt.% as the hydroxyapatite crystals grow [10]. Apatite nanocrystals grow with their *c*-axis preferentially oriented parallel to the long axes of the microribbons [26]. Dental enamel does not contain cells and as a conse-

Fig. 7.3. Scanning electron micrograph of enamel carbonated hydroxyapatite "spaghetti shaped nanocrystals" arranged in bundles oriented along three different directions. Scale bar = 10 μm. (Reproduced by permission from Ref. [10].)

Biological apatites

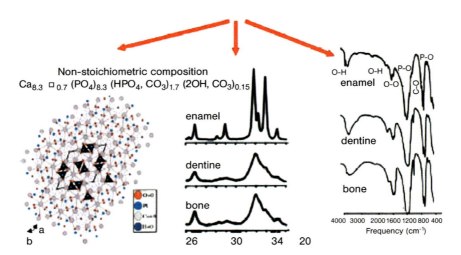

Fig. 7.4. Crystal structure of natural carbonate hydroxyapatite. Powder X-ray diffraction patterns and infrared spectra of enamel, dentine and bone are given. (Reproduced by permission of Ref. [27].)

quence of it any degradation or damage cannot be biologically repaired, evidencing the need for synthetic enamel biocompatible materials repairing tooth deterioration.

Biogenic hydroxyapatite exhibits special behavior due to its nano-size dimensions, low degree of crystallinity, non-stoichiometric composition, crystalline surface disorder and presence of carbonate ions in the crystal lattice. Figure 7.4 shows the crystal structure of biological hydroxyapatite, together with the X-ray diffraction patterns and infrared spectra of enamel, dentine and bone [27].

7.4
Biomimetic Hydroxyapatite: Porous and Substituted Apatites

In the last decade advanced technology has been utilized to synthesize a new generation of inorganic alloplastic materials that mimic natural bone and can function as a scaffold upon which cells can be seeded, developing "*in vitro* autologous bone" [28–30].

The aim is to provide synthetic 3D scaffolds onto which cells are cultured and the artificial piece can be colonized both under *in vitro* and *in vivo* conditions. To control the interaction with biological entities much attention has been devoted to charactering the surface properties of the scaffolds. The use of nanosized hydroxyapatite to prepare apatitic bioceramics with tailored microstructure and porosity opens up many possibilities. In fact poorly crystalline HA nanocrystals, in addition to the excellent biological properties of HA such as non-toxicity and lack of inflammatory and immunitary responses, have a high bioresorbability. This further property allows this kind of bioceramic to be completely replaced by neo-formed bone.

The bioreabsorbability of hydroxyapatite under physiological conditions can be modulated by modifying its degree of crystallinity which can be obtained by implementation of innovative synthesis with a nanodimensional control. Sol–gel, spray-drying, hydrothermal synthesis, recently reviewed [31], have been developed to obtain nanoparticles with high specific surface area, high bioreabsorbability and containing foreign ions, mimicking bone hydroxyapatite chemical–physical and physiological behavior [32–35]. Mg^{2+} ions which cause acceleration of nucleation kinetics of hydroxyapatite and inhibit its crystalline growth [36], are present in high concentration in bone tissue during the initial phases of osteogenesis and tend to disappear when the bone is mature [37]. Similar behavior can be reproduced in synthetic Mg^{2+}-substituted hydroxyapatites characterized by a low degree of crystallinity and an high bioreabsorbability. Carbonate apatite containing different amounts of Mg^{2+} can be synthesized by dropping solutions of $NaHCO_3$ and H_3PO_4 simultaneously into a Ca^{2+} and Mg^{2+} basic suspension [38], or, under aqueous conditions, by employing a cyclic pH variation technique [36]. The replacement of magnesium for Ca^{2+} in the hydroxyapatite structure and its destabilizing effect on hydroxyapatite crystallization has been widely investigated [39–41]. The effect of magnesium in limiting the degree of hydroxyapatite crystallinity has been utilized to synthesize powders that form agglomerates of about 10–25 μm from primary particles about 30–50 nm in size. The contemporary substitution of 0.25–0.30 of Mg^{2+} relative to Ca^{2+} and a carbonation fraction between 0.07 and 0.13 relative to PO_4^{3-} optimizes the most favorable and biomimetic stoichiometry, reflecting as much as possible the composition of bone tissue. An *in vivo* experiment performed filling defects in rabbit femur with granules of the carbonate Mg-HA showed that a period of one month is enough for this filler to be completely replaced by new bone tissue [42]. After preparing powders that preserve the composition of natural apatite, the next step is to use them to design implants having a biomimetic porosity. Porous HA with a morphology simulating that of spongy bone (Fig. 7.5) (porosity varying from a microporosity of >1 μm to a macroporosity ranging from 300 to 2000 μm) has been prepared using various technologies to control pores dimension, shape, distribution and interconnections. HA ceramics processed by high-temperature treatment [43] present a significant reduction of reactivity and growth kinetics of new bone. New formation methods at lower temperatures have been developed, allowing one to obtain porous bioceramics with a low degree of crystallinity. Colloidal processing [44], starch consolidation [45, 46], gel casting and foams out [47] have yielded excellent results, producing bioceramics with a bimodal distribution of the pores size that can be modified as a function of the sintering conditions.

Porous coralline HA can be synthesized by a hydrothermal method for HA formation directly from natural sea corals [48] and HA replaces aragonite whilst preserving its porous structure. The biaxial strength of coralline apatite could be

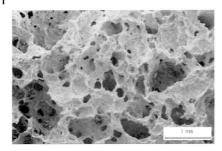 

Fig. 7.5. SEM images of a porous apatitic bioceramic mimicking spongy bone porosity, having a macroporosity ranging from 300 to 2000 μm (scale bar = 1 mm) (a) and microporosity of >1 μm (scale bar = 10 μm) (b).

improved due to a unique double treatment that includes a nano-coating layer to cover meso- and nanopores. In this two-stage process, the coral is fully converted into hydroxyapatite and then coated with a sol–gel-derived apatite (Fig. 7.6a–c). This new material can be applied to bone graft applications where high strength requirements and longevity are pertinent [49, 50].

The interconnected network of pores promotes bone in-growth, but also allows bioceramics to be utilized as drug delivery agents, by inserting different bioactive molecules. Many studies have demonstrated that hydroxyapatite ceramics can be used to deliver steroids, antibiotics, proteins, hormones, anticancer drugs. The aim has been to develop devices able to deliver the appropriate drug amount for a relatively long period of time, whilst minimizing the concentration of the drug in the bloodstream and other organs and the potential side effects produced by systemic administration. Porous ceramics closely mimicking spongy bone morphology have been synthesized by impregnation of cellulosic sponges with poorly crystalline HA water suspension [51]. These porous ceramics have been tested as controlled drug delivery bone grafts to evaluate the fundamental parameters that control release kinetics. A theoretical approach, based on the use of the Finite Element Method, was adopted to describe the Ibuprofen-lysine and Hydrocortisone Na-succinate release kinetics, comparing the numerical results with the experimental ones [52].

An alternative approach to tissue engineering, which uses cells seeded onto macropores of these HA scaffolds to promote bone ingrowth, is represented by filling the macro and micropores with gelatine, which can act as cell nutrient and/or delivery agent of bioactive molecules [53].

If bioceramics are used in powder form for bone filling applications, the HA powder is usually mixed with a polymeric carrier matrix to avoid migration out of the implant region. Both non-absorbable [poly(methyl methacrylate) [54], polyethylene [55] and polysulfone] and biodegradable (poly(lactic acid) [56], polyglycolic acid, collagen, cellulose and starch [57, 58]) polymeric matrices can be used, even

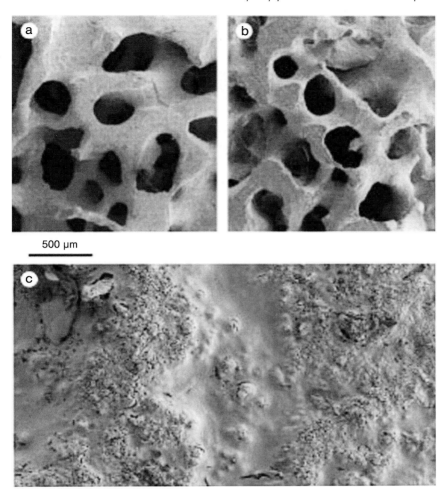

500 μm

Fig. 7.6. SEM images of (a) Goniopora coral morphology,
(b) Goniopora coral converted into pure HA, and (c) sol–gel
derived nanocoated surface of previously converted Goniopora
coral (scale bar = 500 μm). (Reproduced by permission of Ref.
[49].)

if the non-biodegradability reduces drastically the bioactivity of the HA crystals.
Biodegradable polymers are widely used in the medical field as scaffolds in tissue
engineering. Among natural polymers, those from the vegetable kingdom are
sometimes preferred to those of animal origin to avoid immunogenic reactions. Al-
ginates, a family of unbranched binary copolymers are extracted from marine
brown algae. They form insoluble hydrogels for binding with bivalent cations like

Ca^{2+} and, as a consequence, of it calcium crosslinked alginate hydrogels have been utilized in many biomedical applications, including cell transplantation and drug delivery [59–61]. A biomimetic composite has been prepared by direct nucleation of HA on alginate copolymer, exploiting a self-assembly process and a typical egg-box structure can be obtained in samples with a 60/40 HA/Alg. wt ratio [62].

HA nucleation on alginate leads to the nucleation of HA nanocrystals – the polymer prevents further growth. In fact the X-ray diffraction pattern of the HA-alginate scaffold shows the characteristic diffraction maxima of poorly crystalline bone-like apatitic crystals, and TG analysis reveals that HA nanocrystals are 4 wt.%-carbonated [63]. An alginate-HA nanocrystals scaffold seeded with MG63 cells, cultured for 7 days and then subjected to morphological analysis has been shown to favor cell growth while maintaining their osteoblastic functionality [64]. An alginate-Mg carbonated HA nanocrystals bone filler has been patented for orthopedic and odontoiatric implantations [65].

7.5
Biologically Inspired Hydroxyapatite: HA–Collagen Composites and Coatings

Biological organisms are the only true intelligent systems capable of acting as a real "functional material". However, recently, synthetic biologically inspired materials have been prepared with the aim of reproducing, even if only partially, some specific functionalities of biological tissue. Presently, the induction of spontaneous self-assembling of molecules by building intelligent interfaces capable of interactive responses is a real possibility that resembles what occurs in living organisms. The driving forces governing the self-assembly are essentially hydrogen bonding, Van der Waals electrostatic forces and electron-transfer interactions [66, 67].

New methods of synthesis have been developed to utilize the capacity of biological systems to store and transfer information at a molecular level to obtain the spontaneous self-assembly of these entities into a superior architectural arrangement. These syntheses are denoted "biologically inspired" because they reproduce an ordered structure and an environment very close to the biological one. As the complex hierarchic structure of bone originates from the nano-sizes of the first apatite crystals inside the gap regions of collagen fibril structure, nanotechnology concepts and techniques are predominant in preparing biologically inspired bone grafts materials. The *in vitro* self-assembly of collagen molecules induced by thermal or pH variation, to form native fibrils, illustrates that collagen molecules themselves contain all of the structural information necessary for the assembly [68, 69]. Telopeptides-free type I collagen molecules have been utilized as a storehouse of information to nucleate carbonated hydroxyapatite nanocrystals inside the self-assembled collagen fibers (Fig. 7.7a and b) [70].

The two components hydroxyapatite nanocrystals and collagen fibrils, exhibit strong chemical and structural interactions that show a complete analogy of the synthesized composite with natural bone. The apatite crystals have nanometric dimensions, acicular-shaped morphology, and preferential orientation of their *c*-axis

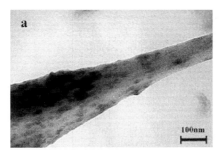

Fig. 7.7. TEM micrographs of synthetic biologically inspired self-assembled collagen fibers/HA nanocrystals composites showing HA nanocrystals nucleated inside the self-assembled collagen fibers (scale bar = 100 nm) (a) and acicular-shaped HA nanocrystals subsequently grown onto the collagen fibers surface (scale bar = 100 nm) (b). (Reproduced by permission of Ref. [71].)

parallel to the direction of orientation of collagen fibers [71]. X-ray diffraction analysis revealed not only the crystal preferential orientation, but also a poor degree of crystallinity very close to that of bone apatite reported in Fig. 7.4. Further evidence of the chemical interaction between collagen fibrils and inorganic crystals is furnished by FTIR analysis. In fact the COO^- groups of collagen are the sites for a possible interaction with HA, and in the FTIR spectra of synthetic HA-Collagen composite a shift of the COO^- antisymmetric stretching band from 1340 cm^{-1} of pure collagen to 1337 cm^{-1} can be appreciated, resembling what is observed for naturally calcified collagen [72]. The adsorption band at 870 cm^{-1} present in the FTIR spectra of HA–collagen composite indicates that the nucleation of hydroxyapatite nanocrystals onto collagen implies carbonation of the apatite phase. Moreover, the carbonation can be assigned only to the B position, as confirmed by the absence of the band at 880 cm^{-1} and by EDS analysis, which reveals that the increase in C content, due to $CO_3{}^{2-}$ groups, corresponds to a decrease in P concentration, maintaining almost constant the stoichiometric ratio Ca/(C+P). The interaction of HA with collagen seems to prevent carbonation in the A position, probably by blocking access to OH^- groups [73]. Carbonation of synthetic hydroxyapatite nanocrystals reduces their crystalline degree increasing the similarity with bone inorganic phase and, especially, raising their bioactivity and biodegradability [74, 75]. Histological sections of implants in rabbit femur after two months showed, along with osteoblastomas, osteoclastic-like cells revealing that hydroxyapatite nanocrystals–collagen fibrils composite induces an active remodeling process, characterized by simultaneous osteoclastic phagocytosis and osteoblastic osteogenesis. The bone tissue reaction of hydroxyapatite nanocrystals–self-assembled collagen fibrils composite was examined in rat tibias to clarify the new bone formation mechanism [76]. This biologically inspired composite was implanted into tibias of rats, and observed at different days after implantation. Bone tissue reactions of the composite demonstrated osteoclastic resorption of the biomaterial, followed by new bone formation by osteoblasts. Alkaline phosphatase activity has been observed in the neighborhood of the composite and new bone was actively formed around these sites in a way very similar to the reaction of a transplanted autogenous-bone. Detailed study on these bone reactions is currently progressing, and it has been assumed that the substitution process of the composites to new bone occurs as follows:

1. Formation of the composite debris via erosion by body fluid.
2. Phagocytosis of the debris and the composite surface by macrophages.
3. Induction of osteoclastic cells on the composite surface and resorption of the composite via a similar process to that of bone.
4. Induction of osteoblasts to the resorption lacunae created by osteoclastic cells and formation of new bone in the surroundings of the composite.

These reactions have not been reported on other HA/Collagen composite materials [77, 78] synthesized without using the self-organization process, which causes difference in cell reaction. In fact it is the bone-like nanostructure that can deceive the bone-related cells into starting the remodeling process. The hydroxyapatite nanocrystals–self-assembled collagen fibrils composite can be successfully utilized as an artificial bioinspired material in both the orthopedic and dental fields as an *in vivo* filler and an *in vitro* tissue regenerator.

Nevertheless, calcium phosphate/collagen composites exhibit weak mechanical behavior which can be improved by chemical crosslinking collagen, but at the expense of reducing their bioreabsorbility [79]. Calcium phosphate/collagen composites are limited to non-loading applications, but are particularly suitable to prepare bioactive coatings onto the surface of metallic implants with the aim of accelerating bone formation and implant fixation. Because of the presence of collagen fibers inside this kind of hybrid composite, a coating of it on metallic implants cannot be made by plasma spray or physical vapor deposition. Instead, electrolytic deposition appears to be the most promising technical procedure. Calcium phosphate coatings have been synthesized by electrochemical deposition from different Ca^{2+} and $(PO_4)^{3-}$ solutions, in acid and in basic conditions [80–83], but also without conditioning pH by adding acid or basic reagents (Fig. 7.8a) [84]. Many technical difficulties have arisen in making a homogeneous and uniform collagen/calcium phosphate coating by electrolytic deposition on top of metallic supports. A collagen/calcium phosphate coating has been crystallized on an apposite electrode, but is not surface stable and is spontaneously chipped off forming a precipitate [85]. A bioactive multilayer collagen/calcium phosphate coating has been obtained by electrocrystallization of calcium phosphate on titanium electrode where the collagen was previously physically absorbed on the surface, but the thickness and the collagen/mineral composition ratio are strongly limited [86]. A double-layered collagen fibril/octacalcium phosphate composite coating on silicon substrates has been recently obtained by electrolytic deposition. The Si cathode surface is covered by a 100 nm thin layer of calcium phosphate coating on top of which a 100 μm thick collagen fibril/octacalcium phosphate cluster composite layer is formed [87].

Electrolytic deposition of a biomimetic, bone-like self-assembled collagen fibrils/

Fig. 7.8. SEM micrograph of coatings, electrolytically deposited on titanium plate, consisiting of (a) apatite synthesized from Ca(NO₃)₂ and NH₄H₂PO₄ solutions (scale bar = 10 μm) and (b) self-assembled collagen fibrils/HA nano-crystals composite, both obtained without conditioning solution pH by adding acid or basic reagents (scale bar = 2 μm). (c) TEM image of the collagen/HA composite shown in (b) (scale bar = 200 nm).

HA nano-crystals composite coating on titanium plates has been performed, using Ca(NO₃)₂ and NH₄H₂PO₄ solutions in a purely helical type I collagen molecule suspension. The use of dilute electrolytic solutions, low current density at the cathode and room temperature affords a coating composed of poorly crystalline carbonate hydroxyapatite nano-crystals that nucleate inside and around the reconstituted collagen fibrils distributed in a homogeneous network on Ti plate surface (Fig. 7.8b).

If the telopeptides-free collagen molecule assembling takes place during the crystallization of the apatite phase, the neo-forming collagen fibrils mineralize with a close structural relationship to the inorganic phase. FTIR analysis of this composite shows a shift of COO^- stretching of collagen resembling natural mineralized collagen fibers. This biomimetic feature of nanocrystals that nucleate and grow inside the fibrils is of interest only at the beginning of the calcification process, which continues with the massive growth of HA nanocrystals onto the collagen fibrils surface (Fig. 7.8c).

These two different kinds of collagen fibers mineralization resemble what has been observed in the self-assembled collagen fibers–hydroxyapatite nanocrystals in aqueous solutions [71]. The FTIR adsorption band at 870 cm^{-1} and the absence of that at 880 cm^{-1} observed for the composite coating allows assigning the carbonation of the hydroxyapatite only to the B position, closely resembling the carbonation of bone apatite.

These results [84, 88] suggest auspicious applications in the preparation of medical devices such as biomimetic bone-like composite-coated metallic implants with a loading capability deriving from the metal core and having, at the same time, a bioactive surface that accelerates bone formation and implant fixation.

7.6
Functionalized Hydroxyapatite: HA Nanocrystals – Bioactive Molecules

An important question in biocompatibility is how the device or material communicates its structural make-up to direct or influence the response of proteins, cells, and the organism to it. For devices and materials that do not leach undesirable substances, this communication occurs through the surface structure; the body reads the surface structure and responds. The distribution of ions and molecules on the surface of all condensed matter has a special organization and reactivity and requires novel methods to be tailored above all because these ions and molecules may drive many of the biological reactions that occur in response to the biomaterial.

HA nanocrystals exhibit a high surface area (from 60–130 up to 180 m^2 g^{-1} for natural ones) characterized by an elevated surface disorder in which the ions stoichiometry is not maintained. This property is considered responsible for the higher HA nanocrystals reactivity in respect to the larger ones and can be affected by the crystals shape. In fact the crystals expose different planes to the biological environmental as a function of their shape (plate, acicular or needle shape) even if they exhibit the same surface area. Needle and plate shape (about 100 nm and 15 nm sized) synthetic HA nanocrystals TEM images are shown in Fig. 7.9(a) and (b) respectively. Only in recent years has high-resolution electron microscopy (HREM) been successfully applied to investigate hydroxyapatite structure to discern different calcium phosphate phases that often coexistent in the same crystalline granule. Among the rather few HREM studies on hydroxyapatite, some refer to the material itself [89], others to HA thin films [90]. This kind of analysis allows us to view the surface disorder at the boundaries of the crystallites constituting the single nanocrystals (Fig. 7.10a). Furthermore, HREM at high resolution allows us to observe the HA nanocrystals interplanar spacing (Fig. 7.10b), permitting discernment between HA and OCP, which are both elongated along the c-axis.

Surface functionalization of HA nanocrystals by bioactive molecules is an innovative approach not only to modulate but also to drive their reactivity. In this way HA nanocrystals will not only guarantee, for instance, either osteointegration or osteoinduction enhanced properties but they will also perform, by stimulating specific cellular responses, at the molecular level.

Only in recent years have scientists begun to use biomolecules for the synergistic coupling of crystals synthesis and functionalization. In fact, previous studies have limited the use of biomolecules as simple growth inhibitors of HA crystallization,

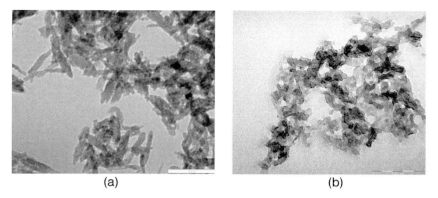

Fig. 7.9. TEM images of synthetic HA nanocrystals that are (a) needle shaped (approx 100 nm in size; scale bar = 200 nm) and (b) plate shaped (approx. 15 nm in size; scale bar 100 nm).

rather than considering their use as a strategy to fine-tune the bioactivity of the nanoparticles [91, 92]. Studies of the effect of biological molecules onto hydroxyapatite crystal growth have been related directly to physiological or pathological calcification processes. Particular interest has been dedicated to amino acids, which are compounds of major importance for living organisms, even because their con-

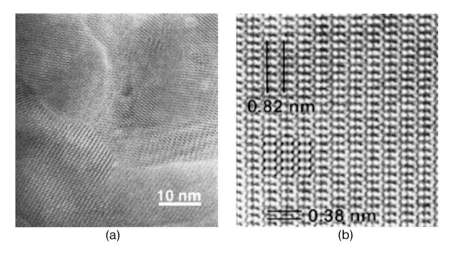

Fig. 7.10. HREM images of HA deposited by laser showing (a) the grain boundaries of crystallites with different orientation at low magnification and (b) in a single crystallite the interplanar spaces $D_{hkl} = 0.82$ and 0.38 that are characteristic of HA. (Reproduced by permission of Ref. [90].)

centration is controlled by physiological mechanisms and they enter inside the cell environment by simple diffusion [93, 94]. The effects of 4-hydroxyproline, tyrosine and serine (amino acids with uncharged polar side groups), glycine, cysteine, cystine, and glutamine (amino acids with neutral polar side groups) and lysine, which is an amino acid with a basic side group, on HA crystallization have been studied by the "constant composition technique" [95–97]. Using this technique, the chemical potentials of the species in the working solution are kept constant during the crystallization process by the stoichiometric addition of reactants and therefore the crystal growth reaction is performed under pseudo-steady-state conditions. Diverse inhibiting activities were observed as a function of the different amino acid side group. These studies have shown that crystallization kinetics are affected by blockage of the active growth sites on the crystal surface.

Considering that the presence of proteins, (and hence amino acids) in biological materials is intrinsic to the bioactivity of HA, amino acids can be considered as agents that can increase the bioactivity of the synthetic HA. Concerning this, aqueous colloids of positively charged amino acid-functionalized HA nano-rods less than 80 nm long and ca. 5 nm wide have been synthesized by hydrothermal crystallization [98]. The colloidal stabilization and its positive zeta potential, together with FTIR spectroscopy data, were consistent with a general model in which the α-carboxylate of the amino acid was preferentially bound to the crystal surfaces, inducing a crystal growth inhibition predominantly onto the Ca-rich surfaces during the initial stages of crystallization. To better define the surface Ca^{2+} role on amino acid linking, calcium deficient hydroxyapatite nanocrystals (CDHA) were synthesized in the presence of differently charged amino acids (alanine, cysteine, arginine, aspartic acid). The data obtained suggest a binding model in which the amino acid amino group is preferentially bound to the CDHA crystals, probably through a phosphorus amiditic linkage, such that the negatively charged carboxylate group exposed at the crystal/solvent interface determines the surface charge. In fact, zeta potential measurements showed that the CDHA/amino acids nanoparticles had an enhanced negative charge with respect to CDHA synthesized in the absence of amino acids [99]. The isoelectric point of the amino acid used as additive during the synthesis strongly affects the nanocrystals morphology and crystallites dimensions.

An innovative improvement of the functionalization process is represented by the synthesis of apatite nano-crystals surface loaded with bioactive molecules that promote a desirable cellular response if released with controlled kinetics. To reach this goal, it is a priority to investigate the chemical mechanisms of the binding and desorption of bioactive molecules by crystals surface. In fact, bioactive ceramics with drug delivery function, previously described, adsorb and release molecules through a mechanism dominated predominantly by physical parameters, such as carrier porosity and dimensions, bioactive molecule steric hindrance and environment diffusive conditions.

Slurries of hydroxyapatite have been successfully used to investigate the link and release processes of cisplatin, one of the most frequently prescribed drugs for the treatment of several solid tumors. Chemotherapeutic treatments applied locally to osteosarcoma can result in tumor inhibition and much less-toxic systemic values of released drugs. Both the adsorption and release process of cisplatin were found to depend significantly on the ionic composition of the aqueous media used [100]. In fact, at a constant pH of 7.4, significantly more cisplatin is adsorbed by the hydroxyapatite crystals from a free buffer solution than from a buffered solution containing chloride ions. The amount of hydroxyapatite-bound cisplatin desorbed into solution was also progressively increased as a function of the increasing concentration of chloride in the equilibrating solution. These results suggest that it is the hydrated positively charged derivatives of cisplatin, whose hydrolysis is inhibited in chloride-rich solution, that are involved in the adsorption of cisplatin by the surface phosphate-rich hydroxyapatite crystals. Further work has demonstrated that subtle variations of the chemical–structural characteristics of the apatitic crystals affect both the adsorption and desorption of the drug cisplatin [101].

Adsorption isotherms obtained in ultrapure water and kinetic release in EPES buffer saline solution (containing chloride to favor release of the adsorbed platinum complex) indicate that cisplatin kinetics increase with decreasing crystallinity, the latter being associated with either smaller nanocrystals dimensions or greater surface area, which is strictly related to the amount of surface lattice defects which create active binding sites [102].

On the other hand, apatitic phase for bone specific delivery of geminal bisphosphonates (BPs) – a class of drugs that have been developed for use in various diseases of bone, tooth and calcium metabolism – are currently under investigation [103]. BPs are compounds characterized by two phosphonate groups attached to a single carbon atom, forming a "P-C-P" structure which is affine to Ca^{2+} ions. As a consequence, they strongly bind to calcium phosphate crystals and inhibit bone resorption by hindering their growth, aggregation and dissolution. However, BPs in vivo action mechanism involves cellular effects that are added to the surface physicochemical outcome of inhibiting HA crystal growth [104, 105]. Interestingly, while the affinity of BPs for bone mineral hydroxyapatite is the basis for their use as inhibitors of bone resorption, at the same time their coordinating abilities towards ions present on apatitic surfaces might be extended to design new delivery system for bone targeted drugs. In this context, calcium phosphate ceramics have been shown to be good biocompatible carriers for BPs, such as 1-hydroxy-2-(imidazol-1-yl-amino)-ethylidenebisphosphonic acid (Zoledronate) that is efficient for the treatment of post-menopausal osteoporosis and bone metastases [106]. A simple mathematical model was designed that correctly described the Zoledronate–calcium phosphate interaction at equilibrium, in simplified media such as ultrapure water or phosphate buffers [107]. However, the chemical link occurring between the BPs and calcium on the surface of HA nanocrystals is so strong that the BPs release kinetics appear affected by the bioresorbability of apatitic nanocrystals [108].

Because of the affinity of bisphosphonate groups to calcium, Keppler has proposed platinum compounds containing aminobisphosphonate ligands, which are expected to have a selective tropism for bone apatite and, at the same time, to present the well-known cytotoxic activity typical of platinum complexes [109]. Bis(ethylenediamino)diplatinum medronate ([$Pt_2(C_2H_8N_2)_2(CH_2P_2O_6)$]) has been linked to two different biomimetic synthetic hydroxyapatite nanocrystals having the same composition but different physicochemical properties such as crystal shape, crystallinity, domain size, and specific surface area. A detailed surface characterization of the drug-loaded HA nanocrystals, necessary to determine the surface stoichiometry, has been carried out on the Pt complex/HA conjugate by Attenuated Total Reflection (ATR) spectroscopy and XPS (X-ray Photoelectric Spectroscopy) analyses. Results have shown a Pt/N ratio very close to that of the not-linked bis(ethylenediamino)diplatinum medronate, confirming that it does not to undergo degradation, as concerned the Pt–N bound, during absorption onto the apatite nanocrystals [102].

Finally, an innovative approach to fine-tune the cellular response to implanted apatite nanoparticles, without using the release of bioactive molecules can be ob-

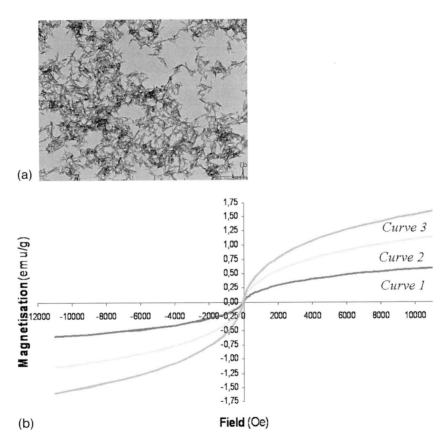

(a)

(b)

Fig. 7.11. (a) TEM image of needle-shaped magnetic HA nanocrystals (scale bar = 100 nm). (b) Magnetic susceptibility of apatite nanocrystals synthesized in the presence of different amounts of the magnetite complex. Curves 2 and 3 are for samples that contain double and triple, respectively, the amount of complex in the sample for curve 1.

tained inducing a specific physical property onto apatite nanocrystals. With this aim a bone-like nanocrystalline apatite showing magnetic properties has been synthesized using a magnetite complex as a co-reagent (Fig. 7.11a) [110, 111]. The magnetic susceptibility of this material shows an increase in magnetization as a function of magnetite complex content (Fig. 7.11b). The well-known bioactive properties of the poorly crystalline apatite, along with the magnetic responsiveness of such HA nanocrystals, make this innovative material an attractive potential candidate for selective heating by external electromagnetic induction and then for bone anticancer hyperthermia treatment [112].

7.7
Conclusion and Future Challenges

Synthesis of substituted, biomimetic, biologically inspired, functionalized hydroxyapatite nano-crystals has been described and their morphological, structural, chemical–physical, surface characterization has been evaluated to explain their high bioactivity and ability to induce bone regeneration and remodeling. The discussion should be enough to describe how size is a crucial factor in determining the HA bone substitutes chemistry in the nano-regime.

Even if spontaneous self-organization is a fascinating and inimitable essence of life material, the biomimetic chemical–physical characteristics of synthetic hydroxyapatite nanocrystals can render themselves *in vitro* and *in vivo* spontaneously mineralizable. The main guides for biomimetic approaches such as supramolecular organization, interfacial molecular recognition and multilevel processing, are strongly connected to the nano-sized dependent chemistry of apatite crystals. Pursuing the aim to mimic more and more the characteristics of bone tissue, the synthesis of biomineral-inspired hydroxyapatite nanocrystals is a reality that is becoming transferable into the preparation of technologically innovative biomedical devices. The synthesis of nanostructured biomimetic bone substitutes resembling the complex hierarchical structure of the biological tissue is an attractive field of research that is achieving the promised results. Moreover, the ability to functionalize the surfaces of apatite nano-crystals with bioactive molecules makes them able to transfer information and to act selectively on the biological environment. Furthermore, this approach offers a promising route to the construction of hierarchically ordered biomaterials based on the principle of controlled nanoparticle self-assembly, allowing us to consider inert mineral materials as templates for life.

Acknowledgments

We thank MIUR (Cofin 2004-032118) and University of Bologna (funds for selected research topics) for financial support, and Bologna University Laboratory of

Environmental and Biological Structural Chemistry (LEBSC) research group and Bristol University Mann's research group for fruitful discussions.

References

1 J. BLACK, G. HASTING, in *Handbook of Biomaterials Properties*, CHAPMAN and HALL (Ed.), American Society, Washington, DC, **1998**.

2 R. LANGER, J. P. VACANTI, Tissue engineering, *Science* **1993**, *260*, 920–926.

3 U. KNESER, D. J. SHAEFER, B. MUNDER, C. KLEMT, C. ANDREE, G. B. STARK, Tissue engineering of bone, *Minimally Invasive Therapy and Allied Technologies*, **2002**, *11*, 107–116.

4 C. T. LAURENCIN, A. M. A. AMBROSIO, M. D. BORDEN, J. A. COOPER, Tissue engineering: Orthopedic application, *Annu. Rev. Biomed. Eng.* **1999**, *1*, 19–46.

5 L. D. K. BUTTERY, J. M. POLAK, in *Learning from Nature: How to Design New Implantable Biomaterials*, R. L. REIS, S. WEINER (Eds.), NATO Science Series, Kluwer Academic Publisher, Dordrecht, **2004**, pp. 199–204.

6 F. M. C. DRIESSENS, Formation and stability of calcium phosphates in relation to the phase composition of the mineral in calcified tissues, in *Bioceramic Calcium Phosphate*, CRC Press, Boca Raton, FL, **1983**, pp. 1–32.

7 C. REY, V. RENUGOPALAKRISHNAN, B. COLLINS, M. GLIMCHER, Fourier transform infrared spectroscopic study of the carbonate ions in bone mineral during aging, *Calcif. Tissue Int.* **1991**, *49*, 251–258.

8 J. M. BURNELL, E. J. TEUBNER, A. G. MILLER, Normal maturational changes in bone matrix, mineral, and crystal size in the rat. *Calcif. Tissue Int.* **1980**, *31*, 13–19.

9 C. REY, B. COLLINS, T. GOEHL, I. R. DICKSON, M. GLIMCHER, The carbonate environment in bone mineral: A resolution-enhanced Fourier transform infrared spectroscopy study, *Calcif. Tissue Int.* **1989**, *45*, 157–164.

10 H. A. LOWESTAN, S. WEINER, *On Biomineralization*, Oxford University Press, New York, **1989**.

11 R. LEGROS, N. BALMAIN, G. BONEL, Age-related changes in mineral of rat and bovine cortical bone, *Calcif. Tissue Int.*, **1987**, *41*, 137–144.

12 R. G. HANDSCHIN, W. B. STERN, Crystallographic and chemical analysis of human bone apatite (Crista Iliaca), *Clin. Rheumatol.*, **1994**, *13*, 75–90.

13 A. S. POSNER, Crystal chemistry of bone mineral, *Physiol. Rev.* **1969**, *49*, 760–762.

14 V. L. SIKAVITSAS, J. S. TEMENOFF, A. G. MIKOS, Biomaterials and bone mechano-transduction, *Biomaterials* **2001**, *22*, 2581–2593.

15 H. C. ANDERSON, Mineralization by matrix vesicles, *Scan. Electron Microsc.* **1984**, 953–964.

16 A. S. POSNER, The mineral of bone, *Clin. Orthop.* **1985**, 87–99.

17 G. A. RODAN, Introduction to bone biology, *Bone* **1992**, *13*, 53–56.

18 H. C. ANDERSON, Molecular biology of matrix vesicles, *Clin. Orthop.* **1995**, *314*, 266–280.

19 H. C. ANDERSON, Matrix vesicles and calcification, *Curr. Rheumatol. Rep.* **2003**, *5*, 222–226.

20 *Collagen: Biochemistry, Biomechanics, Biotechnology*, M. E. NIMNI (Ed.), CRC Press, Boca Raton FL, **1988**.

21 S. WEINER, H. D. WAGNER, The material bone: Structure-mechanical function relations, *Ann. Rev. Mat. Sci.* **1998**, *28*, 271–298.

22 A. ASCENZI, E. BONUCCI, A. RIPAMONTI, N. ROVERI, X-ray diffraction and electron microscope study of osteons during calcification, *Calcif. Tiss. Res.* **1978**, *25*, 133–143.

23 A. ASCENZI, E. BONUCCI, P. GENERALI, A. RIPAMONTI, N. ROVERI, Orientation of apatite in single osteon samples as

studied by pole figures, *Calcif. Tissue Int.* **1979**, *29*, 101–105.

24 A. ASCENZI, A. BIGI, A. RIPAMONTI, N. ROVERI, X-ray diffraction analysis of transversal osteonic lamellae, *Calcif. Tissue Int.* **1983**, *35*, 279–283.

25 A. ASCENZI, A. BIGI, M. H. J. KOCH, A. RIPAMONTI, N. ROVERI, A low-angle X-ray diffraction analysis of osteonic inorganic phase using synchrotron radiation, *Calcif. Tissue Int.* **1985**, *37*, 659–664.

26 C. DU, G. FALINI, S. FERMANI, C. ABBOTT, J. MORADIAN-OLDAK, Supramolecular assembly of amelogenin nanospheres into birefringent microribbons, *Science*, **2005**, *307*, 1450–1454.

27 M. VALLET-REGI, J. M. GONZALES CALBET, Calcium phosphates as substitution of bone tissue, *Progr. Solid State Chem.* **2004** *32(1–2)*, 1–31.

28 K. SATO, Y. KUMAGAI, J. TANAKA, Apatite formation on organic monolayers in simulated body environment, *J. Biomed. Mater. Res.* **2000**, *50(1)*, 16–20.

29 A. L. BOSKEY, Will biomimetics provide new answers for old problems of calcified tissues?, *Calcified Tissue Int.* **1998**, *63(3)*, 179–182.

30 CHANG, MYUNG CHUL, T. IKOMA, M. KIKUCHI, J. TANAKA, Preparation of a porous hydroxyapatite/collagen nanocomposite using glutaraldehyde as a crosslinkage agent, *J. Mater. Sci. Lett.* **2001**, *20(13)*, 1199–1201.

31 M. P. FERRAZ, F. J. MONTEIRO, C. M. MANUEL, Hydroxyapatite nanoparticles: A review of preparation methodologies, *J. Appl. Biomater. Biomech.* **2004**, *2*, 74–80.

32 E. BOUYER, F. GITZHOFER, M. I. BOULOS, Morphological study of hydroxyapatite nanocrystal suspension, *J. Mater. Sci. Mater. Med.* **2000**, *11*, 523–531.

33 L. YUBAO, K. DE GROOT, J. DE WIJN, C. P. A. T. KLEIN, S. V. D. MEER, Morphology and composition of nanograde calcium phosphate needle-like crystals formed by simple hydrothermal treatment, *J. Mater. Sci. Mater. Med.* **1994**, *5*, 326–331.

34 G. BEZZI, G. CELOTTI, E. LANDI, T. M. G. LA TORRETTA, I. SOPYAN, A. TAMPIERI, A novel sol-gel technique for hydroxyapatite preparation. *Mater. Chem. Phys.* **2003**, *78(3)*, 816–824.

35 D. JANACKOVIC, I. PETROVIC-PRELEVIC, L. KOSTIC-GVOZDENOVIC, R. PETROVIC, V. JOKANOVIC, D. USKOKOVIC, Influence of synthesis parameters on the particle size of nano-structured calcium-hydroxyapatite, *Key Eng. Mater.* **2001**, *192-5(Bioceramics)*, 203–206.

36 A. BIGI, F. MARCHETTI, A. RIPAMONTI, N. ROVERI, Magnesium and strontium interaction with carbonate-containing hydroxyapatite in acqueous medium, *J. Inorg. Biochem.* **1981**, *15*, 317–327.

37 A. BIGI, E. FORESTI, R. GREGORIANI, A. RIPAMONTI, N. ROVERI, J. S. SHAH, The role of magnesium on the structure of biological apatites, *Calcif. Tissue Int.* **1992**, *50(5)*, 439–444.

38 R. Z. LE GEROS, J. P. LE GEROS, in *Phosphate Minerals*, J. O. NRIAGU, P. B. MOORE (Eds.), Springer, New York, **1984**, pp. 351–385.

39 S. BARAVELLI, A. BIGI, E. FORESTI, A. RIPAMONTI and N. ROVERI, Thermal behavior of bone and synthetic hydroxyapatites submitted to magnesium interaction in aqueous media, *J. Inorg. Biochem.* **1984** *20*, 1–12.

40 A. BIGI, G. FALINI, E. FORESTI, M. GAZZANO, A. RIPAMONTI, N. ROVERI, Rietveld structure refinements of calcium hydroxylapatite containing magnesium, *Acta Crystallogr., Sect. B: Struct. Sci.* **1996**, *52(1)*, 87–92.

41 A. BIGI, G. FALINI, E. FORESTI, M. GAZZANO, A. RIPAMONTI, N. ROVERI, Magnesium influence on hydroxyapatite crystallization, *J. Inorg. Biochem.* **1993**, *49(1)*, 69–78.

42 A. TAMPIERI, G. CELOTTI, E. LANDI, From biomimetic apatite to biologically inspired composites, *Anal. Bioanal. Chem.* **2005**, *381*, 568–572.

43 L. M. RODRIGUEZ-LORENZO, M. VALLET-REGI, J. M. F. FERREIRA, Fabrication of hydroxyapatite bodies by uniaxial pressing from a precipitated

powder, *Biomaterials* **2001**, *22(6)*, 583–588.

44 D. Tadic, F. Beckmann, K. Schwarz, M. Epple, A novel method to produce hydroxyapatite objects with interconnecting porosity that avoids sintering. *Biomaterials* **2004**, *25(16)*, 3335–3340.

45 D. Tadic, F. Beckmann, K. Schwarz, M. Epple, A novel method to produce hydroxyapatite objects with interconnecting porosity that avoids sintering. *Biomaterials* **2004**, *25(16)*, 3335–3340.

46 L. M. Rodriguez-Lorenzo, M. Vallet-Regi, J. M. F. Ferreira, M. P. Ginebra, C. Aparicio, J. A. Planell, Hydroxyapatite ceramic bodies with tailored mechanical properties for different applications, *J. Biomed. Mater. Res.* **2002**, *60(1)*, 159–166.

47 S. Padilla, J. Roman, M. Vallet-Regi, Synthesis of porous hydroxyapatites by combination of gelcasting and foams burn out methods, *J. Mater. Sci.: Mater. Med.* **2002**, *13(12)*, 1193–1197.

48 D. M. Roy, S. K. Linnehan, Hydroxyapatite formed from coral skeletal carbonate by hydrothermal exchange, *Nature* **1974**, *247(438)*, 220–222.

49 B. Ben-Nissan, Natural bioceramics: From coral to bone and beyond, *Curr. Opin. Solid State Mater. Sci.* **2003**, *7(4–5)*, 283–288.

50 B. Ben-Nissan, A. Milev, R. Vago, Morphology of sol-gel derived nano-coated coralline hydroxyapatite, *Biomaterials* **2004**, *25(20)*, 4971–4975.

51 A. Tampieri, G. Celotti, S. Sprio, A. Delcogliano, S. Franzese, Porosity-graded hydroxyapatite ceramics to replace natural bone. *Biomaterials* **2001**, *22(11)*, 1365–1370.

52 B. Palazzo, M. C. Sidoti, N. Roveri, A. Tampieri, M. Sandri, L. Bertolazzi, F. Galbusera, G. Dubini, P. Vena, R. Contro, Controlled drug delivery from porous hydroxyapatite grafts: An experimental and theoretical approach, *Mater. Sci. Eng., C: Biomimetic Supramol. Systems* **2005**, *C25(2)*, 207–213.

53 A. Tampieri, G. Celotti, E. Landi, M. Montevecchi, N. Roveri, A. Bigi, S. Panzavolta, M. C. Sidoti, Porous phosphate-gelatin composite as bone graft with drug delivery function, *J. Mater. Sci.: Mater. Med.* **2003**, *14(7)*, 623–627.

54 J. Pena, M. Vallet-Regi, J. San Roman, TiO$_2$-polymer composites for biomedical applications, *J. Biomed. Mater. Res.* **1997**, *35(1)*, 129–134.

55 M. Wang, T. Kokubo, W. Bonfield, in *Bioceramics*, T. Kokubo, T. Nakamura, F. Miyaji (Eds.), Elsevier, Osaka, Japan **1996**, Vol. 9, p. 387.

56 M. Kikuchi, K. Sato, Y. Suetsugu, J. Tanaka, in *Bioceramics*, R. Z. Le Geros, J. P. Le Geros (Eds.), World Scientific, New York, **1998**, Vol. 11, p. 153.

57 D. Bakos, M. Soldan, I. Hernandez-Fuentes, Hydroxyapatite-collagen-hyaluronic acid composite, *Biomaterials* **1999**, *20*, 191–195.

58 A. Cherng, S. Takagi, L. C. Chow, Effects of hydroxypropyl methyl cellulose and other gelling agents on the handling properties of calcium phosphate cement, *J. Biomed. Mater. Res.* **1997**, *35(3)*, 273–277.

59 P. Eiselt, J. Yeh, R. K. Latvala, L. D. Shea, D. J. Mooney, Porous carriers for biomedical applications based on alginate hydrogels, *Biomaterials* **2000**, *21(19)*, 1921–1927.

60 B. Amsden, N. Turner, Diffusion characteristics of calcium alginate gels, *Biotechnol. Bioeng.* **1999**, *65(5)*, 605–610.

61 K. H. Bouhadir, K. Y. Lee, E. Alsberg, K. L. Damm, R. W. Anderson, D. J. Mooney, Degradation of partially oxidized alginate and its potential application for tissue engineering, *Biotechnol. Progr.* **2001**, *17(5)*, 945–950.

62 S. C. N. Chang, J. A. Rowley, G. Tobias, N. G. Genes, A. K. Roy, D. J. Mooney, C. A. Vacanti, L. J. Bonassar, Injection molding of chondrocyte/alginate constructs in the shape of facial implants, *J. Biomed. Mater. Res.* **2001**, *55(4)*, 503–511.

63 E. Landi, A. Tampieri, G. Celotti,

L. VICHI, M. SANDRI, Influence of synthesis and sintering parameters on the characteristics of carbonate apatite, *Biomaterials* **2004**, *25(10)*, 1763–1770.

64 S. YANG, K. F. LEONG, Z. DU, C. K. CHUA, The design of scaffolds for use in tissue engineering. Part I. Traditional factors, *Tissue Eng.* **2001**, *7(6)*, 679–689.

65 A. TAMPIERI, M. SANDRI, E. LANDI, G. CELOTTI, N. ROVERI, M. MATTIOLI-BELMONTE, L. VIRGILI, F. GABBANELLI, G. BIAGINI, HA/alginate hybrid composites prepared through bio-inspired nucleation, *Acta Biomater.* **2005**, *1*, 343–351.

66 E. RUIZ-HITZKY, Functionalizing inorganic solids: Towards organic-inorganic nanostructured materials for intelligent and bioinspired systems, *Chem. Record* **2003**, *3*, 88–100.

67 B. DIETRICH, P. VIOUT, J. M. LEHN, in *Macrocyclic Chemistry: Aspects of Organic and Inorganic Supra Molecular Chemistry*, VCH, Weinheim, **1993**, pp. 384.

68 L. DONALD, A. VEIS, Collagen self-assembly in vitro, *J. Biol. Chem.* **1981**, *256*, 7118–7128.

69 G. FALINI, S. FERMANI, E. FORESTI, B. PARMA, K. RUBINI, M. C. SIDOTI, N. ROVERI, Films of self-assembled purely helical type I collagen molecules. *J. Mater. Chem.* **2004**, *14*, 2297–2302.

70 N. ROVERI, G. FALINI, M. C. SIDOTI, A. TAMPIERI, E. LANDI, M. SANDRI, B. PARMA, Biologically inspired growth of hydroxyapatite nanocrystals inside self-assembled collagen fibers, *Mater. Sci. Eng., C: Biomimetic Supramol. Systems* **2003**, *C23*, 441–446.

71 A. TAMPIERI, G. CELOTTI, E. LANDI, M. SANDRI, N. ROVERI, G. FALINI, Biologically inspired synthesis of bone-like composite: Self-assembled collagen fibers/hydroxyapatite nanocrystals, *J. Biomed. Mater. Res., A* **2003**, *67*, 618–625.

72 M. KIKUCHI, S. ITOH, S. ICHINOSE, K. SHINOMIYA, J. TANAKA, Self organization mechanism in a bone-like hydroxyapatite/collagen nanocomposite synthesized in vitro and its biological

reaction in vivo, *Biomaterials* **2001**, *22*, 1705–1711.

73 A. TAMPIERI, G. CELOTTI, E. LANDI, N. ROVERI, *Eur. Patent* 04001417.7, **2004**.

74 S. A. REDEY, S. RAZZOUK, C. REY, D. BERNACHE-ASSOLANT, G. LEROY, M. NARDIN, G. COURNOT, Osteoclast adhesion and activity on synthetic hydroxyapatite, carbonated hydroxyapatite, and natural calcium carbonate: Relationship to surface energies. *J. Biomed. Mater. Res.* **1999**, *45*, 140–147.

75 G. GUILLEMIN, S. J. HUNTER, C. V. GAY, Resorption of natural calcium carbonate by avian osteoclasts in vitro, *Cells Mater.* **1995**, *5(2)*, 157–165Y.

76 M. KIKUCHI, T. IKOMA, S. ITOH, H. N. MATSUMOTO, Y. KOYAMA, K. TAKAKUDA, K. SHINOMIYA, J. TANAKA, Biomimetic synthesis of bone-like nanocomposites using the self-organization mechanism of hydroxyapatite and collagen, *Composites Sci. Technol.* **2004**, *64(6)*, 819–825.

77 MIYAMOTO, K. ISHIKAWA, M. TAKECHI, T. TOH, T. YUASA, M. NAGAYAMA et al., Basic properties of calcium phosphate cement containing atelocollagen in its liquid or powder phases, *Biomaterials* **1998**, *19*, 707–715.

78 C. DU, F. Z. CUI, X. D. ZHU, K. DE GROOT, Three-dimensional nano-HAp/collagen matrix loading with osteogenic cells in organ culture, *J. Biomed. Mater. Res.* **1999**, *44*, 407–415.

79 M. KIKUCHI, H. N. MATSUMOTO, T. YAMADA, Y. KOYAMA, K. TAKAKUDA, J. TANAKA, Glutaraldehyde cross-linked hydroxyapatite/collagen self-organized nanocomposites, *Biomaterials* **2004**, *25*, 63–69.

80 M. SHIRKHANZADEH, Direct formation of nanophase hydroxyapatite catohodically polarised electrodes, *J. Mater. Sci. Mater. Med.* **1998**, *9*, 67–72.

81 H. SCHLIEPHAKE, D. SCHARNWEBER, M. DA, S. ROSSELER, A. SEWING, C. HUTTMANN, Biological performance of biomimetic calcium phosphate coating of titanium implants in the dog

mandible, *J. Biomed. Mater. Res.* **2003**, *64*, 225–234.

82 M. MANSO, C. JIMENEZ, C. MORANT, P. HERRERO, J. M. MARTINEZ DUART, Electrodeposition of hydroxyapatite coating in basic conditions, *Biomaterials* **2000**, *21*, 1755–1761.

83 K. DUAN, Y. W. FAN, R. WANG, Electrochemical deposition and patterning of calcium phosphate bioceramics coating, *Ceram. Trans.* **2003**, *147*, 53–61.

84 S. MANARA, F. PAOLUCCI, B. PALAZZO, M. MARCACCIO, E. FORESTI, N. PARMA, N. ROVERI, Electrochemical deposition of hydroxyapatite-collagen fibrils composites" Submitted to *Advanced Functional Materials.*

85 H. OKAMURA, M. YASUDA, M. OHTA, Synthesis of calcium-deficient hydroxyapatite-collagen composite, *Electrochemistry* **2000**, *68*, 486–488.

86 H. SCHLIEPHAKE, D. SCHARNWEBER, M. DARD, S. ROSSLER, A. SEWING, C. HUTTMANN, Biological performance of biomimetic calcium phosphate coating of titanium implants in the dog mandible. *J. Biomed. Mater. Res.* **2003**, *64*, 225–234.

87 Y. FAN, K. DUAN, R. WANG, A composite coating by electrolysis-induced collagen self-assembly and calcium phosphate mineralization, *Biomaterials* **2004**, *26*, 1623–1632.

88 G. FALINI, E. FORESTI, S. MANARA, B. PALAZZO, M. MARCACCIO, F. PAOLUCCI, N. ROVERI, Electrochemically co-deposition of hydroxyapatite-collagen composite on Ti surface, *19th European Conference on Biomaterials*, Sorrento 11–15 September 2005.

89 E. I. SUVOROVA, P. A. BUFFAT, Electron diffraction from micro and nano-particles of hydroxyapatite, *J. Microsc. (Oxford)* **1999**, *196(1)*, 46–58.

90 L. C. NISTOR, C. GHICA, V. S. TEODORESCU, S. V. NISTOR, M. DINESCU, D. MATEI, N. FRANGIS, N. VOUROUTZIS, C. LIUTAS, Deposition of hydroxiapatite thin films by Nd:YAG laser ablation: A micro structural study, *Mater. Res. Bull.* **2004**, *39*, 2089–2101.

91 L. ADDADI, S. WEINER, Stereochemical and structural relations between macromolecules and crystals in biomineralization, in *Biomineralization: Chemical and Biochemical Prospective*, S. MANN, J. WEBB, R. J. P. WILLIAMS (Eds.), VCH Verlagsgesellschaft, Weinheim, **1980**, pp. 132–156.

92 A. L. BOSKEY, Phospholipids and calcification: An overwiew, in *Cell Mediated Calcification and Matrix Vesicles*, S. Y. ALI (Ed.), Elsevier, Amsterdam, **1986**, pp. 175–179.

93 A. MEISTER, *Biochemistry of Amino Acids*, 2nd edn., Academic Press, New York, **1965**.

94 A. L. LEHNINGER, *Biochemistry*, 2nd edn., Worth Publishing, New York, **1975**.

95 S. KOUTSOPOULOS, E. DALAS, Hydroxyapatite crystallization in the presence of serine, tyrosine and hydroxyproline amino acids with polar side groups, *J. Crystal Growth*, **2000**, *216*, 443–449.

96 S. KOUTSOPOULOS, E. DALAS, Hydroxyapatite crystallization in the presence of amino acids with uncharged polar side groups: glycine, cysteine, cystine, and glutamine, *Langmuir* **2001**, *17*, 1074–1079.

97 S. KOUTSOPOULOS, E. DALAS, The crystallization of hydroxyapatite in the presence of lysine, *J. Colloid Interface Sci.* **2000**, *231*, 207–212.

98 R. GONZALEZ-MCQUIRE, J. Y. CHANE-CHING, JEAN-YVES, E. VIGNAUD, A. LEBUGLE, S. MANN, Synthesis and characterization of amino acid-functionalized hydroxyapatite nanorods, *J. Mater. Chem.* **2004**, *14(14)*, 2277–2281.

99 B. PALAZZO, S. MANN, N. WALSH, N. ROVERI, G. FALINI, E. FORESTI, G. NATILE, M. LAFORGIA, Apatite surface functionalization with aminoacids and platinum complexes, *IV Symposium on Pharmaco-Biometallics*, CIRCMSB (ed.), Tecnomack, Bari, Italy, p. 23.

100 A. BARROUG, J. M. GLIMCHER, Hydroxyapatite crystals as a local delivery system for cisplatin: Adsorption and release of cisplatin in vitro, *J. Orthopaedic Res.* **2002**, *20*, 274–280.

101 A. BARROUG, L. T. KUHN, L. C.

GERSTENFELD, M. J. GLIMCHER, Interaction of cisplatinum with calcium phosphate nanoparticles: In vitro controlled adsorption and release, *J. Orthopaedic Res.* **2004**, *22*, 703–708.

102 B. PALAZZO, M. IAFISCO, M. LAFORGIA, N. MARGIOTTA, G. NATILE, C. L. BIANCHI, D. WALSH, N. ROVERI, "Biomimetic hydroxyapatite nanocrystals as bone substitute with anti-tumour drugs delivery function" Submitted to *Advanced Functional Materials*.

103 S. JOSSE, C. FAUCHEUX, A. SOUEIDAN, G. GRIMANDI, D. MASSIOT, B. ALONSO, P. JANVIER, S. LAIB, P. PILET, O. GAUTHIER, G. DACULSI, J. GUICHEUX, B. BUJOLI, J. M. BOULER, Novel biomaterials for bisphosphonate delivery, *Biomaterials* **2005**, *26(14)*, 2073–2080.

104 R. G. G. RUSSELL, M. J. ROGERS, Bisphosphonates: From the laboratory to the clinic and back again, *Bone (New York)* **1999**, *25(1)*, 97–106.

105 A. EZRA, G. GOLOMB, Administration routes and delivery systems of bisphosphonates for the treatment of bone resorption, *Adv. Drug Deliv. Rev.* **2000**, *42*, 175–195.

106 L. S. ROSEN, D. GORDON, M. KAMINSKI, A. HOWELL, A. BELCH, J. MACKEY, J. APFFELSTAEDT, M. HUSSEIN, R. E. COLEMAN, D. J. REITSMA, J. J. SEAMAN, B. L. CHEN, Y. AMBROS, Zoledronic acid versus pamidronate in the treatment of skeletal metastases in patients with breast cancer or osteolytic lesions of multiple myeloma: A phase III, double-blind, comparative trial, *Cancer J. (Sudbury, Mass.)*, **2001**, *7(5)*, 377–387.

107 H. ROUSSIERE, G. MONTAVON, S. LAIB, P. JANVIER, B. ALONSO, F. FAYON, M. PETIT, D. MASSIOT, J. M. BOULER, B. BUJOLI, Hybrid materials applied to biotechnologies: Coating of calcium phosphates for the design of implants active against bone resorption disorders, *J. Mater. Chem.* **2005**, *15(35–36)*, 3869–3875.

108 B. PALAZZO, M. IAFISCO, M. LAFORGIA, N. MARGIOTTA, G. NATILE, N. ROVERI, Hydroxyapatite nanocrystals as local anticancer platinum complexes delivery system. *V Symposium on Pharmaco-Biometallics*, 11–12 November 2005 Bertinoro (Bologna), Italy, CIRCMSB (ed.), Tecnomack, Bari, Italy.

109 T. KLENNER, P. VALENZUELA-PAZ, F. AMELUNG, H. MUNCH, H. ZAHN, B. K. KEPPLER, H. BLUM, in *Metal Complexes in Cancer Chemotherapy*, B. K. KEPPLER (Ed.), VCH, Weinheim, **1993**, pp. 85–127.

110 B. PALAZZO, D. WALSH, S. MANN, N. ROVERI, Magnetic nanoapatite for bone tissue cancer treatment by locally induced hyperthermia, *V Symposium on Pharmaco-Biometallics*, 11–12 November 2005, Bertinoro (Bologna), Italy, CIRCMSB (ed.), Tecnomack, Bari, Italy.

111 B. PALAZZO, D. WALSH, S. MANN, N. ROVERI, Magnetic hydroxyapatite nanocrystals for bone tissue cancer treatment by locally induced hyperthermia, submitted to *J. Mater. Chem.*

112 J. N. WEINSTEIN, R. L. MAGIN, M. B. YATVIN, D. S. ZAHARKO, Liposomes and local hyperthermia: Selective delivery of methotrexate to heated tumors, *Science*, **1979**, 204, 188–191.

8
Magnetic Nanoparticles for Tissue Engineering

Akira Ito and Hiroyuki Honda

8.1
Introduction

Magnetic nanoparticles, from nanometer to submicron size, have been developed for medical applications by many researchers [1, 2]. Since magnetic nanoparticles have unique magnetic features not present in other materials, they can be applied to special medical techniques. Magnetic particles are attracted to high magnetic flux density and this feature is used for drug targeting [3] and bioseparation [4–6], including cell sorting [7]. A comprehensive review on the use of magnetic particles in clinical applications was given by Häfeli et al. [8]. In these applications, magnetic particles are given various names, e.g. magnetic microspheres, magnetic nanospheres, and ferrofluids, among others. Here, we use the general term "magnetic nanoparticles", and review and discuss their new applications in tissue engineering.

Magnetite (Fe_3O_4) particles are being used in an increasing number of biological and medical applications, including cell sorting. To add an affinity and target ability to cells, we applied the concepts involved in drug delivery systems (DDSs) to magnetite nanoparticles and developed functionalized magnetite nanoparticles. We developed magnetite cationic liposomes (MCLs), which are cationic liposomes containing 10-nm magnetite nanoparticles, to improve accumulation of magnetite nanoparticles in target cells by taking advantage of the electrostatic interaction between MCLs (positively charged) and the cell membrane (negatively charged) [9–11]. Since MCLs are designed to interact with target cells via electrostatic interaction, there is the risk of non-specific interaction between MCLs and various non-target cell types. A promising technique is the use of antibodies raised against target cells to isolate them from the other cells. We developed antibody-conjugated magnetoliposomes (AMLs) for use in cancer-targeted therapy [12–14]. The AMLs were made from magnetoliposomes (MLs) consisting of neutral lipids, to reduce electrostatic interaction with target cells. Target cells can be magnetically labeled by the functionalized magnetite nanoparticles (MCLs or AMLs), via uptake of magnetite nanoparticle into the cells (Fig. 8.1).

The loss or failure of an organ or tissue is one of the most frequent, devastating,

Nanotechnologies for the Life Sciences Vol. 9
Tissue, Cell and Organ Engineering. Edited by Challa S. S. R. Kumar
Copyright © 2006 WILEY-VCH Verlag GmbH & Co. KGaA, Weinheim
ISBN: 3-527-31389-3

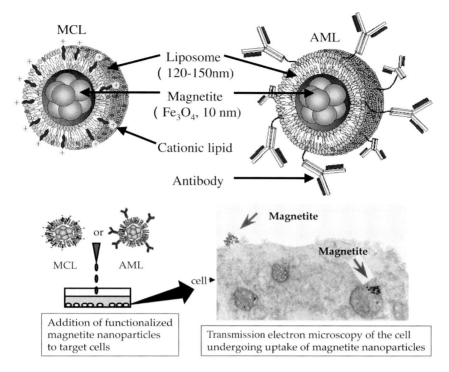

Fig. 8.1. Schematic of functionalized magnetic nanoparticles.

and costly problems in human health care. Tissue engineering applies the principles of biology and engineering to the development of functional substitutes for damaged tissue. Langer and Vacanti laid the foundations of this interdisciplinary field in 1993 [15], and worldwide interest in tissue engineering was initiated. The following decade saw growing enthusiasm for tissue engineering, and tissue engineering has been a promising technology for overcoming the organ transplantation crisis resulting from donor organ shortage. In general, tissue engineering consists of the following processes: (a) Target cells are isolated and expanded to the required cell number; (b) cells are harvested and reseeded into three-dimensional (3D) biodegradable scaffolds, allowing 3D cell culture; and (c) the cultured 3D constructs are implanted into patients.

Although an overall technology of these processes in tissue engineering has been established, there is still room for improvement in each process. From the point of view of bioprocess engineering, development of a methodology of "physical manipulation of target cells" is essential for tissue engineering in the next few decades. We selected magnetic force as a physical force, and manipulated target cells labeled with magnetite nanoparticles using DDS techniques. Thus, we have developed a novel cell-manipulating technology using functionalized magnetite nanoparticles and magnetic force, which we have designated "Mag-TE". The present chapter focuses on Mag-TE that has been applied to the processes in tissue engineering: (a) Magnetic isolation and expansion of mesenchymal stem cell (Section 8.2); (b) magnetic cell-seeding into 3D biodegradable scaffolds (Mag-seeding, Section 8.3); and (c) construction of 3D tissue-like structure by using magnetic force (Section 8.4).

8.2
Mesenchymal Stem Cell Isolation and Expansion

Bone marrow-derived mesenchymal stem cells (MSCs) can differentiate into osteoblasts, chondrocytes, adipocytes, muscle cells or nerve cells *in vitro* and *in vivo* [16–19]. Since MSCs can easily be obtained by bone marrow aspiration, transplantation of bone marrow MSCs may provide a new treatment for regeneration of mesenchymal tissues [18]. However, marrow aspiration of too great a volume causes damage and pain to the donor. Thus, it is difficult to obtain the large number of MSCs required for regeneration of injured tissues. Expansion of MSCs *in vitro* is a necessary step in the clinical application of MSCs.

Despite the great interest in MSCs, there is still no well-defined protocol for isolation and expansion of MSCs in culture. Most experiments have been conducted using MSCs isolated primarily from bone marrow aspirates by their tight adherence to plastic dishes, as described by Friedenstein et al. [20]. However, this method for isolating cells does not help increase the number of MSCs, as there is only a small number of MSCs in bone marrow aspirate. Pittenger et al. reported that only a small percentage (0.0001–0.01%) of bone marrow aspirate cells that attached to the culture dishes were MSCs [16]. In addition, culture volume inevitably becomes large, since there are numerous nonadherent cells such as hematopoietic cells that must be diluted and removed by washing and changes in medium, resulting in low-density culture of MSCs. Under such low-density culture conditions, MSCs do not proliferate immediately, and require much time to develop into colonies as growth factors, including autocrine [21–24], paracrine [24], and juxtacrine [25] factors, play an important role in cell growth. Therefore, we developed a new method for producing high-density culture, using magnetic nanoparticles to promote expansion of MSCs.

8.2.1
MSC Expansion using MCLs

We have developed MCLs that have been used as carriers to introduce DNA into cells [26], and as heat generating mediators for cancer therapy [27–31]. Here, in this study [32], magnetic forces were used to move MSCs labeled with MCLs, to hold them *in situ*, and to culture them at high density. First, we investigated the applicability of this combined methodological approach to the enrichment and proliferation of MSCs using MCLs. Figure 8.2 illustrates a schematic of MSC expansion using MCLs.

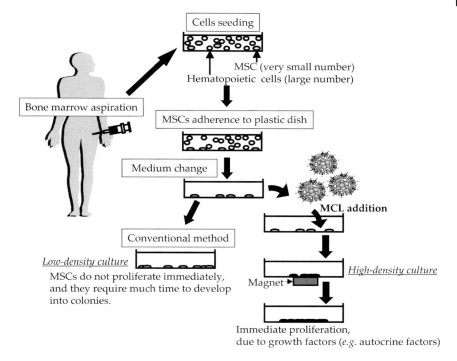

Fig. 8.2. Schematic of the expansion of MSCs using MCLs.

Expansion of MSCs using MCLs was investigated as follows. At 4 h after addition of the MCLs, 1000 human MSCs (a very small cell number corresponding to the MSCs in 1 mL of bone marrow aspirate) were seeded into a 100-mm cell culture dish with 10 mL of cell culture medium; a cylindrical neodymium magnet (diameter, 2.2 cm; height, 1 cm; 4000 Gauss) was then placed under the dish to provide magnetic force vertical to the dish.

When the MCLs were added to human MSCs at 100-pg magnetite per cell, uptake of magnetic nanoparticles began rapidly and maximum uptake (20 pg cell^{-1}) was achieved at 4 h after addition. Subsequently, the amount of magnetite per cell decreased due to dilution as a result of cell growth. Growth of MSCs in medium containing MCLs (MCL(+)MSCs) was compared with growth of MSCs in medium without MCLs (MCL(−)MSCs). MCLs did not inhibit nor stimulate growth of MSCs. Furthermore, no effects of MCL addition on MSC differentiation were observed. Osteogenic differentiation of MCL(+)MSCs transferred into osteogenic medium was attained 17 days after transfer. MSCs incubated in osteogenic medium changed shape from fibroblastic (undifferentiated MSC) to polygonal, and formed calcium nodules. Adipogenic induction of MCL(+)MSCs was attained after culturing in adipogenic medium, as indicated by accumulation of lipid-rich vacuoles within cells and the presence of Oil red-O-positive cells. Thus, we observed no tox-

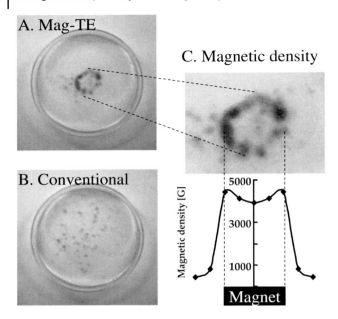

Fig. 8.3. Cultures of MSCs produced by Mag-TE (A), compared with conventional means (B), have a high density (C).

icity against MSC proliferation or differentiation when the magnetite concentration was 100 pg cell^{-1}.

MCL(+)MSCs accumulated circumferentially along the edge of the magnet at the center of dish where the magnet was positioned (Fig. 8.3A), while MCL(−)MSCs formed scattered colonies throughout the dish (Fig. 8.3B). Since magnetic particles are attracted to high magnetic flux density, MCL(+)MSCs were attracted to where the magnetic density was highest (Fig. 8.3C). The growth area regulated by the magnetic force was approximately 1 cm^2, as indicated by image analysis. These results suggest that the MCL(+)MSCs were seeded at 1000 cells cm^{-2}, which was a much higher concentration than that of MCL(−)MSCs (18 cells cm^{-2}; 1000 cells per 55-cm^2 dish). A total of 2.2×10^4 cells were counted on day 7, corresponding to the number of cells in 1 cm^2 of confluent culture (approximately 2×10^4 cells cm^{-2}). This was five-fold greater than the number obtained for MCL(−)MSCs using the conventional method.

Methods for expansion of MSCs are in great demand for clinical applications. One approach to expansion of MSCs is identification of growth factors involved in self-renewal of MSCs. Tsutsumi et al. reported that fibroblast growth factor-2 (FGF-2) increases the growth rate of rabbit, canine and human bone marrow MSCs [33]. However, the mechanism by which FGF-2 maintains proliferation without differentiation is unknown, and clinical applications of cytokines should be pursued with caution due to the unknown functions of cytokines. Alternatively, growth factors such as cytokines are produced by cells themselves and act as autocrine factors. Huss et al. reported that autocrine factors such as stem cell factor (SCF) and interleukin-6 (IL-6) are involved in proliferation and differentiation of a canine bone marrow-derived cell line, and that an increase in local cell concentration is associated with cell viability [34]. In addition, Gregory et al. reported that conditioned medium from cultures of human MSCs increases the rate of proliferation of freshly plated cultures of human MSCs; secretion of high levels of dickkopf-1 (Dkk-1), an inhibitor of the canonical Wnt signaling pathway, was involved in the reentry of MSCs into the cell cycle in lag phase because of low cell density [35]. These results prompted us to use magnetic force for MSC separation. As mentioned above, when MSCs were seeded at high density using MCLs, the number of cells obtained was five-fold greater than the number obtained from culture without MCLs. These results suggest that growth factors, including autocrine, paracrine, and juxtacrine factors, are involved in the proliferation of MSCs. "High-density culture" using magnetic nanoparticles provides a new methodology for expansion of MSCs.

8.2.2
MSC Isolation and Expansion using AMLs

As mentioned in Section 8.2.1, we used magnetic force to move MSCs labeled with MCLs and to culture them at high density, and demonstrated the feasibility of magnetic force-based high-density culture for proliferation of MSCs *in vitro*. Because MCLs were designed to interact with target cells via electrostatic interaction, there is the risk of non-specific interaction between MCLs and various non-target cell types. The MCLs were mixed with purified MSCs that had been isolated from bone marrow aspirates.

A promising technique is the use of antibodies raised against MSCs to isolate MSCs from bone marrow aspirates. We previously developed antibody-conjugated magnetoliposomes (AMLs) for use in cancer therapy [12–14]. The AMLs were made from magnetoliposomes (MLs) that consisted of neutral lipids, to reduce electrostatic interaction with target cells. Recently, many researchers have used antibodies to isolate or characterize MSCs [36–38]. Barry et al. reported that CD105 (endoglin), which is the TGF (transforming growth factor)-β receptor I/III present on endothelial cells, macrophages, and connective tissue stromal cells, is a useful surface antigen for isolation of MSCs [36]. Therefore, we investigated whether magnetic force-based high-density culture of MSCs, using anti-CD105 antibody-conjugated magnetoliposomes, is an effective method for expansion of MSCs [39]. Figure 8.4 illustrates a strategy for MSCs expansion using AMLs.

Magnetic separation of human MSCs from bone marrow aspirates using anti-CD105 antibody-conjugated magnetoliposomes was performed as follows. After obtaining informed consent, an average of 10 mL of iliac and maxillofacial bone marrow was collected from donors. These bone marrow aspirates were mixed with 20 mL of cell growth medium, and 3 mL of that mixture was combined with 7 mL of

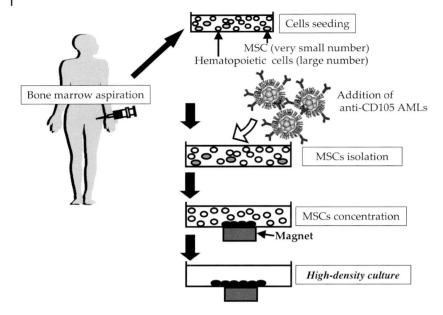

Fig. 8.4. Schematic of the separation and expansion of MSCs using AMLs.

medium containing AMLs, after which bone marrow aspirates with AMLs were placed in 100-mm tissue culture dishes. A disk-shaped neodymium magnet (diameter, 2.2 cm; 4000 Gauss) was then placed under the 100-mm dish to expose the aspirate to magnetic force vertical to the bottom of the dish. After every 3 to 4 days of culture, the nonadherent cells were removed and the medium was replaced.

The immobilization density and average particle size of αCD105-AMLs were 47.7 μg-antibody/mg-magnetite and 136.7 nm, respectively. As shown in Fig. 8.5, 85% of the human MSCs were magnetically separated using αCD105-AMLs, which was a significantly higher percentage than the values obtained using magnetoliposomes not conjugated to antibody (MLs; 30%).

In the initial state of cell culture after the magnet was placed under the culture dish, the AMLs accumulated at the center of the dish along the edge of the magnet. At 1 day after the start of culture, when the medium was replaced, nonadherent cells and excess AMLs were removed, and very few cells with fibroblastic morphology were found to have adhered to the dish. In the control dishes, cultured using the conventional method, none or very few colonies were observed. In contrast, in the αCD105-AML cultures, cells accumulated at the center of dish along the edge of the magnet, followed by development of those cells into variously sized colonies at the periphery of the magnet, resulting in high-density cell culture. On day 7, the number of viable cells was assayed (Fig. 8.6). In αCD105-AML culture of iliac bone marrow aspirate, an average of 246 cells were detected per dish, which

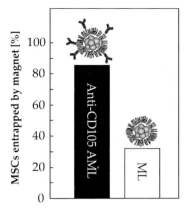

Fig. 8.5. Separation of MSCs using anti-CD105 AML or non-antibody-conjugated MLs.

was 24-fold greater than the numbers obtained with control cultures. No viable cells were detected in the control maxillofacial bone marrow culture. In contrast, an average of 138 cells per dish was detected in the αCD105-AML maxillofacial bone marrow culture.

In this study, we used AMLs to create a "high-density MSC culture" from bone marrow aspirate. A net bone marrow aspirate of 1 mL was used as a model for expansion of MSCs from an extremely small number of cells. In this model, few or no cells were detected in cultures obtained from iliac and maxillofacial bone marrow aspirates using conventional methods based on the plastic-adherent tendencies of cells. In contrast, after 7 days of culture with αCD105-AMLs, 246 cells were

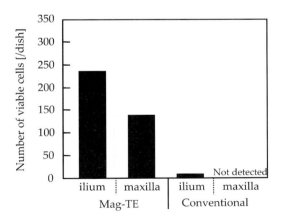

Fig. 8.6. MSC separation and expansion from bone marrow aspirates using anti-CD105 AMLs (Mag-TE) or without AMLs (a conventional method).

detected in iliac bone marrow culture and 138 cells were detected in maxillofacial bone marrow (Fig. 8.6). These results suggest that this method, which we have termed "Mag-TE", is an effective method for expansion of MSCs from very small initial numbers of cells. Typically, only a small number of cells can be obtained from bone marrow aspirates; this is especially true for maxillofacial bone marrow. However, we achieved efficient expansion of MSCs from maxillofacial bone marrow using αCD105-AMLs and a magnet.

8.3
Mag-seeding

One approach to tissue engineering is to seed the cells in 3D porous biodegradable scaffolds that allow the cells to form a continuous structure via cell adhesion, proliferation, and deposition of ECM. Although cell-seeding into scaffolds at a high density is closely associated with enhancement of tissue formation in 3D constructs (e.g., cartilage [40], bone [41] and cardiac tissue [42]), effective and high-density cell seeding into 3D scaffolds is difficult to achieve. Technical difficulties in cell-seeding are caused by the complicated structure of the scaffold [43–46], and insufficient migration into the scaffolds due to their pore sizes and materials, which may cause crucial prolongation of the culture period due to a shortage of initially seeded cells. Therefore, numerous methodologies for effective cell-seeding into 3D scaffolds have been investigated [47], and novel methodologies have been also highly sought.

One conventional cell-seeding methodology (static-seeding) involves cell suspension being seeded onto small scaffolds using small volumes of highly concentrated cell suspension, as such a large number of cells cannot be obtained from Petri dish culture [47]. Here, the inevitable problem arises, namely that the cell suspension seeded onto the scaffold flows away along with medium flow, and only a small number of cells remains on the scaffolds. Mag-TE could help overcome this problem, as magnetic force would attract seeded cells that had been magnetically labeled to prevent them from flowing away, with the result that numerous cells could be seeded onto the scaffolds.

We applied a Mag-TE technique to a tissue engineering process in cell-seeding, (termed "Mag-seeding"), and investigated whether Mag-seeding enhances cell-seeding efficiency into 3D porous scaffolds for tissue engineering [48]. NIH/3T3 fibroblast cells (FBs) were used as the model, along with six kinds of scaffold: five were collagen sponges, each of a different pore size (50–600 μm), while the other was a D,D-L,L-polylactic acid (PLA) sponge. Mag-seeding was performed as follows: FBs were incubated with culture medium containing MCLs, while scaffolds were hydrated in culture medium and then placed in the well of tissue culture plates. A magnet was placed on the reverse side of the tissue culture plate to provide magnetic force vertical to the plate, and aliquots (100 μL) of the magnetically labeled cell suspension then poured onto the top of the hydrated scaffolds. Figure 8.7(A) illustrates a schematic of Mag-seeding.

A. Schematic of Mag-seeding

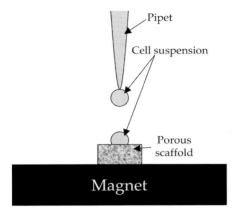

B. Cell-seeding effect of Mag-seeding by pore size

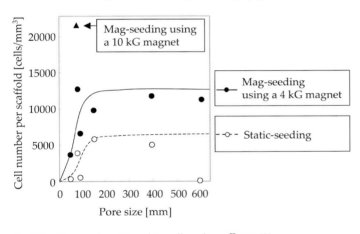

Fig. 8.7. Mag-seeding (A) and its cell-seeding efficacy (B).

The cell-seeding efficiency of Mag-seeding was compared with that of static-seeding. As shown in Fig. 8.7(B), the cell-seeding efficiency for all scaffolds was enhanced by Mag-seeding. Accompanying an increase in pore size, the number of cells in the scaffold increased, and then reached a plateau; approximately 5000 cells mm^{-3} for static-seeding and 12 000 cells mm^{-3} for Mag-seeding. Since the apparent size of the scaffold was particularly small in a collagen sponge with a pore size of 600 μm, very few cells were detected when the cells were seeded by static-seeding. This is because the poured cell suspension onto the scaffold flowed away. Conversely, a substantial number of cells was detected when the cells were seeded by Mag-TE, suggesting that the magnetic force facilitated cell-seeding.

The effects of magnetic force intensity on cell-seeding efficiency were examined. Two kinds of magnet (magnetic field intensity; 4 and 10 kG) were used in this experiment. The seeding efficiency by Mag-seeding using the 10-kG magnet was significantly higher than the efficacy by static-seeding or by Mag-seeding using the 4-kG magnet (Fig. 8.7B).

In conclusion, we demonstrated that a large number of cells can be seeded into scaffolds by Mag-seeding. Magnetically labeled cells are attracted by magnetic force and consequently numerous cells remain in the scaffold across the medium flow. The cell-seeding efficiency depends on the pore size of the scaffolds. In static-seeding, cells seeded onto scaffolds may enter into the scaffold due to natural precipitation by gravity. For scaffolds with small pore size, most of the cells poured onto the scaffold cannot enter, and thus only a small number of cells are seeded within the scaffold. By Mag-seeding, a significantly larger number of cells enter the scaffold, regardless of pore size (50–600 μm tested). Moreover, cell-seeding efficiency is enhanced by the use of a magnet with higher magnetic induction. These results indicate that Mag-seeding is an easy and reliable cell-seeding technique. In addition, since cells labeled with MCLs can be easily manipulated using magnetic force, automation of cell-seeding for tissue engineering is a distinct possibility, which could aid industrial engineered-tissue production.

8.4
Construction of 3D Tissue-like Structure

Currently, tissue engineering is based on seeding cells onto 3D biodegradable scaffolds, which allows the cells to reform their original structure [15]. However, some problems remain with this approach, e.g., insufficient cell migration into the scaffolds and inflammatory reactions due to scaffold biodegradation, and thus novel approaches for achieving 3D tissue-like constructs are required. We applied Mag-TE to the construction of "scaffold-less" 3D tissue-like structures.

8.4.1
Cell Sheet Engineering using RGD-MCLs

A major difficulty obstructing the fabrication of *in vivo*-like 3D constructs without the use of artificial 3D scaffolds is a lack of cell adherence in the vertical direction via cell–cell junctions. This non-adherence is caused by enzymatic digestion of adhesive proteins. To overcome this, Okano et al. employed a thermo-responsive culture surface grafted to poly (*N*-isopropylacrylamide) (PIPAAm) [49–51]. Cells adhered to and proliferated on the thermo-responsive surface, as well as on tissue culture polystyrene dishes. Furthermore, confluent cells on the PIPAAm dishes were expelled as intact contiguous sheets by decreasing the temperature to below the lower critical solution temperature (LCST) of PIPAAm. Kushida et al. recovered monolayer cell sheets from a surface grafted with PIPAAm and then deposited extracellular matrices (ECMs); digestive enzymes were not used and the ECMs

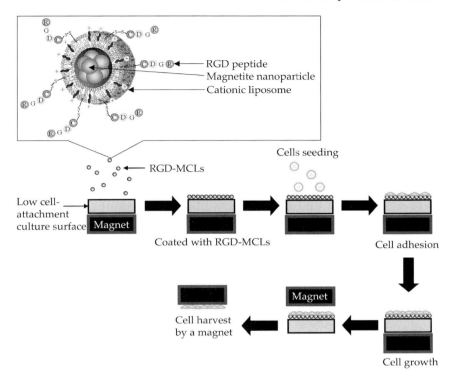

Fig. 8.8. Schematic for Mag-TE using RGD-MCLs.

remained to enhance cell–cell attachment [52]. These engineered cell sheets could be layered to construct multi-layered 3D cell sheets, and Okano et al. proposed the concept of "cell sheet engineering".

We have developed a new biomaterial and methodology for Mag-TE. The RGD (Arg-Gly-Asp) sequence, an integrin recognition motif found in fibronectin [53, 54] and one of the most extensively studied cell adhesion peptides, was conjugated with magnetite cationic liposomes (RGD-MCLs) [55]. Figure 8.8 illustrates the RGD-MCLs.

RGD-MCLs, which contained magnetite nanoparticles that possessed a positively charged cationic lipid surface and were coupled covalently with the cell adhesion peptide RGD (Arg-Gly-Asp-Cys), were constructed according to the method of Gyongyossy-Issa et al. [56] with slight modifications. The average particle size of the RGD-MCLs was 243 nm. The density of the immobilized peptides in RGD-MCLs was 0.226 mg peptide/mg magnetite.

The new technique using RGD-MCLs consists of the following processes. RGD-MCLs are pre-seeded onto a low cell-attachment culture surface, consisting of a hydrophilic and neutrally charged covalently bound hydrogel layer, and a magnet is set on the underside of the well to attract the RGD-MCLs to the well surface. Target

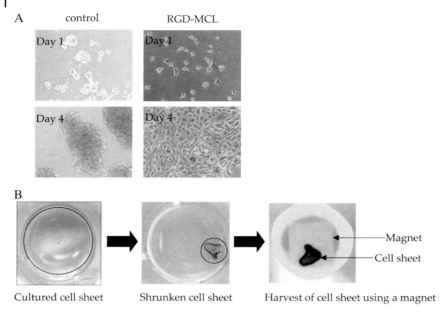

A   control   RGD-MCL

Fig. 8.9. (A and B) Formation of cell sheets by Mag-TE using RGD-MCLs.

cells are then seeded into the wells, which are coated with RGD-MCLs, and are then incubated until confluency. The magnet is removed to detach the cell sheets from the well, after which the cell sheets are harvested using a magnet inserted into the well. Figure 8.9 gives a schematic of Mag-TE using RGD-MCLs.

When NIH/3T3 cells, which are anchorage dependent, were cultured in low-attachment plates, the cell number did not change during the 4 day incubation period. For RGD-MCLs, apparent cell proliferation was observed and the cells proliferated to confluency during the 4 day incubation period. Figure 10.8(A) shows photomicrographs of NIH/3T3 cells cultured on the uncoated surface (Control), and that coated with RGD-MCLs for 1 day or 4 days. When the cells were cultured in low-attachment plates, the cells floated in the media and no cells were observed to be attached to the plate. Subsequently, the cells aggregated and formed spheroid constructs on day 4. Conversely, when the cells were cultured on the surface coated with RGD-MCLs, NIH/3T3 cells had adhered and spread by day 1, and had proliferated to confluency by day 4.

The NIH/3T3 cells grown to confluency on RGD-MCLs were subjected to a novel fabricating and harvesting method using a magnet. After 4 days of incubation on RGD-MCLs, the magnet placed on the underside of the low-attachment plate was removed. When the medium was gently pipetted to cause medium flow from the periphery of the cell sheets, the cell sheet detached from the bottom of the plate and shrunk, thus resulting in the formation of a contiguous cell sheet-like construct. The cell sheets fabricated by Mag-TE were black-brown due to the magnetite nanoparticles, and we investigated whether these cell sheets could be harvested using magnets. The cylindrical neodymium magnet positioned on the underside of the low-attachment plate was removed. A hydrophilically treated membrane was the pasted to the head of a cylindrical alnico magnet, and the magnet moved into the well toward the surface of the culture medium. Due to the magnetic force, the cell sheets floated up to the surface of the culture medium and without disruption, stuck to the membrane (Fig. 8.9B).

In this study, cell integrin ligand RGD peptides were covalently coupled to cationic liposomes containing magnetite nanoparticles, and we used these RGD-MCLs as a novel biomaterial for tissue engineering. RGD-MCLs have two possible active effects on cell adherence; one is positive charge of the cationic liposomes, and the other is the RGD peptide on their surface. When cells were cultured on the surface of neutrally charged magnetoliposomes, the cells floated in the media and no cells were observed to be attached to the plate. Conversely, cells attached onto the culture surface coated with MCLs, thus suggesting that the positive charge due to cationic liposomes facilitated cell attachment to the culture surface. However, few NIH/3T3 cells spread and proliferated on MCLs. In contrast, when NIH/3T3 cells were cultured on a surface coated with RGD-MCLs, the cells adhered, spread, and proliferated, thus suggesting that RGD peptides strongly promote cell adhesion. These data encourage further development of tissue engineering techniques using magnetite cationic liposomes coupled with cell adhesion peptides, such as KQAGDV (Lys-Gln-Ala-Gly-Asp-Val) peptide for smooth muscle cells [57] and YIGSR (Tyr-Ile-Gly-Ser-Arg) peptide for neurons [58].

In cell sheet engineering, methodologies for handling cell sheets are needed because grafts fabricated by cell sheet engineering are easily damaged by handling. We proposed a novel methodology for handling cell sheets using RGD-MCLs and magnetic force; the Mag-TE method enabled us to handle cell sheet-engineered grafts. Because cell sheets constructed by Mag-TE contain magnetite nanoparticles, NIH/3T3 cell sheets can be recovered and handled using a magnet. This feature is not present in cell sheets produced using other methodologies, such as the use of a PIPAAm-grafted surface. These results also suggest the possibility of developing a "tissue-engineered graft delivery system" using Mag-TE, while Mag-TE could also be applied to industrial tissue engineering because magnetic force could greatly simplify the recovery step of cell sheet production. This recovery step could be automated by substituting the magnet used to harvest cell sheets with an electromagnet, which can immediately release the sheets from the surface of the magnet via electrical control. These findings indicate that RGD-MCLs are potent tools for industrial tissue engineering.

8.4.2
Construction of a Keratinocyte Sheet using MCLs

The epidermis is one of only a few tissues for which it is possible to culture its principal cell (the keratinocyte) and to use these cultured cells to reconstitute stratified and differentiated human tissue [59, 60]. These skin equivalents have been

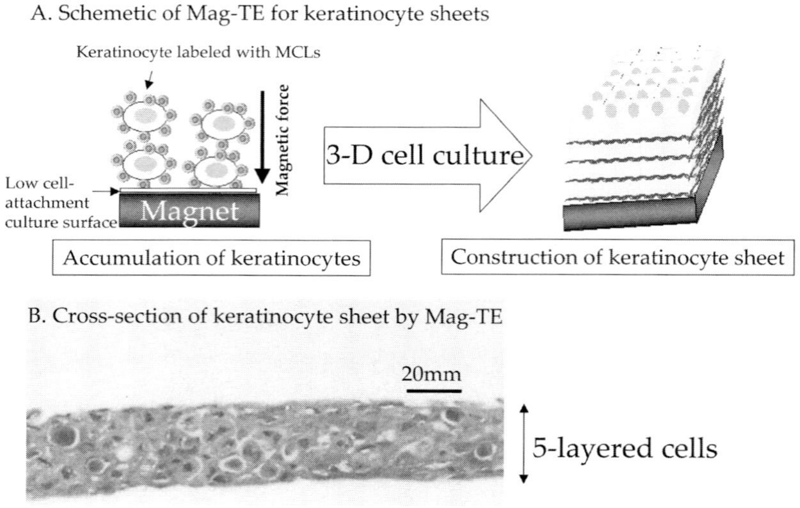

A. Schemetic of Mag-TE for keratinocyte sheets

Keratinocyte labeled with MCLs

Accumulation of keratinocytes

3-D cell culture

Construction of keratinocyte sheet

B. Cross-section of keratinocyte sheet by Mag-TE

20mm

5-layered cells

Fig. 8.10. (A) Scheme for the construction of a keratinocyte sheet by Mag-TE; (B) cross-section of a keratinocyte sheet so-produced.

used clinically to repair burns and wounds [61, 62]. In this study [63], MCLs were used to label human keratinocytes magnetically, and we investigated whether magnetically labeled keratinocytes could be accumulated using a magnet, and whether stratification is promoted by magnetic force to form a sheet-like 3D construct by Mag-TE (Fig. 8.10A).

In this case, MCLs were added to keratinocytes, to manipulate cells magnetically in a similar way to that described in Section 8.2.1. MCLs-labeled keratinocytes (2×10^6 cells, corresponding to five-fold confluency in 24-well plates) were seeded into 24-well low-attachment plates, to investigate whether five-layered keratinocyte sheets can be constructed by the accumulation of cells by magnetic force. A 30 mm neodymium magnet (4000 G) was placed under the plate. Keratinocytes without MCLs or with MCLs in the absence of a magnet did not attach to the plates. In contrast, keratinocytes labeled with MCLs at 50 pg cell^{-1} accumulated on the low-attachment plates in the presence of the magnet. Keratinocytes labeled with MCLs formed a sheet-like construct in the presence of the 30 mm magnet. Phase-contrast microscopy of the cross-sections of the sheets (Fig. 8.10B) revealed that keratinocytes labeled with MCLs formed five-layered sheets.

Notably, when keratinocytes were seeded onto monolayer keratinocytes cultured on tissue culture plates they did not form a sheet-like construct. Moreover, when keratinocytes were seeded onto the low-attachment plates, they did not attach to the surface of the plates and did not form sheets. When protease (trypsin) was used for preparation of keratinocyte suspension, the ECMs may have been digested. As a way to enhance layered cell–cell interactions, we took a physical approach, using magnetic attraction. Keratinocytes magnetically labeled with MCLs evenly accumulated onto the low-attachment plates in the presence of a magnetic force. Keratinocytes magnetically labeled with 50 pg cell^{-1} MCLs formed 5-layered keratinocyte sheets. We speculate that the cell–cell adhesion was caused by the very close placement of cells by the magnetic force.

In our experience, if epithelial sheets fabricated by the method of Rheinwald and Green [64] have five or more cellular layers, they are sufficiently strong for recovery and transplantation [65]. Therefore, in this study, magnetically labeled keratinocytes of five-fold confluency against the culture area were seeded to construct five-layered keratinocytes. As a result, keratinocyte sheets with five layers were constructed. The sheets fabricated by Mag-TE consisted of undifferentiated keratinocytes, which apparently differs from the epidermal sheets fabricated by the method of Rheinwald and Green [64]. To the best of our knowledge, this is the first time that multilayered "undifferentiated" keratinocyte sheets have been constructed. Undifferentiated keratinocytes in keratinocyte sheets produced by Mag-TE may have greater effects on wound healing than cornified and anucleate keratinocytes fabricated by inducing terminal differentiation.

Yamato et al. have reported that thermo-responsive culture dishes grafted with PIPAAm allow the intact harvest of keratinocyte sheets without damage caused by protease (e.g. dispase) treatment [66]. Dispase, a neutral protease from *Bacillus polymyxa*, is widely used to harvest multilayered keratinocyte sheets from culture dishes [67]. In clinical use, extensive washing to remove dispase from keratinocyte sheets is required before they can be applied to wounds, because residual dispase is harmful to the wound site. In industrial production of keratinocyte sheets, this washing is laborious, and is a technological barrier to automation of the process. In our method, we used low-attachment plates, to harvest keratinocytes from the plate without enzymatic treatment after removing the magnet, because the keratinocytes did not adhere to the plate surface. Bioreactors automated for successive culture for epithelial sheet generation have been developed [68, 69]. Because Mag-TE does not require washing of the keratinocyte sheets, which is a laborious step, it allows automation of tissue engineering. Moreover, we used magnetic force to make the recovery step easier, which could aid industrial production of keratinocyte sheets. For automation of this recovery step, the magnet used to harvest keratinocyte sheets may be substituted with an electromagnet, which can release the sheets from the surface of the magnet when switched off. Together, these findings indicate that MCLs are potent tools for industrial engineered-tissue production.

8.4.3
Delivery of Mag-tissue Engineered RPE Sheet

Age-related macular degeneration (AMD) is an eye disease and a major cause of blindness [70]. The most severe form of AMD is characterized by choroidal neovascularization (CNV), which causes rapid visual loss. When CNV is surgically excised from patients with AMD, the retinal pigment epithelium (RPE) is also removed along with CNV. Since RPE cells play a critical role in assisting photoreceptors in vision, RPE cell transplantation into the sub-retinal space, from where these cells are surgically excised, is one possible approach [71]. RPE cells in suspension have been injected into the sub-retinal space of animal models and AMD patients. While some success has been achieved in animal models, with recovery of vision [72, 73], the results have been limited in humans [74, 75]. The poor outcomes are due to two major factors; limitations in obtaining autologous RPE cells and difficulty in RPE cell delivery.

Tissue engineering is a possible technology for solving the above-mentioned problems. Conventionally, tissue engineering has been based on two steps: expanding autologous cells *in vitro*, and seeding the cells onto 3D biodegradable scaffolds to reform their native structure [15]. First, a method for cultivating RPE cells from CNV specimens that have been surgically removed from patients with AMD has been developed [76]. The use of autologous RPE cells can overcome immune rejection, and the expansion of RPE cells may resolve donor shortage. Conversely, the use of biodegradable scaffolds poses problems, such as insufficient cell migration into the scaffolds and inflammatory reaction due to the biodegradation of the scaffolds. Particularly for tissue engineering of RPE, since RPE constructs desired for transplantation are of very small size (1–4 mm^2) and also inflammation is a complicating factor in eye diseases, an autologous cell sheet-like structure without artificial scaffolds may be more suitable. However, since cell–cell interactions are difficult to manipulate, assembly of a 3D cell construct without scaffolds remains a challenge. Moreover, the difficulty faced constructing RPE cell sheets is in the handling of such small tissue-engineered grafts for transplantation. To the best of our knowledge, there are few reports on tissue-engineered graft delivery systems.

In one study [77] we investigated whether RPE cell sheets could be constructed using Mag-TE. Thus, MCLs were used to label RPE cells magnetically, and we investigated whether magnetically labeled RPE cells could be accumulated to form RPE cell sheets, instead of artificial scaffolds, using a magnet. Moreover, since the RPE cell sheets constructed by Mag-TE contained magnetite nanoparticles, we investigated whether these cell sheets could be handled with a magnet, and assessed the feasibility of using a tissue-engineered graft delivery system by Mag-TE *in vitro*.

ARPE-19 cells, used in this study, are a spontaneously arising human RPE cell line with normal karyology [78]. Figure 8.11 illustrates a schematic for the construction and transplantation of RPE cell sheets using MCLs and magnetic force (Mag-TE).

To produce small RPE cell sheets, less than 4 mm^2, a cloning ring of 2.4 mm caliber (inner area, 4 mm^2) was used. A cloning ring was placed at the center of a 24-well low-attachment plate and a 22 mm neodymium magnet was placed under the plate. Magnetically labeled ARPE-19 cells (8×10^3 cells mm^{-2}, which corresponds to a 10-fold concentration of confluency) were seeded into the cloning ring to investigate whether multilayered RPE cell sheets less than 4 mm^2 can be constructed via accumulation of cells by magnetic force. ARPE-19 cells without MCLs or with MCLs in the absence of the magnet did not form cell sheets, and the cells were dispersed when the cloning ring was removed. In contrast, ARPE-19 cells labeled with MCLs formed an approximately 1 mm^2 sheet-like construct in the

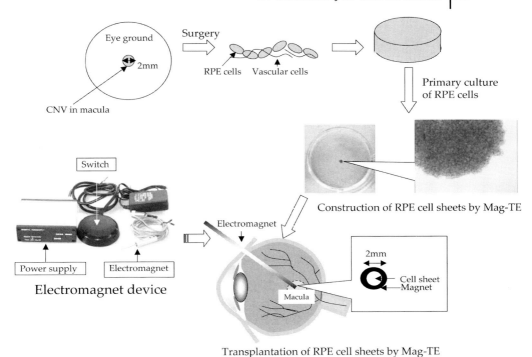

Fig. 8.11. Scheme for the construction and transplantation of RPE cell sheets by Mag-TE (the apparatus used is shown on the bottom left of the figure).

presence of a magnet. The cross-sections of the layers revealed that the ARPE-19 cells labeled with MCLs formed 15-layered sheets with a thickness of 60 μm each.

Next, we investigated whether these magnetically labeled RPE cell sheets could be harvested and delivered using a magnet. When the cloning ring and the cylindrical neodymium magnet positioned on the reverse side of the 24-well low-attachment plate were removed, the RPE cell sheets detached from the bottom of the well. As a device for delivery of the magnetically labeled tissue, an iron wire was attached magnetically to the cylindrical neodymium magnet. The magnetic flux density at the tip of the iron wire was 1100 Gauss. The tip of the iron wire was then positioned at the surface of the culture medium. Due to the magnetic force, the RPE cell sheets floated up to the surface of the culture medium without disruption, and stuck to the tip of the iron wire. As a model for transplantation, we investigated whether the RPE cell sheets constructed by Mag-TE could be delivered onto the bottom of the tissue culture dish containing 10 mL of medium. When the magnet was removed from the iron wire and the wire was tapped gently, the cell sheet detached from the tip of wire and sank onto the tissue culture surface. After incubation for 1 day, the cell sheets had attached to the tissue culture dish. Subsequently, the cell sheets were further incubated. ARPE-19 cells growing from the cell sheets were observed after a 16 day incubation period, suggesting that the transplanted cells had retained their activity.

The Mag-TE method enabled us to handle the tissue-engineered grafts using magnetic force. We showed that the small RPE cell sheets constructed by Mag-TE could be recovered and delivered using magnetic force. Due to the magnetic force, the RPE cell sheets labeled with MCLs floated up to the surface of the culture medium and stuck to the tip of the iron wire on the magnet. To deliver the RPE cell sheet into the sub-retinal space of AMD patients, the cell sheet must be sufficiently strong so as not to break in the intraocular space, in which a balanced salt solution is added during surgery to adjust intraocular pressure. Furthermore, the cell sheets should not disperse into unintended spaces in the eye during transplantation. In our study, the RPE cell sheets were transplanted into the tissue culture dish containing 10 mL medium, as an *in vitro* model. In this experimental model, the RPE cell sheets were successfully transferred into another dish and released from the tip of iron wire by detaching the magnet and nullifying the magnetic effect of the iron wire. In the near future, we intend to study the feasibility of Mag-TE in animals and clinical trials. Presently, we have developed a tissue delivery device for Mag-TE, which is an electromagnet device that enables on-off control, to manipulate Mag-tissue engineered grafts (Fig. 8.11).

8.4.4
Construction of a Liver-like Structure using MCLs

Tissues and organs *in vivo* often consist of several types of cell layers. Cell–cell interactions among these layers are important in maintaining the normal physiology of organ systems, such as the vasculature (smooth muscle and endothelial cells [79]), skeletal muscle (myocytes and peripheral nerves [80]), and liver (hepatocytes and sinusoidal endothelial cells [81]). However, cell–cell interactions are difficult to manipulate in co-culture systems with two or more cell types, even in 2D cultures. Moreover, the assembly of 3D tissues containing various cell types remains a challenge.

Heterotypic interactions play a fundamental role in liver function. Since the liver is formed from endodermal foregut and mesenchymal vascular structures, it may be functionally mediated by heterotypic interactions [82, 83]. Liver-specific functions in isolated hepatocytes that require nonparenchymal cells disappear in homotypic cultures [84]. Various 2D co-culture systems of hepatocytes and nonparenchymal cells have been investigated, including those using microfabrication [85] and 2D patterning [86]. However, novel technologies are required to reconstruct the liver to function as it does *in vivo*. This would require a 3D construct containing various types of cells that could thrive beyond the cell type limitations of co-culture. This study [87] uses magnetic force to precisely place magnetically labeled cells onto target cells and to promote heterotypic cell–cell adhesion to form a 3D construct. Here, we magnetically labeled human aortic endothelial cells (HAECs) using MCLs. We then investigated whether the labeled HAECs could be placed onto

A. Cross-section of 3-D co-culture by Mag-TE

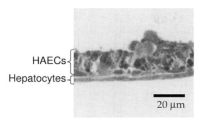

HAECs
Hepatocytes

20 μm

B. Albumin secretion of 3-D co-culture on day 7

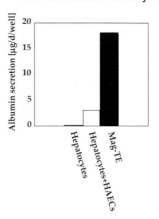

Fig. 8.12. (A and B) Construction of liver-like structure by Mag-TE.

a rat hepatocyte layer using magnetic force, and whether magnetic force promotes the adhesion of heterotypic cells.

We seeded 1.2×10^5 HAECs (the number of cells required to reach confluence in 24-well cell culture plates) onto hepatocyte monolayers and manipulated a 22 mm magnet to form double cell layers throughout the wells. HAECs labeled or not with MCLs did not attach to hepatocytes in the absence of a magnet. Conversely, in the presence of the magnet, HAECs evenly attached to the hepatocyte layer throughout the wells. The hepatocyte layer expressed albumin under the HAEC monolayer that stained positively with Berlin blue, indicating the presence of MCLs (Fig. 8.12A). We measured hepatic albumin expression to determine cellular function in the layered co-culture system by Mag-TE (Fig. 8.12B). Albumin secretion in homotypic hepatocyte culture was undetectable on day 7. Hepatic albumin secretion was slightly enhanced in co-cultures of HAECs, even when the magnet was not present. Conversely, layered co-cultures in the presence of a magnetic force maintained a high level of albumin secretion on day 7.

Two major difficulties obstruct fabrication of an *in vivo*-like 3D construct for 3D tissue engineering including heterotypic cells. One is non-adherence to heterotypic cells caused by the cell type limitation of co-culture. Okano et al. have developed double layered co-cultures using a thermo-responsive surface of grafted PIPAAm to overcome this difficulty [49–51]. Harimoto et al. [88] reported that whereas trypsinized single endothelial cells do not attach to hepatocytes, endothelial cell sheets fabricated by cell sheet engineering would attach to hepatocyte monolayers. However, the assembly of heterotypic cells into complex tissues such as duct-like constructs also presents a challenging technological barrier, even for cell sheet engineering. Another difficulty is to spatially control the positioning of target cells. Mironov et al. [89] have developed a computer-aided jet-based cell printer that could place cells at specific sites on thermo-responsive gels, and termed this "organ printing". This technology seemed to overcome the difficulties with spatial control. However, printed cells can only form monolayers, which must be stratified layer-by-layer to assemble 3D organs. Here, we developed a new method using magnetite nanoparticles and magnetic force, which we refer to as Mag-TE, to overcome these difficulties.

Co-culture systems with nonparenchymal cells maintain hepatocyte functions for long periods [90–92]. To examine the feasibility of the layered co-culture system using magnetic force, we used the liver model of hepatocytes and endothelial cells. We seeded 1.2×10^5 HAECs, which corresponds to the number at confluence in 24-well cell culture plates, onto hepatocyte monolayers and positioned a 22 mm magnet that could apply 4000 G uniformly throughout the plates. This caused an almost uniform upper layer of HAECs. Albumin secretion by hepatocytes was enhanced in this double-layered co-culture system. The precise mechanisms that regulate increases in liver-specific function in hepatocyte co-cultures have not been elucidated. Bhandari et al. [93] have reported that 3T3 fibroblast cells persist in co-cultures with hepatocytes, but 3T3 cell conditioned medium could not substitute for viable co-cultured 3T3 cells in preserving hepatocyte function, suggesting that cell–cell interaction is essential for modulating hepatocyte functions. Potential mediators of cell–cell interactions include soluble factors such as cytokines [94] and insoluble cell-associated factors such as ECMs [95]. We surmise that the tight and close interaction of overlaying HAECs and monolayer hepatocytes using magnetic force caused ECMs and cytokines to be deposited between the layers, thus powerfully enhancing liver function. In contrast, cellular interaction was very weak in co-cultures without the magnetic force and HAECs hardly attached to the hepatocytes. The mechanism by which close cell–cell distance enhances cell to cell interaction, such as gap junction formation, remains to be elucidated.

8.4.5
Construction of Tubular Structures using MCLs

Tissues and organs *in vivo* have unique shapes, and are often composed of several types of cell layers required to maintain the normal physiology of organ systems. In one study [96] we investigated whether we could use Mag-TE to form tubular structures such as ureters (which consist of monotypic urothelial cell layers) and blood vessels (which consist of heterotypic layers of aortic endothelial cells, smooth muscle cells, and fibroblasts).

First of all, Mag-tissue engineered cell sheets were constructed. After incubation with MCLs, human aortic endothelial cells (ECs), human aortic smooth muscle cells (SMCs), mouse NIH/3T3 fibroblasts (FBs), or canine urothelial cells (UCs) were seeded into low-attachment plates. A cylindrical neodymium magnet (diameter, 30 mm; magnetic induction, 4000 G) was then placed on the reverse side of the low-attachment plates to provide magnetic force vertical to the plate, and the cells were cultured for 24 h. Figure 8.13(A) illustrates the procedure used to construct tubular structures. A cylindrical magnet with magnetic poles on its curved surface

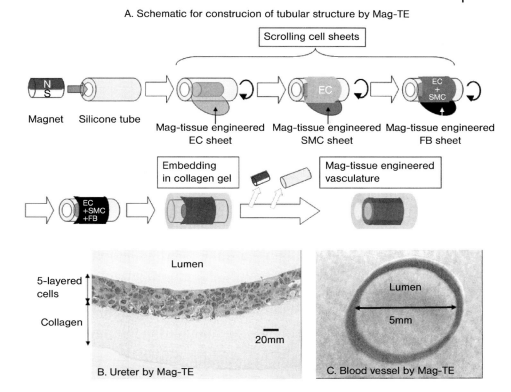

A. Schematic for construcion of tubular structure by Mag-TE

Fig. 8.13. (A) Scheme for the construction of tubular structures by Mag-TE. (B and C) Ureter and blood vessel, respectively, produced by Mag-TE.

was used (diameter, 3 mm; length, 10 mm; magnetic induction, 1300 G). The magnet was inserted into a silicone tube (outer diameter, 5 mm; inner diameter, 3 mm; length, 25 mm). The 5 mm outer diameter was chosen to construct small-diameter blood vessels (inner diameter, <6 mm). The cylindrical magnet in the silicone tube was then rolled over the magnetically labeled cell sheet. To construct a tissue-engineered ureter, a UC sheet only was used. To construct a blood vessel, an EC sheet was used first, followed by addition of an SMC sheet and an FB sheet. Collagen was injected and allowed to gelate around the cell sheets, to support the cell constructs.

When the cylindrical magnet in the silicone tube was rolled over the magnetically labeled UC sheet, UCs were attracted onto the silicone tube by magnetic force. UCs ultimately covered the entire surface of the silicone tube. The sheets formed from ECs, SMCs and FBs in the construction of blood vessels were thicker than the UC sheets. When the silicone tube and magnet were removed, we obtained tubular constructs of tissue engineered urothelial and vascular tissue. Histological observation revealed that the constructed urothelial and vascular tissues formed contiguous tubular structures with a lumen of 5 mm; Fig. 8.13(C) shows a macroscopic view of the vascular tissue. The urothelial structure consisted of five- to seven-layer monotypic urothelial tissue that formed on the collagen gel (Fig. 8.13B).

In this study, for the construction of magnetically labeled cell sheets using magnetic force, four types of cells (UCs, ECs, SMCs, FBs) were used. In general, it is difficult to fabricate 3D tissue constructs without using 3D scaffolds, due to the lack of cell adherence via cell–cell junctions, particularly in the vertical direction. We developed a novel methodology for fabrication of cell sheets, which we have termed "Mag-TE".

For clinical applications, tubular constructs of urothelial and vascular tissue can be cultured for long periods in tubular tissue-oriented bioreactors, such as those that provide biomechanical stimuli similar to those that occur *in vivo* [97–99], to induce these cells to express differentiated phenotypes. For cultivation using bioreactors, because the tubular tissues constructed by Mag-TE contain a sufficient amount of suitably placed cells, culturing for cell proliferation may not be required. Consequently, only culturing for expression of differentiated phenotypes and deposition of ECMs (e.g., collagens and elastins of vascular tissues [100]) would be required. This suggests that use of tubular constructs created using Mag-TE can shorten the culture period required before treatment. In this study, collagen gel was used as a support to reinforce the mechanical strength of tubular tissue constructed by Mag-TE; there has been no quantitative analysis of the mechanical properties of such tissue constructs. Several groups are working to improve the mechanical properties of collagen-gel-based vascular grafts. Control of collagen gel fibril orientation has been investigated in an effort to maximize the mechanical strength of collagen-gel-based scaffolds [101, 102].

8.5
Conclusion

In this chapter, we have only been able to highlight magnetic nanoparticles and Mag-TE as "metallic nanoparticles for tissue engineering". These techniques for tissue engineering are based on biochemical engineering, cell biology, magnetics, nanotechnology, and many other scientific fields. Currently, magnetic techniques complement other methods presently used in medical applications, and this combination therapy should result in more effective medical treatment, including tissue engineering and regenerative medicine. Greater understanding of the properties of magnetic particles will increase their potential for medical application.

When considering the use of metallic nanoparticles, toxicity is an important issue in clinical applications. For magnetite nanoparticles, no toxic effects against proliferation of several cell types were observed within the range of magnetite concentrations tested (e.g., human keratinocytes, <50-pg magnetite/cell; human endothelia cells, human smooth muscle cells, mouse fibroblast cells, canine urothelial cells, human mesenchymal stem cells [MSCs], <100 pg cell^{-1}), and no effect on the differentiation of MSCs was observed. Moreover, in a preliminary study [103], we investigated the toxicity of systemically administered MCLs (90 mg, i.p.) in mice; none of the 10 mice injected with MCLs died during the study. Transient accumulation of magnetite was observed in the liver and spleen of the mice, but the magnetite nanoparticles had been cleared from circulation by hepatic Kupffer cells and/or fixed macrophages in the spleen by the 10th day after administration [103]. However, further study is required to assess the toxicity of residual magnetic nanoparticles in grafts before Mag-TE can be used for clinical applications.

References

1 GUPTA, A.K., GUPTA, M. Synthesis and surface engineering of iron oxide nanoparticles for biomedical applications. *Biomaterials*, **2005**, 26, 3995–4021.

2 SHINKAI, M., ITO, A. Functional magnetic particles for medical application. *Adv. Biochem. Eng. Biotechnol.* **2004**, 91, 191–220.

3 LUBBE, A.S., ALEXIOU, C., BERGEMANN, C. Clinical applications of magnetic drug targeting. *J. Surg. Res.* **2001**, 95, 200–206.

4 DUNNILL, P., LILLY, M.D. Letter: Purification of enzymes using magnetic bio-affinity materials. *Biotechnol. Bioeng.* **1974**, 16, 987–990.

5 MOSBACH, K., ANDERSSON, L. Magnetic ferrofluids for preparation of magnetic polymers and their application in affinity chromatography. *Nature* **1977**, 270, 259–261.

6 SAFARIK, I., SAFARIKOVA, M. Magnetic techniques for the isolation and purification of proteins and peptides. *Biomagn. Res. Technol.* **2004**, 2, 7.

7 DAVID, R., GROEBNER, M., FRANZ, W.M. Magnetic cell sorting purification of differentiated embryonic stem cells stably expressing truncated human CD4 as surface marker. *Stem Cells* **2005**, 23, 477–482.

8 HÄFELI, U., SHÜTT, W., TELLER, J., ZBOROWSKI, M. *Scientific and Clinical Applications of Magnetic Carriers*, Plenum Publishing Corporation, New York, **1997**.

9 SHINKAI, M., YANASE, M., HONDA, H., WAKABAYASHI, T., YOSHIDA, J., KOBAYASHI, T. Intracellular hyperthermia for cancer using magnetite cationic liposomes: in vitro study. *Jpn. J. Cancer Res.* **1996**, 87, 1179–1183.

10 YANASE, M., SHINKAI, M., HONDA, H., WAKABAYASHI, T., YOSHIDA, J., KOBAYASHI, T. Intracellular hyperthermia for cancer using magnetite cationic liposomes: An in vivo study. *Jpn. J. Cancer Res.* **1998**, 89, 463–469.

11 ITO, A., TANAKA, K., KONDO, K., SHINKAI, M., HONDA, H., MATSUMOTO, K., SAIDA, T., KOBAYASHI, T. Tumor regression by combined immunotherapy and hyperthermia using magnetic nanoparticles in an experimental subcutaneous murine melanoma. *Cancer Sci.* **2003**, 94, 308–313.

12 LE, B., SHINKAI, M., KITADE, T., HONDA, H., YOSHIDA, J., WAKABAYASHI, T., KOBAYASHI, T. Preparation of tumor-specific magnetoliposomes and thier application for hyperthermia. *J. Chem. Eng. Jpn.* **2001**, 34, 66–72.

13 SHINKAI, M., LE, B., HONDA, H., YOSHIKAWA, K., SHIMIZU, K., SAGA, S., WAKABAYASHI, T., YOSHIDA, J., KOBAYASHI, T. Targeting hyperthermia for renal cell carcinoma using human MN antigen-specific magnetoliposomes. *Jpn. J. Cancer Res.* **2001**, 92, 1138–1145.

14 ITO, A., KUGA, Y., HONDA, H., KIKKAWA, H., HORIUCHI, A.,

WATANABE, Y., KOBAYASHI, T. Magnetite nanoparticle-loaded anti-HER2 immunoliposomes for combination of antibody therapy with hyperthermia. *Cancer Lett.* **2004**, 212, 167–175.

15 LANGER, R., VACANTI, J.P. Tissue engineering. *Science* **1993**, 260, 920–926.

16 PITTENGER, M.F., MACKAY, A.M., BECK, S.C., JAISWAL, R.K., DOUGLAS, R., MOSCA, J.D., MOORMAN, M.A., SIMONETTI, D.W., CRAIG, S., MARSHAK, D.R. Multilineage potential of adult human mesenchymal stem cells. *Science* **1999**, 284, 143–147.

17 MAKINO, S., FUKUDA, K., MIYOSHI, S., KONISHI, F., KODAMA, H., PAN, J., SANO, M., TAKAHASHI, T., HORI, S., ABE, H., HATA, J., UMEZAWA, A., OGAWA, S. Cardiomyocytes can be generated from marrow stromal cells in vitro. *J. Clin. Invest.* **1999**, 103, 697–705.

18 DEANS, R.J., MOSELEY, A.B. Mesenchymal stem cells: biology and potential clinical uses. *Exp. Hematol.* **2000**, 28, 875–884.

19 KOPEN, G.C., PROCKOP, D.J., PHINNEY, D.G. Marrow stromal cells migrate throughout forebrain and cerebellum, and they differentiate into astrocytes after injection into neonatal mouse brains. *Proc. Natl. Acad. Sci. U.S.A.* **1999**, 96, 10711–10716.

20 FRIEDENSTEIN, A.J., GORSKAJA, J.F., KULAGINA, N.N. Fibroblast precursors in normal and irradiated mouse hematopoietic organs. *Exp. Hematol.* **1976**, 4, 267–274.

21 MARUYAMA, Y., TOHDA, S., NAGATA, K., SUZUKI, T., MUROHASHI, I., NARA, N. Role of humoral and cellular factors on the growth of blast progenitors of acute myeloblastic leukemia in serum-free culture. *Hematol. Pathol.* **1990**, 4, 115–123.

22 REILLY, I.A., KOZLOWSKI, R., RUSSELL, N.H. The role of cell contact and autostimulatory soluble factors in the proliferation of blast cells in acute myeloblastic leukemia. *Leukemia* **1989**, 3, 145–150.

23 GLOECKNER, H., LEMKE, H.D. New miniaturized hollow-fiber bioreactor for in vivo like cell culture, cell expansion, and production of cell-derived products. *Biotechnol. Progr.* **2001**, 17, 828–831.

24 KIMURA, A., KATOH, O., HYODO, H., KUSUMI, S., KURAMOTO, A. Autocrine and/or paracrine mechanism operate during the growth of human bone marrow fibroblasts. *Br. J. Haematol.* **1991**, 78, 469–473.

25 EHMANN, U.K., TERRIS, M.K. Juxtacrine stimulation of normal and malignant human bladder epithelial cell proliferation. *J. Urol.* **2002**, 167, 735–741.

26 NAGATANI, N., SHINKAI, M., HONDA, H., KOBAYASHI, T. Development of a new transformation method using magnetite cationic liposomes and magnetic selection of transformed cells. *Biotechnol. Techniques* **1998**, 12, 525–528.

27 TANAKA, K., ITO, A., KOBAYASHI, T., KAWAMURA, T., SHIMADA, S., MATSUMOTO, K., SAIDA, T., HONDA, H. Intratumoral injection of immature dendritic cells enhances antitumor effect of hyperthermia using magnetic nanoparticles. *Int. J. Cancer* **2005**, 116, 624–633.

28 ITO, A., MATSUOKA, F., HONDA, H., KOBAYASHI, T. Heat shock protein 70 gene therapy combined with hyperthermia using magnetic nanoparticles. *Cancer Gene Ther.* **2003**, 10, 918–925.

29 ITO, A., MATSUOKA, F., HONDA, H., KOBAYASHI, T. Antitumor effects of combined therapy of recombinant heat shock protein 70 and hyperthermia using magnetic nanoparticles in an experimental subcutaneous murine melanoma. *Cancer Immunol. Immunother.* **2004**, 53, 26–32.

30 ITO, A., SHINKAI, M., HONDA, H., YOSHIKAWA, K., SAGA, S., WAKABAYASHI, T., YOSHIDA, J., KOBAYASHI, T. Heat shock protein 70 expression induces antitumor immunity during intracellular hyperthermia using magnetite nanoparticles. *Cancer Immunol. Immunother.* **2003**, 52, 80–88.

31 ITO, A., SHINKAI, M., HONDA, H., KOBAYASHI, T. Heat-inducible TNF-alpha gene therapy combined with hyperthermia using magnetic nanoparticles as a novel tumor-targeted therapy. *Cancer Gene Ther.* **2001**, 8, 649–654.

32 ITO, A., HIBINO, E., HONDA, H., HATA, K., KAGAMI, H., UEDA, M., KOBAYASHI, T. A new methodology of mesenchymal stem cell expansion using magnetic nanoparticles. *Biochem. Eng. J.* **2004**, 20, 119–125.

33 TSUTSUMI, S., SHIMAZU, A., MIYAZAKI, K., PAN, H., KOIKE, C., YOSHIDA, E., TAKAGISHI, K., KATO, Y. Retention of multilineage differentiation potential of mesenchymal cells during proliferation in response to FGF. *Biochem. Biophys. Res. Commun.* **2001**, 288, 413–419.

34 HUSS, R., HOY, C.A., DEEG, H.J. Contact- and growth factor-dependent survival in a canine marrow-derived stromal cell line. *Blood* **1995**, 85, 2414–2421.

35 GREGORY, C.A., SINGH, H., PERRY, A.S., PROCKOP, D.J. The Wnt signaling inhibitor dickkopf-1 is required for reentry into the cell cycle of human adult stem cells from bone marrow. *J. Biol. Chem.* **2003**, 278, 28 067–28 078.

36 BARRY, F.P., BOYNTON, R.E., HAYNESWORTH, S., MURPHY, J.M., ZAIA, J. The monoclonal antibody SH-2, raised against human mesenchymal stem cells, recognizes an epitope on endoglin (CD105). *Biochem. Biophys. Res. Commun.* **1999**, 265, 134–139.

37 HAYNESWORTH, S.E., BABER, M.A., CAPLAN, A.I. Cell surface antigens on human marrow-derived mesenchymal cells are detected by monoclonal antibodies. *Bone* **1992**, 13, 69–80.

38 QUIRICI, N., SOLIGO, D., BOSSOLASCO, P., SERVIDA, F., LUMINI, C., DELILIERS, G.L. Isolation of bone marrow mesenchymal stem cells by anti-nerve growth factor receptor antibodies. *Exp. Hematol.* **2002**, 30, 783–791.

39 ITO, A., HIBINO, E., SHIMIZU, K., KOBAYASHI, T., YAMADA, Y., HIBI, H., UEDA, M., HONDA, H. Magnetic force-based mesenchymal stem cell expansion using antibody-conjugated magnetoliposomes. *J. Biomed. Mater. Res. B Appl. Biomater.* **2005**, 75, 320–327.

40 FREED, L.E., LANGER, R., MARTIN, I., PELLIS, R.R., VUNJAK-NOVAKOVIC, G. Tissue engineering of cartilage in space. *Proc. Natl. Acad. Sci. U.S.A.* **1997**, 94, 13 885–13 890.

41 HOLY, C.E., SHOICHET, M.S., DAVIES, J.E. Engineering three-dimensional bone tissue in vitro using biodegradable scaffolds: Investigating initial cell-seeding density and culture period. *J. Biomed. Mater. Res.* **2000**, 51, 376–382.

42 CARRIER, R.L., PAPADAKI, M., RUPNICK, M., SCHOEN, F.J., BURSAC, N., LANGER, R., FREED, L.E., VUNJAK-NOVAKOVIC, G. Cardiac tissue engineering: cell seeding, cultivation parameters, and tissue construct characterization. *Biotechnol. Bioeng.* **1999**, 64, 580–589.

43 KIM, B.S., PUTNAM, A.J., KULIK, T.J., MOONEY, D.J. Optimizing seeding and culture methods to engineer smooth muscle tissue on biodegradable polymer matrices. *Biotechnol. Bioeng.* **1998**, 57, 46–54.

44 BRUININK, A., SIRAGUSANO, D., ETTEL, G., BRANDSBERG, T., BRANDSBERG, F., PETITMERMET, M., MULLER, B., MAYER, J., WINTERMANTEL, E. The stiffness of bone marrow cell-knit composites is increased during mechanical load. *Biomaterials* **2001**, 22, 3169–3178.

45 LI, Y., MA, T., KNISS, D.A., LASKY, L.C., YANG, S.T. Effects of filtration seeding on cell density, spatial distribution, and proliferation in nonwoven fibrous matrices. *Biotechnol. Progr.* **2001**, 17, 935–944.

46 XIAO, Y.L., RIESLE, J., VAN BLITTERSWIJK, C.A. Static and dynamic fibroblast seeding and cultivation in porous PEO/PBT scaffolds. *J. Mater. Sci. Mater. Med.* **1999**, 10, 773–777.

47 VUNJAK-NOVAKOVIC, G., RADISIC, M. Cell seeding of polymer scaffolds. *Methods Mol. Biol.* **2004**, 238, 131–146.

48 SHIMIZU, K., ITO, A., HONDA, H.

Enhancement of cell-seeding efficiency into 3D porous scaffolds by magnetic force for tissue engineering. *J. Biomed. Mater. Res. B Appl. Biomater.* **2006**, 77, 265–272.

49 HIROSE, M., YAMATO, M., KWON, O.H., HARIMOTO, M., KUSHIDA, A., SHIMIZU, T., KIKUCHI, A., OKANO, T. Temperature-responsive surface for novel co-culture systems of hepatocytes with endothelial cells: 2-D patterned and double layered co-cultures. *Yonsei Med. J.* **2000**, 41, 803–813.

50 SHIMIZU, T., YAMATO, M., ISOI, Y., AKUTSU, T., SETOMARU, T., ABE, K., KIKUCHI, A., UMEZU, M., OKANO, T. Fabrication of pulsatile cardiac tissue grafts using a novel 3-dimensional cell sheet manipulation technique and temperature-responsive cell culture surfaces. *Circ. Res.* **2002**, 90, 40–48.

51 SHIMIZU, T., YAMATO, M., KIKUCHI, A., OKANO, T. Cell sheet engineering for myocardial tissue reconstruction. *Biomaterials* **2003**, 24, 2309–2316.

52 KUSHIDA, A., YAMATO, M., KONNO, C., KIKUCHI, A., SAKURAI, Y., OKANO, T. Decrease in culture temperature releases monolayer endothelial cell sheets together with deposited fibronectin matrix from temperature-responsive culture surfaces. *J. Biomed. Mater. Res.* **1999**, 45, 355–362.

53 RUOSLAHTI, E., PIERSCHBACHER, M.D. New perspectives in cell adhesion: RGD and integrins. *Science* **1987**, 238, 491–497.

54 PIERSCHBACHER, M.D., RUOSLAHTI, E. Cell attachment activity of fibronectin can be duplicated by small synthetic fragments of the molecule. *Nature* **1984**, 309, 30–33.

55 ITO, A., INO, K., KOBAYASHI, T., HONDA, H. The effect of RGD peptide-conjugated magnetite cationic liposomes on cell growth and cell sheet harvesting. *Biomaterials* **2005**, 26, 6185–6193.

56 GYONGYOSSY-ISSA, M.I., MULLER, W., DEVINE, D.V. The covalent coupling of Arg-Gly-Asp-containing peptides to liposomes: purification and biochemical function of the lipopeptide. *Arch. Biochem. Biophys.* **1998**, 353, 101–108.

57 MANN, B.K., WEST, J.L. Cell adhesion peptides alter smooth muscle cell adhesion, proliferation, migration, and matrix protein synthesis on modified surfaces and in polymer scaffolds. *J. Biomed. Mater. Res.* **2002**, 60, 86–93.

58 SANEINEJAD, S., SHOICHET, M.S. Patterned glass surfaces direct cell adhesion and process outgrowth of primary neurons of the central nervous system. *J. Biomed. Mater. Res.* **1998**, 42, 13–19.

59 YANNAS, I.V., BURKE, J.F., ORGILL, D.P., SKRABUT, E.M. Wound tissue can utilize a polymeric template to synthesize a functional extension of skin. *Science* **1982**, 215, 174–176.

60 BELL, E., EHRLICH, H.P., BUTTLE, D.J., NAKATSUJI, T. Living tissue formed in vitro and accepted as skin-equivalent tissue of full thickness. *Science* **1981**, 211, 1052–1054.

61 HANSBROUGH, J.F., BOYCE, S.T., COOPER, M.L., FOREMAN, T.J. Burn wound closure with cultured autologous keratinocytes and fibroblasts attached to a collagen-glycosaminoglycan substrate. *JAMA* **1989**, 262, 2125–2130.

62 FALANGA, V., MARGOLIS, D., ALVAREZ, O., AULETTA, M., MAGGIACOMO, F., ALTMAN, M., JENSEN, J., SABOLINSKI, M., HARDIN-YOUNG, J. Rapid healing of venous ulcers and lack of clinical rejection with an allogeneic cultured human skin equivalent. Human skin equivalent investigators group. *Arch. Dermatol.* **1998**, 134, 293–300.

63 ITO, A., HAYASHIDA, M., HONDA, H., HATA, K., KAGAMI, H., UEDA, M., KOBAYASHI, T. Construction and harvest of multilayered keratinocyte sheets using magnetite nanoparticles and magnetic force. *Tissue Eng.* **2004**, 10, 873–880.

64 RHEINWALD, J.G., GREEN, H. Serial cultivation of strains of human epidermal keratinocytes: The formation of keratinizing colonies from single cells. *Cell* **1975**, 6, 331–343.

65 HATA, K., KAGAMI, H., UEDA, M., TORII, S., MATSUYAMA, M. The characteristics of cultured mucosal cell sheet as a material for grafting; comparison with cultured epidermal cell sheet. *Ann. Plast. Surg.* **1995**, 34, 530–538.

66 YAMATO, M., UTSUMI, M., KUSHIDA, A., KONNO, C., KIKUCHI, A., OKANO, T. Thermo-responsive culture dishes allow the intact harvest of multi-layered keratinocyte sheets without dispase by reducing temperature. *Tissue Eng.* **2001**, 7, 473–480.

67 GREEN, H., KEHINDE, O., THOMAS, J. Growth of cultured human epidermal cells into multiple epithelia suitable for grafting. *Proc. Natl. Acad. Sci. U.S.A.* **1979**, 76, 5665–5668.

68 KINO-OKA, M., UMEGAKI, R., TAYA, M., TONE, S., PRENOSIL, J.E. Valuation of growth parameters in monolayer keratinocyte culture based on a two-dimensional cell placement model. *J. Biosci. Bioeng.* **2000**, 89, 285–287.

69 UMEGAKI, R., MURAI, K., KINO-OKA, M., TAYA, M. Correlation of cellular life span with growth parameters observed in successive cultures of human keratinocytes. *J. Biosci. Bioeng.* **2002**, 94, 231–236.

70 LEE, P., WANG, C.C., ADAMIS, A.P. Ocular neovascularization: An epidemiologic review. *Surv. Ophthalmol.* **1998**, 43, 245–269.

71 BOULTON, M., ROANOWSKA, M., WESS, T. Ageing of the retinal pigment epithelium: Implications for transplantation. *Graefes Arch. Clin. Exp. Ophthalmol.* **2004**, 242, 76–84.

72 GOURAS, P., KONG, J., TSANG, S.H. Retinal degeneration and RPE transplantation in Rpe65($-/-$) mice. *Invest. Ophthalmol. Vis. Sci.* **2002**, 43, 3307–3311.

73 LUND, R.D., ADAMSON, P., SAUVE, Y., KEEGAN, D.J., GIRMAN, S.V., WANG, S., WINTON, H., KANUGA, N., KWAN, A.S., BEAUCHENE, L., ZERBIB, A., HETHERINGTON, L., COURAUD, P.O., COFFEY, P., GREENWOOD, J. Subretinal transplantation of genetically modified human cell lines attenuates loss of visual function in dystrophic rats.

Proc. Natl. Acad. Sci. U.S.A. **2001**, 98, 9942–9947.

74 ALGVERE, P.V., GOURAS, P., DAFGARD KOPP, E. Long-term outcome of RPE allografts in non-immunosuppressed patients with AMD. *Eur. J. Ophthalmol.* **1999**, 9, 217–230.

75 BINDER, S., STOLBA, U., KREBS, I., KELLNER, L., JAHN, C., FEICHTINGER, H., POVELKA, M., FROHNER, U., KRUGER, A., HILGERS, R.D., KRUGLUGER, W. Transplantation of autologous retinal pigment epithelium in eyes with foveal neovascularization resulting from age-related macular degeneration: a pilot study. *Am. J. Ophthalmol.* **2002**, 133, 215–225.

76 SCHLUNCK, G., MARTIN, G., AGOSTINI, H.T., CAMATTA, G., HANSEN, L.L. Cultivation of retinal pigment epithelial cells from human choroidal neovascular membranes in age related macular degeneration. *Exp. Eye. Res.* **2002**, 74, 571–576.

77 ITO, A., HIBINO, E., KOBAYASHI, C., TERASAKI, H., KAGAMI, H., UEDA, M., KOBAYASHI, T., HONDA, H. Construction and delivery of tissue-engineered human retinal pigment epithelial cell sheets, using magnetite nanoparticles and magnetic force. *Tissue Eng.* **2005**, 11, 489–496.

78 DUNN, K.C., AOTAKI-KEEN, A.E., PUTKEY, F.R., HJELMELAND, L.M. ARPE-19, a human retinal pigment epithelial cell line with differentiated properties. *Exp. Eye Res.* **1996**, 62, 155–169.

79 FILLINGER, M.F., O'CONNOR, S.E., WAGNER, R.J., CRONENWETT, J.L. The effect of endothelial cell coculture on smooth muscle cell proliferation. *J. Vasc. Surg.* **1993**, 17, 1058–1067.

80 STREIT, J. Mechanisms of pattern generation in co-cultures of embryonic spinal cord and skeletal muscle. *Int. J. Dev. Neurosci.* **1996**, 14, 137–148.

81 GUGUEN-GUILLOUZO, C., CLEMENT, B., BAFFET, G., BEAUMONT, C., MOREL-CHANY, E., GLAISE, D., GUILLOUZO, A. Maintenance and reversibility of active albumin secretion by adult rat hepatocytes co-cultured with another

liver epithelial cell type. *Exp. Cell Res.*
1983, 143, 47–54.

82 HOUSSAINT, E. Differentiation of the
mouse hepatic primordium. I. An
analysis of tissue interactions in
hepatocyte differentiation. *Cell Differ.*
1980, 9, 269–279.

83 DOUARIN, N.M. An experimental
analysis of liver development. *Med.
Biol.* **1975**, 53, 427–455.

84 BHATIA, S.N., BALIS, U.J., YARMUSH,
M.L., TONER, M. Effect of cell-cell
interactions in preservation of cellular
phenotype: Cocultivation of hepatocytes and nonparenchymal cells.
FASEB J. **1999**, 13, 1883–1900.

85 BHATIA, S.N., YARMUSH, M.L., TONER,
M. Controlling cell interactions by
micropatterning in co-cultures: Hepatocytes and 3T3 fibroblasts. *J. Biomed.
Mater. Res.* **1997**, 34, 189–199.

86 YAMATO, M., KWON, O.H., HIROSE, M.,
KIKUCHI, A., OKANO, T. Novel
patterned cell coculture utilizing
thermally responsive grafted polymer
surfaces. *J. Biomed. Mater. Res.* **2001**,
55, 137–140.

87 ITO, A., TAKIZAWA, Y., HONDA, H.,
HATA, K., KAGAMI, H., UEDA, M.,
KOBAYASHI, T. Tissue engineering
using magnetite nanoparticles and
magnetic force: Heterotypic layers of
cocultured hepatocytes and endothelial
cells. *Tissue Eng.* **2004**, 10, 833–840.

88 HARIMOTO, M., YAMATO, M., HIROSE,
M., TAKAHASHI, C., ISOI, Y., KIKUCHI,
A., OKANO, T. Novel approach for
achieving double-layered cell sheets
co-culture: Overlaying endothelial cell
sheets onto monolayer hepatocytes
utilizing temperature-responsive
culture dishes. *J. Biomed. Mater. Res.*
2002, 62, 464–470.

89 MIRONOV, V., BOLAND, T., TRUSK, T.,
FORGACS, G., MARKWALD, R.R. Organ
printing: Computer-aided jet-based 3D
tissue engineering. *Trends Biotechnol.*
2003, 21, 157–161.

90 GREGORY, P.G., CONNOLLY, C.K.,
GILLIS, B.E., SULLIVAN, S.J. The effect
of coculture with nonparenchymal
cells on porcine hepatocyte function.
Cell Transplant. **2001**, 10, 731–738.

91 SHIMAOKA, S., NAKAMURA, T.,

ICHIHARA, A. Stimulation of growth of
primary cultured adult rat hepatocytes
without growth factors by coculture
with nonparenchymal liver cells.
Exp. Cell Res. **1987**, 172, 228–242.

92 VILLAFUERTE, B.C., KOOP, B.L., PAO,
C.I., GU, L., BIRDSONG, G.G.,
PHILLIPS, L.S. Coculture of primary
rat hepatocytes and nonparenchymal
cells permits expression of insulin-like
growth factor binding protein-3 in
vitro. *Endocrinology* **1994**, 134,
2044–2050.

93 BHANDARI, R.N., RICCALTON, L.A.,
LEWIS, A.L., FRY, J.R., HAMMOND,
A.H., TENDLER, S.J., SHAKESHEFF,
K.M. Liver tissue engineering: A role
for co-culture systems in modifying
hepatocyte function and viability.
Tissue Eng. **2001**, 7, 345–357.

94 MORIN, O., GOULET, F., NORMAND, G.
Liver sinusoidal endothelial cells:
Isolation, purification, characterization
and interaction with hepatocytes. In:
Cell Biology Reviews, MORIN, O.,
GOULET, F., NORMAND, G. (Eds.),
Springer International, New York.
1988, Vol. 15, pp. 1–73.

95 GOULET, F., NORMAND, C., MORIN, O.
Cellular interactions promote tissue-
specific function, biomatrix deposition
and junctional communication of
primary cultured hepatocytes.
Hepatology **1988**, 8, 1010–1008.

96 ITO, A., INO, K., HAYASHIDA, M.,
KOBAYASHI, T., MATSUNUMA, H.,
KAGAMI, H., UEDA, M., HONDA, H. A
novel methodology for fabrication of
tissue-engineered tubular constructs
using magnetite nanoparticles and
magnetic force. *Tissue Eng.* **2005**, 11,
1553–1561.

97 NIKLASON, L.E., GAO, J., ABBOTT,
W.M., HIRSCHI, K.K., HOUSER, S.,
MARINI, R., LANGER, R. Functional
arteries grown in vitro. *Science* **1999**,
284, 489–493.

98 NARITA, Y., HATA, K., KAGAMI, H.,
USUI, A., UEDA, M., UEDA, Y. Novel
pulse duplicating bioreactor system
for tissue-engineered vascular
construct. *Tissue Eng.* **2004**, 10,
1224–1233.

99 SODIAN, R., LEMKE, T., FRITSCHE, C.,

HOERSTRUP, S.P., FU, P., POTAPOV, E.V., HAUSMANN, H., HETZER, R. Tissue-engineering bioreactors: a new combined cell-seeding and perfusion system for vascular tissue engineering. *Tissue Eng.* **2002**, 8, 863–870.

100 MITCHELL, S.L., NIKLASON, L.E. Requirements for growing tissue-engineered vascular grafts. *Cardiovasc. Pathol.* **2003**, 12, 59–64.

101 TRANQUILLO, R.T., GIRTON, T.S., BROMBEREK, B.A., TRIEBES, T.G., MOORADIAN, D.L. Magnetically orientated tissue-equivalent tubes: application to a circumferentially orientated media-equivalent. *Biomaterials* **1996**, 17, 349–357.

102 BAROCAS, V.H., GIRTON, T.S., TRANQUILLO, R.T. Engineered alignment in media equivalents: Magnetic prealignment and mandrel compaction. *J. Biomech. Eng.* **1998**, 120, 660–666.

103 ITO, A., NAKAHARA, Y., TANAKA, K., KUGA, Y., HONDA, H., KOBAYASHI, T. Time course of biodistribution and heat generation of magnetite cationic liposomes in mouse model. *Jpn. J. Hyperthermic Oncol.* **2003**, 19, 151–159.

9
Applications and Implications of Single-walled Carbon Nanotubes in Tissue Engineering

Peter S. McFetridge and Matthias U. Nollert

9.1
Introduction

The discovery of single-walled carbon nanotubes (SWNT) by Iijima in 1991 [1] initiated a surge of interest that continues to grow as our understanding of these unique materials evolves. The extraordinary mechanical [2, 3] and electrical properties [4] of SWNT have fueled continued interest as the spectrum of potential applications grows that can take advantage of these unique nano-scaled materials. Applications encompass a wide range of composite materials from electrically conducting polymers and biopolymers, and nanoelectronics, to biosensors and materials that interface directly with biological systems [5–7].

A rapidly growing discipline that may take advantage of SWNT unique properties is the area of regenerative medicine in the field of tissue engineering [8]. The broad concept, as the name implies, is the regeneration of diseased or damaged tissues such that complications of using transplant tissue can be avoided. Central to this theme is the use of 3D scaffolds that provide support for adhesion-dependent cells in a manner that guides tissue regeneration by the host, or other transplanted cells. The aim is that the preformed scaffold (in the shape of the end product) will guide cellular regeneration to result in the development of neo-tissue. Importantly, not only the physical parameters of the scaffold are critical to allow the correct structural and biochemical development, but also the environment in which the cells are cultured. It is the combination of these factors that are fundamentally important if fully functional tissue is to be developed.

Although the number of peer reviewed published articles on SWNT has been increasing dramatically (both in terms of material processing and its applications) as interest grows, there has been little published in the area of tissue engineering. In this chapter we will discuss general applications of SWNT in the field of tissue engineering, paying particular attention to the preparation of these nano-materials and how they interact with cells. We first review some of the history behind the electrical stimulation of cells and the reasoning behind the use of SWNT as a conductive material to support or promote organ regeneration. We then review

Nanotechnologies for the Life Sciences Vol. 9
Tissue, Cell and Organ Engineering. Edited by Challa S. S. R. Kumar
Copyright © 2006 WILEY-VCH Verlag GmbH & Co. KGaA, Weinheim
ISBN: 3-527-31389-3

purification and preparation processes, and then take a closer look at cellular interactions, and how functionalizing or modifying SWNT surface chemistry can modulate cell function. The chapter concludes by reviewing specific applications of
SWNT in tissue engineering and the current pertinent literature.

9.2
Electromagnetic Fields for Tissue Regeneration

The particular attributes of carbon nanotubes that are of special interest to regenerative medicine are their mechanical and electrical properties. The importance of
electrical properties in modulating biological processes has been recognized since
the mid-1700s. In fact, by the end of the eighteenth century the injury potential,
which is a direct current voltage gradient induced within an injured tissue space
by current flowing into and around an injured nerve, had been recognized and
measured. In recent times, these currents have been shown to be essential for regeneration [9, 10]. Within a hundred years of the early work of Galvani and Volta
in describing and characterizing electric currents and fields, many others promoted bioelectricity for a wide variety of dubious medical applications [11], including insomnia, migraine and baldness among others. The dubious scientific work
and downright charlatanism from this period tainted perceptions that last to this
day regarding bioelectric phenomena.

There are now several systems that clearly demonstrate that electrical field modulation promotes tissue regeneration, for which the unique properties of carbon
nanotubes may be exploited to develop improved therapeutic options. One such
area is in nerve regeneration. The seminal work demonstrating the influence of
electrical fields on nerve growth was by Borgens et al. [12]. They showed that
injury currents carried by Na^+ and Ca^{2+} ions were driven into the cut ends of severed lamprey spinal cord axons. Increased branching and faster regeneration of the
axons was observed in the presence of a steady electric field of opposite polarity.
This work has been extended to a guinea pig model in which the spinal cord was
severed and significantly greater axonal growth was observed in the presence of an
applied electrical field compared with the case where no electrical field was present
[13]. In another study, improvements in nerve function were also observed with an
applied electrical field [14].

Another area that has seen extensive work on the interaction between electrical
fields and biomedical applications is in bone growth. Pulsed electromagnetic fields
(PEMF) stimulation has been in clinical use for nearly 30 years on patients with
delayed fracture healing and nonunion and has been demonstrated in a multitude
of clinical case reports [15–17]. Double-blinded studies have confirmed the clinical
effectiveness of pulsed electromagnetic fields stimulation on osteotomy healing
[18, 19] and delayed union fractures [20]. Brighton et al. [21] conducted a multicenter study of the nonunion and reported an 84% clinical healing rate of non-
union with direct current treatment. Recently, Schaden et al. [22] reported 76% of

non-union or delayed union patients treated with one time extracorporeal shock wave therapy resulted in bony consolidation with a simultaneous decrease in symptoms.

Both of these applications have led investigators to develop electrically conducting materials that have favorable biocompatibility characteristics. The most widely studied conducting biomaterial is the polymer polypyrrole. The biocompatibility properties of this material were first studied by Wong et al. [23]. They showed that extracellular matrix proteins could adsorb on a thin film of polypyrrole and could support attachment and spreading of cells. In the presence of an electrical potential, cell growth and DNA synthesis were altered. A range of inherently conducting polymers has been developed and they have been reviewed [24]. Potential applications of polypyrrole based electrically conducting biomaterials include glucose biosensors [25, 26], neural prosthetics [27, 28], and bone tissue engineering [29–31]. Taken together, these studies demonstrate that there is great potential in developing novel biomaterials and biomedical devices with the capacity to be electrically conducting. Since carbon nanotubes can act as conductors, there is immense interest in exploring their potential in biomedical applications.

9.3
Tissue Engineering

As a discrete discipline, tissue engineering is little more than a decade old; however, growth of cells on synthetic materials as a means to improve the performance of implantable materials dates back to the 1970s [32–35]; in particular, in the development of small diameter vascular prosthetics where it became obvious that these materials need to be more than just immunologically inert, rather they need to functionally integrate with surrounding tissues to enhance performance. As such, there has been continuous development of these materials to understand and improve interactions with cell systems.

In addition to the many biological molecules that enhance cellular interactions, such as growth factors and adhesion peptides, magnetic and electrical stimulation (as described above) has a strong influence on cellular growth, cellular phenotype, and tissue regeneration [36–42]. As such, designer materials that allow modulation of mechanical, electrical and thermal properties may produce materials that can be fine-tuned to suit specific applications. For example, when a diseased and occluded blood vessel is to be excised and a prosthetic tube implanted, if the mechanical compliance between a patient's existing blood vessel and the new implant do not match, the chance of graft failure increases significantly [43]. By varying the concentration of SWNT within an appropriate polymer a wide range of compliance values can be obtained, then specifically matched to each patient. Although most polymer additives will allow control over the material's bulk mechanical properties, few materials offer the combined influence of mechanical and electrical modulation.

Incorporating conductive additives into a bulk material allows the gross conductive properties to be modulated such that electrical stimulation can be used to enhance tissue repair and/or regeneration [44]. One distinct advantage that SWNT have over other conductive additives is the dual capability of varying the material's conductivity *and* enhancing the material's mechanical properties. One biopolymer that has seen a wide range of clinical successes is the extracellular matrix (ECM) molecule collagen [45–49]. Typically collagen is hydrolyzed and stored in an acidic solution that is returned to a neutral pH to initiate polymerization. Unlike most synthetic polymers this method allows living cells to be added during the polymerization step, to produce preformed, homogenous, cell dense materials. This is particularly important as it reduces the time and cost of construct development. By starting with a cell dense material one can avoid having to wait several weeks for the cell populations to proliferate and fully populate the matrix. The earlier work of Weinberg and Bell, who developed the first collagen-based *de novo* blood vessel in 1986 [50], had shown the potential of these natural polymer scaffolds; however, the major draw back was poor mechanical strength and stability. Collagen in its natural form is composed of similarly repeating amino acids, where glycine is repeated every third amino acid, with many of the remaining amino acids being proline and hydroxyproline. As collagen is hydrolyzed, collagen bundles dissociate creating a mass of disorganized fibers that absorbs a significant amount of water to form what is called a hydrogel. Cells remodel the collagen and these disorganized fibers are realigned back into their original form as functional tissue is created. As the fibers realign, a significant amount of shrinkage occurs as the fibers condense and water is expelled.

The most common method used to enhance mechanical properties of collagen-based hydrogels is the addition of chemical crosslinks that bind, strengthen and minimize enzymatic digestion of the long-chain polymer [51–54]. Many crosslinkers, such as glutaraldehyde, are toxic and often leave residual toxins in the scaffold after processing. A further drawback of this approach is that it renders the material impervious to cellular infiltration. The barrage of ECM degrading and remodeling enzymes, secreted by cells, have little effect on many of these chemically induced crosslinks, and as such there is little cell penetration or remodeling of the scaffold. The result is stable, but inert, materials. From a tissue engineering perspective, cell migration into the scaffold and subsequent remodeling are fundamentally important if functional neo-tissue is to be formed. By using conductive additives, which also enhance the material's mechanical characteristics, the opportunity exists to develop off-the-shelf, high-performance materials that may improve repair and regeneration of diseased tissues.

However, with the current state of technology, single-walled carbon nanotubes (SWNT) cannot be used for medial applications straight from the production plant. In general the production of SWNT results in a high percentage of contaminants that must be removed prior to application. Purification technologies are diverse, and often change the surface chemistry of the SWNT, which in turn changes the way biological systems respond to the presence on SWNT. As often is the case with nano-scale materials, the particle's surface chemistry dominates the interactions between the nanotube and the surrounding environment. Changing functional groups on the surface of SWNT can dramatically alter the way in which they interact with hydrophobic and hydrophilic materials; we can use this property to improve dispersion and function within various different polymers. In the next section we discuss the purification and dispersion of SWNT in relation to tissue engineering scaffolds for tissue regeneration.

9.4
SWNT Preparation: Purification and Functionalization

As a potentially implantable material, not only will the SWNT be exposed to the body, but also any contaminants from the manufacturing process not removed through purification processes. If a SWNT composite is non-degradable with nanotubes bound within the material, any negative biological effects of the nanotubes will likely be minimized compared with materials that degrade over time. As in the case with degradable tissue engineering scaffolds, SWNT and contaminants would come into direct contact with host tissue at higher concentrations. Consequently, compatibility with cells and surrounding tissue is crucial, particularly if the SWNT are not eliminated through normal bodily function. For these new materials to be successful they must be biologically, mechanically, and chemically compatible with the host tissue, and, as such, a comprehensive understanding of how SWNT interact with living systems is required. This is important from not only a materials perspective, but also as an airborne contaminant, or any other interaction these nano-materials may have with biological systems.

Various different methods are used to produce SWNT, including chemical vapor deposition, laser pyrolysis, pulse laser vaporation, and carbon arc-discharge. Each of these methods results in differing qualities of nanotubes with yields containing different types and amounts of production contaminants. Different production methods will result in a different elemental composition, and as such the efficiency of purification techniques will have a strong bearing on biological interactions. By understanding these processes we can gain insight into what the dominant mode in which SWNT (or production contaminants) interfere with cell function. This may lead to determining the critical parameters of SWNT preparations that must be modulated to improve cell interactions and thus develop enhanced biomaterials.

To date, few investigations have assessed SWNT–cell interactions and those that have shown a toxic response by cells and tissues [55–59]. A study by Warheit et al. (2004) observed the formation of a series of multifocal granuloma in rats when SWNT were instilled intratracheally [59]. Interestingly these results were not dose-dependent as other studies have shown. Oberdörster et al. (2004) found that oxidative stress was induced when juvenile Largemouth Bass were exposed to C_{60} fullerenes within brain tissue, with the fullerenes localizing in lipid-rich regions such as cell membranes [57]. Schuler et al. (2004) expressed concern that, due to the small dimensions of nanoparticles, they might penetrate the skin and possibly reach the brain by eluding the immune system [60]. Similarly, in a study by Shvedova et al. (2003), using unpurified SWNT produced by the HiPCO process with a

30% iron mass, (NASA-JSC, Houston TX), caused changes in cell ultrastructure and morphology, a loss of cell integrity and apoptosis, accelerated oxidative stress, an accumulation of peroxidative products, and depletion of antioxidants in immortalized human epidermal keratinocytes (HaCaT) [61].

Although the exact mechanism that causes these deleterious interactions with cell systems is unknown, a comprehensive understanding of how SWNT interact with living systems is, clearly, required. By understanding the mechanism(s) that causes these effects on cell function (such as production contaminants, mass transfer limitations, hydrophobic interactions etc.) it will be possible to modulate SWNT chemistry to improve SWNT–cell interactions, and as such develop improved materials.

A recent publication by Nimmagadda et al. (2006) describes how purification and SWNT surface chemistry variation effects the function and viability of 3T3 fibroblasts [62]. In these investigations SWNT were obtained from CarboLex Inc. (Lexington, KY), manufactured by a modified electric-arc technique. In this method a plasma discharge is induced at high temperatures onto a cathode containing graphite powder, metal catalysts, and adhesives. The deposit that forms contains the SWNT. This process results in 40–60% amorphous carbon, as well as non-carbon contaminants including, Ni, Y, Co, and others [63–65]. The study aimed to determine if SWNT themselves, or the production contaminants, were responsible for the toxic responses observed, and to assess if variation in SWNT surface hydrophobicity would improve cellular interactions. Three different preparations of SWNT were assessed, including "as purchased" nanotubes (AP-NT), nanotubes purified by acid washing to remove catalyst residue (PUR-NT), and purified nanotubes functionalized with the small hydrophilic carbohydrate, glucosamine (GA-NT). SWNT that were purified by nitric acid oxidation resulted in dissolution of much of the metal contaminants, but also functionalized the damaged (or defect) ends of the SWNT with carboxylic acid groups reducing the material's hydrophobicity. With the additional functionalization step of added glucosamine, a small sugar molecule, to the SWNT, the overall hydrophobic nature of these NT is further reduced.

An assessment of the effects of these different SWNT preparations has shown a dose-dependent relationship between the viability of mouse 3T3 fibroblasts and concentration of SWNT, where increasing concentrations of SWNT resulted in decreased cell viability. The effects of "as-purchased" SWNT on 3T3 mouse fibroblasts are similar to other reported data, where cell viability was significantly diminished [55, 58, 66]. The observed effects are strongly influenced by contaminants in the preparation, as well as by soluble and non-soluble components. Improved cell viability occurs with PUR-NT and GA-NT over AP-NT preparations (at a given concentration), emphasizing that purity alone (and thus production method) plays an important role in biocompatibility of SWNT. These investigations have shown significant variation of 3T3 fibroblasts viability and metabolic function (after three days in culture) compared with control values. At the lowest concentration of AP-NT (0.001% w/v), cell viability was reduced to 55% of control values (Fig. 9.1a). In the same cell cultures 3T3 fibroblasts demonstrated dramatic changes in their met-

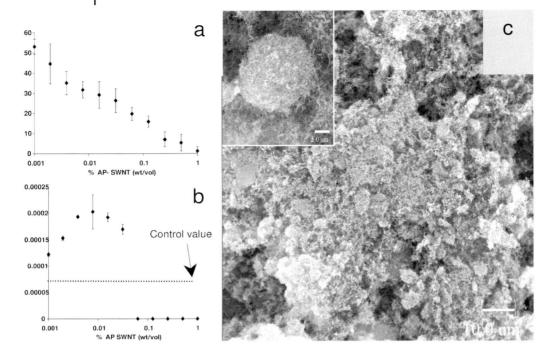

Fig. 9.1. (a) AP-NT/3T3 cultures demonstrated a negative dose-dependent relationship with increasing AP-NT concentration. At the lowest tested concentration of AP-NT (0.001%) only 55% of fibroblasts remained viable ($p < 0.05$). (b) At lower concentrations, cells in AP-NT preparation reduced Alamar Blue dye as much as 200% more than the cells under control conditions (0.78×10^{-4} fluorescence units). No metabolic activity was observed at and above 0.0625% NT ($n = 3$). (c) SEM displayed spherical bodies dispersed throughout AP-NT/3T3 cultures. These were hypothesized to be detached cells encased with nanotubes, where cells may have coalesced to form the larger spherical bodies. In this analysis 3T3 cells were cultured with an AP-NT concentration of 0.0625% NT (wt./vol.), with a cell viability of 19.65% compared with control values (100%) [62].

abolic activity (Fig. 9.1b), illustrating a nonlinear effect dependent on NT concentration. The variation observed in metabolic activity is consistent with results from a recent study by Cui et al. (2005) [55]. In their study, the ability of human HEK293 cells to adhere and proliferate when cultured in the presence of SWNT was also diminished. Cui et al. also assessed gene expression profiles for several specific genes and showed that variation in metabolic activity could in part be explained by the variation in gene expression profiles. Unfortunately, neither the elemental composition nor the purification process (if any) of the nanotubes purchased from Carbon Nanotechnologies, Inc. was described. As such the effect, or comparative effect, of residual contaminants was not assessed.

SEM analysis of 3T3 cells cultures in AP-NT, gives further insight into cellular interactions with NT, where spherical bodies ranging from 5 to 15 μm in size have been observed (Fig. 9.1c) [62]. These spherical bodies were hypothesized to be detached, or semi-detached, cells encased in nanotubes that may, in some cases, have coalesced to form the larger spherical structures.

Purification and functionalization of AP-NT reduces the negative effect on cell viability and metabolic activity, demonstrating that as-purchased NT contain impurities/contaminants that may be responsible for the variation in cell viability and function seen in Fig. 9.1. The effect of these components was assessed by acid washing and chemical surface modification of AP-NT. This suggests that the removal of production contaminants and SWNT hydrophobicity both play an important role in cell-SWNT interactions. The overall trend shows cells incubated in GA-NT preparations having a higher cell viability with less deviation in metabolic activity (from control values) compared with cell/PUR-NT cultures. This suggests the surface properties of SWNT may have a profound effect on cell function. However, the dose-dependant relationship still exists as increasing concentrations of both PUR-NT and GA-NT preparations still reduce cell viability (Fig. 9.2).

At the concentration of 0.0625% w/v 3T3 fibroblasts cultured in the presence of glucosamine functionalized SWNT (GA-NT) for 6 days were found to migrate to the upper surface of the GA-NT to form sheets of cells over the nanotube preparation. Figure 9.3(a) shows an enlargement of Fig. 9.2(c) (incubated for 3 days), compared with sheets of cells seen spreading over the surface of the nanotube preparation after 6 days in culture (Fig. 9.3b).

Because of the small size of these particles, the surface properties dominate the overall chemistry of the molecule and, as such, the addition of functional moieties can dramatically alter the manner in which these particles interact with the surrounding solution. The basic structure of SWNT is a graphene sheet of carbon atoms linked by non-polar covalent bonds. Since electrons are equally shared there is no charge separation and these molecules cannot enter into a charge interaction with water [67]. They are therefore hydrophobic, resulting in minimal solubility in aqueous media. In an investigation by Nimmagadda et al. all NT preparations aggregated to some degree, although the purified nanotubes with carboxylic acid groups and the SWNT functionalized with glucosamine displayed varying degrees of solubility [62]. SEM analysis of AP-NT (the most hydrophobic of the three NT preparations) cultured with 3T3 fibroblasts displayed spherical structures encased in NT-like particles. This led us to the hypothesis that the hydrophobic NT were penetrating, and aggregating within the hydrophobic cell membrane leading to cell rupture (Fig. 9.1c). These effects were not seen with NT functionalized with glucosamine where the hydrophobic nature of SWNT was reduced.

Evidence for the translocation of nanotubes through cell membranes has been confirmed by fluorescence imaging, with [68, 69] and without [70] functionalized markers or molecules. These studies clearly show the nanotubes entering cells, and the potential of nanotubes localizing in hydrophobic regions of the cell membranes. A study by Cherukuri et al. (2004) using a pluronic surfactant that disperses the nanotubes suggests an active mechanism of SWNT uptake. In this study, mouse peritoneal macrophage-like cells were used whose function is the active uptake of targeted particles. However, this may be a secondary uptake path-

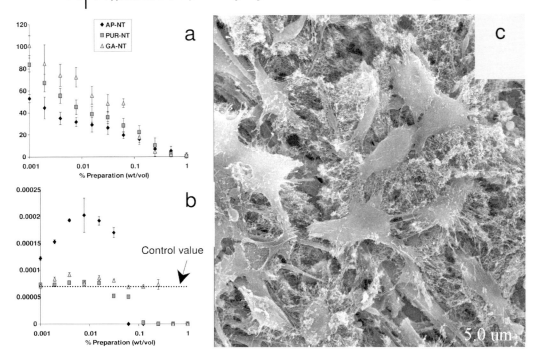

Fig. 9.2. (a) Cells incubated with GA-NT showed the highest viability followed by cells/PUR-NT cultures, and cell/AP-NT cultures have the lowest viability at SWNT concentrations below 0.125%. Preparations and concentrations below 0.125% NT (wt./vol.) show a statistically significant difference ($p < 0.05$), with the exception of concentrations above 0.125%, where no significant difference between each preparation was found ($n = 3$). (b) 3T3's in GA-NT and PUR-NT preparations show metabolic activity similar to 3T3's in control cultures. There was no significant difference ($p < 0.05$) between preparations or concentrations between 0.5 and 1.0% NT (wt./vol.). (c) SEM displays 3T3 cells dispersed throughout the 3D GA-NT matrix. No spherical bodies were observed as seen with 3T3 cells incubated with AP-NT [62].

way, whereas in normal stromal cells uptake may be passive translocation driven by hydrophobic/hydrophilic interactions. The results presented by Cherukuri et al. (2004) do not show any toxicity effects at SWNT concentrations of 3.8 µg mL^{-1}, where they estimated an uptake rate of 1 nanotube per cell per second (total of 70 000 nanotubes ingested) over the incubation period of approximately 20 h [70]. By comparison, at the lowest SWNT concentrations assessed by Nimmagadda et al. [62], 0.000976 w/v, or 9.76 µg mL^{-1}, fibroblast viability was reduced to 50.08 ± 3.74% for "as purchased" nanotubes (AP-NT), with 83.58 ± 6.22% viability with "purified" nanotubes (PUR-NT). Only glucosamine functionalized nanotubes (GA-NT) had no effect on cell viability at these concentrations (Fig. 9.2).

Cherukuri et al. have shown a 40% reduction of SWNT internalization when incubated at 27 °C compared with 37 °C, which is consistent with the temperature

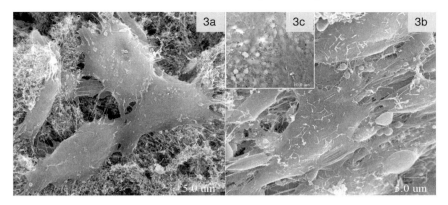

Fig. 9.3. (a) 3T3 fibroblasts cultured in the presence of GA-NT at 0.0625% (w/v) for 3 days, showing an enlargement over Fig. 9.2(c). (b) Similar magnification of 3T3 cells cultured under the same conditions for 6 days that have migrated to the upper surface of the GA-NT to form sheets, with (c) (inset) showing a lower magnification of cells covering the surface of the nanotube preparation [62].

dependence of phagocytosis [71]. By contrast, Pantarotto et al. (2004) used amino-modified SWNT functionalized with the fluorescent probe fluorescein isothiocyanate (FITC) and SWNT functionalized with a peptide responsible for G-protein function (also covalently bonded to FITC) to show that uptake by human 3T6 and murine 3T3 fibroblasts was not affected by temperature variation or the use of endocytosis inhibitors. They concluded that the method of translocation appeared similar to cell penetrating peptides, which is not associated with the endocytosis pathway, but the exact mechanism was unclear [72, 73]. Recently, Monteiro-Riviere et al. (2005) have shown that multi-walled carbon nanotubes (MWCNT) were present in cytoplasmic vacuoles of human epidermal keratinocytes [56]. Our investigations have shown that, depending on the hydrophobicity state of the SWNT, these nano-particles appear to aggregate on the surface of cell-like structures. Although several different mechanisms of SWNT translocation are likely, we hypothesize that due to the hydrophobic nature of these nano-particles they tend to aggregate within hydrophobic regions of the lipid bilayer to form focal points. As yet there is no clear evidence that hydrophobic nanotubes move from the lipid bilayer and back into the aqueous environment (either cytoplasmic or extracellular) and as such there is some question as to whether these nanotubes remain within hydrophobic lipid membranes or as postulated enter the cells aqueous environment. The fluidity of these lipid membranes, which interact with other cell membranes, may transport SWNT into the cell, though not necessarily out of the lipid membranes. From our observations, as concentrations increase, the likelihood of the small focal aggregations enlarging in size increases; it is therefore quite possible that, at sufficiently high concentrations, SWNT interfere with membrane integrity, resulting in rupture and cell death.

In our investigations SWNT within each preparation settled out of solution over

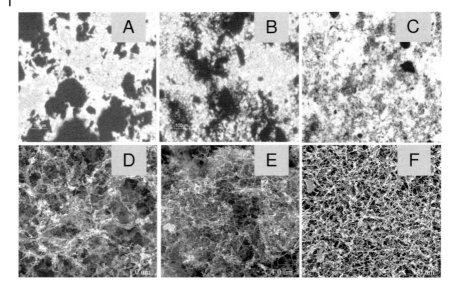

Fig. 9.4. Images (A–C light microscopy, D–F SEM) displaying SWNT preparations dispersed in aqueous cell culture media inoculated on 3T3 fibroblasts after 3 days incubation. (A) AP-NT displaying large aggregates of nanotubes, (B) SWNT aggregates are progressively dispersed through purification steps (PUR-NT), and (C) further reduced hydrophobic interactions of SWNT within the aqueous environment by functionalization with glucosamine. Lower SEM images show (D) AP-NT, (E) PUR-NT and (F) GA-NT surface topography as the SWNT become less hydrophobic, with improved dispersion.

the 3 day incubation period to form layers directly on the cell cultures. Light microscopy illustrates how the differentially treated nanotubes tend to aggregate as a function of hydrophobicity, GA-NT layers appeared more dispersed compared with AP-NT and PUR-NT preparations (Fig. 9.4a–c), with SEM images showing variable surface topography as NT dispersion increases (Fig. 9.4d–f). The presence of glucosamine molecules functionalized on the surface of SWNT improved SWNT interactions with the aqueous media, allowing SWNT bundles to disperse more uniformly with less aggregation [74].

It is unlikely that tissue engineered scaffolds would be made solely of SWNT, rather they will be formed as a composite material where the bulk polymer will be chosen for selected characteristics pertinent to its endpoint application. Materials designed for use in soft tissue repair applications will be vastly different from bone, or artificial limbs, and as such properties such as elasticity, biocompatibility, porosity, charge, etc. need to be considered.

From a materials perspective, carbon nanotubes used as an additive within a bulk polymer need to be dispersed uniformly, ideally as aligned, individual tubes rather than large random aggregates. Without uniform dispersion we will not be able to take full advantage of the unique properties of SWNT, and as such is a critical issue. Significant effort has focused on full-length SWNT and MWNT dispersion, with several approaches being taken to improve dispersion and alignment. Both chemical and mechanical dispersion methods have been used, including, but not limited to, sonication, surfactants, and functionalization with different moieties or macromolecules. Improving alignment of SWNT has been equally challenging and several methods have been investigated, including mechanical shear [75, 76], anisotropic flow [77], and magnetic fields [78, 79].

As discussed, the surface chemistry of these nano-particles dictates the manner in which SWNT are prepared to interface appropriately with the bulk polymer. Clearly, SWNT dispersion within aqueous hydrogels requires a different approach than dispersion in non-polar polymers. Sennett et al. (2003) mixed both SWNT and MWCT in molten polycarbonate (PC) using a twin-screw extruder to disperse the nanotubes, then used a melt-spinning technique to align the NT with the PC fibers. MWCT and SWNT displayed different dispersion characteristics, with larger individual MWCT dispersing uniformly; however, SWNT dispersion was limited to bundles of nanotubes rather than single tubes [76]. This effect will in part be due to the mixing characteristics of different sized particles but dispersion will again be dominated by the nano-particle's surface chemistry. MWCT and SWNT that are produced using different production methods will have different contaminants and will require different purification strategies, most of which can alter the surface charge/chemistry. As such the two materials will likely behave very differently under similar solution conditions.

Other methods of dispersion include functionalization of nanotubes to alter their surface chemistry and/or provide secondary chemical bonding sites. These functional moieties range in chemistry and function, and include (amongst others) biological polymers, including DNA and other large macromolecules. One of the more common methods to functionalize nanotubes is the formation of carboxylic acid groups as either a final process or as an intermediate in a process called defect functionalization using a strong oxidant such as nitric acid (HNO_3) [80–83]. Fu et al. (2002) functionalized SWNT with oligomeric poly(ethylene glycol) to solubilize SWNT and MWNT in aqueous solutions. They then added serum albumin using surface-bound ester linkages, citing an ester to amide transformation as the mechanism of adsorption [83]. These preparations were shown to have a high solubility in water and suggested this method to conjugate fragile biological species. Of the biological polymers used, Zheng et al. (2003) have shown effective dispersion of SWNT in water when sonicated in the presence of single stranded DNA (ssDNA), suggesting binding through π-stacking of ssDNA wrapping in a helical fashion around the SWNT [84]. Baker et al. (2002) oxidized SWNT to form carboxylic acid groups then reacted with thionyl chloride and then ethylenediamine to generate a terminal amine group to which (after several intermediate steps) single-stranded DNA oligonucleotides were functionalized to the SWNT. These investigations have shown the ability of DNA to be covalently bound to SWNT, then for complementary DNA sequences to hybridize with minimal interaction from non-complementary sequences [81]. Bianco et al. (2005) have discussed several strategies for using, and functionalizing, nanotubes in biomedical applications. The paper focuses on nanotube functionalization for specific applications in drug, vaccine, and gene delivery and provides an excellent overview of different functionalization strategies [85].

As a material with potential biomedical applications the absorption pattern of these and other molecules to the surface of SWNT is important to further our understanding of how SWNT are likely to interact with not only the bulk polymer but with surrounding cells and tissues. The inherent hydrophobic nature of SWNT has been hypothesized to play a dominant role in the non-specific molecular binding of various molecules to the surfaces of SWNT [69, 72, 86–88]. In addition to these direct effects of SWNT on cells, it is also possible that SWNT aggregates may indirectly alter cell function by limiting mass transfer of substances either to or from the cell surface. We investigated this possibility by examining the interactions of differentially functionalized SWNT on glucose and protein transfer through SWNT membranes [62]. These investigations were designed to assess any mass transfer limitation that might occur as the SWNT settled onto the cell layer and potentially affected cell function. At sufficiently high concentrations, all SWNT preparations (AP-NT, PUR-NT and GA-NT) resulted in a reduced rate of glucose transfer, compared with controls. However, because the glucose mass transfer rate was still in excess of cellular demand it was concluded that there was no inhibition of cell function. Although no conclusions can be drawn regarding glucose adsorption from the data available, the three preparations clearly had a different topography and porosity (Fig. 9.4), which would be the likely cause of variation in mass transfer rate. Similarly, protein (balanced salts with 10% serum) transfer through the membrane displayed mass transfer variation, indicating inhibition. At lower concentrations of SWNT there was no significant variation in mass transfer across any of the three SWNT preparations. At higher SWNT concentrations (0.25% and 1%) a significant decrease in protein transfer was noted across the PUR-NT preparation compared with AP-NT and GA-NT preparations (Fig. 9.5b) [62].

The results of this study clearly demonstrate that chemical composition and nature of nanotube preparation is of fundamental importance for biological applications. Furthermore, choice of supplier, manufacturing method, and the effectiveness of NT purification will strongly influence the overall composition, including contaminants within the raw material. These impurities will have a range of effects on cellular viability and metabolism, depending on the contaminants involved and their concentration; however, the mechanism by which either the contaminants or the SWNT alter cell viability or metabolism is yet to be resolved. SWNT purification is therefore a critical step and it is important to note that these processes can alter the surface chemistry of the nanotubes and thus their interactions with cells and tissues. We have hypothesized that when SWNT remain strongly hydrophobic (AP-NT and PUR-NT preparations) these nanoparticles interfere with the lipid bilayer of the cell membrane by aggregating within these hydrophobic regions and disrupting cell membrane. Although other mechanisms presumably also occur, SEM evidence points strongly to these hydrophobic interactions as playing a dominant role in early cell failure.

The nanotubes used in the above investigations showed significant effects on *in vitro* cellular function that cannot be attributed to one factor alone, but are likely

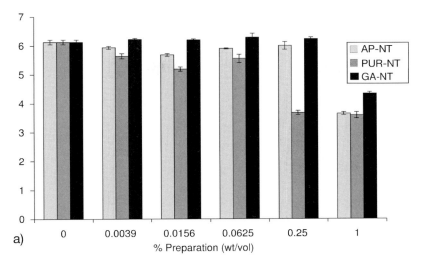

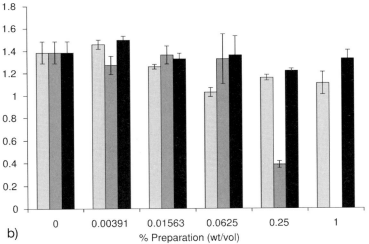

Fig. 9.5. (a) A decrease in glucose transfer rate was observed at higher concentrations of NT preparations (1%). From the transfer rates observed and the cell consumption rate (1.368×10^{-6} mg dL^{-1} h^{-1} cell^{-1}), no apparent mass transfer limitation is likely to have occurred. (b) Analysis of total protein (human plasma) transfer through NT layers showed no significant difference ($p < 0.01$) between AP-NT, GA-NT and control values, whereas PUR-NT layers displayed a significant reduction at higher concentrations [62].

the result of several unfavorable interactions. These interactions were significantly reduced by comprehensive purification processes, and by modifying NT surface chemistry to reduce hydrophobicity and/or the introduction of functional groups. With high purity SWNT the end application will dominate the requirement to modify NT surface chemistry.

9.5
Specific Applications of Carbon Nanotubes in Tissue Engineering

Carbon nanotubes offer a unique potential to develop smart materials that can be tailored to meet specific mechanical and electrical properties. These properties can be optimized to suit specific applications in areas such as bone development where materials must be rigid yet retain a degree of elasticity. The ability to alter the material's conductivity also provides the opportunity to modulate cell interactions/ function by applying electrical fields. The same process can be used for most tissues, especially those that respond in a positive way to electrical stimulation. Some of the more obvious applications include bone, nerve, and cardiac tissue engineering; however, other areas, including skin wound healing, have been shown to respond to electrical fields and are thus candidates for SWNT scaffolds.

There are three primary drivers for the use of SWNT as additives in tissue engineering scaffolds: first is to fine tune scaffold mechanical properties (application specific); the second is to use the NT conductive properties to modulate the materials electrical properties so electrical fields can be used to stimulate the growth and development of tissue engineered constructs; and the third is to promote phenotype specific cellular regeneration by varying the scaffold's conductivity and modulating an applied electrical field. Clearly, a significant amount of basic research is required to verify these goals, but the potential is there given the current status of our knowledge.

In this last section we review several recent publications that focus directly on tissue applications using NT as a component of scaffold materials. Due to the relative infancy of this research area, publications are limited but they clearly show the direction and potential of these novel materials. In a series of investigations interfacing structural biological polymers with SWNT, MacDonald et al. (2005) have assessed the interactions between cell-loaded collagen hydrogels dispersed with varying concentrations of SWNT (0.2, 0.4, 0.8, and 2.0 wt.%) (Fig. 9.6). Figure 9.7 shows progressive enlargements of the composite scaffolds using SEM image analysis of the cartoon given in Fig. 9.6 [89].

The overall goal of these investigations is to enhance the mechanical and functional properties of these hydrogels for potential use in areas such as orthopedic, cardiac and neuronal tissue engineering. Collagen hydrogels have been used in a wide variety of medial implants with excellent results; however, from a purely structural perspective these gels lack the mechanical stability that more demanding applications such as blood vessels and bone replacement require.

The potential to improve the material's mechanical attributes while at the same time adding functionality to the matrix will undoubtedly prove to be a productive avenue of research for both hard and soft tissue replacement or repair. As one of the few investigations that use a degradable polymer with NT dispersed at concentrations as high as 2.0 wt.% it is promising that the authors show no loss in cell viability (rat aortic smooth muscle cells) over the 7 day culture period. Although there was a general reduction in cell number at day 3, by day 7 there was no statis-

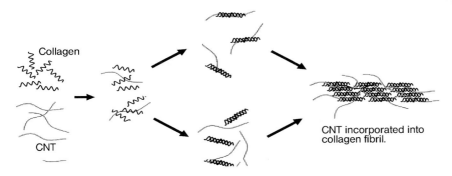

Fig. 9.6. Diagrammatic representation of the process by which CNT can be incorporated firstly into collagen fibrils then form a collagen–CNT composite material [89].

tical difference in total cell number between controls (hydrogels less NT), and NT treated constructs [89].

This data contrasts the reduced cell viability seen in our own investigations [62] where at 1.0 wt.% SWNT no cells were viable. In our investigations the SWNT were dispersed directly in the cell culture media, eventually coming out of suspension, and as such where in direct physical contact with cells. The increased cell viability in the studies by MacDonald et al. (2005) is likely due to the interaction between NT and collagen fibers, where collagen fibers may bind the NT, reducing the direct interactions/contact with cells seen in our investigations. This adds weight to our hypothesis that non-bound hydrophobic NT localize in hydrophobic regions of the cell membrane, resulting in mechanical disruption of the cell membrane (Fig. 9.1).

Not surprisingly, bone and neuronal engineering have been an early focus for several tissue engineering groups. In one of the earliest publications using NT in a tissue engineering application, Supronowicz et al. (2002) assessed cellular interactions with NT composite materials. In these investigations the interactions of

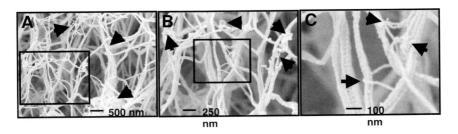

Fig. 9.7. Progressive enlargement of the same section of a composite scaffold (A–C), using scanning electron microscopy, showing examples of nanotubes and nanotube bundles incorporated into the collagen fibrils, identified by their typical banding pattern [89].

murine osteoblasts seeded onto poly(lactic acid) (PLA)/MWNT composites were assessed while stimulating cells with alternating current. PLA, normally an insulator, was made conductive with the addition of 20% (w/w) NT, resulting in a resistivity of 0.2 ohms (Ω). Osteoblasts seeded at a density of 7500 cells cm^{-2} were exposed to an alternating current of 10 µA at a frequency of 10 Hz with a 50% duty cycle. Results show a significant increase in cellular proliferation, type I collagen deposition, and calcium content although no expression of alkaline phosphatase or bone sialoprotein was recorded. Using only the MWNT/PLA composite, samples were either subjected to the alternating current, or not. Unfortunately, there was no control using PLA alone (no MWNT additive), and as such we cannot determine whether the effects noted were the result of the applied current or the presence of MWNT [90]. The translation of this polymer composite to a tissue engineering scaffold requires a degree of porosity be built into the initial material; however, at such high MWNT concentrations (20% w/w) the structural properties may be negatively effected.

A different approach by Zhao et al. (2005) used differentially functional SWNT as a scaffold template aiming to mimic collagen fibrils at the nanoscale such that nucleation is initiated and crystallization of hydroxyapatite (HA) occurs to encourage bone formation. Various functional groups were explored to expose negatively charged surfaces that are believed to promote supersaturation of local ions leading to mineralization. Unlike carboxylic acid functionalized SWNT, the paper shows that sulfonate- and phosphonate-functionalized nanotubes lead to nucleation of HA on their surfaces [91]. This is a promising approach to a complex issue where the defined orientation of HA in natural mineralization processes may be mimicked using SWNT to enhance the regeneration capacity of newly formed bone.

The mechanical and rheological properties of poly(propylene fumerate) (PPF)/ SWNT composites were investigated by Shi et al. (2005) as a potential injectable scaffold for bone repair. This work again draws attention to the importance of SWNT dispersion within any polymer matrix. Although the mechanical properties were enhanced at lower concentrations (0.02, 0.03, 0.05 wt.%), higher concentrations (0.1 and 0.2 wt.%) resulted in air bubble formation and SWNT aggregates within the polymer matrix [92]. Interestingly, functionalization of the NT or the addition of surfactants did not improve dispersion in PPF, underscoring our earlier observation that NT surface chemistry may need to be specifically tailored for improved dispersion in this particular polymer.

A second area of specific interest in SWNT composite materials is as a substrate for neuronal growth and development [93, 94]. Lovat et al. (2005) describe the use of MWNT neural signal transfer whilst supporting cell adhesion and dendrite elongation. These results show adhered hippocampal neuronal cells spontaneously fire action potentials with an increased frequency. The authors suggest that growing neuronal circuits on NT substrates promotes an increase in network operation [95]. Also using NT as substrates for neuronal growth, Hu et al. (2005) have prepared polyethyleneimine (PEI) functionalized SWNT films deposited on glass surfaces as a copolymer to assess hippocampal neuron function. These investigations assessed neurite outgrowth and branching as a comparative investigation to substrates of PEI (a widely used substrate for neuronal culture) and as-purchased MWNT. Results show that neuronal cells cultured on SWNT-PEI substrates produce out growths and branching intermediate between PEI (alone) and as-purchased MWNT, suggesting that the ratio of NT to PEI could be titrated to produce neurons with specific characteristics [96]. In an earlier paper, Webster et al. (2004) showed the potential of carbon nanofiber reinforced polycarbonate urethane (PU) as a scaffold to promote adhesion and growth of cell-specific phenotypes compared with cells grown on glass or 100% PU substrates. Using different cell types on increasing concentrations of NT (NT with a diameter 60 nm), it was shown that the adhesion of astrocytes and fibroblasts generally decreased with increasing concentrations of NT (ratio of PU:NT wt.%); conversely, osteoblast adhesion increased under the same conditions [97]. The authors see the potential of either PU/NT coated or bulk materials being used that offer reduced fibroblast adhesion, and thus reduce the soft tissue formation on orthopedic implants, while at the same time increasing the opportunity for osteoblast adhesion leading to new bone growth. Correa-Duarte et al. (2004) have discussed the generation of thin films of MWNT to form 3D sieve architectures [98]. Using specific treatments and lengths of NT, SEM analysis shows NT forming in perpendicular alignments, pyramid-like formations, and also honeycomb structures. In addition to scaffold surface treatments to modulate cell function, this technology would also be useful in determining the effect of physical patterning and cell adhesion at the nanoscale with identical materials with different structures.

9.6
Conclusions

Clearly, cell survival on a biomaterial is not adequate, in itself, for functional tissue regeneration. The use of NT offers several unique opportunities to develop nano-textured surfaces, variably conductive materials, and materials with adjustable mechanical properties that can be tailored to meet specific design considerations. As these investigations continue to test and further understand cell interactions at macro, micro, and nano-scales the promise of smart, functional materials will become a reality. Ultimately, it is likely to be a combination of multiple stimuli, including mechanical, chemical, and electrical factors that will enhance the development of new functional tissue engineered devices.

References

1 IIJIMA S. Helical microtubules of graphitic carbon. *Nature* **1991**, 354, 56–58.
2 WONG EW, PAUL E. SHEEHAN, CHARLES M. LIEBER. Nanobeam mechanics: Elasticity, strength, and toughness of nanorods and nanotubes. *Science* **1997**, 277, 1971–1975.
3 DRESSELHAUS MS, DRESSELHAUS G, CHARLIER JC, HERNANDEZ E. Electronic, thermal and mechanical properties of carbon nanotubes. *Philos.*

Transact A Math. Phys. Eng. Sci. **2004**, 362, 2065–2098.

4 SANDER TJ, DEVORET MH, DAI H, THESS A, SMALLEY RE, GEERLIGS LJ, DEKKER C. Individual single-wall carbon nanotubes as quantum wires. *Nature* **1997**, 386.

5 LIN Y, SHELBY TAYLOR, HUAPING LI, SHIRAL FERNANDO KA, LIANGWEI QU, WEI WANG, LINGRONG GU, BING ZHOU, YA-PING SUN. Advances toward bioapplications of carbon nanotubes. *J. Mater. Chem.* **2004**, 14, 527–541.

6 WEBSTER TJ, WAID MC, MCKENZIE JL, PRICE RL, EJIOFOR JU. Nano-biotechnology: Carbon nanofibres as improved neural and orthopaedic implants. *Nanotechnology* **2004**, 15, 48–54.

7 CHEN RJ, BANGSARUNTIP S, DROUVALAKIS KA, WONG SHI KAM N, SHIM M, LI Y, KIM W, UTZ PJ, DAI H. Noncovalent functionalization of carbon nanotubes for highly specific electronic biosensors. *Proc. Acad. Natl. Sci. U.S.A.* **2003**, 100(9), 4984–4989.

8 MATTSON MP HR, RAO AM. Molecular functionalization of carbon nanotubes and use as substrates for neuronal growth. *J. Mol. Neurosci.* **2000**, 14(3), 175–182.

9 ALTIZER AM, STEWART SG, ALBERTSON BK, BORGENS RB. Skin flaps inhibit both the current of injury at the amputation surface and regeneration of that limb in newts. *J. Exp. Zool.* **2002**, 293(5), 467–477.

10 MCGINNIS ME, VANABLE JW, JR. Voltage gradients in newt limb stumps. *Prog. Clin. Biol. Res.* **1986**, 210, 231–238.

11 VANABLE JW, JR. A history of bioelectricity in development and regeneration. In: *A History of Regeneration Research*, DINAMORE CE (Ed.), Cambridge University Press, Cambridge, **1991**, pp. 151–177.

12 BORGENS RB, ROEDERER E, COHEN MJ. Enhanced spinal cord regeneration in lamprey by applied electric fields. *Science* **1981**, 213(4508), 611–617.

13 BORGENS RB. Restoring function to the injured human spinal cord. *Adv.*

Anat. Embryol. Cell Biol. **2003**, 171, III–IV, 1–155.

14 BORGENS RB, BLIGHT AR, MCGINNIS ME. Functional recovery after spinal cord hemisection in guinea pigs: The effects of applied electric fields. *J. Comp. Neurol.* **1990**, 296(4), 634–653.

15 BASSETT CA. Fundamental and practical aspects of therapeutic uses of pulsed electromagnetic fields (PEMFs). *Crit. Rev. Biomed. Eng.* **1989**, 17(5), 451–529.

16 EYRES KS, SALEH M, KANIS JA. Effect of pulsed electromagnetic fields on bone formation and bone loss during limb lengthening. *Bone* **1996**, 18(6), 505–509.

17 RYABY JT. Clinical effects of electro-magnetic and electric fields on fracture healing. *Clin. Orthop. Relat. Res.* **1998**(355 Suppl), S205–215.

18 BORSALINO G, BAGNACANI M, BETTATI E, FORNACIARI F, ROCCHI R, ULUHOGIAN S, CECCHERELLI G, CADOSSI R, TRAINA GC. Electrical stimulation of human femoral intertrochanteric osteotomies. Double-blind study. *Clin. Orthop. Relat. Res.* **1988**(237), 256–263.

19 MAMMI GI, ROCCHI R, CADOSSI R, MASSARI L, TRAINA GC. The electrical stimulation of tibial osteotomies. Double-blind study. *Clin. Orthop. Relat. Res.* **1993**(288), 246–253.

20 SHARRARD WJ. A double-blind trial of pulsed electromagnetic fields for delayed union of tibial fractures. *J. Bone Joint Surg. Br.* **1990**, 72(3), 347–355.

21 BRIGHTON CT, BLACK J, FRIEDENBERG ZB, ESTERHAI JL, DAY LJ, CONNOLLY JF. A multicenter study of the treatment of non-union with constant direct current. *J. Bone Joint Surg. Am.* **1981**, 63(1), 2–13.

22 SCHADEN W, FISCHER A, SAILLER A. Extracorporeal shock wave therapy of nonunion or delayed osseous union. *Clin. Orthop. Relat. Res.* **2001**(387), 90–94.

23 WONG JY, LANGER R, INGBER DE. Electrically conducting polymers can noninvasively control the shape and growth of mammalian cells. *Proc.*

Acad. Natl. Sci. U.S.A. **1994**, 91(8), 3201–3204.

24 WALLACE GG, INNIS PC. Inherently conducting polymer nanostructures. *J. Nanosci. Nanotechnol.* **2002**, 2(5), 441–451.

25 RUBIO RETAMA J, LOPEZ CABARCOS E, MECERREYES D, LOPEZ-RUIZ B. Design of an amperometric biosensor using polypyrrole-microgel composites containing glucose oxidase. *Biosens. Bioelectron.* **2004**, 20(6), 1111–1117.

26 LIU X, NEOH KG, CEN L, KANG ET. Enzymatic activity of glucose oxidase covalently wired via viologen to electrically conductive polypyrrole films. *Biosens. Bioelectron.* **2004**, 19(8), 823–834.

27 LAKARD S, HERLEM G, VALLES-VILLAREAL N, MICHEL G, PROPPER A, GHARBI T, FAHYS B. Culture of neural cells on polymers coated surfaces for biosensor applications. *Biosens. Bioelectron.* **2005**, 20(10), 1946–1954.

28 KIM DH, ABIDIAN M, MARTIN DC. Conducting polymers grown in hydrogel scaffolds coated on neural prosthetic devices. *J. Biomed. Mater. Res. A* **2004**, 71(4), 577–585.

29 CASTANO H, O'REAR EA, MCFETRIDGE PS, SIKAVITSAS VI. Polypyrrole thin films formed by admicellar polymerization support the osteogenic differentiation of mesenchymal stem cells. *Macromol. Biosci.* **2004**, 4(8), 785–794.

30 DE GIGLIO E, GUASCITO MR, SABBATIN L, ZAMBONIN G. Electropolymerization of pyrrole on titanium substrates for the future development of new biocompatible surfaces. *Biomaterials* **2001**, 22(19), 2609–2616.

31 MATTIOLI-BELMONTE M, GIAVARESI G, BIAGINI G, VIRGILI L, GIACOMINI M, FINI M, GIANTOMASSI F, NATALI D, TORRICELLI P, GIARDINO R. Tailoring biomaterial compatibility: In vivo tissue response versus in vitro cell behavior. *Int. J. Artif Organs* **2003**, 26(12), 1077–1085.

32 HERRING M, GARDNER A, GLOVER J. Seeding endothelium on to canine arterial prostheses. *Arch. Surgery* **1979**, 114, 679–682.

33 GRAHAM LM, VINTER DW, FORD JW, KAHN RH, BURKEL WE, STANLEY JC. Cultured autogenous endothelial cell seeding of prosthetic vascular grafts. *Surg. Forum* **1979**, 30, 204–206.

34 ESKIN SG, TREVINO L, CHIMOSKEY JE. Endothelial cell culture on dacron fabrics of different configurations. *J. Biomed. Mater. Res.* **1978**, 12(4), 517–524.

35 ESKIN SG, ARMENIADES CD, LIE JT, TREVINO L, KENNEDY JH. Growth of cultured calf aortic smooth muscle cells on cardiovascular prosthetic materials. *J. Biomed. Mater. Res.* **1976**, 10(1), 113–122.

36 CHENG WL, LIN CC. The effects of different electrical stimulation protocols on nerve regeneration through silicone conduits. *J. Trauma* **2004**, 56(6), 1241–1246.

37 KERNS JM, PAVKOVIC IM, FAKHOURI AJ, WICKERSHAM KL, FREEMAN JA. An experimental implant for applying a DC electrical field to peripheral nerve. *J. Neurosci. Methods* **1987**, 19(3), 217–223.

38 KOTWAL A, SCHMIDT CE. Electrical stimulation alters protein adsorption and nerve cell interactions with electrically conducting biomaterials. *Biomaterials* **2001**, 22(10), 1055–1064.

39 MCDEVITT L, FORTNER P, POMERANZ B. Application of weak electric field to the hindpaw enhances sciatic motor nerve regeneration in the adult rat. *Brain Res.* **1987**, 416(2), 308–314.

40 SHEN N, ZHU J. Experimental study using a direct current electrical field to promote peripheral nerve regeneration. *J. Reconstr. Microsurg.* **1995**, 11(3), 189–193.

41 SIDZHANOV ZH M, SHABANOV AM, PAMURZIN LG, ZHARMAGAMBETOV S, SADYKOV RG. Use of an electrical current for accelerating bone healing. *Ortop. Travmatol. Protez.* **1976**(10), 64–66.

42 WEIGERT M. Stimulation of bone formation using electrical current. *Hefte Unfallheilkd* **1973**, 115(Suppl 115) 1–10.

43 Seifalian AM, Giudiceandrea A, Schmitz-Rixen T, Hamilton G. Non-compliannce: The silent acceptance of a villain. In: *Tissue Engineering of Vascular Prosthetic Grafts*. Zilla P, Greisler HP (Eds.), R.G. Landers Company, Austin, **1999**, p. 621.

44 Robinson KR. The response of cells to electrical field: A review. *J. Cell Biol.* **1985**, 101, 2023–2027.

45 Crispin B, Weinberg CB, Bell E. A blood vessel model constructed from collagen and cultured vascular cells. *Science* **1986**, 231(4736), 397–400.

46 Guidoin R, Marceau D, Couture J, Jian Raio T, Merhi Y, Roy PE, De la Faye D. Collagen coatings as biological sealants for textile arterial prostheses. *Biomaterials* **1989**, 10(April), 156–165.

47 Kuntz E. Ethicon, Inc., assignee. Preparation of collagenous materials. *US Patent 3 152 976*, **1964**.

48 Lee CH, Singla A, Lee Y. Biomedical applications of collagen. *Int. J. Pharm.* **2001**, 221(1–2), 1–22.

49 Ramshaw JAM, Werkmeister JA, Glattauer V. Collagen-based biomaterials. *Biotechnol. Genetic Eng. Rev.* **1995**, 13, 335–382.

50 Weinberg CB, Bell E. A blood vessel model constructed from collagen and cultured vascular cells. *Science* **1986**, 231(4736), 397–400.

51 Elbjeirami WM, Yonter EO, Starcher BC, West JL. Enhancing mechanical properties of tissue-engineered constructs via lysyl oxidase crosslinking activity. *J. Biomed. Mater. Res. A* **2003**, 66(3), 513–521.

52 Mou SS, Ma AD, Tu M, Li LH, Zhou CR. Preparation and biocompatibility of tissue-engineered scaffold materials based on collagen. *Di Yi Jun Yi Da Xue Xue Bao* **2002**, 22(10), 878–879.

53 Orban JM, Wilson LB, Kofroth JA, El-Kurdi MS, Maul TM, Vorp DA. Crosslinking of collagen gels by transglutaminase. *J. Biomed. Mater. Res. A* **2004**, 68(4), 756–762.

54 Pieper JS, van der Kraan PM, Hafmans T, Kamp J, Buma P, van Susante JL, van den Berg WB, Veerkamp JH, van Kuppevelt TH. Crosslinked type II collagen matrices: Preparation, characterization, and potential for cartilage engineering. *Biomaterials* **2002**, 23(15), 3183–3192.

55 Cui D, Tian F, Ozkan CS, Wang M, Gao H. Effect of single wall carbon nanotubes on human HEK293 cells. *Toxicol. Lett.* **2005**, 155(1), 73–85.

56 Monteiro-Riviere NA, Nemanich RJ, Inman AO, Wang YY, Riviere JE. Multi-walled carbon nanotube interactions with human epidermal keratinocytes. *Toxicol. Lett.* **2005**, 155(3), 377–384.

57 Oberdörster E. Manufactured nanomaterials (fullerenes, C_{60}) induce oxidative stress in the brain of juvenile Largemouth Bass. *Environ. Health Perspectives* **2004**, 112(10), 1058.

58 Shvedova AA, Castranova V, Kisin ER, Schwegler-Berry D, Murray AR, Gandelsman VZ, Maynard A, Baron P. Exposure to carbon nanotube material: Assessment of nanotube cytotoxicity using human keratinocyte cells. *J. Toxicol. Environ. Health A* **2003**, 66(20), 1909–1926.

59 Warheit DB, Laurence BR, Reed KL, Roach DH, Reynolds GA, Webb TR. Comparative pulmonary toxicity assessment of single-wall carbon nanotubes in rats. *Toxicol. Sci.* **2004**, 77(1), 117–125.

60 Schuler E. Perception of risks and nanotechnology. In: Discovering the Nanoscale, Baird D, Nordmann A, Schummer J (eds.), IOS Press. **2004**, 219.

61 Shvedova AA. Exposure to carbon nanotube material: Assessment of nanotube cytotoxicity using human keratinocyte cells. *J. Toxicol. Environ. Health, Part A* **2003**, 66, 1909–1926.

62 Nimmagadda A, Thurston K, Nollert UM, McFetridge PS. Chemical modification of SWNT alters in vitro cell-SWNT interactions. *J. Biomed. Mater. Res. Part A* **2006**, 76A(3), 614–625.

63 Bethune DS, Klang CH, Devries MS, Gorman G, Savoy R, Vazquez J, Beyers R. Cobalt-catalysed growth of carbon nanotubes with single-atomic-layer wall. *Nature* **1993**, 363, 605–607.

64 CHEN J, HAMON MA, HU H, CHEN Y, RAO AM, EKLUND PC, HADDON RC. Solution properties of single-walled carbon nanotubes. *Science* **1998**, 282, 95–98.

65 IIJIMA S. Single shell carbon nanotubes of one nanometer diameter. *Nature* **1993**, 363, 503–605.

66 MAYNARD AD, BARON PA, FOLEY M, SHVEDOVA AA, KISIN ER, CASTRANOVA V. Exposure to carbon nanotube material: Aerosol release during the handling of unrefined single-walled carbon nanotube material. *J. Toxicol. Environ. Health A* **2004**, 67(1), 87–107.

67 DRESSELHAUS MS. Nanotechnology – New tricks with nanotubes. *Nature* **1998**, 391, 19.

68 WU W, WIECKOWSKI S, PASTORIN G, BENINCASA M, KLUMPP C, BRIAND JP, GENNARO R, PRATO M, BIANCO A. Targeted delivery of amphotericin B to cells by using functionalized carbon nanotubes. *Angew. Chem. Int. Ed.* **2005**, 44(39), 6358–6362.

69 WONG SHI KAM N, JESSOP TC, WENDER PA, DAI H. Nanotube molecular transporters: Internalization of carbon nanotube-protein conjugates into mammalian cells. *J. Am. Chem. Soc.* **2004**, 126, 6850–6851.

70 CHERUKURI P, BACHILO SM, LITOVSKY SH, WEISMAN RB. Near-infrared fluorescence microscopy of single-walled carbon nanotubes in phagocytic cells. *J. Am. Chem. Soc.* **2004**, 126(48), 15 638–15 639.

71 MATSUI H, ITO T, OHNISHI S. Phagocytosis by macrophages. III. Effects of heat-labile opsonin and poly(L-lysine). *J. Cell Sci.* **1983**, 59, 133–143.

72 PANTAROTTO D, BRIAND JP, PRATO M, BIANCO A. Translocation of bioactive peptides across cell membranes by carbon nanotubes. *Chem Commun (Camb)* **2004**(1), 16–17.

73 PANTAROTTO D, SINGH R, MCCARTHY D, ERHARDT M, BRIAND JP, PRATO M, KOSTARELOS K, BIANCO A. Functionalized carbon nanotubes for plasmid DNA gene delivery. *Angew. Chem. Int. Ed.* **2004**, 43(39), 5242–5246.

74 POMPEO F, RESASCO DE. Water solubilization of single-walled carbon nanotubes by functionalization with glucosamine. *Nano Lett.* **2002**, 2(4), 369–373.

75 JIN YP, GUPTA D, DZIARSKI R. Endothelial and epithelial cells do not respond to complexes of peptidoglycan with soluble CD14 but are activated indirectly by peptidoglycan-induced tumor necrosis factor-alpha and interleukin-1 from monocytes. *J. Infectious Dis.* **1998**, 177(6), 1629–1638.

76 SENNETT M, WELSH ER, WRIGHT JB, LI WZ, WEN JG, REN ZF. Dispersion and alignment of carbon nanotubes in polycarbonate. *Appl. Phys. A: Mater. Sci. Proc.* **2003**, 76, 111–113.

77 HAGGENMUELLER R, GOMMANS HH, RINZLER AG, FISCHER JE, WINEY KI. Aligned single-wall carbon nanotubes in composites by melt processing methods. *Chem. Phys. Lett.* **2000**, 330(3–4), 219–225.

78 FISCHER JE, ZHOU W, VAVRO J, LLAGUNO MC, GUTHY C, HAGGENMUELLER R, CASAVANT MJ, WALTERS DE, SMALLEY RE. Magnetically aligned single wall carbon nanotube films: Preferred orientation and anisotropic transport properties. *J. Appl. Phys.* **2003**, 94(4), 2157–2163.

79 WALTERS DA, CASAVANT MJ, QIN XC, HUFFMAN CB, BOUL PJ, ERICSON LM, HAROZ EH, O'CONNELL MJ, SMITH K, COLBERT DT et al. In-plane-aligned membranes of carbon nanotubes. *Chem. Phys. Lett.* **2001**, 338(1), 14–20.

80 POMPEO F, RESASCO D. Water solubilization of single-walled carbon nanotubes by functionalization with glucosamine. *Nano Lett.* **2002**, 2(4), 369–373.

81 BAKER SE, CAI W, LASSETER TL, WEIDKAMP KP, HAMERS RJ. Covalently bonded adducts of deoxyribonucleic acid (DNA) oligonucleotides with single-wall carbon nanotubes: Synthesis and hybridization. *Nano Lett.* **2002**, 2(12), 1413–1417.

82 CHEN J, RAO AM, LYUKSYUTOV S, ITKIS ME, HAMON MA, HU H, COHN RW, EKLUND PC, COLBERT DT,

SMALLEY RE and others. Dissolution of full-length single-walled carbon nanotubes. *J. Phys. Chem. B* **2001**, 105(13), 2525–2528.

83 FU K, HUANG W, LIN Y, ZHANG D, HANKS TW, RAO AM, SUN YP. Functionalization of carbon nanotubes with bovine serum albumin in homogeneous aqueous solution. *J. Nanosci. Nanotechnol.* **2002**, 2(5), 457–461.

84 ZHENG M, JAGOTA A, SEMKE ED, DINER BA, MCLEAN RS, LUSTIG SR, RICHARDSON RE, TASSI NG. DNA-assisted dispersion and separation of carbon nanotubes. *Nat. Mater.* **2003**, 2(5), 338–342.

85 BIANCO A, KOSTARELOS K, PARTIDOS CD, PRATO M. Biomedical applications of functionalised carbon nanotubes. *Chem. Commun. (Camb)* **2005**(5), 571–577.

86 MING Z, JAGOTA A, SEMKE ED, DINER BA, MCLEAN RS, LUSTIG SR, RICHARDSON RE, TASSI NG. DNA-assisted dispersion and separation of carbon nanotubes. *Nat. Mater.* **2003**, 2, 338.

87 ERLANGER BF, BI-XING CHEN, MIN ZHU, LOUIS BRUS. Binding of an anti-fullerene IgG monoclonal antibody to single wall carbon nanotubes. *Nano Lett.* **2001**, 1(9), 465–467.

88 MATARREDONE O, HEATHER RHOADS, ZHONGRUI LI, JEFFREY H. Harwell, Leandro Balzano, Daniel E. Resasco. Dispersion of single-walled carbon nanotubes in aqueous solutions of the anionic surfactant NaDDBS. *J. Phys. Chem. B* **2003**, 107, 13 357–13 367.

89 MACDONALD RA, LAURENZI BF, VISWANATHAN G, AJAYAN PM, STEGEMANN JP. Collagen-carbon nanotube composite materials as scaffolds in tissue engineering. *J. Biomed. Mater. Res. A* **2005**, 74(3), 489–496.

90 SUPRONOWICZ PR, AJAYN PM, ULLMANN KR, ARULANADAM BP, METZGER DW, BIZIOS R. Novel current-conducting composite substrates for exposing osteoblasts to alternating current stimulation. *J. Biomed. Mater. Res.* **2002**, 59, 499–506.

91 ZHAO B, HU H, MANDAL SK, HADDON RC. A bone mimic based on the self-assembly of hydroxyapatite on chemically functionalized single-walled carbon nanotubes. *Chem. Mater.* **2005**, 17, 3235–3241.

92 SHI X, HUDSON JL, SPICER PP, TOUR JM, KRISHNAMOORTI K, MIKOS AG. Rheological behaviour and mechanical characterization of injectable poly(propylene fumarate)/single-walled carbon nanotube composites for bone tissue engineering. *Nanotechnology* **2005**, 16, S531–S538.

93 HU H, NI Y, MONTANA V, HADDON RC, PARPURA V. Chemically functionalized carbon nanotubes as substrates for neuronal growth. *Nano Lett.* **2004**, 4, 507–511.

94 MATTSON MP, HADDON RC, RAO AM. Molecular functionalization of carbon nanotubes and use as substrates for neuronal growth. *J. Mol. Neurosci.* **2000**, 14(3), 175–182.

95 LOVAT V, PANTAROTTO D, LAGOSTENA L, CACCIARI B, GRANDOLFO M, RIGHI M, SPALLUTO G, PRATO M, BALLERINI L. Carbon nanotube substrates boost neuronal electrical signaling. *Nano Lett.* **2005**, 5(6), 1107–1110.

96 HU H, NI Y, MANDAL SK, MONTANA V, ZHAO B, HADDON RC, PARPURA V. Polyethyleneimine functionalized single-walled carbon nanotubes as a substrate for neuronal growth. *J. Phys. Chem. B: Lett.* **2005**, 109, 4285–4289.

97 WEBSTER TJ, WAID MC, MCKENZIE JL, PRICE RL, EJIOFOR JU. Nano-biotechnology: Carbon nanofibres as improved neural and orthopaedic implants. *Nanotechnology* **2004**, 15(1), 48–54.

98 CORREA-DUARTE MA, WAGNER N, ROJAS-CHAPANA J, MORSCZECK C, THIE M, GIERSIG M. Fabrication and biocompatibility of carbon nanotube-based 3D networks as scaffolds for cell seeding and growth. *Nano Lett.* **2004**, 4(11), 2233–2236.

10

Nanoparticles for Cell Engineering –
A Radical Concept

Beverly A. Rzigalinski, Igor Danelisen, Elizabeth T. Strawn,
Courtney A. Cohen, and Chengya Liang

10.1
Introduction and Overview

Nanotechnology holds great promise for the treatment of disease through cellular engineering. Engineering has made substantial advances in materials science and construction of nanoscale devices over the last decade. Movement of this emerging technology into the realm of cell biology and medicine involves applying these advances to biological systems, providing novel opportunities to intervene in the progression and pathology of disease. To accomplish this, scientists must develop the ability to work and reason in a manner that merges previously distinct disciplines; applying the principles of physics and engineering to cell biology and physiology. This chapter is directed towards achieving this goal.

It is of critical importance to address this merging of disciplines in our written discussions of nanotechnological inroads into the biomedical sciences. The present chapter discusses such a progression in an interdisciplinary manner for the first time, by examining the role of engineered nanoparticles as free radical scavengers at the cellular level and their potential use as pharmacological agents for free radical-mediated disorders.

We begin with a brief discussion of free radicals and their role in cellular function and dysfunction, followed by examination of the role of oxidative stress in aging and disease. The current pharmacology of oxidative stress will be addressed along with its limited efficacy. Comparisons will be made regarding free radical processes in biological vs. industrial and chemical processes. Next, we will examine an alternative, nanotechnological approach to oxidative stress at the cellular level and examine the emerging potential for the utilization of different nanoparticles in the treatment of disease.

Nanotechnologies for the Life Sciences Vol. 9
Tissue, Cell and Organ Engineering. Edited by Challa S. S. R. Kumar
Copyright © 2006 WILEY-VCH Verlag GmbH & Co. KGaA, Weinheim
ISBN: 3-527-31389-3

10.2
Free Radicals and Oxidative Stress

Free radicals are highly reactive compounds, lacking the full complement of electrons necessary for molecular stability. In biological organisms, unstable free radicals can strip electrons from cellular macromolecules, rendering them dysfunctional. Free radical species encountered within the cell include superoxide (O_2^-), the hydroxyl radical ($OH^{\cdot}$), nitric oxide (NO), peroxynitrite ($ONOO^-$), lipid hydroperoxides, and others [1–5]. Importantly, many free radical reactions in the cell are self-propagating [6]. An example of this is the lipid peroxidation chain reaction, initiated by the hydroxyl radical ($OH^{\cdot}$). The reaction is initiated by interaction of the hydroxyl radical with a membrane lipid (L):

$$LH_2 + OH^{\cdot} \rightarrow LH^{\cdot} + OH^- + H^+$$

In the second phase of the propagation process:

1. $LH^{\cdot} + O_2 \rightarrow LHOO$

2. $LHOO^{\cdot} + LH_2 \rightarrow LHOOH + LH^{\cdot}$

Thus, the free radical species is regenerated to propagate the peroxidation reaction. In hydrophobic areas such as the plasma membrane, chain reactions of lipid peroxidation can damage membrane structure and induce cell death [6–8]. One of the end products of peroxidative damage to cellular lipids is the destructive aldehyde 4-hydroxy-2-nonenal (4HNE), a highly reactive electrophile that interacts with protein constituents such as cysteine, histidine, and lysine [9, 10]. 4-HNE has been implicated as a damaging factor in many human disease states, by virtue of disruption of protein function. Formation of prostaglandin-like compounds, termed isoprostanes, by free radical-mediated lipid peroxidation has also been implicated in numerous pathological conditions [11]. In addition to lipid peroxidation, cellular sugars and proteins can also undergo free radical mediated oxidation, which may induce loss of catalytic function, destruction of cellular architecture, increased degradation or accumulation of damaged cellular constituents, as occurs in many neurodegenerative, ischemic, diabetic, and other disorders (Table 10.1). Engineering and materials science readers should note that these biological free radical processes are similar to those encountered in industrial and materials science applications – the players are somewhat different but the game remains the same.

10.2.1
Sources of Intracellular Free Radicals

Free radicals arise from many sources within the cell, as described in several good reviews [5, 12–16]. Mitochondria and the oxidative phosphorylation process are considered to be a primary site of free radical production, giving rise to superoxide

Tab. 10.1. Human disease states that generate oxidative stress or of which oxidative stress is a known component.

Organ system	Disease
Brain, CNS, PNS	Alzheimer's, Parkinson's, Huntington's, Multiple Sclerosis Trauma, Stroke, Ischemia, Other Neurodegenerative disorders
Circulatory	Atherosclerosis, Hyperlipidemia, Cardiovascular diseases Vascular disorders, Hypertension
Endocrine	Diabetes, Metabolic Syndrome
Musculoskeletal	Arthritis, Physical injury, Joint disorders
Immune	Allergic disorders, Autoimmune disorders, Inflammatory disorders
Respiratory	Asthma, emphysema, COPD, Bronchitis
Digestive	Inflammatory bowel disease, Crohn's disease

radicals during the passage of electrons through the electron transport chain [17–19]. Free radicals are also formed at sites of inflammation, released by neutrophils and macrophages to destroy invading bacteria [1, 18]. However, cells of the immune system also generate free radicals in inflammatory disorders such as Alzheimer's disease, arthritis, and cardiovascular disease, where they initiate or propagate tissue damage [5].

Free radical production is also a by-product of certain normal cellular enzymatic reactions. The cyclooxygenase enzyme produces free radicals during prostaglandin synthesis [20, 21]. NAPDH oxidase and xanthine/xanthine oxidase are also intracellular enzyme systems associated with generation of free radicals during normal metabolic function [1, 6, 21]. NO is a free radical produced by nitric oxide synthase (NOS) and is commonly utilized in cell signaling in the nervous and cardiovascular systems [22]. However, overproduction of NO in inflammatory and disease states such as atherosclerosis can result in formation of the highly reactive peroxynitrite radical ($ONOO^{\cdot}$), which can subsequently induce nitrosylation of critical cellular proteins and damage their function.

In addition to endogenously generated free radicals, organisms are exposed to numerous exogenous sources of free radicals and free radical generating agents during a typical life span. These include environmental toxins such as pesticides, radiation, pharmacological agents, industrial pollutants, food additives, cigarette smoke, and many others. In summary, the typical cell is exposed to numerous free radicals from both endogenous and exogenous sources during a lifetime. Such free radical exposure, if unchecked or excessive, may induce considerable damage to cellular constituents necessary for survival and function.

10.2.2
Oxidative Stress

Healthy cells are able to counteract the damaging effects of free radicals via innate, natural, enzymatic or molecular mechanisms. Superoxide dismutase (SOD) catalyzes the conversion of the superoxide radical into $H_2O_2 + O_2$. Although H_2O_2 is also damaging to biological systems, catalase and glutathione peroxidase enzymes convert H_2O_2 into $O_2 + 2H_2O$. Other endogenous reductants that ameliorate cellular free radicals include *n*-acetyl cysteine, vitamin E, vitamin C, carotenes, melatonin, and lipoic acid derivatives, and are the subject of many recent reviews [1, 23–27].

Although excessive free radical production can be damaging, production of free radicals in low amounts is necessary for normal cell function and signaling. For example, the free radical NO is necessary for normal signal transduction pathways in the vasculature [28] and neurotransmission in the brain [23]. Phagocytic cells routinely utilize free radicals for destruction of invading bacteria or cells infected with viruses. Regulation of cell signaling pathways such as phosphatase activity, calcium signaling, and transcription factors have all been implicated in having regulatory links with free radical production within the cell [22, 29, 30]. Hence a low amount of free radicals appears to be important to normal cell function and the healthy cell achieves a biological balance between free radical production and degradation.

Cellular havoc arises when the production of free radicals exceeds the capacities of cellular defense mechanisms or when cell defense mechanisms are compromised by aging and other pathological states, disturbing the biological balance. In this case, production of free radicals far exceeds their normal signaling functions and damage to cellular macromolecules ensues. The burden of free radicals imposed upon a cell in such a situation is termed oxidative stress, and often arises as a prerequisite or result of aging or disease states. For an example of this, let us return to the subject of the NO free radical. NO is an important neurotransmitter in regulated, low, amounts. However, when produced in overabundance due to neuronal damage or in the presence of hydroxyl radicals, oxidative stress can be imposed upon the brain. The damaging $ONOO^-$ radical is formed under these conditions, which can induce lipid peroxidation and nitration of protein tyrosine residues [31]. Tyrosine nitration inhibits phosphorylation of tyrosine residues in proteins, an important signaling mechanism in cell regulation. Other free radicals may catalyze formation of carbonyl groups on proteins, which can interfere with tertiary conformation and inhibit or reduce their function [30, 32, 33]. The end result of such cumulative damage is cell dysfunction or demise.

Within the cell, there is a balance between the rate of free radical production and the capacity of cellular defense mechanisms to mitigate free radical damage. When cellular mitigating systems are compromised, high levels of oxidative stress can do extensive cell damage. Cellular pathologies associated with free radicals include lysosome and proteasome dysfunction leading to cell death [34–36], loss of signaling cascades mediated by phosphorylation [32, 37], oxidation of DNA bases [38, 39], destruction of the membrane lipid environment [40, 41], loss of vascular tone and responsivity [32, 42], aberrant neurotransmission [16, 43, 44], mitochondrial dysfunction [1, 13, 14], and initiation of cell death pathways [12, 16, 45].

10.2.3
Oxidative Stress and Disease

A multitude of disease pathologies are caused or otherwise related to oxidative stress, including Alzheimer's disease, inflammatory disorders, Parkinson's disease, diabetes, atherosclerosis, hypertension, and mechanical injury, which are summarized in Table 10.1 and are the subject of numerous works [1, 18, 25, 43, 46–49]. As with Alzheimer's disease, free radical production is often not the ultimate causative factor, but is a byproduct of cellular malfunction initiated by amyloid beta peptide. Aberrant free radical production by damaged cells can lead to destruction of uninjured "bystander" tissue, thereby propagating the damage. In Alzheimer's disease, excessive free radical production in damaged neurons and inflammatory activation of microglia leads to free radical injury of healthy, neighboring neurons [37, 44, 50]. Similar pathologies have been associated with atherosclerosis, arthritis, and other inflammatory disorders that are prevalent in aging organisms [1, 6, 51–54].

Of particular interest to us all has been the role of free radicals in aging and age-associated disorders. As described by Harman [55], aging began with the origin of life, and represents the accumulation of diverse deleterious changes in cells and tissues that increase the risk of disease and death. Although there are several theories on aging, the "Free Radical Theory of Aging" appears to have the most biochemical support [23, 50, 55]. According to this theory, aging is associated with accumulation of macromolecular free radical damage to lipid membranes, mitochondria, protein, and DNA, promoting cell dysfunction and death. Numerous lines of evidence support this hypothesis and are the subject of several excellent reviews [1, 45, 49, 50]. Aging is accompanied by an increase in the level of oxidative damage such as carbonyl groups and tyrosine nitration of cellular proteins, a general index of free radical damage [32, 33, 56]. Late onset Alzheimer's is characterized by an accumulation of oxidatively modified proteins [32, 57]. The antioxidant spin trap n-t-butyl-α-phenylnitrone improved spatial memory of aged rats relative to untreated controls, and also decreased the level of oxidative damage to cellular macromolecules in the brain [32]. Lipid peroxidation products have also been shown to increase with age [58], suggesting that membrane damage accumulates over the course of a lifespan.

Quality of life, not only life span, is important in the aging process. Once again, the free radical hypothesis maintains that free radical production is associated with cognitive deficits, neuronal dysfunction, atherosclerotic conditions, and inflammatory conditions associated with aging. Neural tissue has the highest ATP and oxygen utilization of all the organs in the body. It follows that neural tissue is therefore exposed to some of the highest levels of oxidative stress in neurodegenerative diseases and aging. Several reviews [59–61], suggest that free radical scavengers may be useful therapeutic adjuncts in treatment and prevention of age-associated disorders, improving the quality of life.

Given the widespread involvement of free radicals in aging and disease, the use of free radical scavengers or antioxidants as pharmacological therapy has been reported in many disorders (reviewed in Refs. [25, 27, 52, 54, 59–61]). Free radical scavengers such as vitamin E, vitamin C, *n*-acetyl cysteine, melatonin, SOD mimetics, 21-aminosteroids, 2-methylaminochromans and nitrosone spin traps have been utilized to lessen the severity of disorders associated with free radical production. However, many of these free radical scavengers have met with only limited clinical success. For example, SOD and poly(ethylene glycol)-conjugated SOD showed promise in the treatment of traumatic head injury, where superoxide radical production contributes to disease pathology [59, 61, 62]. Yet clinical trials of these agents yielded far from promising results, likely due to lack of cell or tissue penetration. On the whole, pharmacological use of free radical scavengers has several problematic aspects. The first is lack of penetration to the site of radical production, as may be the case with exogenous SOD mimetics. Similarly, vitamin C is also a free radical scavenger, yet it is thought to function only in aqueous environments and has difficulty crossing the lipid membrane of the cell. Additionally, due to its chemical structure, high concentrations of vitamin C are also reported to act as free radical generators, so proper dose is critical to antioxidant function. The second problematic issue is the requirement for repeated dosage. Many of these scavengers require repetitive daily dosing to replace molecular species that were utilized in free radical reduction. In the case of vitamins C or E and other antioxidants, one molecule of antioxidant is utilized for each free radical detoxified, and regeneration is low or non-existent in the case of exogenous agents such as spin traps and 21-aminosteroids. In summary, although supplementation with free radical scavengers has shown to be beneficial in several disease states, the overall success has been limited at the clinical level.

10.3
A Nanotechnological Approach to Oxidative Stress

Up to this point, our discussion of free radicals and oxidative stress have focused on events from a biological standpoint. However, if we examine the basic chemistry of free radical mechanisms, the engineer will realize that the cellular processes underlying biological oxidative stress are encountered in many non-biological instances. Let us take, for example, the processes involved in three-way catalysis for improved combustion and removal of environmental contaminants from engine exhaust. These processes have much in common with the biological concept of redox reactions and antioxidants from a chemical and physical standpoint. Likewise, the role of coatings in reduction of metal oxidation involves chemical principles similar to those associated with prevention of oxidation of biomolecules, intracellularly. Nanoparticle formulations added to such industrial catalytic systems often serve the role of a catalysis promoter or a stabilizer, to improve the efficiency of the catalyst. However, looking at this issue from a cross-disciplinary standpoint, the actions of enzymes and cofactors involved in biochemical reactions can be defined in a similar manner, in that they act at catalysis promoters that are unchanged by the reaction they catalyze. From an engineering standpoint, nanotechnology has provided dramatic improvement in industrial applications related to efficiency of redox reactions. Cross-disciplinary application of these processes to cellular redox reactions occurring on the nano-scale intracellularly may provide a new basis for pharmacological treatment of many diseases related to oxidative stress.

Three of the most-studied nanoparticle redox reagents, at the cellular level, are rare earth oxide nanoparticles (particularly cerium), fullerenes and their derivatives, and carbon nanotubes. In the following sections, we discuss the chemistry and biological properties of these nanoparticles and their potential use in preservation of cellular redox status and treatment of disease.

10.3.1
Rare Earth Oxide Nanoparticles – Cerium

Rare earth oxides of cerium, neodymium, praseodymium, and ruthenium have been utilized in three-way catalysts for enhancement of fuel combustion, removal of soot from engine exhaust, removal of environmental free radical contaminants such as NO and SO from fuel exhaust, removal of organics from wastewater, promoters of catalysis in environmental clean up, and in development of fuel cell technologies [75]. These applications are all related to the high redox capabilities of rare earth oxides; that being their ability to serve as free radical scavengers. Out of the lanthanide series of elements, oxides of cerium (ceria, CeO_2) nanoparticles are reported to have several unique properties that make them highly efficient redox reagents (Fig. 10.1). To date, on the intracellular level, ceria has been the most-studied of this nanoparticle group.

As a rare earth element of the lanthanide series, cerium oxide has a characteristic fluorite lattice structure. Lanthanides exist in the earth's crust at about 100 ppm,

Fig. 10.1. Electron micrograph of cerium oxide nanoparticles.
(Courtesy of Nanophase Inc., Romeoville, Illinois.)

with cerium making up approximately 24 ppm [63]. However, cerium has several unique properties among the rare earths, established by examining microcrystalline cerium oxide structures [65]. First, cerium has two partially filled subshells of electrons, 4f and 5d, with 14 excited sub-states of cerium predicted [66]. The cerium atom can exist in either the +3 (fully reduced) or +4 (fully oxidized) state, and may readily flip-flop between the two in a redox reaction [66–68]. Both the crystal lattice size and the bond length change when alterations in redox state occur. It is hypothesized that cerium oxides make excellent oxygen buffers with high oxygen storage capacities, because of their high redox capabilities [64]. In addition to alterations in cerium redox state, cerium oxide also exhibits oxygen vacancies, or defects, in the lattice structure, by loss of oxygen and/or its electrons [66, 67]. Formation of each defect removes a subgroup of ions from the lattice structure. Thus, creation and annihilation of oxygen vacancies occurs in the cerium oxide lattice during redox reactions [65]. Cerium and its oxides are also unique, among the rare earths, in that there is a high hydrogen absorbing capacity on the surface, providing for ease of reactions with H_2, O_2, or H_2O, as compared with other rare earths.

The valence and defect structure of cerium oxide is dynamic and may change spontaneously or in response to physical parameters such as temperature, pH, presence of other ions, and oxygen partial pressure [67–69]. Several groups have examined the use of cerium oxide particles as metal coatings to reduce oxidation and as coatings for catalytic converters, to enhance oxidation of carbon monoxide and hydrocarbons, and reduce nitrogen oxide emissions [70, 71]. A primary role of cerium oxide under these conditions is to act as an oxygen storage and release component [69]. The chemical reactions of cerium oxide have been reviewed [68, 72–75] and include oxygen atom transfer and absorption, oxidation of unsaturated hydrocarbons, electron transfer to hydrocarbon radicals, high catalytic activity in redox reactions, and reduction of nitrogen oxide. One cannot avoid noting the similarities between these uses and the properties necessary for a good cellular antioxidant. Thus, the chemistry of cerium oxide further supports a potential role as either a biological free radical scavenger or antioxidant.

The atomic and molecular properties of materials at the nanometer scale change in ways we are just beginning to comprehend. The chemical behavior of cerium oxide nanoparticles may reflect the chemical properties of its macromolecular structure to some extent, but altered characteristics are likely at the nanoscale. For example, Raman spectra of cerium oxide nanoparticles are very different from microparticles [68]. Creation and annihilation of oxygen vacancies is reported to occur more rapidly in nano-ceria, making it a more efficient redox center in industrial applications. Guo and Waser have reported that the electrical conductivity of nanocrystalline CeO_2 is several orders of magnitude higher than its microcrystalline form [76]. Several other groups report that nanocrystalline cerium oxide shows enhanced electrical conductivity as grain size decreases below 100 nm, due to a decrease in the enthalpy of oxygen defect formation [67–69]. Thus, oxygen vacancies are likely to form more readily at the nanoscale, increasing the number of sites available for free radical scavenging. Additionally, the surface area of cerium oxide particles (grain boundary) is dramatically increased at the nanoscale, so greater oxygen exchange and redox reactions may occur over the increased surface area [71]. It is precisely these proposed alterations in cerium oxide nanoparticles that may impart cellular biological effects. Several reports indicate that ceria and its oxides are the only rare earth elements capable of undergoing redox transformations at ambient temperatures, such as those encountered in the intracellular environment [77, 78]. However, the chemical properties of cerium oxide nanoparticles under physiologically relevant parameters remain unknown, as most studies to date utilize temperatures and pressures well above the physiological range. Thus, there is a distinct need to further study the chemistry and physics of ceria nanoparticles at physiological temperatures and in medium with ionic composition similar to the intracellular and interstitial fluid environment.

10.3.1.1 Biological Effects of Cerium

Biologically, cerium has been utilized most often in the form of nitrates or chlorides of cerium. In China, cerium chlorides and nitrates are used in fertilizers, to increase harvest [74]. Aged rice seed treated with cerium nitrate showed an enhanced respiratory rate and increased activities of superoxide dismutase, catalase, and peroxidase, enzymes associated with reduction of oxidative stress and free radicals [79].

Cerium has been found in humans and other animals, with amounts varying from several parts per billion to parts per million. In a feeding study conducted in pigs, feed containing rare earth elements (38% La, 52% Ce, 10% other) increased weight gain and feed conversion ratios [80], thereby enhancing biomass production. No negative health effects were observed. Serum cholesterol, triglycerides, total protein, albumin and ionic composition were normal and concentration of rare earths in the tissues was below 52 µg per kg dry weight. Cerium nitrate has been used in treatment of burns, and improved the outcome of burn patients [81, 82]. Improved outcome was associated with a reduction in the systemic inflammatory response observed in burn victims and improved capacity for tissue repair. Rare earths in general have potent anti-atherosclerotic, anti-arthritic, and anti-inflammatory activity. Histologically, the chloride form of cerium has been used to identify subcellular sites of free radical (superoxide, H_2O_2) production [83, 84]. It has also been used to study the location of xanthine/xanthine oxidase in rat liver cells, an enzyme associated with free radical production [22]. Although these studies involve the use of nitrates and chlorides of cerium, they support a distinct role for cerium in interaction with cellular free radicals.

From the few existing reports, cerium is relatively inert from the biological standpoint of toxicity – but reports are scarce and do not address oxides of cerium, much less nanoparticles. However, some negative biological effects have been reported, particularly at high concentrations. One report of inhalation pneumoconiosis was identified, occurring in a lens grinder exposed to prolonged high inhaled concentrations or rare earths [85]. Millimolar solutions have been shown to aggregate erythrocyte membranes *in vitro*, through disulfide bond crosslinks [86]. At high concentrations and non-physiological pH, cerium chloride served as a catalyst for DNA and peptide hydrolysis [87]. Cerium chloride given to rats at 0.2–20 mg kg^{-1} increased the oxygen affinity of hemoglobin by alteration of the secondary structure [87]. Based on its redox activities, it appears likely that cerium interacts with proteins such as hemoglobin and cytochrome P450, which contain heme-iron centers at the active site. However, once again, these reports represent the activity of cerium in the chloride or nitrate form, at millimolar concentrations. The biological properties of cerium oxide nanoparticles are likely to substantially differ.

10.3.1.2 Biological Effects of Cerium Oxide Nanoparticles

The previous section describes the biological effects of cerium in general; however, cerium oxide nanoparticles, although having some similar properties, appear to be quite different on the whole. At the cellular level, cerium oxide nanoparticles have several potent properties, notably at extension of cellular longevity and as regenerative antioxidants.

In several studies, the Rzigalinski laboratory has demonstrated that nanoparticles of cerium oxide, of the size range of 6–20 nm, prolong the lifespan of mixed brain cell cultures, by 6–8-fold [88–90]. Ceria nanoparticles of greater than 20 nm were without effect at prolongation of life span, as were ruthenium, titanium, and neodymium oxide nanoparticles. In these experiments, cerium oxide nanoparticles were added directly to the tissue culture medium on day 10 *in vitro*, allowed to remain in the medium for 48 h, followed by replacement with fresh medium. Thus, a 48 h period was allowed for particle uptake. The optimal particle dose was 10 nM in the tissue culture medium. Figure 10.2 shows representative light micrographs of mixed brain cell cultures treated with cerium oxide. Importantly, neurons present in these aged cultures demonstrated signaling characteristics similar to their younger, untreated counterparts, demonstrating that normal neuronal function was also preserved in ceria nanoparticle-treated cultures [88–90].

Based on the industrial uses of ceria nanoparticles in redox reactions, it was hypothesized that the cellular effects of ceria nanoparticles were due to a free radical scavenging activity. To examine the mechanism of action further, cell cultures were exposed to free radical generating conditions of UV light, γ-radiation, H_2O_2, and trauma [88, 90–93]. Cultures treated with ceria nanoparticles showed dramatic resistance to these forms of oxidative stress, and survival exceeded that observed with the biological free radical scavengers vitamin E, melatonin, and *n*-acetyl cysteine.

Of critical importance is the dosing regime used in these studies. First, cultures were only treated once, on day 10 *in vitro*. At this time in culture, a confluent astrocyte monolayer covers the bottom of the tissue culture well, with neurons, microglia, and oligodendrocytes loosely attached to the astrocyte layer. Little cell division occurs at this time, and most of the cells in culture are terminally differentiated. A single dose of ceria nanoparticles was responsible for the dramatic extension of lifespan observed in these cultures. Further, a single dose on day 10 demonstrated enhanced protection against oxidative stressors through the extended lifespan of the cultures. This suggests that the antioxidant capacity of ceria nanoparticles in the cell may be, unlike many free radical scavengers, regenerative. The known

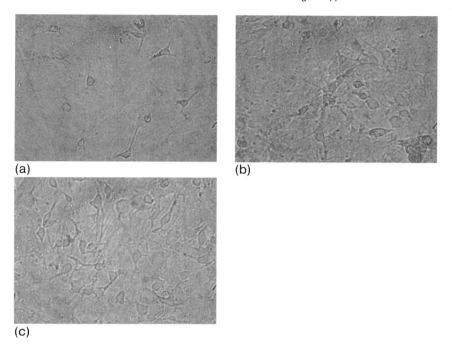

(a)

(b)

(c)

Fig. 10.2. Light micrograph of control and nanoparticle treated rat cortical mixed cell cultures. (A) 27-day-old organotypic culture near the end of its *in vitro* life span. Note the lack of confluent astrocyte monolayer in the background and low numbers of remaining, processed neurons. (B) 27-day-old culture treated with 10 nM cerium oxide nanoparticles once, on day 10 *in vitro*. Note the healthy, confluent astrocyte monolayer and abundant neurons with robust processes. (C) 123-day-old culture treated with 10 nM cerium oxide nanoparticles on day 10 *in vitro*. Total magnification 800×.

chemistry of ceria nanoparticles appears to support this hypothesis. Ceria undergoes rapid, reversible reduction and oxidation and can readily take up and release oxygen, alternating between CeO_2 and CeO_{2-x} [64, 68, 94]. Therefore, it is a redox cycling agent that does not in itself generate free radicals in the process. Based on reports from the materials science field, we can only speculate on the actions of ceria nanoparticles in the physiological milieu. In aqueous medium, it appears likely that formation of adsorbed OH groups occurs on the surface of the ceria, with a concomitant reduction of the Ce atom from +4 to +3 (Fig. 10.3). This change in cerium valence alters the structure of the oxide lattice, creating oxygen vacancies and expansion of the lattice [95, 96]. This electron shuffling in the lattice, along with the oxygen vacancies generated, provide redox potential for free radical scavenging [76, 97]. After the scavenging event, the original lattice structure may be regenerated by H_2O and the cerium atom returned to the +4 state [98]. However, we would point out that the precise mechanisms responsible for lattice regeneration in the intracellular environment can only be speculated (Fig. 10.3). Thus,

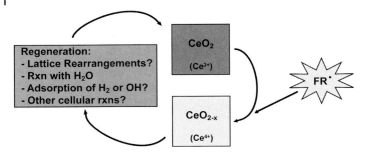

Fig. 10.3. Hypothesized intracellular mechanism(s) of action of ceria nanoparticles (see text for description). FR = free radical.

the redox properties of ceria appear to make it possible for the nanoparticles to both oxidize and reduce molecules, depending on the conditions in the surrounding milieu and the level of oxidative stress.

Why then, are micro-sized ceria particles ineffective or weakly effective as biological free radical scavengers? First, and probably most obvious, micro-ceria is unlikely to enter the cell due to its size. As shown in Fig. 10.4, nano-ceria readily enter mixed brain cell cultures and other cells [99], and hence are able to carry out redox reactions. Figure 10.4 shows brain macrophage or microglia that has phagocytosed one nanoparticle, with possibly a second in the process of consumption. Figure 10.5 is an electron micrograph taken from mixed brain cell cultures, demonstrating the presence of nanoparticles in the cytoplasmic space. Second, nano-ceria has a dramatically increased surface area as compared with micro-sized particles, thereby increasing the available redox sites. Last, nano-ceria is capable of redox behavior at far lower temperatures than its micro-sized counterparts, and has more oxygen vacancies for scavenging electrons, which may allow for redox cycling in physiological environs [96, 98, 100].

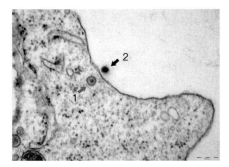

Fig. 10.4. Electron micrograph of intracellular cerium oxide. Note that one particle appears to already have entered the cell (first arrow), while a second may be in the early stages of phagocytosis (second arrow).

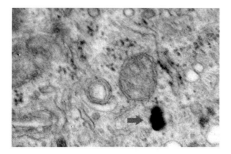

Fig. 10.5. Electron micrograph of mixed cortical cultures, showing intracellular accumulation of ceria nanoparticles (arrow).

10.3.1.3 Other Oxide Nanoparticles

Materials science has utilized other oxides of the lanthanide series for redox reactions, including neodymium, ruthenium, praseodymium, lanthanum, and titanium. In many cases, ceria nanoparticles are doped with these compounds to enhance redox activity or stability. Although pure nano-oxides of these rare earths were without the life extension and radical scavenging abilities of ceria, mixtures of ceria-doped nanoparticles remain to be further examined for biological effects. The stability of ceria nanoparticles within the cell and the persistence during repetitive challenges of oxidative stress are presently unknown, and dopants may improve the intracellular half-life. Toxicities of these oxides have not been noted, with the exception of ruthenium, which can interfere with mitochondrial function [101]. However, there remains much work to be done regarding biological properties of ceria-doped nanoparticles.

Nanoparticles of zinc oxide and iron oxide have also received much attention at the biological level. Zinc oxide nanoparticles are routinely used as UV-absorbing agents in sunscreens. Their effects as intracellular antioxidants and lifespan extenders remain to be examined. Iron oxide nanoparticles have several emerging applications for drug delivery and biomedical imaging. However, problematic issues arise in that iron oxides may produce free radicals via the Fenton reaction, and hence may have damaging effects on cells. In this case, the addition of ceria nanoparticles to iron oxide nanoparticle preparations utilized in drug delivery may serve to abrogate deleterious effects of Fenton-reaction generated intracellular free radicals.

10.3.1.4 Fullerene Derivatives and Carbon Nanotubes

Ceria nanoparticles are not the only nanoparticles that have future potential as free radical scavengers. Although experiments on cell lifespan are lacking, numerous reports suggest that fullerenes and carbon nanotubes may also act intracellularly as potent free radical scavengers. Corona-Morales et al. [102] have demonstrated that the water soluble carboxylic acid derivative of fullerene, carboxyfullerene, prolonged survival of adrenal chromaffin cells exposed to the free radical generating neurotransmitter dopamine. In a series of papers, Dugan et al. [103–105] showed that carboxyfullerenes acted as free radical scavengers and dramatically reduced neuronal death induced by excitotoxic doses of NMDA, amyloid beta peptide, and H_2O_2, suggesting a role in neuroprotection. Carboxyfullerenes out-performed several other groups of free radical scavengers, including 21-aminosteroid compounds, vitamin E analogs, and spin trap agents. The extensive double bond system in fullerenols has led Krusic et al. to characterize them as a "free radical sponge" [106]. In rats, carrying the mutant SOD gene responsible for amyotropic lateral sclerosis, carboxyfullerenes infusion delayed both functional deterioration and death [105]. Additionally, carboxyfullerene prevented the iron-induced oxidative injury in the nigrostriatal dopaminergic system of rats by acting as an antioxidant and suppressing free radical damage [107], suggesting potential use in Parkinson's disease-like disorders. The mechanism of action of fullerenes has been hypothesized by Jain [108], in which a free radical forms a bond with fullerene creating a stable, nonreactive fullerene radical. Taken together, these reports suggest that like ceria nanoparticles, fullerene derivatives hold great potential in treating oxidative-stress related degenerative conditions, such as those associated with aging. Their role in aging, however, remains to be investigated.

The free radical scavenging and generating activities of carbon nanotubes have also been a subject of great debate. Unpurified single-walled carbon nanotubes (SWCNT) were shown to stimulate production of free radicals in human keratinocytes and bronchial epithelial cells [109, 110]. However, the free radical generating activities of SWCNT appeared to be due, in part, to metal contamination during particle synthesis. Muller et al. [111] reported that multi-walled carbon nanotubes (MWCNT) induced inflammatory and fibrotic reactions in rat lung after intratracheal instillation. They hypothesized that the mechanism of action was via free radical generation. In a recent subsequent study, Fenoglio et al. [112] directly examined the effect of MWCNT on free radical generation via electron spin resonance and found that MWCNT did not generate free radicals in physiological buffers. In contrast, they found that MWCNT were excellent scavengers of free radicals, including hydroxyl and superoxide radicals. In support of a role for CNT as free radical scavengers, one must consider that functionalization and polymer grafting onto CNT involves reactions similar to radical addition to the carbon framework [113, 114].

10.4
Nano-pharmacology

Substantial evidence suggests that nanoparticles comprise a class of materials that may be potent antioxidants and intracellular redox reagents. However, treatment at the organismal level is confounded by the need for establishing the behavior of nanoparticles within the pharmacological parameters of ADME – absorption, distribution, metabolism, and excretion. Given the novelty of nanotechnology and the requirement for alternative ways of thinking about chemistry and physics in the realm of the very small, traditional pharmacological parameters will no doubt require additional adjustments.

10.4.1
Absorption

For a nanoparticle to be utilized in pharmacotherapy, absorption into the system is generally one of the first hurdles to be met. Several reviews address the study of nanoparticle-assisted drug delivery [115, 116] but much work remains in this realm, particularly for rare earth nanoparticles such as ceria and carbon derivatives such as CNT and fullerene compounds. It has been well established that various nanoparticles can enter an organism via skin absorption, digestive tract, and parenteral administration. Many even cross the blood–brain barrier. Our laboratory has conducted experiments injecting ceria nanoparticles into the tail vein of the mouse (Fig. 10.6). We found that injection of a 90 nmol total over 3 days via tail vein resulted in an increased accumulation of tissue cerium in brain, heart, and lung at 3 months post injection (Fig 10.6). Interestingly, these three tissues are the most highly oxidative organs of the body. During the 3 month post injection period, no overt toxicological effects were noted in these experiments; however, histological assessments were not conducted (most of the tissue available was utilized to mea-

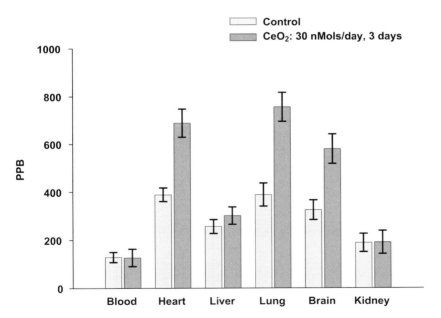

Fig. 10.6. Distribution of cerium in mouse tissues. Mice were injected with CeO_2 nanoparticles in saline via the tail vein (10 µL, 3 mM stock), every other day for a total of three injections. Mice were sacrificed 3 months post injection. Tissues were collected, ashed, and analyzed for cerium content by inductively coupled mass spectrometry.

sure tissue cerium levels). As measured by ashing followed by inductively coupled plasma mass spectrometry, tissue cerium levels were approximately doubled, but remained in the ppb range (Fig. 10.6). Six months post injection, tissue cerium levels were similar. These studies suggest that ceria nanoparticles are readily absorbed by tissues and tend to accumulate and persist in highly oxidative areas. One particular difficulty in administration was noted during intravenous injection in that injection of stock solutions higher than 3 mM appeared to agglomerate at the site of injection, blocking circulation and resulting in necrosis at the tail. Thus, agglomeration may be problematic in nanoparticle delivery. Additionally, notably, the long-term effects of accumulated ceria have not yet been examined and metabolic parameters such as excretion in urine and feces are unknown.

10.4.1.1 **Agglomeration**

Agglomeration appears to be a primary issue in dosing, as mentioned above. Ceria, as well as other nanoparticles, have a high propensity for agglomeration, particularly in physiological medium. Agglomeration increases particle size and may block cellular or tissue uptake mechanisms. At the organismal level, systemic agglomeration of nanoparticles may block capillary circulation, lymphatic vessels, alveoli, or kidney perfusion. Thus, the use of dilute, well dispersed solutions of nanoparticles is critical. In our studies, care was taken to eliminate pre-administration agglomeration of nano-ceria by avoiding phosphate buffers and the use of rigorous sonication prior to delivery [88]. Poorly dispersed ceria solutions or solutions prepared in phosphate buffer, result in poor intracellular delivery of nanoparticles. However, as can be seen in Figs. 10.4 and 10.5, it appears likely that some agglomeration occurs in the cells themselves, since the particle size in electron micrographs appears to be somewhat larger than that in the stock solutions and intracellular particles shown in Fig. 10.5 appear to be associated. In lung cells, Stark et al. also found agglomerates of ceria [99], which appeared to be localized to vacuole-like structures.

In addition to agglomeration, contaminants from synthesis of nanoparticle preparations may also cause a range of effects and must be carefully controlled in cellular and animal studies. Metal contaminants from synthesis of carbon nanotubes may promote free radical generation and toxic effects. Likewise, tailing surfactant in nano-ceria preparations synthesized via micro-emulsion techniques have also proved damaging to the cell. Therefore, an assessment of preparation purity is a critical step that must be considered prior to biological use.

10.4.1.2 **Dose**

Dose is also a parameter that needs to be addressed in a somewhat non-traditional pharmacological manner. Traditional pharmacology generally expresses dose in a molar or mg kg^{-1} manner. However, nanoparticles are engineered "mini-reactors", and dose via mass or molar number may be an inappropriate descriptor, since each individual particle appears to be an entity unto itself. For example, in nano-ceria or fullerenes, the radical scavenging effect is primarily due to surface area reactions, which are enhanced at the nano-scale. Based on this premise, Oberdorster et al.

[117] propose that dose be described in terms of surface area, rather than mass. Other groups have utilized the descriptor of number of nanoparticles delivered per cell for *in vitro* experiments, which works well for nanoparticles of discrete size. Until further experimentation provides adequate guidelines, both these types of dose descriptors appear useful for experimental analysis and comparison. In the cases of carbon nanotubes, additional parameters of particle length and circumference need to be considered.

10.4.2
Distribution, Metabolism, and Excretion

Distribution, metabolism, and excretion are parameters that also must be addressed as nanotechnology moves into the realm of conventional pharmacology. This area is one in which research is decidedly scarce. Studies with ultrafine particulates have shown that inhaled doses reach the systemic circulation and penetrate the blood–brain barrier, and this scenario is likely for many forms of nanoparticles [118, 119]. However, progress in this area has been hindered by our lack of ability to detect nanoparticle distribution and excretion. In the case of fullerenes, chemical detection in tissues is confounded by the fact that carbon is a ubiquitous backbone of all organic molecules. With cerium oxide and other rare earth nanoparticles, assessment of distribution is hindered by our ability to extract cerium oxide from tissues and the lack of chemical assessment of its presence. Of course, it is certainly possible to measure rare earths in tissue by inductively coupled mass spectrometry, but this only measures the rare earth element itself, and not the presence of oxides or other derivatives. Current research directed to the use of fluorescent tags may improve our ability to track the distribution of nanoparticles in living cells and tissues. Additionally, nanoparticles synthesized with heavy metal cores may permit localization via CT or MRI in the whole animal.

10.5
Nanoparticle Antioxidants and Treatment of Disease

Research to date suggests that nanoparticles may provide the field of medicine with effective, long lasting antioxidants for the treatment of diseases associated with free radical production (Table 10.1). Neurodegenerative disorders such as Alzheimer's disease, Parkinson's disease, multiple sclerosis, amyotropic lateral sclerosis, and traumatic and ischemic brain injury all have free radical mediated components [1, 3, 5, 6, 120, 121]. Efficient nanoparticle free radical scavengers may hold great potential in abrogating the functional deficits associated with many of these neurological disorders. Using an *in vitro* model for traumatic brain injury [122–125], our group has shown that pre-treatment of brain cell cultures with ceria nanoparticles prior to injury, reduces cell death and neuronal dysfunction after injury [91, 92]. Importantly, ceria nanoparticles were effective at reducing neuronal death and dysfunction even when delivered up to 3 h post injury – an effect not reproduced by other free radical scavengers. These effects were due, in part, to a reduction in the inflammatory reactions in brain microglia [92]. Other work also supports a role for fullerene derivatives in mediation of free radical damage in the nervous system [102–106].

Cardiovascular disorders are another realm in which nanoparticle free radical scavengers may hold great promise. Development of atherosclerotic lesions is promoted by free radical-mediated vascular damage and free radical generation to the vessel wall increases and promotes the subsequent inflammatory processes. Hence, effective administration of nanoparticle antioxidants may blunt this process, particularly in susceptible populations. Percutaneous transluminal angioplasty (PTCA) is a widely used procedure in treatment of acute and chronic ischemic heart disease. One of the most serious and reoccurring complications of PTCA is re-stenosis, leading to further blockage and occurrence of new ischemic events. The pathophysiology of re-stenosis is multifactorial and consists of inflammation, smooth muscle cell migration, proliferation, extracellular matrix remodeling, and free radical damage. Poly(ethylene glycol)-based core shell nanoparticles (NK911) loaded with the antineoplastic drug doxorubicin were able to prevent re-stenosis after balloon angioplasty in animal models [126]. It appears the core shell nanoparticles directed the doxorubicin to the damaged vascular epithelium. Incorporation of free radical scavenging nanoparticles into the core shell of such preparations may lead to further enhancement of endothelial function and reduction in future damage. Similarly, the use of coronary stents in treatment of chronic and acute coronary occlusion is also complicated by the occurrence of in-stent re-stenosis [127] and is a clinical problem treatable only by coronary artery by-pass. Stent re-stenosis is a result of continuing inflammatory reactions within the vascular walls, of which free radical production is an important component. Coating of vascular stents with ceria or other nanoparticles may aid in reducing inflammation around the site of stent insertion and decrease the incidence of re-stenosis.

Oxidative stress and free radical production are also a primary component of inflammatory disorders such as arthritis. Accumulation of activated inflammatory cells within the arthritic joint release high levels of free radicals that promote destruction of tissue within the joint. An as-yet unexplored possibility lies in the use of ceria and other antioxidant nanoparticles in reduction of damage to the arthritic joint, by direct delivery to the joint itself or via coating of implantable materials. In support of this potential usage, our laboratory has found that treatment of activated microglia and macrophages with ceria nanoparticles reduces release of free radicals from these cells and protects bystander cells from damage from activated immune cells [88, 92, 93]. Further, work by our collaborators, Drs. S. Olgun and C. Reilly, suggests that treatment of lupus mice with ceria nanoparticles decreases disease severity and inflammatory cell activation (Olgun, Reilly and Rzigalinski, unpublished results).

Of importance to us all is the association between oxidative stress and aging and age-related disorders. Can nanoparticles be the fountain of youth long sought by Ponce de Leon? Although promising, the utility of nanoparticles in retarding aging are far more complex than to venture an answer to such a question. Ceria nanoparticles, and possibly others, certainly hold promise for extension of lifespan and ab-rogation of age related disorders. However, much work is necessary to move the current work in cellular studies, to the level of organism longevity. In preliminary studies in the fruit fly, we have found a 30% increase in lifespan in flies fed cerium oxide nanoparticles, dosed at 10–100 nM in the food. However, doses above 1 mM appeared to result in a decrease in lifespan. Hence, total cumulative dose to the or-ganism is likely to be an important parameter. Also, the effect of dosing through-out the lifespan vs. dosing for a limited time period needs to be examined. Al-though excessive free radical production contributes to cell damage, free radicals, as discussed in Section 10.1, also have beneficial effects on survival – particularly in the vascular and nervous systems. Thus, blockade of free radical production may reach the level of "too much of a good thing". For example, NO radicals are critical to maintenance of vascular tone and endothelial function. Satoh et al. [128] have shown that fullerene nanoparticles interfere with NO-mediated vasodilation and may have deleterious effects on the vasculature at high concentrations. For ceria nanoparticles, the vasoactive actions of NO may also be rendered non-functional at certain concentrations, resulting in disruption of the normal vascular response *in vivo*. Thus the question arises as to how much free radical scavenging is too much, and at which point these highly efficient nanoparticle scavengers in-terfere with normal cell function. Further engineering of nanoparticle structure may provide a solution to this problem. However, much work is needed to assess these parameters.

An additional paradox arises when one considers the increased incidence of neo-plasia with age. If resistance to free radical damage is conferred to normal cells by nanoparticles, might not cancerous cells or dysfunctional cells destined for apopto-sis also receive a degree of protection when nanoparticles are delivered? Such events might potentially counteract the tumor killing effects of radiation or chemo-therapy, both of which may act through free-radical mediated cell death. Alterna-tively, if such nanoparticles could be delivered to healthy cells surrounding the site of a radiation dose, they may afford protection to healthy bystander cell injury associated with radiation therapy. Again, further engineering and targeting of nanoparticles may provide solutions to these issues. Nonetheless, such avenues re-main to be explored before the full medicinal capabilities of nanoparticles can be put to use.

10.6
Toxicology

Despite the beneficial potential of antioxidant nanoparticles in the treatment of dis-ease, there remains a major hurdle that must be surpassed before progress can be made in this area. Toxicological data is seriously lacking. Presently, there are only scant reports on the effects of nanoparticles in general on organism metabolism and function. As discussed in Section 10.3, some nanoparticles appear to have the potential to induce free radical formation and damage, which may be due to contaminants from synthesis procedures [109, 110, 119]. Several studies have reported that SWCNT and MWCNT induce granulomas and fibrosis in the lung during intratracheal or inhalation exposure [110, 129]. Oberdorster et al. [130] have reviewed the potential toxicological effects of fullerenes as free radical generating agents, particularly after exposure to light. Kamat et al. [131] also report significant lipid peroxidation due to photosensitized C60. Hussain et al. have reported that oxide nanoparticles of molybdenum, aluminum, iron, and titanium were toxic, albeit at high doses [132]. Ueng et al. [133] demonstrated that certain fullerenol derivatives suppressed microsomal enzymes and mitochondrial oxidative phosphorylation *in vitro*, although the compounds were able to scavenge free radicals in brain tissue [134, 135]. Chen et al. report that SiO_2 nanoparticles were taken up into the cell nucleus, where they caused aberrant clustering of nuclear proteins and inhibited replication and transcription [136]. Thus, although highly promising, antioxidant nanoparticles are certainly not a magic bullet as yet and the road ahead is likely long – but equally likely one worth traveling.

10.7
Summary

Free radicals and oxidative stress are associated with a broad spectrum of human diseases. Despite their prominence in pathological conditions, antioxidant therapy has met with only limited success. Nanoparticle technology encompassing rare earth oxides, fullerene derivatives, and carbon nanotubes show great promise in antioxidant therapy in a myriad of disease conditions. However, much work is needed to fully realize their capacity in nanomedicine.

References

1 BECKMAN, K.B., AMES, B.N. The free radical theory of aging matures, **1998**, *Physiol. Rev.* 78, 547–581.

2 COMPORTI, M., SIGNORINI, C., BUONOCORE, G., CICCOLI, L. Iron release, oxidative stress, and erythrocyte aging, **2002**, *Free Rad. Biol. Med.* 32, 568–576.

3 DROGE, W. Free radicals in the physiological control of cell function, **2002**, *Physiol. Rev.* 82, 47–95.

4 SPITELLER, G. Are changes of the cell membrane structure causally involved in the aging process? **2002**, *Ann. New York Acad. Sci.* 959, 30–44.

5 MARKESBERY, W.R. Oxidative stress hypothesis in Alzheimer's Disease, **1997**, *Free Rad. Biol. Med.* 23, 134–147.

6 MASHIMA, R., WITTING, P.K., STOCKER, R. Oxidants and antioxidants in atherosclerosis, **2001**, *Curr. Opin. Lipidol.* 12, 411–418.

7 CARMODY, R.J., COTTER, T.G. Signalling apoptosis: A radical approach, **2001**, *Redox. Rep.* 6, 77–90.

8 PRATICO, D., LAWSON, J.A., ROKACH, J., FITZGERALD, G.A. The isoprostanes in biology & medicine, **2001**, *Trends Endocrinol. Metab.* 12, 243–247.

9 PETERSEN, D.R., DOORN, J.A. Reactions of 4-hydroxynonenal with proteins and cellular targets, **2004**, *Free Rad. Biol. Med.* 37, 937–945.

10 ESTERBAUER, H., SHAUR, R.J., ZOLLNER, H. Chemistry and

biochemistry of 4-hydroxynonenal, malonaldehyde and related aldehydes, **1991**, *Free Rad. Biol. Med.* 11, 81–128.

11 MONTUSCHI, P., BARNES, P.J., ROBERTS, L.J. II. Isoprostanes: Markers and mediators of oxidative stress, **2004**, *FASEB J.* 18, 1791–1800.

12 KIM, D., WON, S., GWAG, B. Analysis of mitochondrial free radical generation in animal models of neuronal disease, **2002**, *Free Rad. Biol. Med.* 33, 715–726.

13 SOHAL, R.S., SOHAL, B.H. Hydrogen peroxide release by mitochondria increases during aging, **1991**, *Mech. Aging Develop.* 57, 187–202.

14 FRIDOVICH, I. Fundamental aspects of reactive oxygen species, or what's the matter with oxygen?, **1999**, *Ann. New York Acad. Sci.* 893, 13–18.

15 LENAZ, G., BOVINA, C., D'AURELIO, M., FATO, R., FORMIGGINI, G., GENOVA, M.L., GIULIANO, G., PICH, M.M., PAOLUCCI, U., CASTELLI, G.P., VENTURA, B. Role of mitochondria in oxidative stress and aging, **2001**, *Ann. New York Acad. Sci.* 959, 199–213.

16 MATTSON, M.P. Modification of ion homeostasis by lipid peroxidation: Roles in neuronal degeneration and adaptive plasticity, **1998**, *Trends Neurosci.* 21, 53–57.

17 AZBIL, R.D., MU, X., BRUCE-KELLER, A.J., MATTSON, M.P., SPRINGER, J.E. Impaired mitochondrial function, oxidative stress and altered antioxidant enzyme activities following traumatic spinal cord injury, **1997**, *Brain Res.* 765, 283–290.

18 TORTORELLA, C., PIAZOLLA, G., ANTONACI, S. Neutrophil oxidative metabolism in aged humans: A perspective, **2001**, *Immunopharmacol. Immunotoxicol.* 23, 565–572.

19 BRAND, M.D., AFFOURTIT, C., ESTEVES, T.C., GREEN, K., LAMBERT, A.J., MIWA, S., PAKAY, J.L., PARKER, N. Mitochondrial superoxide: Production, biological effects, and activation of uncoupling proteins, **2004**, *Free Rad. Biol. Med.* 37, 755–767.

20 SMITH, W.L. Prostanoid biosynthesis and mechanisms of action, **1991**, *Am. J. Physiol.* 263, F181–F191.

21 FREDERIKS, W.M., VREELING-SINDELAROVA, H. Ultrastructural localization of xanthine oxidoreductase activity in isolated rat liver cells, **2002**, *Acta Histochem.* 104, 29–32.

22 ESPLUGUES, J.V. NO as a signaling molecule in the nervous system, **2002**, *Br. J. Pharmacol.* 135, 1079–1095.

23 HARMAN, D. Aging: Overview, **2001**, *Ann. New York Acad. Sci.* 928, 1–21.

24 MARCHIOLI, R., SCHWEIGER, C., LEVANTESI, G., TAVAZZI, L., VALAGUSSA, F. Antioxidant vitamins and prevention of cardiovascular disease: Epidemiological and clinical trial data, **2001**, *Lipids* 36, S53–S63.

25 KAUL, N., DEVARAJ, S., JIALAL, I. Alpha-tocopherol and atherosclerosis, **2001**, *Exp. Biol. Med.* 226, 5–12.

26 MELDRUM, B.S. Implications for neuroprotective treatments, **2002**, *Prog. Brain Res.* 135, 487–495.

27 MAXWELL, A.J. Mechanisms of dysfunction of the nitric oxide pathway in vascular diseases, **2002**, *Nitric Oxide* 6, 101–124.

28 ABRAMSON, S.B., AMIN, R.A., CLANCY, R.M., ATTUR, M. The role of nitric oxide in tissue destruction, **2001**, *Best Prac. Res. Clin. Rheumatol.* 15, 831–845.

29 TONKS, N.K. Redox redux: Revisiting PTP's and the control of cell signaling. **2005**, *Cell* 121, 667–670.

30 SZWEDA, P.A., FRIGUET, B., SZWEDA, L.I. Proteolysis, free radicals, and aging. **2002**, *Free Rad. Biol. Med.* 33, 29–36.

31 CARNEY, J.M., STARKE-REED, P.E., OLIVER, C.N., LANDUM, R.W., CHENG, M.S., WU, J.F., FLOYD, R.A. Reversal of age-related increase in brain protein oxidation, decrease in enzyme activity, and loss in temporal and spatial memory by chronic administration of the spin-trapping compound n-tert-butyl-alpha-phenylnitrone, **1991**, *Proc. Natl. Acad. Sci. U.S.A.* 88, 3633–3636.

32 SMITH, C.D., CARNEY, J.M., STARKE-REED, P.E., OLIVER, C.N., STADTMAN, E.R., FLOYD, R.A., MARKESBERY, W.R. Excess brain protein oxidation and enzyme dysfunction in normal aging and in Alzheimer's disease, **1991**,

Proc. Natl. Acad. Sci. U.S.A. 88, 10 540–10 543.

33 SOHAL, R.S. Role of oxidative stress and protein oxidation in the aging process, **2002**, *Free Rad. Biol. Med.* 33, 37–44.

34 FRIGUET, B., BULTEAU, A.L., CHONDROGIANNI, N., CONCONI, M., PETROPOULOS, I. Protein degradation by the proteasome and its implications in aging, **2000**, *Ann. New York Acad. Sci.* 908, 143–154.

35 KELLER, J.N., HUANG, F.F., ZHU, H., YU, J., HO, Y.S., KINDY, T.S. Oxidative stress-associated impairment of proteasome activity during ischemia-reperfusion injury, **2000**, *J. Cereb. Blood Flow Metab.* 20, 1467–1473.

36 AKSENOV, M.Y., AKSENOVA, M.V., BUTTERFIELD, D.A., GEDDES, J.W., MARKESBERY, W.R. Protein oxidation in the brain in Alzheimer's disease, **2001**, *Neuroscience* 103, 373–383.

37 GRUNE, T., DAVIES, K.J., Oxidative processes in aging. In: MASORO, E.J., AUSTAD, S.N. (Eds.), *Handbook of the Biology of Aging*, **2001**, San Diego, Academic Press, pp. 25–58.

38 FLOYD, R.A. Oxidative damage to behavior during aging, **1991**, *Science* 254, 1597.

39 PRATICO, D. Lipid peroxidation in mouse models of atherosclerosis, **2001**, *Trends Cardiovasc. Med.* 11, 112–116.

40 CINI, M., MORETTI, A. Studies on lipid peroxidation and protein oxidation in the aging brain, **1995**, *Neurobiol. Aging* 16, 53–57.

41 WEI, E.P., KONTOS, H.A., DIETRICH, W.D., POVLISHOCK, J.T., ELLIS, E.F. Inhbition of free radical scavengers and by cyclooxygenase inhibitors of pial arteriolar abnormalities from concussive brain injury in cats, **1981**, *Circ. Res.* 48, 95–103.

42 NAKAO, N., FRODL, E.M., WIDNER, H., CARLSON, E., EGGERDING, F.A., EPSTEIN, C.J., BRUNDIN, P. Overexpressing Cu/Zn superoxide dismutase enhances the survival of transplanted neurons in a rat model of Parkinson's disease, **1995**, *Nat. Med.* 1, 226–231.

43 GREENLUND, L.J.S., DECKWERTH, T.L., JOHNSON, E.M. JR. Superoxide dismutase delays neuronal apoptosis: A role for reactive oxygen species in programmed neuronal death, **1995**, *Neuron* 14, 303–315.

44 SKULACHEV, V.P. Why are mitochondria involved in apoptosis? Permeability transition proes and apoptosis as selective mechanisms to eliminate superoxide-producing mitochondria, **1996**, *FEBS Lett.* 397, 7–10.

45 BUTTERFIELD, D.A., DRAKE, J., POCERNICH, C., CASTEGNA, A. Evidence of oxidative damage in Alzheimer's disease brain: Central role for amyloid beta-peptide, **2001**, *Trends Mol. Med.* 7, 548–554.

46 UCHIDA, K. Cellular response to bioactive lipid peroxidation products. **2000**, *Free Rad. Res.* 33, 731–737.

47 FORMAN, L.J., LIU, P., NAGELE, R.G., WONG, P.Y. Augmentation of nitric oxide, superoxide, and peroxynitrite production during cerebral ischemia and reperfusion in the rat, **1998**, *Neurochem. Res.* 23, 141–148.

48 BONDY, S.C., LEBEL, C.P. The relationship between excitotoxicity and oxidative stress in the central nervous system, **1993**, *Free Rad. Biol. Med.* 14, 633–642.

49 HALLIWELL, B. Role of free radicals in the neurodegenerative diseases: Therapeutic implications for antioxidant treatment, **2001**, *Drugs Aging* 18, 685–716.

50 WICKENS, A.P. Aging and the free radical theory. **2001**, *Resp. Physiol.* 128, 379–391.

51 SAYRE, L.M., SMITH, M.A., PERRY, G. Chemistry and biochemistry of oxidative stress in neurodegenerative disease, **2001**, *Curr. Med. Chem.* 8, 721–738.

52 CONTESTABILE, A. Oxidative stress in neurodegeneration: Mechanisms and therapeutic perspectives. **2001**, *Curr. Top. Med. Chem.* 1, 553–568.

53 SOHAL, R.S., WEINDRUCH, R. Oxidative stress, caloric restriction, and aging, **1996**, *Science* 273, 59–63.

54 LAURER, H.L., MCINTOSH, T.K. Pharmacologic therapy in traumatic

brain injury: Update on experimental treatment strategies, **2001**, *Curr. Pharm. Des.* 7, 1505–1516.

55 HARMAN, D. Aging: a theory based on free radical and radiation chemistry, **1956**, *J. Gerontol.* 11, 298–300.

56 MARKESBERY, W.R., LOVELL, M.A. 4-Hydroxynonenal, a product of lipid peroxidation, is increased in the brain in Alzheimer's disease, **1998**, *Neurobiol. Aging* 19, 33–36.

57 PAMPLONA, R., BARJA, G., PORTERO-OTIN, M. Membrane fatty acid unsaturation, protection against oxidative stress, and maximum life span, **2002**, *Ann. New York Acad. Sci.* 959, 475–490.

58 LEE, C.K., KLOPP, R.G., WEINDRUCH, R., PROLLA, T.A. Gene expression profile of aging and its retardation by caloric restriction, **1999**, *Science* 285, 1390–1393.

59 SALVEMINI, D., RIPLEY, D.P., CUZZOCREA, S. SOD mimetics are coming of age, **2002**, *Nat. Rev. Drug Discov.* 1, 367–374.

60 LEKER, R.R., SHOHAMI, E. Cerebral ischemia and trauma – different etiologies yet similar mechanisms: Neuroprotective opportunities, **2002**, *Brain Res. Rev.* 39, 55–73.

61 DOPPENBERG, E.M.R., CHOI, S.C., BULLOCK, R. Clinical trials in traumatic brain injury: What can we learn from previous studies?, **1997**, *Ann. New York Acad. Sci.* 825, 305–322.

62 KONTOS, H.A., WEI, E.P., DIETRICH, W.D., NAVARI, R.M., POVLISHOCK, J.T., GHATAK, N.R., ELLIS, E.F., PATTERSON, J.L. JR. Mechanism of cerebral arteriolar abnormalities after acute hypertension, **1981**, *Am. J. Physiol.* 240, H511–H527.

63 EVANS, C.H. *Biochemistry of the Lanthanides*, **1990**, Plenum Press, New York.

64 DAVIS, V.T. Measurement of the electron affinity of cerium, **2002**, *Phys. Rev. Lett.* 88, 1–4.

65 LAND, P.L. Defect equilibria for extended point defects, with application to nanstoichiometric ceria, **1973**, *J. Phys. Chem. Solids* 34, 1839–1845.

66 SUZUKI, K.T., KOSACKI, I., ANDERSON, H.U. Electrical conductivity and lattice defects in nanocrystalline cerium oxide thin films, **2001**, *J. Am. Ceram. Soc.* 84, 2007–2014.

67 HERMAN, G.S. Characterization of surface defects on epitaxial CeO_2 (001) films, **1999**, *Surf. Sci.* 437, 207–214.

68 CONESA, J.C. Computer modeling of surfaces and defects on cerium dioxide. **1995**, *Surf. Sci.* 339, 337–352.

69 MAMONTOV, E. and EGAMI, T. Lattice defects and oxygen storage capacity of nanocrystalline ceria and ceria-zirconia, **2000**, *J. Phys. Chem. B* 104, 11 110–11 116.

70 SEAL, S., SHUKLA, S. Sol-gel derived oxide and sulfide nanoparticles. In: BARATON, M., NALWA, H.S. (Eds.), **2002**, *Functionalization and Surface Treatment of Nanoparticles*, Academic Press, San Diego.

71 HEINEMANN, C., CORNEHL, H.H., SCHRODER, D., DOLG, M., SCHWARZ, H. The CeO_2+ Cation: Gas-phase reactivity and electronic structure, **1996**, *Inorg. Chem.* 35, 2463–2475.

72 TSCHOPE, A., YING, J.Y., TULLER, H.L. Catalytic redox activity and electrical conductivity of nanocrystalline non-stoichiometric cerium oxide, **1996**, *Sensors Actuators, B* 31, 111–115.

73 DU, X., ZHANG, T., LI, R., WANG, K. Nature of cerium(III)- and lanthanum(III)-induced aggregation of human erythrocyte membrane proteins, **2001**, *J. Inorg. Biochem.* 84, 67–75.

74 FASHUI, H. Study on the mechanism of cerium nitrate effects on germination of aged rice seed, **2002**, *Biol. Trace Elem. Res.* 87, 191–200.

75 BERNAL, S., KASPAR, J., TROVARELLI, A. (Eds.), Recent progress in catalysis by ceria and related compounds, *Catal. Today*, **1999**, 50, 173–443.

76 GUO, X., WASER, R. Electrical properties of the grain boundaries of oxygen ion conductors: Acceptor doped zirconia and ceria, **2006**, *Progr. Mater. Sci.* 51, 151–210.

77 FRANKLIN, S.J. Lanthanide-mediated DNA hydrolysis, **2001**, *Curr. Opin. Chem. Biol.* 5, 201–208.

78 YULIATI, L., HAMAJMA, T., HATTORI, T., YOSHIDA, H. Highly dispersed Ce(III) species on silica and alumina as new photocatalysts for non-oxidative direct methane coupling, **2005**, *Chem. Commun. (Camb.)* 4824–4826.

79 HE, M.L., RANZ, D., RAMBECK, W.A. Study on the performance enhancing effect of rare earth elements in growing and fattening pigs, **2001**, *J. An. Phys. An. Nutr.* 85, 263–274.

80 DE GRACIA, C.G. An open study comparing topical silver sulfadiazine and topical silver sulfadiazine-cerium nitrate in the treatment of moderate and severe burns, **2001**, *Burns*, 27, 67–74.

81 ESKI, M., DEVECI, M., CELIKOZ, B., NISANCI, M., TUREGUN, M. Treatment with cerium nitrate bathing modulates systemic leukocyte activation following burn injurys: An experimental study in rat cremaster muscle flap, **2001**, *Burns* 27, 739–746.

82 GRABOWSKI, G.M., PAULAUSKIS, J.D., GODLESKI, J.J. Mediating phosphorylation events in the vanadium-induced respiratory burst of alveolar macrophages, **1999**, *Tox. Appl. Pharmacol.* 156, 170–178.

83 TELEK, G., SCOAZEC, J.-Y., CHARIOT, J., DUCROC, R., FELDMANN, G., ROZE, C. Cerium-based histochemical demonstration of oxidative stress in taurocholate-induced acute pancreatitis in rats: A confocal laser scanning microscopic study, **1999**, *J. Histochem. Cytochem.* 47, 1202–1212.

84 MCDONALD, J.W., GHIO, A.J., SHEEHAN, C.E., BERNHARDT, P.F., ROGGLI, V.L. Rare earth (cerium oxide) pneumoconiosis: Analytical scanning electron microscopy and literature review, **1995**, *Modern Path.* 8, 859–865.

85 DU, X., ZHANG, T., LI, R., WANG, K. Nature of cerium (III)- and lanthanum (III)-induced aggregaton of human erythrocyte membrane proteins, **2001**, *J. Inorg. Biochem.* 84, 67–75.

86 FRANKLIN, S.J. Lanthanide-mediated DNA hydrolysis, **2001**, *Curr. Opin. Chem. Biol.* 5, 201–208.

87 CHENG, Y., LI, Y., LI, R., LU, J., WANG, K. Orally administered cerium chloride induces the conformational changes of rat hemoglobin, the hydrolysis of 2,3-DPG and the oxidation of heme-Fe(II), leading to changes of oxygen affinity, **2000**, *Chem. Biol. Interac.* 125, 191–208.

88 RZIGALINSKI, BEVERLY, A. Nanoparticles & cell longevity, **2005**, *Tech. Cancer Res. Treat.* 4, 651–660.

89 BAILEY, D., CHOW, L., MERCHANT, S., KUIRY, S.C., PATIL, S., SEAL, S., RZIGALINSKI, B.A. Cerium oxide nanoparticles extend cell longevity and act as free radical scavengers, **2003**, *Nat. Biotechnol.* 14, 112.

90 CLARK, A., ELLISON, A., FRY, R., MERCHANT, S., KUIRY, S., PATIL, S., SEAL, S., RZIGALINSKI, B.A. Engineered oxide nanoparticles increase neuronal lifespan in culture and act as free radical scavengers, *Soc. for Neurosci.*, Abs. #878.2, **2003**, published online.

91 FRY, R., ELLISON, A., COLON, J., MERCHANT, S., KUIRY, S., PATIL, S., SEAL, S., RZIGALINSKI, B.A. Engineered oxide nanoparticles protect against neuronal damage associated with in vitro trauma, **2003**, *J. Neurotrauma* 20, 1054.

92 CALLAGHAN, P., COLON, J., MERCHANT, S., KUIRY, S., PATIL, S., SEAL, S., RZIGALINSKI, B.A. Deleterious effects of microglia activated by in vitro trauma are blocked by engineered oxide nanoparticles, **2003**, *J. Neurotrauma* 20, 1053.

93 CLARK, A., STRAWN, E., COHEN, C., RZIGALINSKI, B. Cerium oxide nanoparticles provide superior antioxidant activity and neuroprotection as compared with vitamin E, n-acetyl cysteine, and melatonin, *Soc. for Neuroscience*, Abs. #93.4, **2005**, published online.

94 EVANS, D.E., KU, B.K., RAMSEY, D., MAYNARD, A., KAGAN, V.E., CASTRANOVA, V., BARON, P. Unusual inflammatory and fibrogenic pulmonary responses to single-walled carbon nanotubes in mice, **2005**, *Am. J. Physiol. Lung Cell. Mol. Physiol.* 289, L698–L708.

95 Aneggi, E., Boaro, M., de Leitenburg, C., Dolcetti, G., Trovarellis, A. Insights into the redox properties of ceria-based oxides and their implications in catalysis, **2006**, *J. Alloys Compds*, 408–412, 1096–1102.

96 Trovarelli, A. *Catalysis by Ceria and Related Materials*, **2002**, Imperial College Press, London.

97 Nolan, M., Parker, S.C., Wilson, G.W. The electronic structure of oxygen vacancy defects at the low index surfaces of ceria, **2005**, *Surf. Sci.* 595, 223–232.

98 Sato, S., Takahashi, R., Sodesawa, T., Honda, N. Dehydration of diols catalyzed by CeO_2, **2004**, *J. Mol. Catal. A* 221, 177–183.

99 Limbach, L.K., Li, Y., Grass, R.N., Brunner, T.T., Hintermann, M.A., Muller, M., Gunther, D., Stark, W.T. Oxide nanoparticle uptake in human lung fibroblasts: Effects of particle size, agglomeration, and diffusion at low concentrations, **2005**, *Environ. Sci. Technol.* 39, 9370–9376.

100 Qiu, L., Liu, F., Zhao, L., Ma, Y., Yao, J. Comparative XPS study of surface reduction for nanocrystalline and microcrystalline ceria powder, **2006**, *Appl. Surf. Sci.* 252, 4931–4935.

101 Velasio, I., Tapia, D. Alteration of intracellular calcium homeostasis and mitochondrial function are involved in ruthenium red neurotoxicity in primary cortical cells, **2000**, *J. Neurosci. Res.* 60, 543–551.

102 Corona-Morales, A.A., Castell, A., Escobar, A., Drucker-Colfin, R. Fullerene C60 and ascorbic acid protect cultured chromaffin cells against levodopa toxicity, **2003**, *J. Neurosci. Res.* 71, 121–126.

103 Dugan, L.L., Lovett, E.G., Quick, K.L., Lotharius, J., Lin, T.T., O'Malley, K.L. Fullerene-based antioxidants and neurodegenerative disorders, **2001**, *Parkinsonism Rel. Disord.* 7, 243–246.

104 Dugan, L.L., Gabrielesen, J.K., Yu, S.P., Lin, T.S., Choi, D.W. Buckminsterfullerenol free radical

scavengers reduce excitotoxic and apoptotic death of cultured cortical neurons, **1996, *Neurobiol. Dis. 3*, 129–135.**

105 Dugan, L.L., Turetsky, D.M., Du, C., Lobner, D., Wheeler, M., Almli, C.R., Shen, C.K., Luh, T.Y., Choi, D.W., Lin, T.S. Carboxyfullerenes as neuroprotective agents, **1997**, *Proc. Natl. Acad. Sci. U.S.A.* 94, 9434–9439.

106 Krusic, P.J., Wasserman, E., Keizer, P.N., Morton, J.R., Preston, K.F. Radical reactions of C60, **1991**, *Science* 254, 1183–1185.

107 Lin, A.M.Y., Chyi, B.Y., Wang, S.D., Yu, H.-H., Kanakamma, P.P., Luh, T.-Y., Chou, C.K., Ho, L.T. Carboxyfullerene prevents iron-induced oxidative stress in rat brain, **1999**, *J. Neurochem.* 72, 1634–1640.

108 Jain, K.K. The role of nanobiotechnology in drug discovery, **2005**, *Drug Disc. Today* 10, 1435–1442.

109 Shvedova, A.A., Kisin, E.R., Mercer, R., Murray, A.R., Johnson, V.J., Potapovich, A.I., Tyurina, Y.Y., Gorelik, O., Arepalli, S., Schwegler-Berry, D., Hubbs, A.F., Antonini, J., Shvedova, A.A., Castranova, V., Kisin, E.R., Schwegler-Berry, D., Murray, A.R., Gandelsman, V.Z., Maynard, A., Baron, P. Exposure to carbon nanotube material: Assessment of nanotube cytotoxicity using human keratinocyte cells, **2003**, *J. Toxicol. Environ. Health* 66, 1909–1926.

110 Shvedova, A.A., Kisin, E.R., Murray, A.R., Scjweler-Berry, D., Gandelsman, V.Z., Baron, P., Maynard, A., Gunther, M.R., Castranova, V. Exposure of human bronchial epithelial cells to carbon nanotubes causes oxidative stress and cytotoxicity. In: **2004**, *Proceedings of the Society of Free Radical Research Meeting*, European Section, June 26–29, 2003, Ioannina, Greece, pp. 91–103.

111 Muller, J., Huaux, F., Heilier, J.F., Arras, M., Delos, M., Nagy, B.J., Lison, D. Respiratory toxicity of carbon nanotubes, **2004**, *Toxicol. Appl. Pharmacol.* 197, 305.

112 FENOGLIO, I., TOMATIS, M., LISON, D., MULLER, J., FONSECA, A., NAGY, J.B., FUBINI, B. Reactivity of carbon nanotubes: Free radical generation or scavenging activity?, **2005**, *Free Rad. Biol. Med.*, in the press.

113 MYLVAGANAM, K., ZHANG, L.C. Nanotube functionalization and olymer grafting: An ab initio study, **2004**, *J. Phys. Chem. B* 108, 15 009–15 012.

114 HOLZINGER, M., VOSTROWSKY, Ol, HIRSCH, A., HENRICH, F., KAPPES, M., WEISS, R., JELLEN, F. Sidewall functionalization of carbon nanotubes, **2001**, *Angew. Chem. Int. Ed.* 40, 4002–4005.

115 KALLINTERI, P., HIGGINS, S., HUTCHEON, G.A., ST. POURCAIN, C.B., GARNETT, M.C. Novel functionalized biodegradable polymers for nanoparticle drug delivery systems, **2005**, *Biomacromolecules* 6, 1885–1894.

116 BIANCO, A., KOSTARELOS, K., PRATO, M. Applications of carbon nanotubes in drug delivery, **2005**, *Curr. Opin. Chem. Biol.* 9, 674–679.

117 OBERDORSTER, G., OBERDORSTER, E., OBERDORSTER, A. Nanotoxicology: An emerging discipline evolving from studies of ultrafine particles, **2005**, *Environ. Health Persp.* 113, 823–839.

118 REJMAN, J., OBERLE, V., ZUHORN, I.S., HOEKSTRA, D. Size-dependent internalization of particles via the pathways of clathrin- and caveolae-mediated endocytosis, **2004**, *Biochem. J.* 377, 159–169.

119 DA ROS, T., PRATO, M. Medicinal chemistry with fullerenes and fullerene derivatives, **1999**, *Chem. Commun.* 663–669.

120 HOFFMAN, S.W., RZIGALINSKI, B.A., WILLOUGHBY, K.A., ELLIS, E.F. Astrocytes generate isoprostanes in response to trauma or oxygen radicals, **2000**, *J. Neurotrauma* 28, 844–849.

121 LAMB, R.G., HARPER, C.C., McKINNEY, J.S., RZIGALINSKI, B.A., ELLIS, E.F. Alterations in phosphatidylcholine metabolism of stretch-injured cultured rat astrocytes, **1997**, *J. Neurochem.* 68, 1904–1910.

122 WEBER, J.T., RZIGALINSKI, B.A., ELLIS, E.F. Traumatic injury of cortical neurons causes changes in intracellular free calcium stores and capacitative calcium influx, **2001**, *J. Biol. Chem.* 276, 1800–1807.

123 AHMED, S.M., WEBER, J.T., LIANG, S., WILLOUGHBY, K.A., SITTERDING, H.A., RZIGALINSKI, B.A., ELLIS, E.F. NMDA receptor activation contributes to elevated intracellular calcium and decreased mitochondrial membrane potential in stretch-injured neurons, **2002**, *J. Neurotrauma* 12, 619–629.

124 WEBER, J.T., RZIGALINSKI, B.A., ELLIS E.F. Calcium responses to caffeine and muscarinic receptor agonists are altered in traumatically injured neurons, **2002**, *J. Neurotrauma* 11, 1433–1443.

125 ZHANG, L., RZIGALINSKI, B.A., ELLIS, E.F., SATIN, L.S. Reduction of voltage-dependent Mg2+ blockade of NMDA currents in mechanically injured cortical neurons, **1997**, *Science* 274, 1921–1923.

126 UWATOKU, T., SHIMOKAWA, H., ABE, K., MATSUMOTO, Y., HATTORI, T., OI, K., MATSUDA, T., KATAOKA, K., TAKESHITA, A., Application of nanoparticle technology for the prevention of restenois after balloon injury in rats. **2003**, *Circ. Res.* 92(suppl), e62–e69.

127 SCHIELE, T.M. Current understanding of coronary in-stent restenosis. Pathophysiology, clinical presentation, diagnostic work-up, and management. **2005**, *Z. Kardiol.* 94, 772–790.

128 SATOH, M., MATSUO, K., KIRIYA, H., MASHINO, T., NAGANO, T., HIROBE, M., TAKAYANAGI, I. Inhibitory effects of fullerene derivative dimalonic acid C60, on nitric oxide-induced relaxation of rabbit aorta, **1997**, *Eur. J. Pharm.* 327, 175–181.

129 LAM, C.-W., JAMES, J.T., McCLUSKEY, R., HUNTER, R.L. Pulmonary toxicity of single-wall carbon nanotubes in mice 7 and 90 days after intratracheal instillation, **2004**, *Toxicol. Sci.* 77, 126–134.

130 OBERDORSTER, G., OBERDORSTER, E., OBERDORSTER, A. Nanotoxicology: An emerging discipline evolving from

studies of ultrafine particles, **2005**, *Environ. Health Persp.* 113, 823–839.

131 KAMAT, J.P., DEVASAGAYAM, T.P.A., PRIYADARSINI, K.I., MOHAN, H., MITTAL, J.P. Oxidative damage induced by the fullerene C60 on photosensitization in rat liver microsomes, **1998**, *Chemico.-Biolog. Interact.* 114, 145–159.

132 HUSSAIN, S.M., HESS, K.L., GEARHART, J.M., GEISS, K.T., SCHLAGER, J.J. In vitro toxicity of nanoparticles in BRL 3A rat liver cells, **2005**, *Tox. In Vitro* 19, 975–983.

133 UENG, T.-H., KANG, J.-J., WANG, H.-W., CHENG, Y.-W., CHIANG, L.Y. Suppression of microsomal cytochrome P450-dependent monooxygenases and mitochondrial oxidative phosphorylation by fullerenol, a polyhydroxylated fullerene C60, **1997**, *Toxicol. Lett.* 93, 29–37.

134 CHIANG, L.Y., LU, F.-J., LIN, J.-T. Free radical scavenging activity of water-soluble fullerenols, **1995**, *J. Chem. Soc., Chem. Commun.* 1283–1284.

135 TSAI, M.C., CHEN, Y.H., CHIANG, L.Y. Polyhydroxylated C60 fullerenol, a novel free radical treapper, prevented the hydrogen peroxide and cumene hydroperoxide elicited changes in rat hippocampus in vitro, **1997**, *J. Pharm. Pharmacol.* 49, 438–445.

136 CHEN, M., VON MIKECZ, A. Formation of Nucleoplasmic Protein Aggregates Impairs Nuclear Function in Response to SiO_2 Nanoparticles, **2005**, *Exp. Cell. Res.* 305, 51–62.

11
Nanoparticles and Nanowires for Cellular Engineering

Jessica O. Winter

11.1
Introduction

The use of nanostructures (e.g., nanoparticles and nanowires) has increased dramatically in the last decade. Most applications have exploited the unique optical properties of mesoscale materials for imaging applications, both *in vivo* and *in vitro*. More recently, nanostructures have been investigated as components of actuating systems that can interact directly with the cellular environment. Because of the complementary size-scale of nanostructures and biological components, these systems provide the opportunity to precisely manipulate cells and proteins. This ability will produce greater insight into cellular function and pave the way for novel therapies.

This chapter discusses applications of nanostructures, including nanoparticles and nanowires, to cellular engineering. While several authors discuss specific aspects of nanomaterial interactions with cells (e.g., cell adhesion), this chapter is one of the first comprehensive investigations of the role of nanoparticles and nanowires in manipulating specific cellular components. Most of these approaches are in nascent stages, and therefore much of this chapter is focused on *potential* opportunities. Discussion is confined to semiconductor, metallic, silica, and magnetic materials, as carbon-based materials, polymeric nanostructures, and protein nanoparticles are discussed elsewhere. This chapter is divided into several subsections. Section 11.2 describes the motivation for using nanostructures to explore the biological milieu, their synthesis, and modifications required for biocompatibility. Section 11.3 examines nanostructures that interact with the cell exterior. Methods for modifying cell adhesion, migration and for manipulation of the cytoskeleton are discussed. Section 11.4 explores techniques for nanostructure delivery to the cell interior. Upon contact with cells, nanostructures enter the endocytosis pathway. Methods for eluding this fate are evaluated. Additionally, applications in intracellular tracking and sensing are considered. Section 11.5 describes nanostructure transport using the cytoskeletal system, including actin- and myosin-based approaches. Potential use for internal cytoskeletal manipulation and cargo delivery are evaluated. Section 11.6 considers the use of nanostructures for biomolecule

Nanotechnologies for the Life Sciences Vol. 9
Tissue, Cell and Organ Engineering. Edited by Challa S. S. R. Kumar
Copyright © 2006 WILEY-VCH Verlag GmbH & Co. KGaA, Weinheim
ISBN: 3-527-31389-3

delivery, including drug delivery and gene therapy. As applications in the area of drug delivery are vast, nanotherapeutics for cancer therapy are presented as examples. Section 11.7 examines the potential of nanostructures to manipulate proteins, membrane bound receptors, and ion channels. Finally, concluding thoughts and a look to the future are presented in Section 11.8.

11.2
Biological Opportunities at the Nanoscale

11.2.1
Nanostructures and Cells

Nanoparticles and nanowires have captured the interest of researchers because of their small size and unique characteristics. Materials in this size regime (i.e., <100 nm diameter [1, 2]) display properties that differ from their bulk counterparts. These properties, resulting from quantum confinement, include electron transfer, strong dipole moments, and size-tunable absorbance and fluorescent emission [1, 3]. Applications of these materials have been widespread, particularly in the electronics field, where there is a constant impetus to uphold Moore's law [4] and produce ever smaller electrical components. Nanoparticles have been integrated into LEDs, solar cells, and lasers [5], whereas nanowires have been used primarily to create electronic and computational components, including p-n diodes, field effect transistors, and logic gates [6].

Although the applications have not been as obvious, nanostructures also show great potential in the biological sciences [7]. Many of life's most basic functions occur in the nanometer regime. For example, the diameter of DNA is ∼2 nm, of a cell surface receptor is ∼10 nm, and of a virus is ∼50 nm [8] (Fig. 11.1). Because of their similar size scales, nanocomponents easily interface with biological molecules [9], a feature that has already been exploited to create new sensing technologies, diagnostics, and therapeutics. Nanoparticles have been utilized as fluorescent dyes for cell labeling [10, 11], biosensors [12, 13], cell sorting aides [14], and chemotherapy alternatives [15]. Nanowires have been used for biological separations, biosensing, and gene therapy [16]. Thus, the combination of similar size scales with unique optical properties has produced several passive applications for biosensing. Future developments will likely exploit biological machinery to manipulate cells directly.

Nanostructures hold particular promise in the fields of cell and tissue engineering. In addition to possessing favorable size scales to many biological components, nanoparticles and nanowires can be manipulated in external optical [17], electric [1], and magnetic fields [18]. Their high surface to volume ratios present favorable sites for chemical reaction, and their optical properties allow for *in situ* monitoring of local structures and biochemistry [19]. Because of these capabilities, nanostructures provide some of the first opportunities to actively manipulate subcellular components. Some of this work has already been realized *ex vivo*. For example,

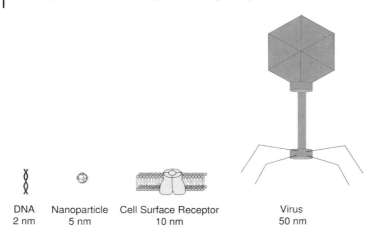

DNA   Nanoparticle   Cell Surface Receptor   Virus
2 nm   5 nm   10 nm   50 nm

Fig. 11.1. Comparative diameters of a nanoparticle and selected biological molecules.

nanoparticles have been harnessed to kinesin and transported along microtubules [20]. However, most possibilities for nanostructures in cellular engineering have yet to be realized.

11.2.2
Nanoparticle and Nanowire Synthesis for Biological Systems

Nanostructures can be synthesized using various techniques, depending on the desired composition and properties. However, developing water-soluble, non-toxic nanostructures has been challenging. Most synthesis methods were originally designed for optical or electronic applications. Particles are commonly made under harsh conditions with high temperatures or organic solvents, incompatible with biological requirements. In fact, the nanomaterials themselves may be toxic (Section 11.2.5). Cadmium, a major component of semiconducting nanostructures, is known to interfere with DNA mismatch repair [21], can inhibit certain types of neuronal firing [22], and is a known carcinogen [23]. Carbon nanotubes and fullerenes can produce asbestos-like lesions in the lung [24]. Carbon [25–27], graphite [27], and iridium [28] particulates can damage tissues. The difficulties of making biological relevant, non-toxic nanostructures are great, yet several synthetic strategies have been developed.

11.2.2.1 **Nanoparticle Synthesis**
Biologically active nanoparticles may be synthesized from semiconducting, metallic, or magnetic materials. Semiconducting particles, also known as quantum dots, are typically composed of CdX (where X is S, Se, or Te), and have been used primarily in optical applications, as a result of their high fluorescence quantum yields [19]. Metallic particles are usually made from gold. They have been used for bio-

sensing because they experience a color change with decreasing inter-particle proximity [13]. Additionally, they have formed the basis of several therapeutic treatments, as particles produce a temperature elevation with near-infrared excitation [15]. Magnetic particles are usually made from iron oxides [18]. Frequently, these particles are embedded in a micron-sized polymer matrix and used as MRI contrast enhancers and for cell and protein separation [18]. The variety of particles used has resulted in several synthetic techniques, each specific to the particle material employed.

Semiconductor Nanoparticles   The most common method of semiconductor nanoparticle synthesis is high temperature precursor decomposition [29]. This procedure allows control of nanoparticle growth and nucleation, and thus size distribution. Depending on the reaction time and final temperature, particles synthesized using precursor decomposition can vary in size from 1 to 11 nm, with size distributions as narrow as 5% [29]. Regular size distributions are critically important for the creation of templated nanostructures, which form the basis of many cellular engineering applications (Section 11.3.2). These structures, created through the ordered self-assembly of nanoparticles, require uniform size distributions to develop patterns over large length scales.

Unfortunately, precursor decomposition utilizes organic solvents, presenting several difficulties to direct application in biology. Particles synthesized with the most common passivating ligands, tri-octyl phosphine (TOP) or tri-octyl phosphine oxide (TOPO), are soluble in chloroform and other organic solvents, but not in alcohols or aqueous solution [29]. Additionally, TOPO is a weakly bound ligand, which can disassociate from the particle over time, producing a cytotoxic product [30]. This loss of passivation exposes the particle core to oxidative attack, which can release toxic ions into the cellular environment [31], eventually resulting in particle dissolution. To protect the particle core from photooxidation and surface reactions, semiconductor particles are often modified with a shell of a wider band gap (i.e., more insulating) material. Most semiconductor particles employed in biological applications contain a shell, with the most common material used being zinc sulfide (ZnS) [32]. ZnS is an ideal shell material as it is substantially less toxic than cadmium, possesses a larger band gap than CdSe, and does not easily form alloys with CdSe [31, 33]. CdSe/ZnS capped particles are stable for months, even in the presence of oxygen and are less toxic than uncapped particles [31].

As an alternative to organic syntheses, nanocrystals have been produced in aqueous solvents through a process known as arrested precipitation. Bulk CdS is relatively insoluble in water ($K_{sp} \sim 10^{-29}$) [34]. However, with the addition of aqueous thiols, nanometer-sized water-soluble colloids are produced. Briefly, upon thiol addition, complexes with cadmium ions are formed. Sulfur ions are added and crystal growth occurs, but it is hindered by the steric presence of the thiol compounds. Eventually, free sulfur anions cannot overcome these steric forces and particle growth is terminated. If the surface is well passivated, the crystals can remain suspended in water.

This process is controlled by many factors, including the reactant ratios, the thiolated ligand in question, and the pH of the reacting mixture; all of which contribute to nanoparticle size and uniformity [35, 36]. Aqueous syntheses are very flexible. Various materials can be used as reactants. The first CdS nanocrystals were produced using $Cd(SO_4)$ as a cadmium source [37], although $Cd(ClO_4)_2$ [38] and $CdCl_2$ [35, 36] have also been used. Sulfur can be introduced through various mechanisms, and $(NH_4)_2S$ [37], Na_2S [38], and H_2S [39] have all been investigated. Thiolated ligands can range widely, including mercaptoacetic acid, mercaptopropionic acid, mercaptoethylamine, and β-mercaptoethanol [35, 40, 41]. Although thiolated ligands are used most frequently, any ligand that binds to Cd^{2+} ions and promotes water solubility of the particle may be employed.

Nanoparticles produced using aqueous syntheses offer many benefits for biological studies. For example, they can be manufactured using simple bench-top chemistry, requiring only a fume hood. Particle surfaces may be readily altered through the use of cysteine- or thiol-terminated biomolecules, or by performing post-synthetic conjugation chemistry on a ligand functional group. The main limitation of this procedure is the large particle size distribution, which can increase with time, developing a tail at larger sizes, indicating particle instability [42]. This is likely a result of Ostwald ripening, the growth of larger particles at the expense of the thermodynamically less stable smaller ones [41]. However, we have observed that only solutions of larger particles exhibit this behavior; solutions containing smaller particles (i.e., ∼3 nm and less) can be stable in aqueous solution for months. Thus, the particles can exhibit remarkable stability, comparable only to silica-capped CdSe/ZnS [10, 43].

Metallic Nanoparticles Gold or silver nanoparticles, the most widely investigated, have a long history, originally being described by Michael Faraday in 1857 [44]. They are normally synthesized by the reduction of metal salts in citrate solution [45]. Similar to aqueous syntheses for semiconductor nanoparticles, various capping agents can be added to control particle surface properties and growth [46]. Gold nanoshells can also be created through the reduction of metal salts onto a pre-existing substrate [47]. Shell growth occurs by the formation of nanoparticles on a surface (i.e., gold [47] or silica [48]). As nanoparticle growth continues, particles coalesce to form a shell. These shells exhibit interesting plasmon resonance shifts, which can produce local heating in the presence of IR irradiation. As an alternative to aqueous routes, gold particles have also been produced in inverse micelle preparations [49]. In this synthesis technique, particle growth is limited by the size of the micelles formed. Unfortunately, most inverse micelle preparations utilize organic solutions, which may necessitate ligand exchange to create biologically stable particles.

Gold particles, in particular, have found application in the biomedical field because of their easily altered surface chemistry and optical properties. Gold readily binds thiol-containing molecules [50], offering a wide-range of surface chemistry options. Additionally, gold colloid solutions change color, from red to purple, upon particle aggregation [13], and particles can be deposited in self-assembled monolayers, facilitating integration into biosensors [51]. The presence of gold particles can be detected using various methods, including optical and electron microscopy and Raman spectroscopy, and gold particles have been used in various diagnostic methods [46]. Finally, gold particles and nanoshells experience a temperature increase upon near-infrared irradiation that has been used for cancer therapy, e.g., hyperthermia (part of Section 11.6.2.2).

Magnetic Nanoparticles   Most magnetic nanoparticles are composed of iron oxides. In general, particles are prepared by the addition of base to an aqueous solution of iron ions, forming iron oxide [18]. To prevent oxidation, particles are produced under nitrogen. For example, magnetite (Fe_3O_4) may be prepared by Reaction (1) [18].

$$Fe^{2+} + Fe^{3+} + 8OH^- \rightarrow Fe_3O_4 + 4H_2O \tag{1}$$

Crystal growth proceeds through nucleation of small complexes that grow as a result of Ostwald ripening. Particles formed range from 6–15 nm and consist of single magnetic domains. The strategy is very similar to aqueous syntheses for semiconductor and metallic nanoparticles described above. Reactions are controlled by passivating surface molecules, which take on an additional importance because particles are attracted to each other through their magnetic fields in addition to traditional electrostatic and van der Waals forces [18]. Passivating molecules can be added to the reactants, during reaction, which forms microemulsions, or after the reaction. A wide variety of natural and synthetic polymers, chemical stabilizers, and metallic coatings have been investigated. Typically, the passivating coating is not magnetic and can be as much as 20 nm thick, effectively doubling the diameter of the particle [52], which may prove problematic, depending on the application.

Magnetic nanoparticles are characterized by the nature of their magnetic properties. Nanoparticles may be ferromagnetic or superparamagnetic. In ferromagnetic materials, individual atomic dipole magnetic moments align to create permanent magnetization. This occurs because the internal thermal energy of the material is not sufficient to overcome the magnetic attraction of the individual dipoles. Particles in the micron-size range exhibit this behavior [18]. However, as particle size decreases, an interesting phenomenon is observed. Although the particle still possess insufficient thermal energy to overcome individual atomic magnetic attractions, the internal thermal energy of the particle is sufficient to alter the direction of all aligned dipoles. This produces a fluctuation of the magnetic dipole for the entire crystal, resulting in a net magnetic moment of zero. Particles below 15 nm exhibit this behavior [18], known as superparamagnetism. Although they are not "true" paramagnetic materials in that their atomic dipoles are aligned, the entire crystal exhibits paramagnetic behavior (e.g., temporary polarization).

11.2.2.2   Nanowire Synthesis

Nanowires have been manufactured using various techniques, including templated synthesis, directed growth in solution, and vapor phase growth [3, 16]. Because of the range of methods employed, nanowires can created from many substances, including magnetic, semiconducting, and ceramic materials. In templated synthesis, nanowires or hollow nanotubes are grown inside a porous membrane using electrodeposition or sol–gel techniques. Using this method, magnetic [53], metallic [16], and ceramic [16] nanowires with aspect ratios of up to 250 have been manufactured. Templated synthesis has several advantages. Particles of varying composition can be created by changing the electric potential or electroplating solution. The presentation of nanowires may be varied. They can be constructed as supported arrays or, by dissolving the membrane, free-standing entities.

Solution phase synthesis proceeds through several mechanisms [3, 16]. In surfactant driven assembly, surfactants attach to a specific crystal face, limiting growth in those directions. Alternatively, chemical reaction may occur preferentially at a specific crystal face, and reduction at this surface produces nanowire growth and extension. Solution-phase synthesis has the advantage of being scalable (to gram quantities [3]) and occurring at modest temperatures, which are more amenable to biological components. The greatest limitation of this technique is that directed growth in solution is the not well understood and difficult to control.

Vapor-phase synthesis requires high temperatures to create gaseous reactants that are deposited on a substrate as nanowires or nanotubes. High temperature syntheses are primarily used to produce carbon nanotubes, but silicon and several other semiconductor nanowires have also been created using this approach [3, 6]. Nanowire formation is catalyzed by a metallic liquid droplet. As the temperature is increased, gaseous vapor of the desired component begins to form a eutectic mixture within the droplet, entering the liquid phase. Further increases in temperature raise the dissolved concentration, favoring formation of a pure solid composed of the desired component. The size of the droplet restricts crystal growth in a preferred direction and defines the diameter of the nanowires, with an average of 10 nm [6]. Nanowires produced are frequently contaminated by the seed crystal. Because many of the seed metals used (e.g., nickel) are toxic, the development of biocompatible nanowires may be hindered.

As with nanoparticles, surface functionalization is often required to produce biologically compatible, water-soluble nanowires. Many of the techniques employed are similar to those for nanoparticles, including the use of thiols, carboxylic acids, and siloxanes [16]. Because nanowires may be constructed of segments of alternating materials, several passivation strategies may be required. For example, metallic nanowires containing segments of nickel and gold can be labeled with distinct fluorescent agents using differences in surface chemistry (Fig. 11.2) [54]. Nanowires have also been functionalized with biological molecules including DNA and proteins [55, 56].

11.2.3
Surface Passivation Strategies

The main limitation of nanostructures produced in organic solutions is their lack of water solubility. Several techniques have been offered to correct this condition, all relying on modification of the nanostructure surface. This can be accomplished by altering the nature of the passivating ligands or augmenting the crystal with a shell material that is biocompatible. These techniques have both been used to create nanoparticles employed for biological labeling [10, 11]. However, the process for transferring particles into water can be time-consuming and requires equipment outside the reach of most biological laboratories.

The most common method for rendering particles water-soluble includes an additional post-processing step known as ligand exchange. During ligand exchange, the existing chemical surface coating of the particle is replaced with one that can alter the solubility of the colloid [57]. There are several variations in the ligand exchange technique; however, the basic premise remains the same [29]. The particles are dried to remove unreacted ligand and the organic solvent, and are introduced into a solution containing a gross excess of the new ligand. After an incubation of several hours, exchange occurs. Exchange can be confirmed by altered solubility of the particles, or through analytical techniques like FTIR (Fourier-transform Infrared Spectroscopy) and XPS (X-ray Photoelectron Spectroscopy).

The most common ligand exchange used to produce biologically compatible quantum dots exchanges TOPO, bound to CdSe/ZnS core–shell particles, for mercaptoacetic acid (MAA = $HS-CH_2-COOH$) [11, 57]. The binding of MAA occurs through the sulfur group at the molecule's terminus, which adheres to Zn atoms on the ZnS coating [57]. The carboxyl end of MAA provides water solubility. In addition to MAA, several other thiolated chemicals have been examined, including mercaptopropionic acid (MPA = $HS-(CH_2)_2-COOH$), mercaptoethylamine (MEA, $HS-(CH_2)_2-NH_2$), and β-mercaptoethanol (MBE, $HS-CH_2-CHOH$). In fact, any molecule that binds to the surface of the particle and is water-soluble can be used, thus allowing a range of surface chemistries to be explored. However, ligand exchange is a long, multi-step process spanning as many as three days. The particles are only stable in solution for 1–3 weeks [57], indicating possible loss of ligand coverage over time. If the ligands are toxic (e.g., MBE [58]) or bioactive (e.g., MEA [59]), desorption can have deleterious effects on cells exposed to particles.

Another method to create water-soluble nanoparticles alters the surface by applying a biocompatible shell material. An example of this technique, developed by Alivisatos et al. [10, 43], coats TOPO-capped CdSe/ZnS particles with silica. Particles were exposed to a gross excess of mercaptopropyltris(methyloxy)silane (MPS), to which tetramethylammonium hydroxide (TMAH) had been added, producing a 2–5 nm silica shell [43]. This technique is versatile and could be applied to any TOPO-capped particle. Additionally, silica-coated particles meet several of the criteria for biological applications. Unlike particles that undergo ligand exchange, silica-coated nanocrystals exhibit stability for months [10, 43]. Also, biomolecules can be readily conjugated to silica using well-established techniques developed for chromatography. The only potential disadvantage of this technique is that silica is insulting and might alter the properties of the encapsulated particle.

11.2.4
Bioconjugation

Once particles have been prepared for aqueous use, it may be desirable to attach biomolecules to their surface. The range of biomolecules that can be bound to

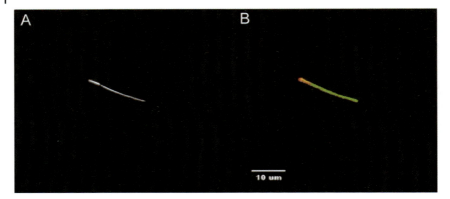

Fig. 11.2. Selective functionalization of multi-component nanowires can be achieved by taking advantage of the selective surface chemistry of the nanowire segments. (A) Reflection image of a two-segment nickel–gold nanowire. The brighter portion of the nanowire represents the more reflective gold surface. (B) The corresponding fluorescence image shows a nanowire that was modified by coupling different fluorescent dyes to each segment, which allowed the Ni segment to emit green and the Au segment to emit red. (Photo courtesy of Gerald Meyer, Johns Hopkins University, Baltimore, MD.)

nanostructures is staggering, and this is one of the factors that have led to their broad appeal for biological applications (see Volume 1 of this series for more detail). There are several methods to link biomolecules to nanostructures [60], the simplest being physisorption on the nanostructure surface (e.g., Zhang et al. [61]). However, most nanostructures are charged, thus electrostatic attraction is more frequently used to mediate attachment. For example, DNA will bind to nanoparticles presenting trimethylammonium ligands [62], and layer-by-layer assembly has been used to create films of alternating-charged nanoparticles and biomolecules [63–65]. Additionally, binding affinities between materials have been exploited to link biomolecules to surfaces. Thiolated compounds bind strongly to Cd-based and gold nanostructures [50], and biomolecules containing cysteine can be attached to nanostructures in this way [66]. Nanostructures can also be functionalized with biomolecules through directed recognition. The avidin–biotin system has been well studied in this regard. Avidin and biotin display strong binding to each other. A molecule that has been modified with avidin can be linked to any biotin-labeled biomolecule. This technique has served as the basis for commercially available fluorescent quantum dot labels [10].

11.2.4.1 Conjugation Strategies to Promote Non-specific Cellular Affinity

Once on a surface, biomolecule-conjugated nanostructures can be directed to a cell or tissue of interest. In some cases, this occurs on the basis of charge, resulting in non-specific adsorption. This recognition strategy is the most straightforward, requiring only adjustment of the surface charge, but has the least specificity. Particles cannot be directed to a specific part of a cell or a specific group of cells. This method was first used by Alivisatos et al. [10], who investigated CdSe–CdS core–shell nanocrystals enclosed in a silica shell (Section 11.2.3) [43]. The shell conferred water-solubility on the particle, but also provided functional groups (e.g., aminosiloxanes) for bioconjugation. To demonstrate non-specific binding, the silica coating was modified with trimethoxysilylpropyl urea (urea = NH_2-C-NH_3^+ at neutral pH), which binds negatively charged acetate groups (-CH_2-COO^- at neutral pH) in the cell nucleus. Nanocrystals introduced to cells were located in the nucleus, as evidenced through fluorescence microscopy measurements. Binding was suppressed with the addition of a negatively-charged reagent or sodium dodecyl sulfate (SDS), indicating that binding was primarily caused by non-specific electrostatic interactions.

11.2.4.2 Biomolecular Recognition

Apart from electrostatic interactions, which are non-specific, particles can be directed to cells and tissue using biomolecular recognition, the strong and highly specific binding that exists between a receptor and its ligand target. This binding is not completely understood, but arises from a combination of electrostatic attractions, hydrogen bonding, and hydrophobic/hydrophilic interactions [67]. The cumulative interaction of these forces, which are individually weak, produces tight binding. Additionally, this affinity is unique to the ligand and its analogs, as only molecules with a certain conformation can access the binding site and form the appropriate bonds. Thus, the interfaces formed through biomolecular recognition are highly specific and controlled.

Several molecules have been identified as biorecognition elements, and can be used to connect small molecules to components of the cell. Typically, these fall into three classes: antibodies, proteins, and peptides [67]. Antibodies recognize components that are foreign to their host system, and can be produced for almost any cellular component. A wide range of antibodies are commercially available, and their large size and high number of functional groups allow for ready conjugation to many surfaces. Antibodies are by far the most commonly used biorecognition element in conjugate chemistry [68]. However, when antibodies cannot be applied, alternatives (e.g., proteins and peptides) may be employed. Because protein or peptide binding will produce a cellular response, these methods are often utilized to impose a specific cellular function (e.g., receptor-mediated endocytosis [11]) to the attached molecule. Proteins can be linked to these molecules through chemistry techniques, the success of which depends widely on the nature of the protein employed. For applications where an entire protein might be too large or expensive, protein fragments, known as peptides, can be employed. Peptides can be synthesized with additional amino acids to allow for conjugation to a chosen molecule.

11.2.4.3 Conjugation Strategies for Antibody-mediated Recognition

Antibody binding is extremely useful for directing nanostructures to particular cell elements. A wide selection of antibodies is commercially available, and they may be easily conjugated to nanostructure surfaces. Antibody–nanoparticle binding was first demonstrated by Nie et al. [11], using CdSe–ZnS core–shell particles. The particle surface was altered through mercaptoacetic acid ligand exchange, placing carboxyl groups on the surface. Using carbodiimide chemistry (i.e., EDC), carboxyl groups were conjugated to the reactive amines of IgG antibodies. Because aggregation is a side-effect of EDC chemistry, conjugates were examined with transmission electron microscopy and were found to be primarily single particles. Fluorescence optical microscopy confirmed that conjugated particles maintained their initial optical properties. To demonstrate biorecognition, IgG conjugates were shown to aggregate in the presence of a specific polyclonal antibody known to bind IgG fragments. Antibody recognition has also been demonstrated on live cells using aqueous CdS quantum dots conjugated to IgG secondary antibodies through EDC chemistry [66]. Primary antibodies recognized and attached to integrin receptors. Quantum dot–IgG conjugates were then able to bind and recognize the primary antibodies with no adverse cellular effects.

11.2.4.4 Conjugation Strategies for Protein- and Peptide-mediated Recognition

Proteins and peptides (i.e., protein fragments) have also been conjugated to particles for cellular recognition. For example, phalloidin protein conjugated to silica-coated quantum dots was used to bind to actin filaments in fixed cells [10]. Phalloidin was attached to the nanoparticle surface using avidin–biotin chemistry. First, the silica coating of the nanocrystal was modified to include primary amines using 3-aminopropyltrimethoxysilane. Then, biotinamidocaproic acid 3-sulfo-*N*-hydroxysuccinimide ester was introduced. The NHS ester group of the activated biotin reacts quickly with primary amines to produce amide bonds, linking the nanocrystals to biotin molecules. The cells were exposed to a separate solution of biotinylated-phalloidin, which adhered to F-actin filaments in the cytoskeleton. Then, streptavidin was introduced, binding the biotin component of biotin–phalloidin. Finally, cells were incubated with the biotinylated-nanocrystals, which bound to the phalloidin–biotin–streptavidin complexes at avidin sites. Additionally, particles have been bound to transferrin, a surface protein that promotes uptake of iron molecules through receptor-mediated endocytosis [11]. EDC chemistry was used to mediate attachment to CdSe/ZnS particles. Transferrin-conjugated particles were incubated with living cells, and were found in intracellular vesicles, as would be expected following receptor-mediated endocytosis.

As an alternative to whole proteins, which can be bulky, a second approach utilizing peptides was developed. Peptides were attached to quantum dots as surface passivation molecules [66], providing a single-step technique for nanocrystal synthesis and bioconjugation. Aqueous CdS nanoparticles were attached to RGDS and YIGSR peptides, which bind to integrin receptors [69, 70] through thiolated ligand-binding using cysteine residues. Peptide–cell binding has several advantages to antibody- or protein-cell binding. The minimum peptide length for recognition depends on the protein selected, but can be as short as 3–5 amino acids (e.g., 1–1.5 nm) [69, 70]. Additionally, peptides can be manufactured to almost any length, with each amino acid roughly corresponding to 3 Å (0.3 nm) for a straight-chain conformation [71]. Thus, the separation distance between the particle and the cell surface can be tuned through altering the length of the peptide. Their small size and tunable length make peptides ideal alternatives to large conjugation molecules. Given these factors, peptide bioconjugation presents a superior approach for forming close-range interfaces between semiconductor quantum dots and cellular receptors.

11.2.5
Toxicity (see also Volume 5 of this Series)

Very few studies have examined the potential health risks of nanostructures. There is limited evidence that suggests risks are present, particularly through inhalation. A recently published report demonstrated that inhaled carbon nanotubes can produce asbestos-like lesions in the lung [24]. Among the highest dose group, the mortality rate was 55% ($N = 9$) after only 7 days of exposure. Although these results were for pulmonary tissue, a separate set of studies examining carbon [25–27], graphite [27], and iridium [28] particulates established the ability of nanomaterials to damage other tissues, including the immune system [26] and brain [25, 27]. Particle effects on the brain included an increase in lipid peroxidation and damage to lipid-rich tissues [25]. The most troubling finding of these studies is that particle clearance proceeds slowly. A low level (i.e., 6% in lung) of particulates remained in tissue a full six months after exposure [28].

However, it is unclear if these results will apply to nanostructures used in biomedical applications. Most of these investigations examined toxicity through inhalation, but semiconducting, metallic, and magnetic particles are primarily produced in solution-phase. A more likely route of exposure is through ingestion or skin contact [72]. Additionally, the materials studied (i.e., carbon particulates and nanotubes) bear little similarity to the nanostructures reviewed here other than size. Several groups have employed semiconducting nanoparticles *in vitro* and *in vivo* without noticeable cell damage [73–78]; however, they did not specifically assess cell viability and function. To date, only a single study [31] has confirmed that semiconducting nanoparticles may be cytotoxic to cells when presented in aqueous solution. This was attributed to the release of free Cd^{2+} ions following oxidation of the particle core. Nanoparticles are susceptible to oxidation and photooxidation [1, 31, 79–82], particularly with UV exposure. Reaction of oxygen with group VI elements (e.g., S, Se, Te) produces oxides [1] that can disassociate from the particle surface. Free Cd atoms are left behind and may eventually enter solution. Elevated free cadmium levels have been reported for poorly-capped particles and particles exposed to prolonged UV excitation [31].

Cd^{2+} ions have well-established health risks [83]. Cadmium is a known carcinogen, and has been linked to lung, testicular, adrenal, liver, and kidney cancer [23]. Although the exact mechanism of carcinogenicity is not understood, cadmium has been shown to interfere with DNA mismatch repair [21]. It also binds strongly to the sulfhydryl groups of mitochondrial proteins [84], and this is the believed source of toxicity observed for the primary hepatocytes studied previously [31]. Additionally, cadmium is a commonly used blocking agent in patch-clamp studies that interferes with neuronal firing [22]. Given all of these factors, the stability of the nanoparticle core is of great concern. Modifying particles with a strongly-bound passivating ligand, an inorganic layer (e.g., ZnS), or a polymer coating can increase particle stability by preventing the entry of oxygen [31, 72]. Most recent reports on semiconducting quantum dot use for living systems utilize some coating of this nature [73–78].

Data for metallic nanoparticles and nanowires is even scarcer. Gold, in the bulk form, is fairly well-tolerated and has been used as an electrode coating in prosthetic implants for some time [85]. However, gold nanostructures have a dramatically higher surface area than bulk material. They are much more reactive and can potentially interfere with biochemical syntheses [86]. Reports on silver nanoparticles are conflicting. Although silver displays toxicity when employed in the nervous system, it has been widely used as a biomedical antimicrobial coating [87]. Silver nanoparticles employed in bone cement exhibited no overt cytotoxicity and a strong antibacterial effect at 1% loading [87]. However, mouse spermatagonia cells demonstrated a reduction in mitochondrial activity, increases in cell lysis and apoptosis, and changes in cell morphology when exposed to silver nanoparticles [88].

Uncapped iron oxide (i.e., magnetic) nanoparticles can also be cytotoxic, reducing cell viability by 20% at concentrations of only 0.05 mg mL^{-1} and 60% at higher concentrations (i.e., 2 mg mL^{-1}) [89]. Reduced viability may result from excessive particle internalization, producing apoptosis and limiting cell migration [90]. This behavior can be altered by the addition of a surface coating. For example, pullulan-conjugated nanoparticles showed no statistical difference in cell viability from control cells at 2 mg mL^{-1} concentrations [89].

Although the data on nanostructure toxicity is slim, it is apparent that coating the particles with a biocompatible shell (e.g., silica) or biomolecule (e.g., pullulan) reduces the risk of adverse reaction. This may result from isolation of a toxic core material, as in the case of Cd-containing particles, or from reduced protein adsorption, which limits particle endocytosis. The most effective coatings are likely those that reduce uptake through the endocytotic pathway [90], allowing the nanoparticle to remain in the extracellular environment as long as possible. As more research is performed, additional modifications to prevent toxicity will likely arise.

11.3
Nanostructures to Modify Cell Adhesion and Migration

An area of cellular engineering with great possibility for nanostructure use is the selective modification of cell adhesion. Cell adhesion is ideal for nanostructure-based evaluations because most of the salient events occur on the extracellular surface. Thus, nanostructure modifications for intracellular delivery (Section 11.4) are not required. Additionally, there is much to be gained from nanoscale investigations of cell adhesion and migration, as these processes are responsible for cancer metastasis [91], angiogenesis [91], tissue repair [92], and synapse formation [93].

Selective cell adhesion is also critical to the development of biocompatible implants and prostheses [94]. Integration with target cells is crucial to the function of the implant. However, in many cases this adhesion must occur while excluding undesirable cell types (e.g., immune cells, glia) that can interfere with device performance and tissue compatibility.

At the micron scale, cell responses to both chemical and physical cues are well established (for a review see Folch and Toner [95]). Surfaces with micron-sized pillars can influence cell adhesion and cytoskeletal protein distributions [96]. Cell extensions (e.g., neurites) are influenced by both the depth and width of surface grooves [97]. Chemical surfaces, containing adhesive islands, can confine cells to regions as small as 10 μm [98]. As area decreases the number of adherent cells, degree of cell spreading, and even the signal for apoptosis (i.e., controlled cell death) can be controlled [99]. Physical and chemical patterns can produce a synergistic response. Neurons cultured on micropatterned substrates align their extensions with regions containing both cues (Fig. 11.3). These patterns have been incorporated into various applications. Physical patterns have been used in biosensors and implants to improve biocompatibility and tissue adherence [100]. Micropatterned adhesive regions can create ordered arrays of muscle cells [101, 102], and guide neurite extension for neural networks [103–106].

11.3.1
Cell Adhesion at the Nanoscale

Although the ability of cells to respond to micron-scale structures has been established, there has been doubt as to whether responses would extend into the nano-

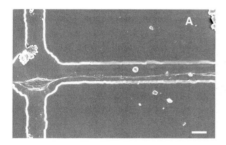
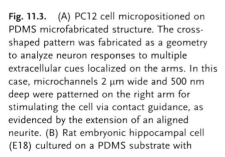

Fig. 11.3. (A) PC12 cell micropositioned on PDMS microfabricated structure. The cross-shaped pattern was fabricated as a geometry to analyze neuron responses to multiple extracellular cues localized on the arms. In this case, microchannels 2 μm wide and 500 nm deep were patterned on the right arm for stimulating the cell via contact guidance, as evidenced by the extension of an aligned neurite. (B) Rat embryonic hippocampal cell (E18) cultured on a PDMS substrate with microchannels and immobilized NGF. Microchannels 2 μm wide and 1 μm deep were patterned in PDMS using replica molding from silicon masters. NGF was immobilized using arylazido photolinkers. The substrate provides a combination of physical and biochemical cues to accelerate polarization of the cell (i.e., determination of an axon). (Photo courtesy of Christine Schmidt, Natalia Gomez University of Texas at Austin, Austin, TX.)

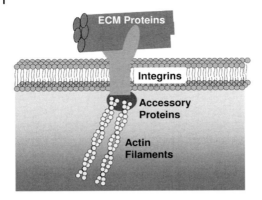

Fig. 11.4. Integrin binding to the ECM and cytoskeleton. Integrins embedded in the plasma membrane are used to form contacts with the extracellular matrix (e.g., collagen, laminin) and cytoskeleton (e.g., actin).

scale [2]. However, the environment surrounding cells is peppered with nanoscale cues. Cells are surrounded by a lipid bilayer ∼5 nm thick [8] (Fig. 11.4). Cell adhesion is mediated by 10 nm diameter integrins embedded in the bilayer [8]. These integrins cluster to create various nanoscale contacts (Table 11.1) [107, 108]. Integrins bind extracellular matrix (ECM) components, which are the primary elements of the basement membrane surrounding most tissues. Some of these components display heights of ∼150 nm and band widths of ∼60–80 nm (Table 11.1) [2, 109], and fibrils and pores in ECM structures as small as 5 nm have been reported [109]. On the cell interior, integrins attach to nanometer-sized cytoskeletal elements, including microtubules, actin, and intermediate filaments (Table 11.1). These elements produce cell shape and structure, and may respond to external forces [110].

11.3.2
Cell Adhesion and Nanoscale Physical Topography

However, creating materials with dimensions that could explore these biomolecules has been challenging. Unlike microfabricated devices, patterning methods are just beginning to extend to the nanoscale. Using lithographic techniques, several features, including grooves, ridges, steps, pillars and pores, have been investigated (see Flemming et al. [109] for a review). Despite being several times smaller than a cell diameter (i.e., ∼10 μm), cell responses were observed. Nanoscale physical features can create networks of aligned cells [111–113], induce macrophage spreading and phagocytotic activity [114], and guide neuronal growth cones [115].

For example, 350 nm wide nanogrooves have been used to evaluate smooth muscle cell (SMC) adhesion *in vitro* [113]. More than 90% of SMCs cultivated on this surface aligned to the grooves, and proliferation of the cells was significantly

Tab. 11.1. Cell–matrix contacts at the nanoscale. (Adapted from Ref. [107].)

Element	Dimensions	Description/purpose
Cell–matrix contacts [107, 198]		
Filopodium	20–200 µm × 200–500 nm	Cell protrusion, actin bundle connected to syndecans (proteoglycans)
Focal contact	1.5 µm × 250 nm	Cell contact, actin microfilaments connected to integrins, substrate adhesion
Hemidesmosome	150 × 40 nm	Cell contact, intermediate filaments connected to integrins
Podosomes	200–400 nm diameter	Cell contact, actin bundle connected to integrins
Microspike	2–10 µm × 200–500 nm	Cell protrusion, actin bundle connected to integrins, NG2 (proteoglycans)
Cell exterior: extracellular matrix proteins [8]		
Collagen	1.5 nm triple helix, 10–300 nm fibers	Structural protein
Fibronectin	2–3 nm arms separated by 100 nm, fibrils 10–1000 nm	Adhesive protein
Laminin	30 × 100 nm	Adhesive protein
Cell interior: cytoskeletal elements [8]		
Actin	5–9 nm diameter	Muscle contraction, cell locomotion, vesicle transport
Microtubules	25 nm diameter	Organelle and protein transport, chromosomal separation, cell division, vesicular transport, primary components of cilia and flagella
Intermediate filaments	10 nm diameter	Mechanical strength, compose nuclear lamina

lower than those grown on control surfaces (Fig. 11.5). Additionally, in a wound healing assay, SMC microtubule organizing centers (MTOCs) were preferentially aligned with the cell axis for grooved surfaces, and the wound location for control surfaces. These studies demonstrate the importance of nanoscale features in guid-

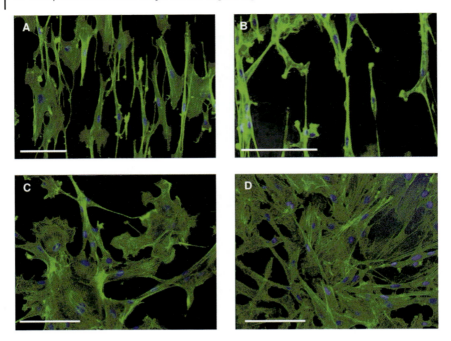

Fig. 11.5. Smooth muscle cell alignment on nanopatterned (A) PMMA and (B) PDMS. Muscle cells orient along the axis of the pattern, when compared with cells cultured on (C) unpatterned PMMA and (D) glass control surfaces. Scale bar = 100 μm. (Photo courtesy of Christopher Chen, Evelyn Yim, University of Pennsylvania, Philadelphia, PA.)

ing cell extension and adhesion. Surfaces with smaller features have also been examined, most notably by direct application of nanoparticles. Silica nanoparticles with dimensions from 7 to 21 nm were deposited directly onto glass coverslips, producing amorphous nanoscale features [116]. Fibroblasts cultured on these surfaces formed clusters and were much less likely to adhere and spread when compared with plain glass controls.

Interest in the creation of nanoscale patterns has led to the development of a new technique, known as colloidal lithography, which uses arrays of ordered self-assembled nanoparticles as a resist (Fig. 11.6) [117]. Metal can be evaporated or sputtered through the colloidal resist, creating a pattern in the crevices between the particles. The shape and dimensions of this pattern can be varied by altering the size and packing of the nanoparticles in question. Particles used are commonly made of gold, silica, or polymers, with diameters on the order of 10–100 nm [118]. The structures formed can display order over length scales of several microns, with increasing order for assemblies of larger particles [51].

Colloidal lithography has been applied to cell adhesion only recently, with much of the work performed by the Sutherland and Curtis research groups (see Dalby et al. [119] for review). Using polystyrene beads roughly 100 nm, structures with

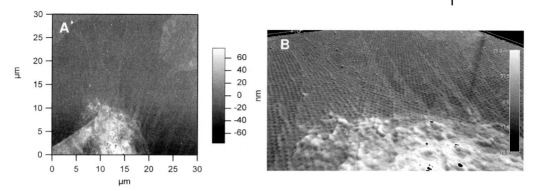

Fig. 11.6. (A) AFM image of colloidal lithography lattice. (B) Pseudo-3D view of same image. (Photo courtesy of Wolfgang Frey, University of Texas at Austin, Austin, TX.)

features ranging from ~50 to 200 nm have been manufactured. Cellular responses to these patterns have been varied. Pancreatic epithelial cells [120] cultured on hemispherical titanium pillars of increasing width (i.e., from 60 to 170 nm) displayed less rounded morphologies and increased spreading with respect to flat control surfaces. However, fibroblasts on 160 nm high by 100 nm wide poly(methyl methacrylate) (PMMA) nanocolumns seemed to prefer smooth surfaces to nanostructures. Vinculin adhesions were smaller on nanocolumn surfaces than in the control group [121], and appeared similar to transient focal adhesion structures observed elsewhere in poor adhesive environments [122]. Cells cultured on nanocolumns contained fewer actin stress fibers [121] and displayed more filopodia per micron of cell perimeter [123] than the controls. These results suggest that the cells were unable to form solid attachments to nanocolumns and continued to explore the local environment (with filopodia) for suitable substrates.

The ability of titanium nanostructures to promote cell spreading, in contrast to PMMA, implies that cells respond not only to the dimensions but also the type of material used and is consistent with previous findings. In bulk conditions, endothelial cells adhere to bare titanium, whereas attachment to PMMA required pre-adsorption of ECM factors [124]. Adhesion to titanium also increases with surface roughness [125], consistent with the finding that cells preferred a rough nanopillared surface to the smooth control. Thus, in addition to physical properties, surface chemistry (e.g., metal vs. polymer) plays an important role in producing nanoscale cell adhesion.

11.3.3
Cell Adhesion and Nanoscale Chemical Patterns

Establishing nanoscale chemical patterns has proven more difficult than creating physical patterns. Many of the methods used at the micron-scale (i.e., microcontact printing) do not readily translate to nanometer dimensions [126]. Several ingenious techniques have been developed to overcome these obstacles. Clusters of adhesive peptide sequences, containing the RGD integrin binding fragment, have been attached to comb polymers with 2 nm monomer chain lengths. Cell adhesion increased with the number of RGD fragments per cluster and the overall density of the clusters [127]. Alternatively, arrays of protein-adsorbed spots have been created using dip-pen nanolithography [128]. An AFM-tip coated with thiolated proteins was exposed to a gold surface. Utilizing the strong attraction of thiols for gold surfaces, proteins are transferred to the gold substrate with tip contact. To examine cell adhesion, spots of the integrin binding protein retronectin 200 nm in diameter with 700 nm spacings were constructed against a non-adhesive background. Fibroblasts adhesion was demonstrated.

Nanoscale chemical patterning is just beginning to extend to nanoparticles. In a rather creative approach, polymer micelles containing gold precursors were used to create ordered arrays of gold nanodots. The micelles were deposited onto substrates and then exposed to oxygen plasma etch to remove the polymer. The remaining gold precursors formed nanodots *in situ*. Dots were then coated with cyclic RGD peptide fragments [108, 129]. RGD spots with diameters < 8 nm and spacings from 28–85 nm were created, with each dot theoretically large enough to accommodate only one integrin binding domain. Cell adhesion and focal contact formation was reduced for spacings > 73 nm, suggesting that integrin clusters separated by <70 nm are required for focal contact formation.

11.3.4
Cytoskeletal Manipulation

Physical and chemical nanopatterns interact with cells passively. They are unable to directly probe specific cellular components. However, as a result of their unique optical, electrical, and magnetic properties, nanostructures *do* possess these capabilities. Most applications of direct cellular manipulation are still in the nascent stages. However, the use of magnetic particles for cell sorting and cytoskeletal manipulation has existed since the 1950s when it was pioneered by Crick (of DNA double helix fame) and Hughes [130]. Manipulation is accomplished using microspheres containing nanometer-sized magnetic colloids encapsulated in a polymer matrix, which are available from several commercial sources (e.g., Dynabeads, Dynal Corp., Brown Deer, WI) (Fig. 11.7).

The most common applications of this technology have been for cell sorting. Biomolecule-labeled magnetic microspheres can be used to isolate specific cells (for reviews see Chalmers et al. [14] and Zborowski et al. [131]), including rare cell lines [14, 132], which could be used for tissue engineering constructs or as a diagnostic tools. More recently, this technique has extended to magnetic nanowires. For example, magnetic nickel nanowires can be constructed using templated electrodeposition. Because of their large aspect ratios, the wires maintain residual magnetism (i.e., are ferromagnetic) in low magnetic field strengths. They can be used in sorting applications that traditional magnetite beads cannot, and in some cases, display superior performance [53, 133].

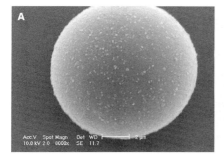

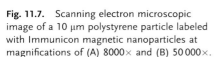

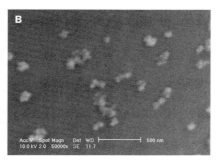

Fig. 11.7. Scanning electron microscopic image of a 10 µm polystyrene particle labeled with Immunicon magnetic nanoparticles at magnifications of (A) 8000× and (B) 50000×.

Note that the bound particles are on the order of 100 nm. (Photo courtesy of Jeffery Chalmers, Ohio State University, Columbus, OH.)

These cell sorting methods have been applied to tissue engineering. For example, magnetite (Fe_3O_4) particles with an average diameter of 10 nm can be encapsulated into cationic liposomes [134]. These liposomes have been used to create tissue engineered cell sheets containing multiple cell types [135, 136]. Liposomes were exposed to cells, accumulating in the cytoplasm. Then, localized magnets were used to precisely position cells on a substrate. This technique has been used to create co-cultures of hepatocytes and endothelial cell [135], sheets of keratinocytes [137], and retinal pigment epithelium [138]. Similarly, nickel magnetic nanowires have also been used to control cell deposition on tissue engineered substrates [139]. Here, nanowires were coupled to cells through non-specific surface protein adsorption. Fibroblasts attached to the nanowires aligned with a magnetic field creating cell–cell chains.

Magnetic structures have also been used to investigate the effects of mechanical stress in cells. For example, the response of individual integrin receptors to external forces was examined using 5.5 µm diameter magnetic spheres [140]. The spheres were coupled to RGD-peptide integrin binding motifs and incubated with cells to produce extracellular integrin binding. Then, spheres were exposed to magnetic fields, applying force to the cytoskeleton. Cytoskeletal stiffness was directly proportional to applied stress, demonstrating that integrins are mechanotransducers, modulating cell behavior in response to mechanical stress. This technique has also been expanded to allow manipulation of multiple cell culture dishes simultaneously, allowing for larger-scale biochemical analyses, including gene expression studies [141].

11.3.5
Future Applications of Nanostructures for Cell Adhesion and Migration

The potential for nanoparticles and nanowires to modulate cell adhesion is great, and many of the most exciting applications have not yet been realized. Most current approaches manipulate whole cells rather than examining the effects of individual receptors and their ligands. Future methods may be able to compare cooperative and individual responses using micro- and nano-scale patterning techniques. Additionally, techniques will become increasingly active, directly manipulating cell receptors and proteins, as opposed to passive patterns on a substrate. These improvements will enhance understanding of the mechanisms for cell adhesion and migration, which may lead to new treatments for spinal cord injury, cancer, and vascular conditions. Increased knowledge of adhesion at the biomaterial–tissue interface will also provide improved prosthetic devices, implants, and tissue-engineered surfaces.

11.3.5.1 Future Physical Nanostructures for Cell Adhesion

Most physical structures investigated to date have features >100 nm. Although this is useful for investigating the cooperative action of cell surface proteins, it is not possible to explore individual receptors, with diameters of ∼10 nm, using these surfaces. Current technologies have been limited by the difficulty of creating regular, physical patterns that have structures on the order of 10 nm and repeat over several microns. Nanoparticle self-assembled monolayers offer several advantages to other physical patterning techniques at the <100 nm length scale. Despite stacking faults, patterns display much more regularity than randomly oriented surfaces (e.g., Fig. 11.8), and at least short-scale order. Synthesis techniques to develop uniformly sized particles are well developed [51], and pattern roughness and dimensions can be easily changed by varying the particle diameter. Although more rigorous surfaces could be created using advanced lithography techniques (e.g., electron beam lithography), many of these techniques are expensive, difficult to reproduce, and inaccessible to the biologist, whereas many nanoparticles are now commercially available. Thus, in future applications, nanoparticle self-assembled monolayers will continue to be used in exploring cell adhesion either directly or as lithographic templates (Section 11.3.2). However, these techniques will probably emphasize directed self-assembly, allowing greater control of the pattern created.

For example, magnetic particles may be organized on a surface using magnetic fields to direct self-assembly [142]. This technique would allow much greater control of the resulting pattern than traditional self-assembly, which relies on thermodynamic forces to drive pattern formation [51]. When combined with techniques for specific cell placement (i.e., micropipette manipulation), magnetic assembly may allow for cells to be aligned with a specific pattern of interest. Extending these patterns to magnetic nanowires will bring further innovations. Because nanowires can be multifunctional, with domains of alternating materials, they offer a unique substrate for investigating cell adhesion. Nanowires produced using templated synthesis may have diameters from 10 nm to 1 μm and lengths of up to 60 μm. This compares favorably with receptor dimensions (∼10 nm) and cell diameters (∼10 μm) [53].

Other methods of directed assembly will also be explored. For example, Koo et al. [127] have demonstrated that star polymers can create structures with regularly-spaced adhesive peptides. The number of peptides bound and their spacing were

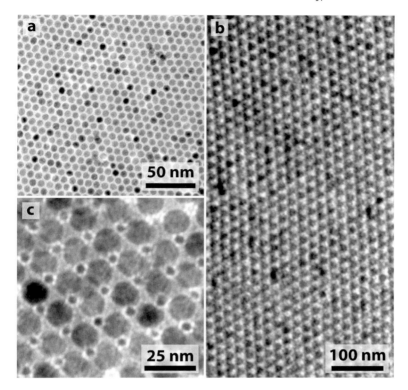

Fig. 11.8. Slow evaporation of an organic solvent containing monodisperse 11.5 nm iron nanocrystals results in the formation of extended regions of (a) monolayer or (b) multilayer superlattices. The superlattice structure can be tuned further through the co-assembly of two different sizes of nanocrystals, and provides a method to combine nanocrystals with different functionality into ordered arrays. A binary nanocrystal lattice (c) with a rock salt (NaCl) structure is formed by evaporating a dispersion containing large iron nanocrystals (13.3 nm) and small gold nanocrystals (5.2 nm). (Photo courtesy of Brian Korgel, Aaron Saunders, University of Texas at Austin, Austin, TX.)

controlled by altering the number of monomers in the star and the monomer length. Similar techniques could be applied to nanostructures, and may employ new classes of polymers. For example, dendritic polymers are well-suited for attachment to spherical nanoparticles. Altering the length of the monomer could be used to control nanoparticle separation distance [143]. The use of various functional groups on the monomer exterior could also provide a method for directed assembly through electrostatic attractions or controlled crosslinking.

11.3.5.2 Future Chemical Patterns for Cell Adhesion

Applications in Cellular Engineering The extension of the physical patterns discussed above to chemical patterning is straightforward. Colloidal lithography (Section 11.3.2) could be used to produce patterned surfaces for chemical adsorption. Gold substrates, in particular, may be created through vacuum deposition and modified with thiolated chemicals and cysteine containing peptides. The dimensions of the pattern will be dependant on the size of the particles used and the method of their assembly. Alternatively, the nanoparticles themselves may be modified with biomolecules and assembled *in situ*. Nanoparticle arrays are an attractive substrate for cell adhesion studies because they provide the opportunity to investigate combined physical and chemical effects as the particles themselves provide shape, in addition to any chemical signatures.

One interesting possibility of nanoparticle arrays is the use of bimodal particle distributions. With a combination of small and large particles, ordered assemblies of alternating particle types can be created (Fig. 11.8) [144]. These assemblies would provide unique physical topography, but could also be used to create alternating adhesive and non-adhesive domains. For example, one particle size could be bound to RGD-containing peptide sequences, whereas another particle size could be passivated with cell resistant molecules (e.g., PEG [145]). Structures of this type would allow researchers to develop new understandings of the distribution requirements of adhesive sites. The greatest challenge in developing such assemblies will probably be controlling populations of particles that may be alternatively hydrophobic and hydrophilic. Micelles or lamellar sheets may be favored over monolayer particle structures. However, the use of amphiphilic peptides or particles of opposing charges may circumvent these difficulties.

Applications in Tissue Engineering Nanoparticles may also be incorporated into existing tissue engineering substrates to alter their chemical properties. Nanoparticles used in this context have the advantage of increasing the useful binding area without requiring a great deal of organic chemistry. One area where this would be of interest is the field of hydrogels. Some hydrogel materials possess good physical properties for tissue engineering, but are resistant to cell adhesion or migration [146, 147]. These substrates are modified through complicated bioconjugation strategies to contain adhesive binding sequences. The amount of biomolecule attached can be difficult to control and may be limited by the number of modifiable sites in the monomer. Scattered gold particles could be incorporated in the substrate to provide a thiol-mediated attachment point for biomolecules. Particles could be included directly in the uncured polymer mixture or could be deposited on the polymer surface and secured through annealing or covalent attachment. A similar approach has already been investigated using micron-sized particles [148]. Polymer templates composed of layer-by-layer assembled polyelectrolyte films were exposed to micron-sized, charged particles, which assembled on the pattern. These particles were then modified with RGD peptides and examined as substrates for cell adhesion.

Another exciting possibility is the incorporation of nanoparticles into smart-substrates. For example, gold nanoparticles have been used to create drug delivery substrates which expand upon exposure to radiation [17]. This technique could be adapted to produce active substrates for the evaluation of cell mechanics. For example, a small number of fluorescent nanoparticles (e.g., CdSe) coupled to cellular integrin-binding domains could be introduced. Upon radiation (preferably in the IR to avoid cell damage), the substrate would expand. This expansion, and the cellular response, could be observed directly by monitoring the movement of the fluorescent particles. This approach has the advantage of potentially binding individual integrins. Expanding substrates might also aid researchers in constructing tissue engineering constructs for contractile tissue (e.g., muscle or heart) *ex vivo*.

11.3.5.3 **Active Investigation of the Cytoskeleton**

To date, magnetic bead examinations of mechanical force in cells have primarily been confined to either the cell surface [140, 149] or to phagocytotic vesicles [150, 151]. However, as particles decrease in size and methods for intracellular delivery improve (Section 11.4.3), it may be possible to manipulate the cytoskeleton directly from the cytoplasm. This would greatly facilitate understanding of the role of the cytoskeleton in cell adhesion and cell mechanics. We already know that the cytoskeleton is a component of an elaborate mechanotransduction system and can influence cell differentiation, apoptosis, and polarity [110]. Many regulatory proteins are physically linked to the cytoskeleton, and changes in external force have been shown to influence local protein expression [110]. Several clinical ailments may have their basis in defective responses to mechanotransduction [110]. Intense study of the relationship between the cytoskeleton, external forces, and gene expression may lead to new treatments for these diseases.

A major challenge of this approach is the ability to develop sufficient magnetic force to manipulate objects. Microbeads encapsulate numerous nanometer-sized colloids that cumulatively generate enough force to produce changes in the cytoskeleton. It is unclear that single, nanometer-sized particles would produce measurable changes in cytoskeletal position, although there is some evidence that they will behave similarly to microparticles [152]. Additionally, cooperative action resulting from the attachment of multiple nanoparticles to the same structure can increase effective force. In some ways, nanoparticles may be particularly appropriate for investigation of the small structures of the cytoskeleton. Along with similar size scales, it is likely that smaller forces will be required to produce movement/tension as the size of the target structure declines (i.e., integrin vs. focal adhesion).

11.4
Nanostructure Cellular Entry

Active cell manipulation requires methods to control nanostructure delivery and targeting. Whereas examination of the extracellular surface is straightforward, additional techniques are required to promote intracellular delivery. Nanostructures may be introduced to cells through endocytosis, but this is a passive technique. Most objects introduced into the endocytosis pathway are confined to lipid covered vesicles in the cell interior. Escape is difficult and materials are given little opportunity to interact with external proteins and biomolecules. Thus, direct delivery of nanostructures is difficult. In this respect, the field of nanocomponent delivery shares many challenges with that of gene therapy, and many of the techniques introduced for the latter can be applied to nanostructures.

Controlled delivery is expected to have many applications. Fluorescent nanostructures could be used for real-time monitoring of intracellular features, which would greatly improve understanding of signal transduction, the endocytotic pathway, and intracellular transport. Magnetic and conducting nanoparticles could be used to control intracellular elements, potentially providing external control of cell adhesion, cell division, differentiation, gene expression, and apoptosis. Nanostructures could also be used as drug delivery agents for drugs, DNA, and proteins providing novel techniques for gene therapy, cancer treatments, and tissue engineering.

11.4.1
Biology of Molecular Delivery

The cell plasma membrane is an effective method of excluding undesired components from the cell interior. This membrane is permeated with several transmembrane proteins, some of which serve as pores for small molecules to pass. However, despite their small size, nanostructures are still too large to enter the cell via this method (e.g., Rosenthal et al. [153]). Nanostructures are instead subject to endocytosis, a process through which objects are transferred from the local environment to the cell interior (Fig. 11.9). Endocytosis [8] is initiated by binding to the cell surface, followed by an invagination of the cell membrane. This membrane surrounds the foreign substance and pinches off from the bilayer, forming a vesicle in the cell interior, known as an endosome. After ~10 min, the endosome migrates from the edge of the membrane to the Golgi apparatus. Endosomes are slightly acidic, with pH values near 6. Some proteins contain targeting sequences that are revealed in the acidic environment, allowing escape from the endosome. Other proteins, most notably membrane bound receptors, release their cargo and are returned to the surface by vesicle fusion. Yet other materials are delivered to lysosomes [8], which have an even lower pH (e.g., pH ~ 5) and are filled with acid hydrolases. Lysosomes are the primary locations for digestion of proteins. Escape of foreign material from endosomes and lysosomes is exceedingly difficult. Endocytosis may be suppressed with temperature decreases (e.g., 4 °C [154]), but this affects other metabolic functions of the cell [155].

11.4.2
Nanostructure Endocytotic Delivery

Internalization of foreign substances occurs through endocytosis, channel transport, or direct passage through the plasma membrane. However, channel transport is usually confined to ions [8], and, given current nanostructure surface treatments and size, it is unlikely that nanostructures could traverse the lipid bilayer unassisted. The endocytosis pathway, with an exclusion size of ~100 nm [156], is the

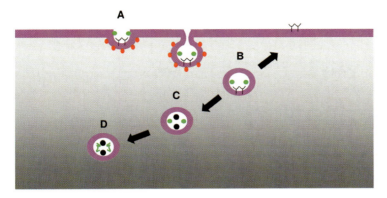

Fig. 11.9. Endocytotic pathway: (A) Following binding, the cell membrane invaginates to encompass membrane-bound substances. The formation of these indentations in the cell membrane is driven by the binding of clathrin protein (red) on the cytoplasmic side. The membrane continues to fold until a vesicle is pinched off from the membrane bilayer. With intracellular entry, the clathrin coat is lost. Vesicles in this stage are known as early endosomes (B) and have a pH of ~6. They are responsible for sorting vesicle contents, e.g., returning membrane-bound receptors (Y-shaped structures) to the cell surface. (C) Next, vesicles migrate toward the Golgi apparatus, where they merge with digestive proteins (black) and become late endosomes (pH ~ 5.5). Late endosomes attempt to digest internalized matter. To complete this process, late endosomes merge with lysosomes (D), which have a pH of ~5.2.

most likely mechanism for nanostructure uptake, and using sugar-coated quantum dot aggregates [156], uptake of particles as small as 5 nm has been demonstrated. Endocytosis can be non-specific, resulting from adsorption on the cell surface, or receptor-mediated, occurring as a consequence of biomolecular recognition binding events between specific receptors and their ligands.

Nanoparticle internalization has been observed for both the non-specific [74, 157] and receptor-mediated [11, 31, 77, 156] pathways. Using the non-specific pathway, particles have been internalized directly from cell culture medium [158], the cell culture substrate [74], and following non-specific binding to the cell surface [76, 159–161]. Internalization likely results from electrostatic attraction between the particle surface and the cell membrane [160]. Interestingly, binding can occur regardless of surface charge, as the plasma membrane exhibits both anion and cationic domains [18]. Receptor-mediated uptake has been observed using the RGD-peptide integrin binding motif [160], epidermal growth factor [31, 162] and transferrin [11, 163]. Remarkably, endocytosis of nanostructures attached to a solid support (e.g., polymeric nanocolumns) has also been observed [164]. In fibroblasts, columns were surrounding by cell membrane and were located near newly-formed intracellular vesicles and clathrin pits. Macrophage-like phagocytotic vesicles, which increased disruption of the cytoskeleton, were also observed.

Regardless of the mechanism, internalization occurs within 30 min, unless specific steps are taken to inhibit internalization pathways (e.g., low temperature [154]). In studies of sugar-coated nanoparticles, the rate of endocytosis was higher with increasing particle size, up to 50 nm. Endocytosis was not observed for aggregates larger than 100 nm [156]. Others have shown that macrophage phagocytosis increases with increasing material size, up to 1.1 μm [165]. Particle endocytosis rates can also be modulated by local biochemistry. For example, uptake of 35-nm dextran coated iron oxide particles was enhanced by high concentrations of interferon-gamma (IFN-γ) and interleukin-4 (IL-4), whereas the application of lovastatin decreased particle uptake [161].

When nanostructures enter the cell, they are found in clathrin-coated pits [162], as would be expected for endocytosis. Co-staining experiments have confirmed that particles are confined to endosomes or lysosomes [159, 162, 163]. Generally, particles remain sequestered in these compartments [18], and are unable to escape. For example, particles internalized by vero cells accumulated in large storage vesicles, similar to those found in macrophages [159]. Vesicles appeared to pass to select daughter cells through asymmetric mitosis. Accumulation of particles can be a significant problem, as aggregations can potentially disrupt the cytoskeleton and cell membrane [166]. It is even possible to overload cells with particles, resulting in apoptosis [18].

Particle endocytosis has been utilized in several applications, the largest area being cell labeling and tracking. Nanostructures, particularly quantum dots, have several advantages in this application. Their fluorescent signals are resistant to bleaching and oxidation over long time scales [11]. Additionally, endocytosed particles remain fluorescent for weeks and are passed to daughter cells [76]. Internalized nanoparticles have been used to investigate several systems that organic dyes could not, including division of HeLa cells [76], formation of aggregate centers in starved *D. discoideum* [76], phagokinetic tracks and metastatic potential [74], as endosomal markers [159], and retrograde transport in filopodia [162]. Particles have also been used to monitor macrophages in atherosclerotic plaques *in vivo* [161].

However, for targeted delivery to cytoplasmic elements, endocytosis is an undesired event, as most nanostructures remain sequestered in lysosomes or endosomes. Endocytosis can be prevented. For example, particles attached to insulin or lactoferrin surface molecules were not internalized as frequently as unmodified controls [157, 167]. Most likely this resulted from binding to "fixed" molecules on the cell surface that are endocytosed at low rates. Also, endocytosis does not occur when particle non-specific binding is low. Mercaptoacetic acid-bound and commercially available streptavidin-conjugated CdSe/ZnS particles with low cell binding potential are not internalized [11, 162] whereas silanized CdSe/ZnS nanocrystals that have shown high cell binding affinities are [10]. Thus, surface chemistry can be used to alter the endocytotic potential of nanostructures. Differences have even been seen for particles with the same surface passivation, but different synthesis techniques (compare, for example, Gomez et al. [160] and Nie et al. [11]). These differences indicate that the most important method of preventing endocytosis is ensuring that the nanocrystal passivation chemistries do not permit non-specific binding to the cell surface. However, notably, methods to prevent endocytosis also prevent *any* cellular entry. Clearly, additional technologies will be required for targeted delivery of nanostructures to the cytoplasm.

11.4.3
Other Methods of Cellular Entry

To circumvent the difficulties associated with endocytosis, several methods of particle entry have been devised. Most of these techniques have been extrapolated directly from the field of gene therapy, the most popular method being the use of translocation peptides. Although there is still much debate as to the exact mechanism of these peptides [168–170] it appears that cell entry begins normally, through the endocytotic pathway. However, the peptides are able to disrupt the endosomal membrane, releasing its contents to the cell interior. This technique has been used with various outcomes. The pep-1 sequence, which forms complexes with quantum dots through its hydrophobic domain, was used to deliver particles to CHO-K1 hamster epithelial cells [171]. However, quantum dots appeared to remain isolated in vesicles, indicating that endosomal escape most likely did not occur. Other peptides have demonstrated greater success. Adenoviral nuclear localization and receptor-mediated endocytosis sequences were used to introduce gold nanoparticles to the cytoplasm [172]. Delivery did not occur unless both sequences were present, with particles targeting the nucleus as expected. Similar results were reported for iron oxide nanoparticles conjugated to the HIV tat peptide [173, 174]. Particles were distributed throughout the cytoplasm with some increase near the nucleus. Quantum dots conjugated to trichosanthin protein [175] and $R_{11}KC$ peptides [176] displayed nuclear targeting with aggregation near the nucleolus [175]. Mitochondrial targeting has also been demonstrated using the Mito-8 peptide [176].

Apart from translocation peptides, other approaches have been examined, with a rather elegant analysis conducted by Derfus et al. [177]. The researchers compared intracellular delivery of quantum dots using translocation peptides, cationic liposomes, dendrimers, electroporation, and microinjection. Of the chemical approaches, cationic liposomes offered the greatest degree of quantum dot delivery, with over 90% of fluorescence distinct from that of separately labeled endocytotic vesicles. In contrast, translocation peptides (i.e., pep-1 as Chariot, Active Motif) displayed fluorescence signals slightly lower than that of the unconjugated control, possibly indicating fluorescence quenching as a result of aggregation, and were localized almost entirely in endosomal compartments. These results are similar to those of Tkachenko et al. [172] who found that pep-1 like peptides produce particle accumulation in lysosomes/endosomes. Additional translocation peptides (i.e., TAT or HA-TAT) were not investigated and may yield different results (e.g., Tkachenko et al. [178]). All three chemical techniques yielded particle aggregates that were on the order of a hundred nanometers in diameter. These aggregates can diminish nanoparticle fluorescence, and may interfere with intracellular manipulation and targeting.

As an alternative, Derfus et al. also investigated physical delivery methods, including electroporation and microinjection [177]. Electroporation produced significant intracellular delivery to the cytoplasm, but particles displayed aggregates with diameters as high as 500 nm. This effect was not related to a loss of surface passi-

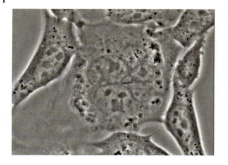 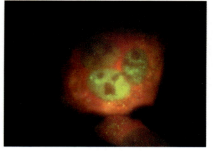

Fig. 11.10. Subcellular localization of quantum dots. A nuclear localization sequence (NLS) peptide was conjugated to the surface of PEGylated green quantum dots (emission maxima 550 nm). These particles were microinjected into the cytoplasm of 3T3 fibroblast cells, along with a 70 kDa rhodamine-dextran. After several hours, green emission from the quantum dots was observed in the nucleus of injected cells, while the red signal from high molecular weight dextran remained in the cytoplasm. (Photo courtesy of Sangeeta Bhatia and Austin Derfus, MIT, Boston, MA.)

vation, as highly crosslinked BSA coated particles exhibited similar results [177]. Similar aggregation has been reported with negatively-charged DNA plasmids [177], and may result from alterations of particle surface charge induced by the electric field. Alternatively, microinjection techniques delivered monodisperse particles that were not confined to endocytotic vesicles (Fig. 11.10) [177]. Previously, these particles were shown to pass to daughter cells during mitosis and retain fluorescence for up to 4 days [73]. The primary disadvantage of the microinjection technique is that it is difficult to scale-up to large populations, as injection is required for each cell.

11.4.4
Nanoparticle Intracellular Sensing

A direct impact of intracellular delivery has been the ability to visualize internal components of a cell. Nanostructures have many optical properties that make them excellent candidates for intracellular labeling. For example, quantum dots display high quantum yields, narrow emission bandwidths, resist photobleaching, and can be readily modified with biomolecules to allow targeting [19]. Fluorescence persists for days [73] or weeks [76]. Additionally, magnetic nanostructures can be monitored with existing hospital equipment (e.g., MRI). Using nanostructures for sensing, it is possible that internal cell functions, including cell signaling and gene expression could be monitored in real time. This ability would represent a tremendous advance for diagnostics and treatment.

11.4.4.1 **Semiconductor Quantum Dots**
Because of their unique optical properties, quantum dots have been widely used in sensing applications. Unfortunately, delivery of quantum dots to living cells has

Fig. 11.11. (A) Transmission images of lung cancer (H1299) tumor. (B) Fluorescence image of same tumor incubated for 1 h with MPA-PEG-capped CdTe nanocrystals injected intratumorally and imaged at an excitation wavelength of 545 nm. (C) Fluorescence image of tumor, reverse angle. (Photo courtesy of Brian Korgel, Felice Shieh, University of Texas at Austin, Austin, TX.)

been challenging. Quantum dots can be introduced through endocytosis, and have been used to monitor cells during growth and development [76]. Particles introduced in this way remained in the cell for up to a week; however, targeting was non-specific, not directed to a particular intracellular component, and particles remained in endosomes. In another study, particles, introduced through microinjection, were used to follow development of *xenopus* embryos [73]. In this particular application, quantum dots were far superior to previous methods. Because quan-

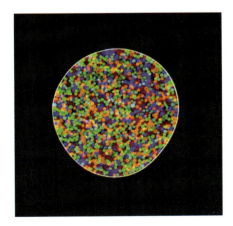

Fig. 11.12. True-color fluorescence images of mesoporous silica beads (5 µm diameter) doped with single-color quantum dots emitting light at 488 (blue), 520 (green), 550 (yellow), 580 (orange), or 610 nm (red). A large population of the QD-beads is confined in a water droplet. (Photo courtesy of Xiaohu Gao, University of Washington, Seattle, WA.)

tum dots remained fluorescent in the daughter cell population for weeks, this technique allowed researchers to examine the fate of cells in the blastula, an achievement not possible using fluorescent dyes that photobleach in minutes. This technique was successful because only a few cells needed to be targeted for delivery; it would be difficult to use microinjection, a serial technique, in applications that require observation of multiple cells (e.g., organ labeling).

Both of these studies examined whole cells, and not specific targets. Most subcellular analyses have focused on extracellular targets, highlighting the difficulties of controlled intracellular delivery. For example, Dahan et al. [179] monitored the diffusion of glycine receptors in the synaptic cleft using semiconductor quantum dots. Quantum dots are particularly well-suited for this application because of their resistance to photobleaching. Traditional dyes only allow monitoring for short time periods, on the order of minutes, whereas quantum dots allowed receptor tracking for up to 20 min. Additionally, their small size allowed access to the synaptic cleft, a region that cannot be reached by microbeads (e.g., gold) that might be used in single particle studies. Alternatively, the diffusion of epidermal growth factor receptors (EGFR) was followed using quantum dots bound to EGF through avidin–biotin interactions [162]. Because of their resistance to photobleaching, receptor dynamics could be followed for long time periods, and a previously unreported mechanism of retrograde receptor transport was discovered. Quantum dots have also been conjugated to aptamers, DNA or RNA molecules that have been selected from populations for their specific binding affinities [180]. Aptamer-labeled particles were used to identify cancer cells *in vitro* and in lung cancer tumors (Fig. 11.11).

Applications for intracellular tracking are still in development. In particular, researchers are interested in creating tags that could be used for multiplexed intracellular measurements. Quantum dots are good candidates for these tags because they can exhibit a range of emission wavelengths under the same excitation, allowing for observation of multiple cues simultaneously (Fig. 11.12) [181]. These tags have great potential *ex vivo* as diagnostic tools and for high throughput screening. However, it may also be possible to observe the expression of several intracellular proteins concurrently through optical signals. In fixed cells, quantum dots conjugated to specific DNA sequences have been used to label the Y chromosome [182]. They have also been used to examine the intracellular expression of eNOS, a cardiovascular enzyme that converts nitrates into NO [183]. If successfully introduced into the cytoplasm, labeling of individual DNA strands or mRNAs may be possible. One limitation of quantum dots for intracellular tracking is the high blinking interval [184], which may prevent the tracking of intracellular components with a high diffusion rate. Nevertheless, with their persistent fluorescence and range of emission wavelengths, quantum dots will likely increase our understanding of intracellular signaling and could potentially revolutionize the field of diagnostics.

11.4.4.2 **Magnetic Nanoparticles**

Magnetic nanoparticles have also been employed for whole cell labeling and intracellular tracking. Particles have been conjugated to the tat peptide, a nuclear localization sequence, and utilized to determine the fate of stem cells [185]. With MRI, particles were visualized in blood and neuronal stem cells, allowing the location of those cells to be determined following injection in the rat tail vein. Alternatively, particles were coated with a combination of tat and Ha2 nuclear localization sequences and fluorescent dyes [174]. These particles responded to magnetic fields *in vivo* and were used to detect expression of the DEVD peptide, an apoptosis inhibitor. Particles have also been used for extracellular protein detection. For example, nanoparticles conjugated to transferrin have been used to monitor the expression of the transferrin receptor *in vitro* [186].

11.4.5
Future Directions

11.4.5.1 Nanostructure Intracellular Delivery
Existing methods of particle delivery have not yet yielded a technique for large-scale introduction of monodisperse particles. Biochemical methods can be used to evade endocytotic vesicles, but reports indicate [171, 172, 177] significant aggregation of particles. Alternatively, microinjection can produce disperse particle delivery, but is a serial technique, requiring cell-by-cell injections. Taking cues from research in gene therapy, researchers will continue to develop methods of particle delivery. Several new techniques are being explored as possible drug or DNA delivery methods, and could be extrapolated to nanoparticle delivery.

For example, polymersomes of PEG-polyester have been shown to foster endosome rupture and have been used to deliver two anti-cancer drugs [187, 188]. The low pH of endosomes encourages polymersome degradation. Soluble polymer components are then free to disrupt the lipid membrane, facilitating cargo escape into the cytoplasm. Similar polymersomes might be constructed to delivery nanoparticles to the cytoplasm. Likewise, fusogenic liposomes have been created to encapsulate polymeric nanoparticles [189]. Liposomes were created by encapsulating polymeric particles in traditional liposomes whose surface was modified by the addition of inactivated Sendai virus. The liposomes were able to deliver DNA to the cytoplasm, technology that could easily be adapted for delivery of magnetic or semiconducting nanoparticles.

Another potential delivery method uses carbon nanotubes to promote rupture of endocytotic vesicles [190] or directly penetrate the cell membrane [191, 192]. For example, nanotubes can be internalized through endocytosis, and then irradiated in the near-infrared to produce endosomal release. The primary limitation of this method is ensuring that heating levels remain low enough to prevent cell death. It is also possible that magnetic particles in a magnetic field could be used for this purpose (see, for example, Won et al. [174]), and a similar technique has been developed by Cai et al. using the Ni catalyst at a carbon nanotube tip for magnetic manipulation [192]. A rotating magnetic field allowed the nanotubes to penetrate the cell plasma membrane, where they remained. Soluble factors (i.e., DNA) bound to the surface were released from the nanotube surface through an unknown mechanism. Alternatively, carbon nanotubes functionalized with ammonium permeate the cell membrane without the aid of magnetic fields [191]. The hydrophobic nanotubes can insert into the lipid bilayer of the plasma membrane, possibly transversing it. Any of these techniques might be adapted for nanostructure delivery; however, adjustment may be challenging as the surface chemistry and size of nanostructure cargos are significantly different from the DNA normally transported.

11.4.5.2 Intracellular Sensing

An interesting possibility for sensing subcellular components is the use of fluorescence resonance energy transfer (FRET). This simple method for detection employs fluorescent molecules to produces an on or off signal upon binding. FRET occurs when an excited electron dissipates its energy by transferring it to a second molecule rather than through photon emission. This quenches the fluorescence of the first molecule. FRET is a distance dependant event. As the FRET donor and acceptor move away from each other, transfer no longer occurs and fluorescence in the donor is restored [193]. If a binding-induced conformational change separates two fluorescent molecules, e.g., attached to different portions of a protein, a fluorescent on/off signal can be produced.

This has been accomplished in several systems. A combination of gold quantum dots (which do not fluoresce) and organic fluorophores have been utilized to create DNA sensors [194]. Gold quantum dots were bound to single-stranded DNA sequences, which were terminated with fluorescent molecules, or fluorophores. The fluorophores formed an arched structure that interacted with the nanoparticle surface, quenching fluorescence (Fig. 11.13). However, when the complement DNA was introduced, binding occurred and the conformation changed. The fluorophore migrated away from the quantum dot surface and its fluorescence was restored. This technique is very accurate, the mismatch of a single base pair in a 30 amino acid DNA sequence could be detected [194].

A second, similar system utilized fluorescent CdSe/ZnS quantum dots to detect sugar binding [195]. The initial configuration contained cyclodextrin molecules

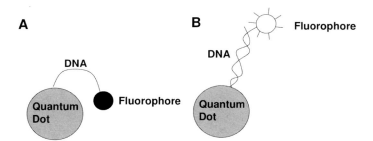

Fig. 11.13. Detection of DNA binding through FRET. (A) Fluorophore-conjugated single-stranded DNA is bound to a quantum dot, forming an arched structure with the nanoparticle surface. Fluorophore fluorescence is quenched through FRET with the nanoparticle. (B) When the complement DNA is added, the DNA structure changes, moving the fluorophore away from the nanoparticle and restoring fluorescence.

conjugated to a non-fluorescent FRET acceptor. The cyclodextrin adhered to maltose (i.e., sugar) binding protein, which was present on the quantum dot surface. FRET between the nanoparticle and the acceptor molecule quenched quantum dot fluorescence. However, when maltose was introduced, the cyclodextrin was displaced, restoring fluorescence to the quantum dot. Although maltose sensing was demonstrated, these sensors could eventually be adapted to detect glucose, a crucial molecule to monitor in the management of diabetes.

Finally, systems need not be hybrids, incorporating nanoparticles and fluorescent dyes; homogeneous systems have also been constructed [196, 197]. Two types of CdTe quantum dots were bound to model proteins. Red-emitting quantum dots were conjugated to bovine serum albumin (BSA), while green-emitting quantum dots were conjugated to an antibody that binds BSA (i.e., anti-BSA). When the two quantum dots were incubated together, the green fluorescence was quenched, as excited electrons transferred their energy to the red-emitting particles. Further, the red fluorescence was enhanced because energy transferred from the green-emitting quantum dots was dissipated through the red-emitting quantum dot. Similarly, the fluorescence of biotinylated-CdTe nanowires has been extinguished by the presence of streptavidin-coated gold nanoparticles (Fig. 11.14). Biotinylated-wires exposed to streptavidin alone displayed no change in fluorescence over time (Fig. 11.14A); however, upon addition of streptavidin-coated gold, fluorescence was increased in the CdTe acceptors (Fig. 11.14B).

FRET-based quantum dot sensing is elegant, and in theory can provide analyte detection of as few as 10 parts per trillion [196]. Additionally, the signal is easy to interpret: fluorescence indicates the presence of an analyte. FRET-based sensing can also provide quantitative information, as fluorescent intensity can be correlated to the number of molecule binding events. Although FRET systems could be constructed entirely from fluorescent dye molecules, quantum dots provide an excellent alternative. They display high quantum yield; and their acceptor energy and emission wavelengths can be tailored by changing the size of the particle.

11.5
Intracellular Transport of Nanostructures

Nanostructures that evade endocytosis have the potential to interact with intracellular structures, including components of the cytoskeleton. The cytoskeleton is responsible for the movement of proteins and vesicles through the cytoplasm, as well as cell mitosis, locomotion, and signaling. Nanostructures have been used to manipulate and interact with the cytoskeleton from the cell exterior (Section 11.3.4), but numerous additional possibilities exist for intracellular interactions. Perhaps the most intriguing potential application is the use of cytoskeletal elements to control intracellular nanostructure transport. This line of inquiry will not only improve biologists understanding of motion through the cell interior, but may aid in assembly of nanoscale structures *ex vivo*. To date, nanostructure-cytoskeleton research has focused on *ex vivo* applications, including sensing and molecular assembly.

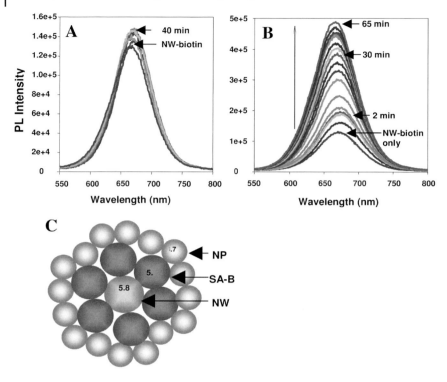

Fig. 11.14. (A, B) Transient luminescence spectra of bioconjugated nanowires in solution. (A) Nanowire-biotin with streptavidin (without Au) monitored for 40 min after mixing, (B) Nanowire-biotin with streptavidin-Au monitored for 65 min. Excitation wavelength for both (A) and (B) is 420 nm. (C) Cross section of closely packed model arrangement of CdTe nanowires and Au nanoparticles in the bioconjugate. (Photo courtesy of Nicholas Kotov, University of Michigan, Ann Arbor, MI.)

The primary focus has been on kinesin-microtubule structures, but actin–myosin interactions have also been examined. Direct translation of these methods to the clinic is not expected. However, the techniques developed for *ex vivo* studies might be adapted to nanostructures in the cytoplasm. The ability to perform cytoskeletal manipulations would allow researchers to explore internal mechanotransduction and manipulate intracellular transport, providing new insight to gene regulation, protein expression, vesicular transport and mitosis.

11.5.1
Biology of Intracellular Transport

The cytoskeleton is composed of a network that includes actin filaments, intermediate filaments, and microtubules [8]. Intermediate filaments, which are about 10 nm in diameter, provide the cell with mechanical strength and form the nuclear lamina, whereas motion primarily occurs along 5–9 nm actin filaments or 25 nm microtubules. Actin filaments interact with myosin proteins and are responsible for muscle contraction, cell locomotion, and vesicle transport [198]. Myosins contain heavy and light chains, with the heavy chain mediating attachment to actin filaments and the cargo. Motion is driven by the hydrolysis of ATP (i.e., adenosine triphosphate), which initiates conformational changes in the myosin protein, producing a stepping motion along the actin filament. Microtubules interface with kinesins and dyneins and are responsible for organelle and protein transport, chromosomal separation, cell division, vesicular transport, and are the primary components of cilia and flagella [198]. Microtubules are directional, continuously growing at their plus ends and slowly dissociating at their minus ends. Kinesin contains two heavy chains that include microtubule binding sites, whereas the structure of dynein is not well understood, but includes heavy chains and intermediate chains. Both kinesin and dynein movement occur through ATP hydrolysis. However, kinesin moves from the minus to plus end of the protein in discrete 8 nm steps, whereas dynein moves from the plus to minus end of the track [198]. Dyneins are primarily associated with cilia and flagella, cellular elements which produce cell movement.

11.5.2
Actin-based Nanostructure Transport

Movement in actin networks is initiated by ATP-binding to myosin proteins. The myosin II form, which is responsible for muscle contraction, is the most common, but myosin V may be the best suited for hauling cargo because of its processive motion [198]. Several nanostructure cargos have been attached to myosin. For example, quantum dots have been conjugated to myosin V heads [199]. Fluorescence signals from the particles allowed direct visualization of the "hand-over-hand" mechanism of myosin movement down the actin filament. Also, rabbit muscle myosin was conjugated to magnetic nanoparticles obtained from bacteria [152]. Using magnetic force, it was possible to alter the velocity of a myosin molecule along the actin filament. It is envisioned that such a system could be used to analyze single molecule interactions between myosin and actin filaments.

Alternatively, actin has been conjugated directly to nanostructures. Biotinylated phalloidin (which binds actin strongly) was attached to quantum dot surfaces and used to link nanoparticles to actin [200]. Proposed applications of this system include the creation of *in vitro* networks of actin filaments or delivery of specific cargos. Actin has also been attached to the surface of gold nanoparticles, a system used as a template for gold nanowire formation [201]. Gold nanoparticles bound to actin monomers could be connected to form nanowires on either side of an exposed actin protein core. Additionally, when placed on a myosin substrate, actin–myosin interactions directed nanowire movement. This technology is primarily intended for the creation of nanoelectronic networks, but might be used to assemble or transport nanostructures in cells.

11.5.3
Microtubule-based Nanostructure Transport

In contrast to actin filaments, microtubules have been more extensively studied for nanostructure transport, probably because they move a wider variety of natural cargos *in vivo*. Again, much of the work to date has been performed *ex vivo*, with proposed applications in nanoassembly and nanoelectronics. For example, micro-tubules have been organized in microfabricated channels using electric fields [202, 203] and their motion has been directed by absorbed kinesins [204]. Although this method does not apply directly to cellular engineering, many of the tools to adapt these techniques *in vitro* are in place. Silica microbeads were attached to kinesin as early as 1990 [20]. *Ex vivo* binding or transport of microchips [205], 1–10 µm glass, gold, and polystyrene microbeads [206], CdSe quantum dots [207], CdSe nanorods [208], and gold nanowires [202] have all been demonstrated using the kinesin-microtubule system. The wide variety of cargos transported suggests that various nanostructures can be moved with ease. Techniques to control cargo motion have been developed. Photo-caged ATP can be activated with the application of light, permitting control of kinesin movement [209]. Collectively, these advances should allow for intracellular manipulation of the cytoskeleton in the near future.

11.5.4
Future Directions

Future research will focus on adapting *ex vivo* techniques to cell studies. For example, motor proteins without nanostructure modification have already been used *ex vivo* to transport and manipulate DNA [210]. The addition of nanostructures to these systems could provide the ability to control (e.g., magnetic manipulation) and monitor (e.g., fluorescent signal) biomolecule interactions *in situ*. Development of these assays would allow researchers to directly measure the force, speed, and dynamics of protein–cytoskeleton interactions. Additionally, nanostructure transport could be used for controlled delivery of drugs, proteins, vesicles, or even organelles (Section 11.6). A limitation of previous attempts to use nanostructures for this purpose has been the inability to control cargo release. One possibility is the use of photocleavable linkages [211]; however, the use of photo-caged ATP may limit the wavelengths available for this chemistry. An alternative approach might be to use the heat generated by excited magnetic particles to promote release of an attached biomolecule.

In addition to intracellular transport, cytoskeletal systems in whole cells may be harnessed to transport and manipulate objects. For example, the flagella of close-packed bacteria grown on a solid surface can be used to manipulate beads suspended in fluid [212]. Flagellated algae (e.g., *Chlamydomonas reinhartii*) were used to move polymer beads through solution [211]. Algae were directed to specific locations using phototaxis, and bead attachment did not appear to significantly impede algae movement. It is unlikely that flagella or cilia could be harnessed directly to nanostructures, as they are significantly larger ($\sim$1 μm). However, these whole-cell systems could be used for controlled delivery of nanostructures amongst a population of cells. For example, it might be possible to deliver toxic nanostructures to cancer cells using "safe" bacteria to transport particles into a specific organ.

11.6
Biomolecule Delivery Using Nanostructures

One of the most promising applications of nanostructures is their use to deliver biomolecules to specific cells and cellular locations. Nanostructures with diameters from 10 to 100 nm can be introduced intravenously, and remain in the blood for significant periods of time [18]. Their surfaces may be modified with drugs, biomolecule cargos, and sequences to promote recognition by certain cell types (Section 11.2.4), resulting in specific targeting. The magnetic and heat-generating properties of nanostructures may be used to control biomolecule release. Nanostructures with these features have been incorporated into traditional drug delivery constructs to produce "smart" drug delivery devices (e.g., Sershen et al. [17]). Although much work has focused on extracellular delivery, intracellular delivery is also possible (Section 11.4). Nanostructure–DNA binding has been well-established [13], and gene therapy is a logical extension of this work. With their small size and flexible surface chemistry, nanostructures are ideal candidates for biomolecule delivery to cells.

11.6.1
Biology of Controlled Delivery

11.6.1.1 Drug Delivery
Therapeutic drugs typically display half-lives on the order of minutes to hours [213]. To control release and prolong the effectiveness of these molecules, drug delivery systems have been developed. Drug delivery devices may target any component of the cell interior or exterior and have been investigated as treatments for various conditions, including cancer [214], diabetes [215], and neurological disorders [216, 217]. In many cases, the desired treatment requires controlled release of a biomolecule over time, possibly many years [215]. Several systems have been developed to fine tune release, including polymer coatings and degradable hydrogels [218].

Drug delivery systems must operate *in vivo*, where the environment can be harsh. Introducing drugs targeted to specific cell types without direct implantation of a drug delivery system is difficult. Drugs may be administered orally or through direct injection, but the primary method is through intravenous injection [219]. Once inside the body, these molecules must evade the natural defense mechanisms of the body, including the reticulo-endothelial system (RES). Components that are too large ($>$200 nm) will be isolated in the spleen, whereas molecules that are too small ($<$10 nm) are removed from blood vessels by diffusion to surrounding tissues or the kidneys [18, 145]. Some drug delivery applications require release to a specific target (e.g., gene therapy, cancer treatment), and achieving this is challenging. For example, the endocytotic pathway (Section 11.4.1) presents a significant barrier to controlled intracellular delivery (e.g., for cancer treatment). With their readily functionalized surfaces and small size, nanocomponents provide a unique method to address these issues.

11.6.1.2 Gene Therapy

In contrast to most drug delivery applications, gene therapy focuses primarily on nuclear delivery of DNA in a one-time application. The concept behind gene therapy is straightforward. Many diseases result from defective copies of a particular gene. Gene therapy attempts to incorporate functional copies of those genes into the genome of affected individuals. For this to take place, the defective genes must be identified in the individual. New, working copies of these genes must be synthesized, delivered to the nucleus, and incorporated into the existing genome. This is usually achieved with the aid of transcription agents, which may use viral or nonviral mechanisms (for review see Johnson-Saliba et al. [220]). Viral vectors have been the most successful, but can provoke an immune response. Additionally, viruses do not infect all cell types equally, and targeting specific cell types can be difficult [221]. Non-viral delivery has been plagued by the difficulties of traversing the plasma membrane and directed targeting to the nucleus. Transfection efficiencies have remained low (0.01–10% [220]). Some of the greatest barriers to clinical gene transfection have been the long times required to achieve transfection and the low concentrations of available transfection agents [222]. For gene therapy to become a viable clinical option, these barriers must be surmounted.

11.6.2
Drug Delivery

Applications for nanostructures in drug delivery are vast (see Volume 10 of this series for more detail). Although usage of metallic and semiconducting materials has been more recent, nanostructured ferrofluids have been investigated since the early 1970s [223, 224]. Drug delivery systems have been envisioned to treat diabetes [215], cardiovascular diseases [225], and cancer [214]. As a result, this section will focus on several of the challenges facing drug delivery systems, and how nanostructures are poised to meet them. This will be illustrated with a few examples of targeted drug delivery applications, primarily in the area of cancer treatment. These applications are not intended to be comprehensive, but represent general cases where nanostructures improve upon existing technology.

11.6.2.1 Cell Targeting *In Vivo*

For optimal control, nanostructures for drug delivery usually display three separate functionalities: a core material, drug or therapeutic element, and a targeting element [213]. The core material provides physical stability, controls size, and provides a template for drug attachment/encapsulation. In some cases, the core material may serve multiple purposes, incorporating an additional targeting or sensing function. For example, magnetic particles may be manipulated with a magnetic field [18]; gold particles can be detected using Raman spectroscopy [47]; and quantum dots can be seen using fluorescence microscopy [19]. The drug or therapeutic element may be attached directly to the nanostructure, or encapsulated in its interior. Targeting molecules (e.g., biorecognition elements, see Section 11.2.4.2) can be added to the nanostructure surface to direct interactions with the cell of interest.

Many systems also include a component to increase circulation time and reduce RES recognition. Early efforts to use micro- or nano-sized particulates for drug delivery were plagued by short circulation lifetimes. Attachment to a carrier molecule can prevent these difficulties. For example, a significant improvement in delivery occurred with the addition of poly(ethylene glycol) (PEG) to particle surfaces [213]. PEG prevents the adsorption of proteins, which might be recognized by the RES [145]. Additionally, small nanoparticles (<10 nm) coated with specific targeting peptides accumulate in the tissue of interest while evading the RES [75]. Thus, particle surface modification can facilitate drug targeting and prevent RES clearance.

Nanostructures offer several advantages to traditional drug delivery systems. Nanostructures may be loaded with vast quantities of drug through encapsulation or binding to the particle surface. They can be further modified with specific targeting molecules. For each binding event, an entire nanostructure payload may be delivered, in contrast to the single molecules delivered using direct drug–biomolecule conjugation schemes [213]. Additionally, nanostructures can reach some target sites that are not accessible by therapeutic agents administered via traditional routes. For example, they have demonstrated the ability to cross the blood–brain barrier [213]. Unlike traditional systems, nanostructures can also be targeted to specific elements through external manipulation (e.g., magnetic fields), and those same signals can be used to trigger controlled release.

11.6.2.2 Drug Delivery for Cancer Treatment

The possibilities for nanostructures in cancer therapy are staggering (also Volumes 6 and 7 of this series for more detail). Nanostructures can provide imaging capabilities, allowing detection of cancer in early stages [213, 226]. Therapeutics may be administered using nanostructure carriers [213]. If these particles are combined with imaging modalities, drug distribution can be monitored in real time. Delivery vehicles could be coupled to biorecognition elements that target only cancerous, and not healthy, cells [213]. Most research to date has focused on magnetic therapies to deliver cytotoxic agents; however, new materials and new approaches are beginning to be investigated. In some cases, the nanomaterial itself becomes the cytotoxic agent, activated by a remote optical source. Because of the ease of surface modification, size selection, and tunable optical, electrical, and magnetic properties, these drug delivery systems will likely become powerful additions to current clinical cancer treatments.

Molecular Approaches   For recovery, cancer treatment requires complete removal of the cancerous cells [227]. In many cases, tumors cannot be fully excised surgically and traditional chemotherapy is either not possible or undesirable. Targeted cytotoxic drug delivery to the tumor provides one alternative to these standard treatments, and it was for this purpose that therapeutic nanostructures were first investigated [223, 224]. The concept is straightforward. Magnetic particles are conjugated to cytotoxic drugs. External magnetic fields are then used to concentrate the particles near the tumor. Particles are internalized, and the drug payload is released by physiological change (e.g., temperature, pH, enzymatic activity) [228].

Magnetic targeting has been used in many environments. Magnetic microspheres have been utilized to deliver drugs to the brain, with concentrations 100–400× higher than those of mice exposed to drug solution alone [229]. In brain tumor-bearing rats, similar systems administered 41–48% of the dose directly to the tumor, in comparison to 23–31% of dose in the absence of the magnetic field [230]. Particle localization in the vasculature has been controlled with magnetic targeting in a swine model [231]. Starch-coated particles, administered to rabbits intra-arterially, were concentrated in invasive squamous cell carcinomas following applications of magnetic fields up to 1.7 T [232]. Magnetic liposomes were delivered to osteosarcomas, suppressing tumor growth [233].

Nanostructure magnetic therapies have also been applied in the clinic. Lübbe et al. [227, 234] investigated the efficacy of magnetic particle bound-epirubicin to advanced tumors. Tumors were less than 0.5 cm from the magnet source and were subjected to magnetic fields of 0.5–0.8 T. No adverse reactions to the magnetic particles were observed, although toxicity was seen at high epirubicin doses. Results were mixed, as particle accumulation in tumors was not successful in 50% of patients [227]. Physiological parameters played a great role in this variability and must be addressed in future attempts to develop successful therapies.

One difficulty in implementing magnetic therapies is that magnetic field strengths high enough to overcome natural body forces (e.g., blood vessel flow) are difficult to develop and maintain, particularly for deep tissue targets [227]. In addition, optimizing particle size to provide the highest level of particle concentration at a tumor site is difficult. Magnetic susceptibility is inversely related to particle size, but larger particles are more likely to be removed from circulation by the RES [227].

Magnetic fields can be used to produce particle accumulation in a chosen tissue, but specific targeting sequences are required to promote selective internalization by cancerous cells. These surface agents are designed to prevent macrophage endocytosis (i.e., immune reaction), while increasing endocytosis in target cancer cells. Several coatings have been investigated. Poly(ethylene glycol) increased uptake in breast cancer cells, while reducing internalization by macrophages [235]. Modification with folate also produced selective internalization by breast cancer cells, but not other cell types [235]. Amino-poly(vinyl alcohol) (PVA) coatings encouraged endocytosis of magnetic particles by melanoma cells, whereas PVA, carboxylate-PVA, and thiol-PVA did not [236].

Once inside the cell, the drug must be released, enter the cytoplasm, and diffuse to its target (in most cases the nucleus or mitotic spindle). Drug release must be precisely controlled. Once the drug payload has detached from the magnetic particle it can no longer be manipulated by a magnetic field. Drug escape from endosomal/lysosomal compartments is possible. The release of drug payloads from nanoparticles in lysosomal environments has been documented and correlated to observed cytotoxicity in breast and cervical cancer cells [237]. Additionally, oleic acid/pluronic coated-particles containing doxorubicin were internalized by prostate and breast cancer cells, producing dose-dependent cytotoxicity [238].

Apart from magnetic materials, researchers have investigated other nanostructures as nanochemotherapeutics, including colloidal gold [239] and silica [240]. Colloidal gold has been used to deliver tumor necrosis factor (TNF) [239] in a mouse model. Native TNF has been used as an anti-cancer therapy before but, however, was found to be toxic at therapeutic doses [239]. Particles bound to TNF, and containing PEG modifications to reduce RES clearance, accumulated in MC-38 colon carcinoma tumors and produced 50–90% reduction in tumor volume. Additionally, toxicity was drastically reduced. None of the animals examined demonstrated adverse reactions, whereas 25–100% of animals exposed to native TNF died. Development of silica nanoparticle carriers, though, is still in early stages. Silica has been attached to cefradine, a broad spectrum antibiotic [240]. Silica particles demonstrate drug release profiles that are likely tunable with particle size, but have yet to be tested *in vitro* or *in vivo*.

Hyperthermia   Apart from their use as drug delivery vehicles, nanostructures can also provide the toxic moiety, treating cancer directly. Because of their unique optical properties, certain types of nanomaterials can produce local temperature elevations. This process, known as hyperthermia, has been investigated by several groups. As a result, this discussion is limited in scope, for more detailed information the reader is directed to recent review articles (e.g., Moroz et al. [241], Ramachandran et al. [242]). Hyperthermia is a result of particle Nèel relaxation [243] in an AC magnetic field [228] (e.g., 5–30 kA m^{-1} at 100–500 kHz [242]), producing a temperature elevation. When the temperature is maintained at 42 °C or higher for 30 min, cell death, primarily through necrosis, will occur [228, 242]. The exact mechanisms of cell death are not completely understood, but the effects of heat on DNA repair and protein expression are well-documented [244].

The use of magnetic particles to locally heat tumor tissue was first proposed in 1957 by Gilchrist et al. [245]. In that study, 20–100 nm magnetite particles were directly injected into the intestines of dogs. Since that time, several alternatives to direct injection have been developed [241]. Arterial embolization employs the tumor vasculature to deliver particles to the tumor interior. Particles may be surgically implanted, although this technique tends to focus on macroscopic structures (~1 mm) [241]. Finally, particles can be conjugated to appropriate biomolecules and may enter the cell (intracellular delivery).

The most limiting features of magnetic hyperthermia have been poor AC field control, temperature control, and cell targeting. High AC fields are not tolerated well, as they can lead to muscle response (e.g., spasms). The advent of nanometer-sized magnetic particles [246, 247] has allowed for the use of lower, more compatible AC fields [242], which may alleviate this concern. However, a clear understanding of the relationship between AC field strength and the resulting temperature profile has not yet been obtained. Temperature control is critical because if body tissues are heated above 56 °C undesired tissue ablation results [242]. Modeling is helpful to address these concerns, but extensive modeling of magnetic hyperthermia systems has not been performed [242]. Additionally, it is difficult to target nanometer-sized particles to tissues of interest using magnetic fields. In this size regime, Brownian motion becomes dominant [242] and substantial force is required to manipulate particles to regions of interest. For this reason, biomolecular targeting may be a superior approach for directing magnetic particles to particular cells. In addition to these factors, magnetic hyperthermia must compete with other better established, approaches, including whole-body hyperthermia and thermistors, to deliver heat [244].

More recently an exciting new material, gold nanoshells, has demonstrated the potential to produce local hyperthermia (Fig. 11.15) [248]. These nanoshells exhibit unique optical properties, which are tunable, based on their size. Nanoshells absorb light in the near-infrared (NIR), and convert this energy into heat, primarily through vibrations of the crystal lattice (i.e., plasmon-resonance) [249]. Nanoshells offer several advantages to magnetic hyperthermia because of the excellent tissue penetration in the NIR and the small number of particles needed to produce a temperature increase [248]. Small particle size allows facile penetration of tumors through the vasculature [249], which is known to be more permeable in cancerous than healthy tissue. Particles introduced in this way may accumulate in tumors

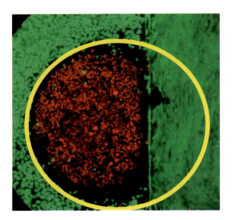

Fig. 11.15. Side-by-side co-culture of SKBR-3 breast carcinoma cells and hamster dermal fibroblasts. Anti-HER2 (a tumor marker overexpressed on the breast carcinoma cells) targeted nanoshells were incubated with the co-culture. After rinsing, the cells were exposed to a near-infrared laser (808 nm, 1 min), then stained for viability with calcein AM (green), indicating live cells, and ethidium homodimer (red), indicating cell death. The laser spot is indicated by the yellow circle. The targeted breast carcinoma cells were ablated while the "normal" cells were unharmed. (Photo courtesy of Jennifer West, Rice University, Houston, TX.)

non-specifically, or through direct, antibody-mediated targeting [250]. This effect has been used to create selective, local tissue ablation targeted specifically to tumors [249]. Additionally, the presence of nanoshells can be confirmed using optical techniques (e.g., using optical coherence tomography), allowing *in situ* monitoring of nanoshell targeting and therapy [15].

Photodynamic Therapy Another interesting application of nanomaterials in cancer therapy is the generation of cytotoxic singlet oxygen. In this process, known as photodynamic therapy (PDT), photosensitizers (PS) are encapsulated in or conjugated to the surface of nanostructures. Additionally, nanostructures can be conjugated to targeting molecules that direct them to local tissues. Upon absorption of light, the photosensitizers excite molecular O_2 to the singlet state, causing cell damage, most notably through disruption of cellular membranes [251]. Singlet oxygen can also cause tissue damage through disruptions of the vasculature, resulting in tissue starvation [251]. Although application of nanomaterials to PDT is recent, several materials have been investigated, including silica, gold, and quantum dot nanoparticles.

Silica nanoparticles are easily modified to allow encapsulation of PS, which are largely hydrophobic. For example, native silica was modified with aminosilanes to promote encapsulation of meta-tetra(3-hydroxyphenyl)-chlorin PS [252, 253]. When excited, the PS-silica nanoparticles produced more singlet oxygen than free PS [253], possibly as a result of increased PS concentration. Silica nanoparticles have also been used to entrain 2-devinyl-2-(1-hexyloxyethyl) pyropheophorbide (HPPH) PS [252]. When exposed to cells, nanoparticles entered the cell interior, most likely through endocytosis. Upon excitation singlet oxygen was produced, resulting in cell death.

Similarly, gold nanoparticles have been prepared as carriers for PS; however, in this case the molecules are linked to the particle surface rather than entrained in particle pores. Using the strong binding affinity between gold and thiol groups, mercaptoalkyl-modified phthalocyanine PS were tethered to the surface of gold nanoparticles [254]. Singlet-oxygen generation was enhanced when compared with that of free phthalocyanine, possibly as a result of the phase transfer agent (i.e., tetraoctylammonium bromide) also tethered to the nanoparticle surface.

An interesting possibility for PDT is the use of semiconductor quantum dots [255]. CdSe quantum dot surfaces (~5 nm) conjugated to Pc4 PS produced excitation and singlet-oxygen generation through fluorescence resonance energy transfer (FRET) [256]. Additionally, unconjugated CdSe quantum dots produced singlet oxygen [256]. Although the efficiency of singlet oxygen generation through this mechanism was very low (5% quantum dots vs. 43% Pc4 PS), quantum dots are extremely resistant to photobleaching and could potentially produce singlet oxygen over longer time periods than traditional PS [255]. The greatest effects on cell viability appear to occur when quantum dots are combined with tethered traditional PS [e.g., trifluoperazine (TFPZ), sulfonated aluminum phthalocyanine (SALPC)] [257]. The exact mechanism for the loss of cell viability in the presence of quantum dots is unclear. It may be attributed to the generation of reactive species, including singlet oxygen, but could also result from the liberation of Cd ions, which are cytotoxic [31, 257].

11.6.3
Gene Therapy

Gene therapy can be thought of as a specific subset of drug delivery. The agent delivered is DNA, instead of a pharmaceutical; and the release of the agent occurs one time rather than at a controlled rate. The ability of nanostructures to bind DNA is well known and was initially investigated for sensing and high throughput screening applications [13]. Recently, several investigators have expanded this research to examine potential applications in gene therapy. There are several potential advantages to nanostructure-based vectors. Like other non-viral systems, they can minimize adverse immune responses. Additionally, nanostructures can be readily modified with biomolecules to promote targeting and uptake. Some nanostructures can also be targeted using external fields and monitored optically.

Initially, nanoparticles were investigated as passive carriers. Silica nanoparticles and gold nanoparticles, in particular, have been examined because of the numerous surface modification strategies available. These systems use nanoparticles to increase presentation of DNA to the cell surface and to couple transfection agents with targeted recognition molecules. Conversely, magnetic nanoparticles have been used as active transfection agents, with magnetic fields directing cell targeting.

11.6.3.1 Silica Nanocarriers

Silica nanocarriers are good candidates for DNA delivery because they are biocompatible, present concentrated DNA to the cell surface, and are easily synthesized. Normally, silica nanoparticles are negatively-charged and will not interact with DNA, which is also negatively-charged. However, their surfaces may be modified with aminosilanes to provide cationic surfaces, and zeta potentials of up to ~50 mV have been observed [258]. These surfaces bind up to 90% of available DNA, primarily through electrostatic attraction [259], and have produced transfection efficiencies 30% as high as those of polyethylenimine (PEI) (60 kDa), a standard cationic carrier. Successful transfection has also been demonstrated *in vivo* (e.g., in the brain), with transfection efficiencies equivalent to those of some viral vectors [260]. Additionally, silica nanocarriers demonstrate far less toxicity than cationic delivery systems [261] or viral vectors [260].

Apart from improvements in biocompatibility, it has been suggested that nanoparticles may enhance transfection efficiencies by concentrating DNA near the cell surface [222]. Traditional transfection systems rely on diffusion to initiate contact between the DNA carrier and cell membrane. Many transfection agents are toxic to cells, and contact time must be limited. Most of the DNA complexes in solution never reach the cell membrane. Because of their increased bulk, nanostructures can overcome these difficulties. In one study, dense silica nanoparticles, coupled to dendrimer transfection agents, sedimented on cell surfaces, increasing transfection efficiencies by almost eight-fold over traditional dendrimers [222]. In another study, the chain length of the dendrimer transfection agent used could be reduced when coupled to silica particles. This is advantageous because long chain (e.g., $G > 5$) dendritic polymers have been found to be effective gene transfection agents; however, they are difficult to synthesize and isolate [262]. When coupled to mesoporous silica, short-chain (i.e., G2) poly-amidoamine (PAMAM) dendrimers were able to deliver gene with transfection efficiencies of 35% [262], comparable to unconjugated long-chain variants.

11.6.3.2 **Gold Nanocarriers**
Like silica, gold nanocarrier surfaces can be readily modified, in this case using thiolated molecules. However, much smaller sizes can be obtained with gold than silica nanocarriers (e.g., 1–10 nm vs. ~30 nm). Smaller nanocarriers potentially increase complex internalization and DNA cargo release [263]. Transfection efficiencies of DNA bound-gold nanocarriers are dependant on the surface charge or hydrophobicity and the chain length of the surface coating. The most successful nanocarriers display efficiencies up to 8× that of PEI (60 kDa) [263].

Gold particles also enhance the efficacy of existing agents. For example, PEI (2 kDa)-bound gold displayed an eight-fold transfection increase over PEI alone [264]. When combined with chemically modified PEI treatments, transfection of up to 50% of cells was possible [264]. The mechanism for this enhancement is uncertain. Physical concentration of DNA molecules through increased sedimentation may play a role, as hypothesized for silica particles [222]. However, the efficiencies of gold nanocarriers are much greater than those of silica (30% of PEI vs. 800% of PEI) despite being almost an order of magnitude smaller. Thus, additional factors likely contribute to transfection enhancements. Studies of endocytosis have shown an increase in internalization for particles in the 50 nm region [156], and it is possible that the smaller size of DNA–gold complexes enhances their uptake.

11.6.3.3 **Magnetic Nanocarriers**
By far the most widely studied non-polymeric nanocarriers have been magnetic particles. Magnetic particles can improve the effectiveness of traditional vectors by promoting specific targeting to the cell surface, and possibly the cell interior [265]. In a technique similar to that used for cell sorting, magnetic nanoparticles have been used to concentrate retroviral vectors, which were isolated in concentrations up to 4200 times that of the control [266]. Using a magnetic field, these particles were then directed to certain portions of a cell culture dish, producing extremely efficient, patterned *in vitro* transfection. In a more comprehensive study, cationic magnetic nanoparticles were examined with non-viral (i.e., PEI, AVET, lipofectamine, geneporter, and DOTAP-cholesterol) and viral (i.e., recombinant adenovirus and a retrovirus) transfection agents *in vitro* and *in vivo* [265]. Using magnetic fields, transfection could be confined to specific regions of a cell culture plate, in agreement with Hughes et al. [266].

The application of a magnetic field produces an extraordinary reduction in transfection times, from 2–4 h to as little as 10 min [265]. This is most likely explained by an increase in vector concentration near the cell surface, resulting from direct magnetic targeting and increased sedimentation [221, 222]. As many vectors demonstrate chronic toxicity in cell culture, a reduction in required transfection time is an important innovation. Additionally, the required DNA dose declined, potentially allowing the use of retroviral transfer agents, which are difficult to isolate in high concentration. Further, targeting of magnetic nanocarriers is extremely specific. Few receptors for the transfection agents employed were required, allowing transfection of cell types that previously did not respond to the viral vectors investigated. A modest increase in transfection was evident for viral agents bound to magnetic particles, and transfection drastically increased (e.g., 20× that of control) the presence of a magnetic field.

Magnetotransfection also shows great promise *in vivo*. Magnetic particle-bound viral agents produced transfection in the gut and stomach, normally inhospitable environments for viral agents and DNA [265]. Transfection has been achieved in primary endothelial cells (i.e., human umbilical vein endothelial cells) that normally display remarkable resistance to DNA and might be used as stem cells [267]. Magnetic nanospheres conjugated to the vascular endothelial growth factor (VEGF) gene have been examined as a treatment for limb ischemia [268]. Using magnetotransfection, overexpression of VEGF was produced, resulting in an increase in capillary formation. This treatment method offers great improvements over either systemic delivery of VEGF or delocalized gene therapy. Systemic delivery has not been very successful because the VEGF half-life is very short, whereas traditional untargeted vector delivery does not target the limb in question and can produce unwanted angiogenesis in alternative locations. Although these studies used external magnets to guide particles, existing MRI technology may also be used for this purpose, reducing one barrier to clinical implementation.

11.6.4
Future Directions

11.6.4.1 Drug Delivery

Nanostructures offer several advantages that will encourage their future use in drug delivery. One benefit is the ability to encapsulate large amounts of drug in a small space. Recent work has highlighted the increased storage potential of magnetic nanotubes/wires in particular. Nanowires possess many of the size advantages of nanoparticles, while having an increased length along which diffusion can occur or drug can be encapsulated. Electrospun nanofibers containing magnetic particles have been manufactured, providing potential MRI targeting ability while increasing available drug storage area [269]. Alternatively, magnetic nanoparticles were deposited on the inner surface of silica nanotubes using templated synthesis, and the controlled release of ibuprofen, 5-FU, and 4-nitrophenol drugs was evaluated [270].

Another interesting system to increase storage is the use of hollow polymeric nanospheres. In a technique similar to colloidal lithography (Section 11.3.2), nanoparticles may be used as templates to self-assemble nanometer-sized degradable

polymer shells [271]. Known as layer-by-layer assembly, this technique systematically adds oppositely charged molecules [e.g., polyelectrolytes, including poly(sodium styrenesulfonate) and poly(allylamine hydrochloride)] to a surface through
electrostatic attraction. Once the shell is assembled, the core material, usually a
polymer nanoparticle (e.g., polystyrene latex [271] or poly(lactic acid)/poly(lactic
acid-*co*-glycolic acid) [272]), can be dissolved. Permeability of the resulting shell is
dependant on the polyelectrolytes selected and on the solution pH [272], allowing
for specific drug release in pH-altered environments (e.g., endosomes, gut, and
stomach). As well as drug, shells can also encapsulate nanostructures, including
magnetic particles [273, 274], CdTe [275, 276], $Cd_xHg_{1-x}Te$ [276] and HgTe [276]
semiconductor quantum dots, and gold-cobalt nanoparticles [277]. The shells provide a protective coating that may facilitate intracellular delivery of nanoparticles
and their cargos.

Nanostructures also provide ideal drug delivery agents because their surfaces
may be altered with multiple agents: drugs, targeting molecules, and chemicals
to facilitate intracellular delivery. For example, one system, conceived by Ferrari
[213], would use nanoparticles to treat large tumor masses. Traditionally, drugs
must access the tumor from its periphery. However, in Ferrari's system, elaborate
mechanisms are used to avoid these barriers. First, the nanoparticle encapsulates
actin monomers and myosin-bound therapeutic agents. The nanoparticle surface
would be bound to molecules that target appropriate cells (e.g., endothelial cells
forming the neovascular tumor barrier). The nanoparticle would also be attached
to molecules that enhance permeation of the endothelium, allowing nanoparticle
access to the tumor interior. Once inside the tumor, the nanoparticle would release
its actin cargo, forming filaments along which the myosin-bound drug payload
can travel. The drug would then be delivered to tumor cells beyond the tumor periphery.

Nanostructures can also be used to create "smart" delivery vehicles, with triggered release produced by an environmental change. For example, the polymeric
nanospheres described above can encapsulate nanoparticles that alter their permeability, actuating release. Au-coated cobalt nanoparticles were able to disrupt polymeric capsules in the presence of a magnetic field, releasing FITC molecules [277].
Another example of selectively triggered release is the use of CdS nanoparticles as
capping agents for mesoporous silica storage spheres [278]. CdS nanoparticles
were used to block surface pores of mesoporous silica. Upon addition of chemicals
that disrupt disulfide bonds, the CdS particles were removed, allowing controlled
release of the drug cargo. Alternatively, gold nanoshells have been encapsulated in
thermoresponsive hydrogels [17]. When the nanoshells are exposed to infrared
light, the temperature increased [Section 11.6.2.2 (Hyperthermia)], causing the
polymer to collapse and release its drug cargo.

Future nanoscale drug delivery devices will likely incorporate many of these techniques. Nanocavities, as found in nanotubes or polymeric nanospheres, may provide space for increased drug storage. Nanostructure surfaces may be altered to target release to a specific location. Additional adaptations may allow for safe passage
of molecules through tissue barriers and into the cell interior. Once in place, optical

or magnetic excitation may be used to actuate drug release. Ultimately, these systems will provide heretofore unseen control in drug delivery.

11.6.4.2 Gene Therapy

Multifunctional Systems Similar to drug delivery, multifunctional systems will likely form the cornerstone of improved nanostructure–DNA delivery. Nanostructures can be attached to several types of biomolecules, creating modular DNA delivery systems that include DNA, cell membrane transporters, and endosomolytic agents. For example, researchers have combined hybrid viruses conjugated to nanoparticle surfaces with magnetic force to create novel gene delivery systems [279]. Previously, the envelope of the hemagglutinating virus of Japan (HVJ) was modified to transfer DNA through cell fusion; however, efficiency remained slow and untargeted. To improve transfection, the virus was coupled with magnetic particles modified with either protamine sulfate or heparin. Successful transfection was achieved using protamine sulfate *in vitro*, whereas heparin was needed to maximize transfection *in vivo*. It is unclear whether these factors enhance direction by promoting cell binding through specific receptors, or simply by altering the zeta potential of the particle surface. In another example of composite systems, researchers combined DNA-covered gold nanoparticles with electromigration and electroporation to enhance delivery. Up to $82\times$ the amount of DNA was delivered when gold particles were subjected to electric fields, promoting electromigration and subsequent electroporation. Although not compared directly with electroporation of free DNA, it is probable that the addition of conductive gold particles enhanced delivery. Although much work is in the early stages, combination approaches can, clearly, merge two mediocre delivery systems to create powerful transfection systems.

Nanorods Polymeric microparticles are by far the most common nanostructures studied for gene therapy. However, nanorods offer some distinct advantages as combination transfection agents. Metallic nanorods can be constructed from alternating material segments (Section 11.2.2.2). Through specific surface chemistry interactions, these segments can be chosen to promote selective attachment of biomolecules to each element [280]. For example, rhodamine dye and DNA were attached to segments of Au-Ni nanorods, respectively. The rhodamine was used to locate nanorods, which were found in acidic organelles, probably cytoplasmic vesicles. Alternatively, DNA and transferrin were attached to nanorods, and transfection efficiencies improved by 2–3-fold [280]. This method represents an advantage to nanoparticle-based systems, which attach multiple elements to the same molecule. In single material systems, kinetic considerations may favor binding of one molecule over that of another.

Another advantage of nanotubes is their potential ability to penetrate the cell membrane directly, reducing dependence on the endocytotic pathway for vector delivery. There is some evidence that magnetic particles may be forced through the cell membrane under the influence of a magnetic field [221]. Also, carbon nanotubes with a magnetic cap can penetrate the cell in a magnetic field [192]. This system was examined as a method of gene delivery, with up to 85% of cells displaying the gene of interest. Transfection was not evidenced in the absence of the magnetic field, demonstrating that the mechanism of infection was active spearing, and not passive diffusion of vectors to the cell surface. However, both rotating and static fields were required for optimal gene delivery. Because of their surface modifiable segments and high aspect ratios favorable for cell penetration, nanotubes may prove to be superior materials for forming composite delivery systems.

Frontiers of DNA Delivery   One of the greatest difficulties in assessing *in vivo* gene therapy success is ascertaining whether the intended target has been transfected. Magnetic particles have been used for some time as diagnostic agents in magnetic resonance imaging (MRI). The combination of magnetic particles as gene delivery agents and imaging modalities may allow researchers to monitor transfection *in vivo* [281, 282]. In addition, it may be possible to monitor gene *expression*. Genes that encode for ferritin formation have been used to synthesize metallic particles directly inside cells [283]. If linked to a transfected gene, ferritin molecules could serve as reporter molecules, indicating the transcription level of the gene of interest. Using existing MRI systems, these exciting developments could allow external observation of transfection and subsequent gene expression *in vivo*.

Also, the potential ability of nanostructures to deliver large quantities of DNA is creating new possibilities for therapeutic treatment. The initial focus of most gene therapy approaches has been delivery of a working copy of a deficient gene. However, in many cases the gene in question *overexpresses* a protein, and it would be desirable to "turn-off" the gene. It has been proposed that complimentary oligonucleotides may bind mRNA, preventing protein synthesis [284]. However, as opposed to the one-shot delivery of vectors, silencing genes will require large amounts of DNA to be delivered continuously. Initial reports indicate that transfection using magnetic particles may provide high levels of these oligonucleotides [285]. Ammonium-modified gold nanoparticles bound DNA, preventing RNA polymerase attachment and subsequent DNA transcription [62]. Reversible hybridization has also been demonstrated on a gold nanocrystal platform [286]. DNA conjugated to gold nanoparticles denatured in the presence of a radio-frequency magnetic field. When the magnetic field was removed, DNA appeared to return to the native configuration. Although the exact mechanism is uncertain, it is likely that heating effects produced by the presence of a magnetic field [Section 11.6.2.2 (Hyperthermia)] caused the loss of hybridization.

11.7
Protein Manipulation

Another potential therapeutic use of nanostructures is the active manipulation of proteins. Enzymes and proteins can be directly conjugated to a nanostructure surface, and, in many cases, activated or deactivated through optical or magnetic effects. Alternatively, substrate molecules can be linked to nanostructures, and used to indirectly manipulate bound proteins, including cell surface receptors or ion channels. Although the applications of nanostructures in this field of cellular engineering are the least developed, they offer the ability to directly control cellular signaling pathways and responses. As techniques continue to advance, nanoscale protein manipulation will lead to new therapies, a greater understanding of cell function, and external systems for large-scale biochemical synthesis.

11.7.1
Biology of Protein Manipulation

Most cell functions, including signaling, replication, and metabolism, are dictated by proteins. Most proteins are globular structures containing positively- and negatively-charged hydrophobic and hydrophilic residues [8]. Upon folding, the hydrophobic residues are confined to the center of the protein, producing secondary structure. The secondary structure frequently contains an active site, a series of charged residues that recognize and bind substrates. Protein–substrate binding affinities result from a combination of electrostatic, hydrogen bonding, and van der Waals interactions [8]. Substrate binding often produces a conformational change in the protein, which activates a signaling pathway or catalyzes a chemical reaction [8].

Proteins may be free or bound to the cell membrane. Proteins embedded in the membrane usually contain several hydrophobic residues, arranged in alpha helices, which span the membrane [8]. These proteins contribute to intercellular signaling, cell motility and adhesion, and the flow of components into and out of the cell. Ion channels, in particular, represent one class of membrane-bound proteins that can respond to electrical, mechanical, or chemical signals [8], and therefore are an interesting target for nanostructure manipulation.

11.7.2
Manipulation of Free Proteins: Enzymes

Nanostructures can be bound to free proteins using many of the techniques discussed previously (Section 11.2.4). The most widely-studied nanostructure system for enzymatic manipulation consists of mesoporous silica containing enzymes immobilized in the particle matrix. This technique has been used to entrap butyrylcholinesterase, an enzyme that cleaves ester bonds, with retention of over 90% of activity [287]. Enzymes bound to silica were more resistant to temperature changes, preserving 100% of activity versus an 85% decrease for free enzyme in solution. Enhanced stability most likely resulted from enzyme interactions with the nanoparticle support, as others have found that enzyme–nanoparticle binding lowers kinetic barriers for conformational rearrangements that may prevent denaturization [288]. Enzymes have also been encapsulated in hollow silica nanoparticles; however, these systems were less successful. Enzyme activity was reduced when compared with free enzyme, most likely as a result of diffusion limitations [289].

Control of the amount of enzyme encapsulated has been obtained using layer-by-layer techniques. Films or shells containing alternative layers of electrolyte and silica nanoparticle or enzyme have been constructed [290, 291]. The amount of enzyme incorporated is controlled by adjusting the number of layers constructed. This technique can also prevent unanticipated enzyme desorption from the particle [292]. Although most of the envisioned applications of this technique are for biosensing or the development of large scale reactors [293], it is possible that layer-by-layer assemblies could be used to construct prosthetic elements that contain a missing or inactive enzyme in the patient.

In addition to silica particles, other materials have been investigated. Trypsin enzyme has been linked to dextran-modified gold through supramolecular interactions [294], and chymotrypsin has been bound to mixed-monolayer protected gold clusters through electrostatic interactions [295]. For chymotrypsin, binding was designed to occur at the active site, and inhibited the enzyme completely with a 1:5 nanoparticle to chymotrypsin ratio. Chymotrypsin has also been attached to CdSe quantum dots with varying surface functionalities. Capping ligands terminated with hydroxyl groups did not interact with enzyme, whereas carboxyl-terminated ligands inactivated enzyme [296]. Enzymes have been coupled to magnetic particles as well, which have the potential benefit of targeting using magnetic fields; however, these are most commonly used in bioreactors, not living systems [297].

Also, nanoparticles have been trapped in the pore of certain proteins, chaperonins, which are responsible for repairing defects in protein folding [298]. Normally, these proteins encompass misfolded proteins, providing a permissible environment for reconfiguration. After folding has been corrected, the protein is released in an ATP-driven process. CdS nanoparticles were complexed with chaperonin proteins with the aid of dimethylformamide. Particles displayed unusually long-lived fluorescence, indicating that chaperonins were able to shield nanoparticles from adverse reactions (e.g., oxidation) in much the same way that misfolded proteins are protected. Further, upon ATP addition, colloids were released and could be isolated through centrifugation.

11.7.3
Manipulation of Bound Proteins: Receptors and Ion Channels

Utilizing nanostructures to manipulate cell surface receptors and ion channels is in many ways easier than the manipulation of free biomolecules. The exterior of the cell is much more accessible than its interior, and introducing nanostructures to an extracellular *in vitro* or *in vivo* environment is straightforward. Although endocytosis may still be an issue, particularly if nanostructures are designed to remain on the cell surface for long periods, performing experiments at low temperatures can alleviate these concerns [154]. This section focuses primarily on nanostructure manipulation of cell surface receptors and ion channels, as nanostructure cytoskeletal manipulation was discussed previously (Section 11.3.4).

Nanoparticle binding to cell surface receptors can influence cell function, particularly the internalization of substances. Magnetic nanoparticles, conjugated to insulin, prevented endocytosis and increased cell proliferation and viability [167]. Gold nanoparticles bound to certain glycoproteins, may inhibit cellular HIV virus transmission [299]. CdSe quantum dots attached to serotonin were used to inactivate serotonin transporters [153], a class of receptors that remove excess transmitter from the extracellular environment following a synaptic event. The particles inhibited serotonin in a dose-dependant manner, most likely because transporters became obstructed when attempting to internalize serotonin-bound nanoparticles.

Nanoparticles have also been employed to investigate ion channels, proteins responsible for signaling in nerve cells. Collagen- and laminin-coated magnetite particles were used to non-specifically coat cell surfaces. When exposed to a magnetic field, an increase in intracellular calcium was observed [300]. These increases result from activation of mechanosensitive ion channels, found often in cardiac tissue. Alternatively, nanoparticles were bound to neuron cell surface receptors, near ion channels of interest. CdS nanoparticles conjugated to integrin-binding peptides were used to explore the potential of electrically exciting ion channels with nanoparticle electrical fields [66, 301]. Ion channels have even been directly linked to nanoparticles. For example, ion channels entrained in silica nanoparticles displayed activity for over a month [302].

11.7.4
Future Directions

The potential of nanostructures to manipulate biomolecules continues to emerge. For example, it is possible that nanostructures could be used to investigate prion proteins, which have recently been identified as potential sources of transmission for mad-cow disease. Defective prions propagate by converting normal prions into the abnormal structure, forming structures similar to amyloid complexes found in Alzheimer's and Parkinson's disease [303]. Gold nanoparticles have been linked to prions, which were used to create self-assembled arrays [304]. Although the main focus of this work was to create stable nanowires, it may be possible to use nanostructures (e.g., magnetite) to manipulate prions and prion propagation, forming a basis for therapeutic treatment. At the very least, manipulation of this new class of biomolecules could provide valuable insight into structure, folding, and prion–prion interactions.

Additionally, future nanostructure–protein systems will expand the ability to manipulate ion channels. For example, organic dyes have been linked to a mechanosensitive channel in such a way that the channels become activated with light [305]. Given their optical and magnetic properties, it is likely that a similar design could be constructed with nanostructures. Although this system was constructed *ex vivo*, it demonstrates the possibility of direct cell manipulation using optical and mechanical elements. Along these lines, gold nanotubes have been used to create artificial ion channels [306]. Nanotubes were embedded in an artificial plasma membrane, composed of polymers, and linked to a DNA molecule. The DNA molecule was used to provide rectification (e.g., on and off states), and was electrophoretically driven to block the mouth of the pore, preventing ion flow. It is likely that nanostructures will continue to be used in this way, possibly being integrated into cells for control of ion flow, signaling, and protein binding.

11.8
Summary and Conclusions

11.8.1
Summary

Nanostructures present several advantages for biological research. They are of similar size to many biological materials. Nanostructure surface areas are high and easily altered through various surface chemistry techniques, allowing for easy biomolecule addition and presentation. Combined with the unique optical and electrical properties of nanostructures, these features have heightened interest in the use of nanostructures for biological investigation. Nanostructures may be composed of many different materials, including semiconductors, metals, and magnetic materials, and may be fashioned into particles, wires, tubes, and shells. Synthesis techniques vary depending on the material employed and the desired shape, but generally include solution-phase synthesis, templated electrodeposition, and deposition from the vapor phase. Once manufactured, nanocomponents can be modified with a range of biomolecules using physisorption, electrostatic attraction, material specific affinities, or biomolecular recognition. These bioconjugates can be used to attach nanostructures to specific components of cells. Although there is some evidence that nanostructures may be cytotoxic, these risks can be mitigated through the application of surface coatings, and nanostructures have already been employed successfully in several *in vivo* and *in vitro* applications.

Nanostructures can be used easily for external investigation of cells. For example, they are ideal for the exploration of cell adhesion, migration, and mechanics, primarily as a result of their small size, similar to that of many components that mediate cell attachment. Interactions with physical topography have chiefly been achieved using structures created through optical lithography. However, colloidal lithography offers a facile technique to create nanostructures much smaller than those produced previously. Additionally, nanoparticles themselves may be organized in self-assembled sheets to create roughened physical surfaces with nanoscale dimensions. Chemical patterns can be produced using many of these same methods. For example, gold-thiol binding affinity may be exploited to assemble proteins on nanoparticles or surfaces created through colloidal lithography. In addition to these passive investigations, magnetic nanostructures can be used to directly manipulate cells. These methods have been implemented for cell sorting and to construct tissue engineering substrates. Also, direct cytoskeletal manipulation is possible, and magnetic nanostructures have been utilized to study the role of integrins in mechanotransduction. Future work will likely focus on integrating physical and chemical cues into multifunctional devices, as well as the creation of active substrates to directly manipulate cells.

Whereas it is straightforward to interface nanomaterials with external structures that govern cell adhesion and migration, controlled intracellular delivery of nanostructures is difficult. Materials can be introduced easily through endocytosis, but remain sequestered in endosomes or lysosomes and are unable to interact with the intracellular environment. Several techniques have been developed to elude this fate, including the addition of translocation peptides, liposomes, dendrimers, electroporation, and microinjection. However, most of these techniques produced significant particle aggregation, which can reduce optical signals and impede interactions with subcellular components. Among these techniques, only microinjection has produced discrete intracellular delivery, but this is a serial technique, difficult to adapt to large numbers of cells or *in vivo* delivery. Despite the limited techniques, several researchers have successfully delivered nanostructures to the cell interior. The primary applications have been for cellular and subcellular tracking. Future research will likely concentrate on improving targeting and delivery of nanostructures, perhaps using biomimetic rather than natural methods.

If controlled delivery can be achieved, it presents the opportunity to directly manipulate internal cellular components, including the cytoskeleton. The cytoskeleton is responsible for cell division, signal regulation, transport, and mechanical properties. Movement along the cytoskeleton is mediated by transport proteins, which bind to specific components, including actin filaments and microtubules. It is possible to conjugate nanostructures to both cytoskeletal components and transport proteins, providing a method for nanostructure movement. These systems have been demonstrated *ex vivo*, primarily for self-assembly and electronics applications; however, it is possible that these techniques may also be explored *in vitro*. Future work will likely adapt *ex vivo* methods for *in vitro* studies. If successful, nanostructures could be used to manipulate the cytoskeleton, increasing understanding of mechanical stimuli in cell signaling and division. Additionally, nanostructures may be transported along cytoskeletal elements, providing a method for directing nanostructures to an organelle of interest, with potential applications in drug delivery and gene therapy.

Additionally, internalized nanostructures could be used for controlled cellular delivery of biomolecules, with applications in drug delivery and gene therapy. The half-lives of drugs are very short, and in many cases, sustained long-term delivery is required for optimal therapeutic effect. Several polymeric release systems have been designed, but these lack targeting ability to specific cells or cell components. Also, many delivery agents are quickly removed from the body through the reticuloendothelial system. Because nanostructure surfaces can be modified easily, they can surmount these challenges. Biorecognition molecules can be added for selective cell targeting; stealth molecules, e.g., poly(ethylene glycol), can be added to increase particle circulation times.

The greatest immediate potential impact lies in the field of cancer therapy, where discrimination between healthy and diseased cells is essential for treatment. Nanostructures have been used for direct molecule delivery, primarily through magnetic targeting. However, they have also been used for triggered release, actuated by an external event or environmental change. For example, gold and magnetic nanostructures experience temperature elevations in the presence of near-infrared irradiation. This effect, hyperthermia, can be used to produce local tissue damage. Alternatively, nanostructures have been linked to photosensitizers for photodynamic therapy. In response to optical excitation, these chemicals catalyze the reaction of nearby molecules to create cytotoxic substances.

Nanostructures have also been used for gene therapy, which is in effect a subset of the field of drug delivery. Nanocarriers for DNA delivery have been created using silica, gold, and magnetic particles. Even in passive systems, increases in transfection efficiency were seen. Improvements most likely resulted from the density of the nanostructures, which promoted sedimentation on the cell surface, concentrating DNA presentation. However, increases were also seen for smaller nanostructures, indicating that the small size of nanocarriers may enhance cellular uptake. The addition of active targeting (e.g., magnetic field) further enhanced DNA delivery.

Future systems for drug and DNA delivery will likely incorporate multiple functionalities. Carriers may contain the cargo, targeting molecules, agents to enhance permeation, and means for optical detection. Many carriers include an intrinsic targeting capability (e.g., magnet or electric field response) that can be incorporated to maximize release. The use of nanorods will likely be investigated, as their larger surface area increases available storage for cargo. Also, there is some evidence that magnetic nanorods can directly penetrate the cell membrane, evading the endocytotic pathway. These enhancements will provide delivery vehicles that can control the location and time of release.

Finally, internalized nanostructures may be used to directly manipulate proteins, an area of huge potential, still in initial stages of investigation. Free proteins, including enzymes, can be encapsulated in or attached to nanostructures, and these composite systems can be used to control enzymatic activity. Most of this work has focused on the creation of biochemical reactor devices, but may serve as the basis for the creation of prosthetic elements, which replace a missing enzymatic function. Additionally, nanostructures may be linked to substrates and used to manipulate membrane-bound proteins. These interactions can modulate the internalization of substances and ion channel activity. Future research will likely continue to examine new classes of proteins (e.g., prions) and expand research in ion channel manipulation.

11.8.2
Conclusions

Nanostructures provide scientists with exciting opportunities to manipulate cells directly. They possess similar sizes to many biological components, including DNA, proteins, peptides, and organelles. Nanostructures are easy to synthesize and can be composed of metallic, semiconducting, ceramic or magnetic materials (in addition to carbon, polymeric, and peptide materials discussed elsewhere). Their surfaces are readily modifiable with a range of biomolecules, which provide a direct means to interact with the cell and its environment. Although there is some concern that certain nanostructures may be cytotoxic, they have nonetheless been employed in several *in vitro* and *in vivo* applications.

Because of their unique optical and electrical properties, nanostructures can be used to manipulate and visualize components of the cell exterior and interior. Most of these investigations have been passive. However, as technologies, particularly for controlled delivery, continue to advance, experiments will shift to more active manipulations of cells and their components. This ability will allow scientists to examine intracellular function with increased control. Additionally, these developments will likely lead to new diagnostic and therapeutic applications, particularly for gene therapy and cancer treatment.

References

1 ALIVISATOS, A. P., Perspectives on the physical chemistry of semiconductor nanocrystals., *J. Phys. Chem.* **1996**, 100, 13 226–13 239.

2 CURTIS, A., WILKINSON, C., Nanotechniques and approaches in biotechnology., *Trends Biotechnol.* **2001**, 19, 97–101.

3 LAW, M., GOLDBERGER, J., YANG, P., Semiconductor nanowires and nanotubes., *Annu. Rev. Mater. Res.* **2004**, 34, 83–122.

4 MOORE, G. E., *Electronics* **1965**, 38.

5 BUKOWSKI, T. J., SIMMONS, J. H., Quantum dot research: Current state and future prospects., *Crit. Rev. Solid State Mater. Sci.* **2002**, 27, 119–142.

6 HUANG, Y., LIEBER, C. M., Integrated nanoscale electronics and opto-electronics: Exploring nanoscale science and technology through semiconductor nanowires., *Pure Appl. Chem.* **2004**, 12, 2051–2068.

7 MURPHY, C. J., COFFER, J. L., Quantum dots: A primer., *Appl. Spectrosc.* **2002**, 56, 16–27.

8 ALBERTS, B. A., BRAY, D., LEWIS, J., RAFF, M., ROBERTS, K., WATSON, J. D., *Molecular Biology of the Cell*, 3rd Ed., Garland Publishing, New York, **1994**, pp. 101, 111, 130–131, 195–198, 275, 477, 486–487, 523–524, 610–611, 618–626, 789, 979, 986, 990, 996–997.

9 MEDINTZ, I. L., UYEDA, H. T., GOLDMAN, E. R., MATTOUSSI, H., Quantum dot bioconjugates for imaging, labeling and sensing., *Nat. Mater.* **2005**, 4, 435–446.

10 BRUCHEZ, M., JR., MORONNE, M., GIN, P., WEISS, S., ALIVISATOS, A. P., Semiconductor nanocrystals as fluorescent biological labels., *Science* **1998**, 281, 2013–2016.

11 CHAN, W. C. W., NIE, S., Quantum dot bioconjugates for ultrasensitive nonisotopic detection., *Science* **1998**, 281, 2016–2018.

12 GAO, X., NIE, S., Molecular profiling of single cells and tissue specimens with quantum dots., *Trends Biotechnol.* **2003**, 21, 371–373.

13 ROSI, N. L., MIRKIN, C. A., Nanostructures in biodiagnositics., *Chem. Rev.* **2005**, 105, 1547–1562.

14 CHALMERS, J. J., ZBOROWSKI, M., SUN, L., MOORE, L., Flow through, immunomagnetic cell separation., *Biotech. Prog.* **1998**, 14, 141–148.

15 LOO, C., LIN, A., HIRSCH, L., LEE, M.-H., BARTON, J., HALAS, N., WEST, J., DREZEK, R., Nanoshell-enabled photonics-based imaging and therapy of cancer., *Technol. Cancer Res. Treat.* **2004**, 3, 33–40.

16 BAUER, L. A., BIRENBAUM, N. S., MEYER, G. J., Biological applications of high aspect ratio nanoparticles., *J. Mater. Chem.* **2004**, 14, 517–526.

17 SERSHEN, S. R., WESTCOTT, S. L., HALAS, N. J., WEST, J. L., Temperature-sensitive polymer-nanoshell composites for photothermally

modulated drug delivery., *J. Biomed. Mater. Res.* **2000**, 51, 293–298.

18 GUPTA, A. K., GUPTA, M., Synthesis and surface engineering of iron oxide nanoparticles for biomedical applications., *Biomaterials* **2005**, 26, 3995–4021.

19 GAO, X., YANG, L., PETROS, J. A., MARSHALL, F. F., SIMONS, J. W., NIE, S., In vivo molecular and cellular imaging with quantum dots., *Curr. Opin. Biotechnol.* **2005**, 16, 63–72.

20 BLOCK, S. M., GOLDSTEIN, L. S. B., SCHNAPP, B. J., Bead movement by single kinesin molecules studied with optical tweezers., *Nature* **1990**, 348, 348–352.

21 HARTWIG, A., SCHWEDTLE, T., Interactions by carcinogenic metal compounds with DNA repair processes: Toxicological implications. *Toxic. Lett.* **2002**, 127, 47–54.

22 LANSMAN, J. B., HESS, P., TSIEN, R. W., Blockade of current through single calcium channels by Cd^{2+}, Mg^{2+}, and Ca^{2+}., *J. Gen. Physiol.* **1986**, 88, 321–347.

23 WAALKES, M. P., Cadmium carcino-genesis., *Mut. Res.* **2003**, 533, 107–120.

24 LAM, C.-W., JAMES, J. T., McCLUSKEY, R., HUNTER, R. L., Pulmonary toxicity of single-wall carbon nanotubes in mice 7 and 90 days after intratracheal instillation., *Toxic. Sci.* **2004**, 77, 126–134.

25 OBERDÖRSTER, E., Manufactured nanomaterials (fullerenes, C60) induce oxidative stress in the brain of juvenile largemouth bass., *Environ. Health Persp.* **2004**, 112, 1058–1062.

26 ELDER, A. C. P., GELEIN, R., AZADNIV, M., FRAMPTON, M., FINKELSTEIN, J., OBERDÖRSTER, G., Systematic effects of inhaled ultrafine particles in two compromised, aged rat strains., *Inhalation Toxic.* **2004**, 16, 461–471.

27 OBERDÖRSTER, G., SHARP, Z., ATUDOREI, V., ELDER, A., GELEIN, R., KREYLING, W., COX, C., Translocation of inhaled ultrafine particles to the brain., *Inhalation Toxic.* **2004**, 16, 437–445.

28 SEMMLER, M., SEITZ, J., ERBE, F., MAYER, P., HEYDER, J., OBERDÖRSTER,

G., KREYLING, W. G., Long-term clearance kinetics of inhaled ultrafine insoluble iridium particles from the rat lung, including translocation into secondary organs., *Inhalation Toxic.* **2004**, 16, 453–459.

29 MURRAY, C. B., NORRIS, D. J., BAWENDI, M. G., Synthesis and characterization of nearly monodisperse CdE (E = S, Se, Te) semiconductor nanocrystallites., *J. Am. Chem. Soc.* **1993**, 115, 8706–8715.

30 Tri-octyl phosphine Oxide MSDS Product Number 223301, version 1.5, 6/28/2004, Sigma Aldrich.

31 DERFUS, A. M., CHAN, W. C. W., BHATIA, S. N., Probing the cytotoxicity of semiconductor quantum dots., *Nano Lett.* **2004**, 4, 11–18.

32 CHAN, W. C. W., MAXWELL, D. J., GAO, X., BAILEY, R. E., HAN, M., NIE, S., Luminescent quantum dots for multiplexed biological detection and imaging., *Curr. Opin. Biotechnol.* **2002**, 13, 40–46.

33 HINES, M. A., GUYOT-SIONNEST, P., Synthesis and characterization of strongly luminescing ZnS-capped CdSe nanocrystals., *J. Phys. Chem.* **1996**, 100, 468–471.

34 WEAST, R. C., ASTLE, M. J. (Eds.), *CRC Handbook of Chemistry and Physics*, 60th Ed., CRC Press, Boca Raton, FL, **1979**, p. B-220.

35 NOSAKA, Y., OHTA, N., FUKUYAMA, T., FUJII, N., Size control of ultrasmall CdS particles in aqueous solution by using various thiols., *J. Colloid Interface Sci.* **1993**, 155, 23–29.

36 WINTER, J. O., GOMEZ, N., GATZERT, S., SCHMIDT, C. E., KORGEL, B. A., Variation of cadmium sulfide nano-particle size and photoluminescence intensity with altered aqueous synthesis conditions, *Colloid Surf. A* **2005**, 254, 147–157.

37 KALYANASUNDARAM, K., BORGARELLO, E., DUONGHONG, D., GRÄTZEL, M., Cleavage of water by visible-light irradiation of colloidal CdS solutions: Inhibition of photocorrosion by RuO_2., *Angew. Chem. Int. Ed.* **1981**, 20, 987–988.

38 HENGLEIN, A., Photo-degradation and

fluorescence of colloidal cadmium sulfide in aqueous solution., *Ber. Bunsenges. Phys. Chem.* **1982**, 86, 301–305.

39 EYCHMÜLLER, A., HÄSSELBARTH, A., KATSIKAS, L., WELLER, H., Fluorescence mechanism of highly monodisperse Q-sized CdS colloids., *J. Luminescence* **1991**, 48–49, 745–749.

40 NOSAKA, Y., YAMAGUCHI, K., MIYAMA, H., HAYASHI, H., Preparation of size-controlled CdS colloids in water and their optical properties., *Chem. Lett.* **1988**, 4, 605–608.

41 KUNDU, K., KHOSRAVI, A. A., KULKARNI, S. K., SINGH, P., Synthesis and study of organically capped ultra small clusters of cadmium sulphide., *J. Mater. Sci.* **1997**, 32, 245–258.

42 ROSSETTI, R., ELLISON, J. L., GIBSON, J. M., BRUS, L. E., Size effects in the excited electronic states of small colloidal CdS crystallites., *J. Phys. Chem.* **1984**, 80, 4464–4469.

43 GERION, D., PINAUD, F., WILLIAMS, S. C., PARAK, W. J., ZANCHET, D., WEISS, S., ALIVISATOS, A. P., Synthesis and properties of biocompatible water-soluble silica-coated CdSe/ZnS semiconductor quantum dots., *J. Phys. Chem. B* **2001**, 105, 8861–8871.

44 FARADAY, M., The Bakerian lecture. Experimental relations of gold (and other metals) to light., *Philos. Trans.* **1857**, 147, 145–181.

45 TURKEVICH, J., STEVENSON, P. C., HILLIER, J., A study of the nucleation and growth processes in the synthesis of colloidal gold., *Disc. Faraday Soc.* **1951**, 11, 55–75.

46 SCHMID, G., Large clusters and colloids. Metals in the embryonic state. *Chem. Rev.* **1992**, 92, 1709–1727.

47 AVERITT, R. D., SARKAR, D., HALAS, N. J., Plasmon resonance shifts of Au-coated Au$_2$S nanoshells: Insight into multicomponent nanoparticle growth., *Phys. Rev. Lett.* **1997**, 78, 4217–4220.

48 WESTCOTT, S. L., OLDENBURG, S. J., LEE, T. R., HALAS, N. J., Formation and adsorption of clusters of gold nanoparticles onto functionalized silica nanoparticle surfaces., *Langmuir* **1998**, 14, 5396–5401.

49 WILCOXON, J. P., WILLIAMSON, R. L., BAUGHMAN, R., Optical properties of gold colloids formed in inverse micelles., *J. Chem. Phys.*, **1992**, 12, 9933–9950.

50 COLVIN, V. L., GOLDSTEIN, A. N., ALIVISATOS, A. P., Semiconductor nanocrystals covalently bound to metal-surfaces with self-assembled monolayers., *J. Am. Chem. Soc.* **1992**, 114, 5221–5230.

51 MURRAY, C. B., KAGAN, C. R., BAWENDI, M. G., Synthesis and characterization of monodisperse nanocrystals and close-packed nanocrystal assemblies., *Ann. Rev. Mater. Sci.* **2000**, 30, 545–610.

52 TOURINHO, F., FRANCK, R., MASSART, R., PERZYNSKI, R., Synthesis and magnetic properties of manganese and cobalt ferrofluids., *Prog. Col. Polym. Sci.* **1989**, 79, 128–134.

53 REICH, D. H., TANASE, M., HULTGREN, A., BAUER, L., CHEN, C. S., MEYER, G. J., Biological applications of multifunctional magnetic nanowires., *J. Appl. Phys.* **2003**, 93, 7275–7280.

54 BAUER, L. A., REICH, D. H., MEYER, G. J., Selective functionalization of two-component magnetic nanowires., *Langmuir* **2003**, 19, 7043–7048.

55 NGUYEN, C. V., DELZEIT, L., CASSELL, A. M., LI, J., HAN, J., MEYYAPPAN, M., Preparation of nucleic acid functionalized carbon nanotube arrays., *Nano Lett.*, **2002**, 2, 1079–1081.

56 BIRENBAUM, N. S., LAI, T. B., REICH, D. H., CHEN, C. S., MEYER, G. J., Selective noncovalent adsorption of protein to bifunctional metallic nanowire surfaces., *Langmuir* **2003**, 19, 9580–9582.

57 CHAN, W. C. W., PRENDERGAST, T. L., JAIN, M., NIE, S., One-step conjugation of biomolecules to luminescent nanocrystals, in *Molecular Imaging: Reporters, Dyes, Markers, and Instrumentation*, BORNHOP, D. J., LICHA, K., Eds., Proceedings of SPIE., **2000**, vol. 3924, pp. 2–9.

58 β-Mercaptoethanol MSDS Product Number M7522, version 1.12, 4/6/2004, Sigma Aldrich.

59 SCHNEIDER, J. A., KATZ, B., MELLES, R. B., Update on nephropathic cystinosis., *Pedia. Nephrol.* **1990**, 4, 645–653.

60 KATZ, E., WILLNER, I., Integrated nanoparticle-biomolecule hybrid systems: Synthesis, properties, and applications., *Angew. Chem. Int. Ed.* **2004**, 43, 6042–6108.

61 ZHANG, J., MATVEEVA, E., GRYCZYNSKI, I., LEONENKO, Z., LAKOWICZ, J. R., Metal-enhanced fluoroimmunoassay on a silver film by vapor deposition., *J. Phys. Chem. B* **2005**, 109, 7969–7975.

62 MCINTOSH, C. M., ESPOSITO, III, E. A., BOAL, A. K., SIMARD, J. M., MARTIN, C. T., ROTELLO, V. M., Gold nanoparticles with biological activity: Disruption of transcription via electrostatic attraction., *J. Am. Chem. Soc.* **2001**, 123, 7626–7629.

63 HUANG, H. Z., YANG, X. R., Chitosan mediated assembly of gold nanoparticles multilayers., *Colloid Surf. A* **2003**, 226, 77–86.

64 LU, L. P., WANG, S. Q., LIN, X. Q., Fabrication of layer-by-layer deposited multilayer films containing DNA and gold nanoparticle for norepinephrine biosensor., *Anal. Chim. Acta* **2004**, 519, 161–166.

65 HE, P. L., HU, N. F., RUSLING, J. F., Driving forces for layer-by-layer self-assembly of films of SiO2 nanoparticles and heme proteins., *Langmuir* **2004**, 20, 722–729.

66 WINTER, J. O., LIU, T. Y., KORGEL, B. A., SCHMIDT, C. E., Recognition molecule directed interfacing between semiconductor quantum dots and nerve cells., *Adv. Mater.* **2001**, 13, 1673–1677.

67 MCCAMMON, J. A., Theory of biomolecular recognition., *Curr. Opin. Struct. Biol.* **1998**, 8, 245–249.

68 HERMANSON, G. T., *Bioconjugate Techniques*, Academic Press, San Diego, **1996**, pp. 456–460.

69 PIERSCHBACHER, M. D., RUOSLAHTI, E., Cell attachment activity of fibronectin can be duplicated by small synthetic fragments of the molecule., *Nature* **1984**, 309, 30–33.

70 GRAF, J., OGLE, R. C., ROBEY, F. A., SASAKI, M., MARTIN, G. R., YAMADA, Y., KLEINMAN, H. K., A pentapeptide from the laminin B1 chain mediates cell adhesion and binds the 67,000 laminin receptor., *Biochemistry* **1987**, 26, 6896–6900.

71 CREIGHTON, T. E., *Proteins: Structures and Molecular Principles*, W. H. FREEMAN and Co., New York, **1983**.

72 COLVIN, V. L., The potential environmental impact of engineered nanomaterials., *Nat. Biotechnol.* **2003**, 21, 1166–1170.

73 DUBERTRET, B., SKOURIDES, P., NORRIS, D. J., NOIREAUX, V., BRIVANLOU, A. H., LIBCHABER, A., In vivo imaging of quantum dots encapsulated in phospholipid micelles., *Science* **2002**, 298, 1759–1762.

74 PARAK, W. J., BOUDREAU, R., LE GROS, M., GERION, D., ZANCHET, D., MICHEEL, C. M., WILLIAMS, S. C., ALIVISATOS, A. P., LARABELL, C., Cell motility and metastatic potential studies based on quantum dot imaging of phagokinetic tracks., *Adv. Mater.* **2002**, 14, 882–885.

75 ÅKERMAN, M. E., CHAN, W. C. W., LAAKKONEN, P., BHATIA, S. N., RUOSLAHTI, E., Nanocrystal targeting in vivo., *Proc. Natl. Acad. Sci. U.S.A.* **2002**, 99, 12617–12621.

76 JAISWAL, J. K., MATTOUSSI, H., MAURO, J. M., SIMON, S. M., Long-term multiple color imaging of live cells using quantum dot bioconjugates., *Nat. Biotechnol.* **2003**, 21, 47–51.

77 WU, X., LIU, H., LIU, J., HALEY, K. N., TREADWAY, J. A., LARSON, J. P., GE, N., PEALE, F., BRUCHEZ, M. P., Immunofluorescent labeling of cancer marker her2 and other cellular targets with semiconductor quantum dots., *Nat. Biotechnol.* **2003**, 21, 41–46.

78 LARSON, D. R., ZIPFEL, W. R., WILLIAMS, R. M., CLARK, S. W., BRUCHEZ, M. P., WISE, F. W., WEBB, W. W., Water-soluble quantum dots for multiphoton fluorescence imaging in vivo., *Science* **2003**, 300, 1434–1436.

79 SPANHEL, L., HAASE, M., WELLER, H., HENGLEIN, A., Photochemistry of colloidal semiconductors. 20. Surface modification and stability of strong luminescing CdS particles., *J. Am. Chem. Soc.* **1987**, 109, 5649–5655.

80 Bowen Katari, J. E., Colvin, V. L., Alivisatos, A. P., X-ray photoelectron spectroscopy of CdSe nanocrystals with applications to studies of the nanocrystal surface., *J. Phys. Chem.* **1994**, 98, 4109–4117.

81 Dabbousi, B. O., Rodrigeuz-Viejo, J., Mikulec, F. V., Heine, J. R., Mattoussi, H., Ober, R., Jensen, K. F., Bawendi, M. G., CdSe(ZnS) core-shell quantum dots: Synthesis and characterization of a size series of highly luminescent nanocrystallites., *J. Phys. Chem. B* **1997**, 101, 9463–9475.

82 Aldana, J., Wang, Y. A., Peng, X., Photochemical instability of CdSe nanocrystals coated with hydrophilic thiols., *J. Am. Chem. Soc.* **2001**, 123, 8844–8850.

83 Patrick, L., Toxic metals and antioxidants: Part II. The role of antioxidants in arsenic and cadmium toxicity., *Alt. Med. Rev.* **2003**, 8, 106–128.

84 Rikans, L. E., Yamano, T., Mechanisms of cadmium-mediated acute hepatotoxicity., *J. Biochem. Mol. Toxic.* **2000**, 14, 110–117.

85 Hambrecht, F. T., Neural prostheses., *Ann. Rev. Biophys. Bioeng.* **1979**, 8, 239–267.

86 Lee, B. I., Qi, L., Copeland, T., Nanoparticles for materials design: Present and future., *J. Ceram. Proc. Res.* **2005**, 6, 31–40.

87 Alt, V., Bechert, T., Steinrücke, P., Wagener, M., Seidel, P., Dingeldein, E., Domann, E., Schnettlet, R., An in vitro assessment of the antibacterial properties and cytotoxicity of nanoparticulate silver bone cement., *Biomaterials* **2004**, 25, 4383–4391.

88 Braydich-Stolle, L., Hussain, S., Schlager, J., Hofmann, M.-C., In vitro cytotoxicity of nanoparticles in mammalian germ-line stem cells., *Toxic. Sci.* **2005**, 88, 412–419.

89 Gupta, A. K., Gupta, M., Cytotoxicity suppression and cellular uptake enhancement of surface modified magnetic nanoparticles., *Biomaterials* **2005**, 26, 1565–1573.

90 Berry, C. C., Wells, S., Charles, S., Aitchison, G., Curtis, A. S. G., Cell response to dextran-derivatised iron oxide nanoparticles post internalization., *Biomaterials* **2004**, 25, 5405–5413.

91 Harlozinska, A., Progress in molecular mechanisms of tumor metastasis and angiogenesis., *Anticanc. Res.* **2005**, 25, 3327–3333.

92 Kirfel, G., Rigort, A., Borm, B., Herzog, V., Cell migration: Mechanisms of rear detachment and the formation of migration tracks., *Eur. J. Cell Biol.* **2004**, 83, 717–724.

93 Clegg, D. O., Wingerd, K. L., Hikita, S. T., Tolhurst, E. C., Integrins in the development, function and dysfunction of the nervous system., *Front. Biosci.* **2003**, 8, d723–d750.

94 Wilson, C. J., Clegg, R. E., Leavesley, D. I., Pearcy, M. J., Mediation of biomaterial-cell interactions by adsorbed proteins: A review., *Tissue Eng.* **2005**, 11, 1–18.

95 Folch, A., Toner, M., Microengineering of cellular interactions., *Annu. Rev. Biomed. Eng.* **2000**, 2, 227–256.

96 Turner, A. M. P., Dowell, N., Turner, S. W. P., Kam, L., Isaacson, M., Turner, J. N., Craighead, H. G., Shain, W., Attachment of astroglial cells to microfabricated pillar arrays of different geometries., *J. Biomed. Mater. Res.* **2000**, 51, 430–441.

97 Rajnicek, A., Britland, S., McCaig, C., Contact guidance of CNS neurites on grooved quartz influence of groove dimensions, neuronal age and cell type., *J. Cell Sci.* **1997**, 110, 2905–2913.

98 Chen, C. S., Mrksich, M., Huang, S., Whitesides, G. M., Ingber, D. E., Geometric control of cell life and death., *Science* **1997**, 276, 1425–1428.

99 Chen, C. S., Jiang, X., Whitesides, G. M., Microengineering the environment of mammalian cells in culture., *MRS Bull.* **2005**, 30, 194–201.

100 BRITLAND, S., PERRIDGE, C., DENYER, M., MORGAN, H., CURTIS, A., WILKINSON, C., Morphogenetic guidance cues can interact synergistically and hierarchically in steering nerve cell growth., *Exp. Biol. Online* **1997**, 1, 1–15.

101 CLARK, P., COLES, D., PECKHAM, M., Preferential adhesion to and survival on patterned laminin organizes myogenesis in vitro., *Exp. Cell Res.* **1997**, 230, 275–283.

102 GOESSL, A., BOWEN-POPE, D. F., HOFFMAN, A. S., Control of shape and size of vascular smooth muscle cells in vitro by plasma lithography., *J. Biomed. Mater. Res.* **2001**, 57, 15–24.

103 CLARK, P., BRITLAND, S., CONNOLLY, P., Growth cone guidance and neuron morphology on micropatterned laminin surfaces., *J. Cell Sci.* **1993**, 105, 203–212.

104 COREY, J. M., BRUNETTE, A. L., CHEN, M. S., WEYHENMEYER, J. A., BREWER, G. J., WHEELER, B. C., Differentiated B104 neuroblastoma cells are a high-resolution assay for micropatterned substrates., *J. Neurosci. Methods* **1997**, 75, 91–97.

105 TURCU, F., TRATSK-NITZ, K., THANOS, S., SCHUHMANN, W., HEIDUSCHKA, P., Ink-jet printing for micropattern generation of laminin for neuronal adhesion., *J. Neurosci. Methods* **2003**, 30, 141–148.

106 VOGT, A. K., BREWER, G. J., DECKER, T., BOCKER-MEFFERT, S., JACOBSEN, V., KREITER, M., KNOLL, W., OFFEN-HAUSSER, A., Independence of synaptic specificity from neuritic guidance., *Neuroscience* **2005**, 134, 783–790.

107 ADAMS, J. C., Cell-matrix contact structures., *Cell. Mol. Life. Sci.* **2001**, 58, 371–392.

108 ARNOLD, M., CAVALCANTI-ADAM, E. A., GLASS, R., BLÜMMEL, J., ECK, W., KANTLEHNER, M., KESSLER, H., SPATZ, J. P., Activation of integrin function by nanopatterned adhesive interfaces., *ChemPhysChem* **2004**, 5, 383–388.

109 FLEMMING, R. G., MURPHY, C. J., ABRAMS, G. A., GOODMAN, S. L., NEALEY, P. F., Effects of synthetic micro- and nano-structured surfaces on cell behavior., *Biomaterials* **1999**, 20, 573–588.

110 INGBER, D. E., Mechanobiology and diseases of mechanotransduction., *Ann. Med.* **2003**, 35, 1–14.

111 CLARK, P., CONNOLLY, P., CURTIS, A. S. G., DOW, J. A. T., WILKINSON, C. D. W., Cell guidance by ultrafine topography in vitro., *J. Cell Sci.* **1991**, 99, 73–77.

112 TEIXEIRA, A. I., ABRAMS, G. A., MURPHY, C. J., NEALEY, P. F., Cell behavior on lithographically defined nanostructured substrates., *J. Vac. Sci. Technol. B* **2003**, 21, 683–687.

113 YIM, E. K. F., REANO, R. M., PANG, S. W., YEE, A. F., CHEN, C. S., LEONG, K. W., Nanopattern-induced changes in morphology and motility of smooth muscle cells., *Biomaterials* **2005**, 26, 5405–5413.

114 WÓJCIAK-STOTHARD, A., CURTIS, A., MONAGHAN, W., MACDONALD, K., WILKINSON, C., Guidance and activation of murine macrophages by nanometric scale topography., *Exp. Cell Res.* **1996**, 223, 426–435.

115 RAJNICEK, A. M., McCAIG, C. D., Guidance of CNS growth cones by substratum grooves and ridges: Effects of inhibitors of the cytoskeleton, calcium channels and signal transduction pathways., *J. Cell Sci.* **1997**, 110, 2915–2924.

116 COUSINS, B. G., DOHERTY, P. J., WILLIAMS, R. L., FINK, J., GARVEY, M. J., The effect of silica nanoparticulate coatings on cellular response., *J. Mater. Sci. Mater. Med.* **2004**, 15, 355–359.

117 FREY, W., WOODS, C. K., CHILKOTI, A., Ultraflat nanosphere lithography: A new method to fabricate flat nanostructures., *Adv. Mater.* **2000**, 12, 1515–1519.

118 STEIN, A., SCHRODEN, R. C., Colloidal crystal templating of three-dimensionally ordered macroporous solids: Materials for photonics and beyond., *Curr. Opin. Solid State Mater. Sci.* **2001**, 5, 553–564.

119 DALBY, M. J., RIEHLE, M. O., SUTHERLAND, D. S., AGHELI, H., CURTIS, A. S. G., Morphological and microarray analysis of human fibroblasts cultured on nanocolumns produced by colloidal lithography., *Eur. Cell Mater.* **2005**, 9, 1–8.

120 ANDERSSON, A.-S., BRINK, J., LIDBERG, U., SUTHERLAND, D. S., Influence of systematically varied nanoscale topography on the morphology of epithelial cells., *IEEE Trans. Nanobiosci.* **2003**, 2, 49–57.

121 DALBY, M. J., RIEHLE, M. O., SUTHERLAND, D. S., AGHELI, H., CURTIS, A. S. G., Fibroblast response to a controlled nanoenvironment produced by colloidal lithography., *J. Biomed. Mater. Res.* **2004**, 69A, 314–322.

122 BERSHADSKY, A. D., TINT, I. S., NEYFAKH, A. A., JR., VASILIEV, J. M., Focal contacts of normal and RSV-transformed quail cells. Hypothesis of the transformation-induced deficient maturation of focal contacts., *Exp. Cell Res.* **1985**, 158, 433–444.

123 DALBY, M. J., RIEHLE, M. O., SUTHERLAND, D. S., AGHELI, H., CURTIS, A. S. G., Changes in fibroblast morphology in response to nano-columns produced by colloidal lithography., *Biomaterials* **2004**, 25, 5415–5422.

124 VAN KOOTEN, T. G., KLEIN, C. L., KÖHLER, H., KIRKPATRICK, C. J., WILLIAMS, D. F., ELOY, R., From cytotoxicity to biocompatibility testing in vitro: Cell adhesion molecule expression defines a new set of parameters., *J. Mater. Sci. Mater. Med.* **1997**, 8, 835–841.

125 KELLER, J. C., STANFORD, C. M., WIGHTMAN, J. P., DRAUGHN, R. A., ZAHARIAS, R., Characterizations of titanium implant surfaces. III., *J. Biomed. Mater. Res.* **1994**, 28, 939–946.

126 MRKSICH, M., WHITESIDES, G. M., Using self-assembled monolayers to understand the interactions of man-made surfaces with proteins and cells. *Ann. Rev. Biophys. Biomol. Struct.* **1996**, 25, 55–78.

127 KOO, L. Y., IRVINE, D. J., MAYES, A. M., LAUFFENBURGER, D. A., GRIFFITH, L. G., Co-regulation of cell adhesion

by nanoscale RGD organization and mechanical stimulus., *J. Cell Sci.* **2002**, 115, 1423–1433.

128 LEE, K.-B., PARK, S.-J., MIRKIN, C. A., SMITH, J. C., MRKSICH, M., Protein nanoarrays generated by dip-pen nanolithography., *Science* **2002**, 295, 1702–1705.

129 SPATZ, J. P., MÖSSMER, S., HARTMANN, C., MÖLLER, M., HERZOG, T., KRIEGER, M., BOYEN, H.-G., ZIEMANN, P., KABIUS, B., Ordered deposition of inorganic clusters from micellar block copolymer films., *Langmuir* **2000**, 16, 407–415.

130 CRICK, F., HUGHES, A., The physical properties of the cytoplasm. I. Experimental., *Exp. Cell Res.* **1950**, 1, 37–80.

131 ZBOROWSKI, M., CHALMERS, J. J., Magnetic cell sorting., *Methods Mol. Biol.* **2005**, 295, 291–300.

132 NAKAMURA, M., DECKER, K., CHOSY, J., COMELLA, K., MELNIK, K., MOORE, L., LASKY, L. C., ZBOROWSKI, M., CHALMERS, J. J., Separation of a breast cancer cell line from human blood using a quadrupole magnetic flow sorter., *Biotechnol. Prog.* **2001**, 17, 1145–1155.

133 HULTGREN, A., TANASE, M., FELTON, E. J., BHADRIRAJU, K., SALEM A. K., CHEN, C. S., REICH, D. H., Optimization of yield in magnetic cell separations using nickel nanowires of different lengths., *Biotechnol. Prog.* **2005**, 21, 509–515.

134 SHINKAI, M., YANASE, M., HONDA, H., WAKABAYASHI, T., YOSHIDA, J., KOBAYASHI, T., Intracellular hyperthermia for cancer using magnetite cationic liposomes: *in vitro* study., *Jpn. J. Canc. Res.* **1996**, 87, 1179–1183.

135 ITO, A., TAKIZAWA, Y., HONDA, H., HATA, K., KAGAMI, H., UEDA, M., KOBAYASHI, T., Tissue engineering using magnetite nanoparticles and magnetic force: Heterotypic layers of cocultured hepatocytes and endothelial cells., *Tissue Eng.* **2004**, 10, 833–840.

136 ITO, A., INO, K., KOBAYASHI, T., HONDA, H., The effect of RGD peptide-conjugated magnetite cationic liposomes on cell growth and cell sheet harvesting., *Biomaterials* **2005**, 26, 6185–6193.

137 ITO, A., HAYASHIDA, M., HONDA, H., HATA, K., KAGAMI, H., UEDA, M., KOBAYASHI, T., Construction and harvest of multilayered keratinocyte sheets using magnetite nanoparticles and magnetic force., *Tissue Eng.* **2004**, 10, 873–880.

138 ITO, A., HIBINO, E., KOBAYASHI, C., TERASAKI, H., KAGAMI, H., UEDA, M., KOBAYASHI, T., HONDA, H., Construction and delivery of tissue-engineered human retinal pigment epithelial cell sheets, using magnetite nanoparticles and magnetic force., *Tissue Eng.* **2005**, 11, 489–496.

139 TANASE, M., FELTON, E. J., GRAY, D. S., HULTGREN, A., CHEN, C. S., REICH, D. H., Assembly of multicellular constructs and microarrays of cells using magnetic nanowires., *Lab Chip* **2005**, 5, 598–605.

140 WANG, N., BUTLER, J. P., INGBER, D. E., Mechanotransduction across the cell surface and through the cytoskeleton., *Science* **1993**, 260, 1124–1127.

141 CHEN, J., FABRY, B., SCHIFFRIN, E. L., WANG, N., Twisting integrin receptors increases endothelin-1 gene expression in endothelial cells., *Am. J. Physiol. Cell. Physiol.* **2001**, 280, C1475–C1484.

142 TANASE, M., SILEVITCH, D. M., HULTGREN, A., BAUER, L. A., SEARSON, P. C., MEYER, G. J., REICH, D. H., Magnetic trapping and self-assembly of multicomponent nanowires., *J. Appl. Phys.* **2002**, 91, 8549–8551.

143 BIELINSKA, A., EICHMAN, J. D., LEE, I., BAKER, J. R., BALOGH, L., Imaging {Au-0-PAMAM} gold-dendrimer nanocomposites in cells., *J. Nanopart. Res.* **2002**, 4, 395–403.

144 SAUNDERS, A. E., KORGEL, B. A., Observation of an AB phase in bidisperse nanocrystal superlattices., *ChemPhysChem* **2005**, 6, 61–65.

145 MOGHIMI, S. M., HUNTER, A. C., MURRAY, J. C., Nanomedicine: Current status and future prospects., *FASEB J.* **2005**, 19, 311–330.

146 BELLAMKONDA, R., RANIERI, J. P., AEBISCHER, P., Laminin oligopeptide derivatized agarose gels allow three-dimensional neurite extension in vitro., *J. Neurosci. Res.* **1995**, 41, 501–509.

147 HERN, D. L., HUBBELL, J. A., Incorporation of adhesion peptides into nonadhesive hydrogels useful for tissue resurfacing., *J. Biomed. Mater. Res.* **1998**, 39, 266–276.

148 ZHENG, H., BERG, M. C., RUBNER, M. F., HAMMOND, P. T., Controlling cell attachment selectively onto biological polymer-colloid templates using polymer-on-polymer stamping., *Langmuir* **2004**, 20, 7215–7222.

149 BAUSCH, A. R., ZIEMANN, F., BOULBITCH, A. A., JACOBSON, K., SACKMANN, E., Local measurements of viscoelastic parameters of adherent cell surfaces by magnetic bead microrheometry., *Biophys. J.* **1998**, 75, 2038–2049.

150 VALBERG, P. A., ALBERTINI, D. F., Cytoplasmic motions, rheology, and structure probed by a novel magnetic particle method., *J. Cell Biol.* **1985**, 101, 130–40.

151 BAUSCH, A. R., MOLLER, W., SACKMANN, E., Measurement of local viscoelasticity and forces in living cells by magnetic tweezers., *Biophys. J.* **1999**, 76, 573–579.

152 TANAKA, T., YAMASAKI, H., TSUJIMURA, N., NAKAMURA, N., MATSUNAGA, T., Magnetic control of bacterial magnetite-myosin conjugate movement on actin cables., *Mater. Sci. Eng. C* **1997**, 5, 121–124.

153 ROSENTHAL, S. J., TOMLINSON, I., ADKINS, E. M., SCHROETER, S., ADAMS, S., SWAFFORD, L., MCBRIDE, J., WANG, Y., DEFELICE, L. J., BLAKELY, R. D., Targeting cell surface receptors with ligand-conjugated nanocrystals., *J. Am. Chem. Soc.* **2002**, 124, 4586–4594.

154 PASTAN, I., WILLINGHAM, M. C. (Eds.), *Endocytosis*, Plenum Press, New York, 1985.

155 GAO, D., CRITSER, J. K., Mechanisms of cryoinjury in living cells. *ILAR J.* **2000**, 41, 187–196.

156 OSAKI, F., KANAMORI, T., SANDO, S., SERA, T., AOYAMA, Y., A quantum dot conjugated sugar ball and its cellular uptake. On the size effects of endocytosis in the subviral region., *J. Am. Chem. Soc.* **2004**, 126, 6520–6521.

157 GUPTA, A. K., CURTIS, A. S. G., Lactoferrin and ceruloplasmin derivatized superparamagnetic iron oxide nanoparticles for targeting cell surface receptors., *Biomaterials* **2004**, 25, 3029–3040.

158 DAHAN, M., LAURENCE, T., PINAUD, F., CHEMLA, D. S., ALIVISATOS, A. P., SAUER, M., WEISS, S., Time-gated biological imaging by use of colloidal quantum dots., *Opt. Lett.* **2001**, 26, 825–827.

159 HANAKI, K.-I., MOMO, A., TAISUKE, O., KOMOTO, A., MAENOSONO, S., YAMAGUCHI, Y., YAMAMOTO, K., Semiconductor quantum dot/albumin complex is a long-life and highly photostable endosome marker., *Biochem. Biophys. Res. Commun.* **2003**, 302, 496–501.

160 GOMEZ, N., WINTER, J. O., SHIEH, F., SAUNDERS, A. E., KORGEL, B. A., SCHMIDT, C. E., Challenges in quantum dot-neuron active interfacing., *Talanta*, **2005**, 67, 462–471.

161 ROGERS, W. J., BASU P., Factors regulating macrophage endocytosis of nanoparticles: Implications for targeted magnetic resonance plaque imaging, *Atherosclerosis* **2005**, 178, 67–73.

162 LIDKE, D. S., NAGY, P., HEINTZMANN, R., ARNDT-JOVIN, D. J., POST, J. N., GRECCO, H. E., JARES-ERIJMAN, E. A., JOVIN, T. M., Quantum dot ligands provide new insights into erbB/HER receptor-mediated signal transduction., *Nat. Biotechnol.* **2004**, 22, 198–203.

163 YANG, P.-H., SUN, X., CHIU, J.-F., SUN, H., HE, Q.-Y., Transferrin-mediated gold nanoparticle cellular uptake., *Bioconj. Chem.* **2005**, 16, 494–496.

164 DALBY, M. J., BERRY, C. C., RIEHLE, M. O., SUTHERLAND, D. S., AGHELI, H., CURTIS, A. S. G., Attempted endocytosis of nano-environment produced by colloidal lithography by human fibroblasts., *Exp. Cell Res.* **2004**, 295, 387–394.

165 PRATTEN, M. K., LLOYD, J. B., Pinocytosis and phagocytosis: The effect of size of a particle substrate on its mode of capture by rat peritoneal macrophages cultured in vitro., *Biochem. Biophys. Acta* **1986**, 881, 307–313.

166 GUPTA, A. K., CURTIS, A. S. G., Surface modified superparamagnetic nanoparticles for drug delivery: Interaction studies with human fibroblasts in culture., *J. Mater. Sci. Mater. Med.* **2004**, 15, 493–496.

167 GUPTA, A. K., BERRY, C., GUPTA, M., CURTIS, A., Receptor-mediated targeting of magnetic nanoparticles using insulin as a surface ligand to prevent endocytosis., *IEEE Trans. Nanobiosci.* **2003**, 2, 255–261.

168 MORRIS, M. C., DEPOLLIER, J., MERY, J., HEITZ, F., DIVITA, G., A peptide carrier for the delivery of biologically active proteins into mammalian cells., *Nat. Biotechnol.* **2001**, 19, 1173–1176.

169 VIVES, E., RICHARD, J. P., RISPAL, C., LEBLEU, B., TAT peptide internalization: Seeking the mechanism of entry., *Curr. Protein Pept. Sci.* **2003**, 4, 125–132.

170 NORI, A., KOPECEK, J., Intracellular targeting of polymer-bound drugs for cancer chemotherapy., *Adv. Drug Deliv. Rev.* **2005**, 57, 609–636.

171 MATTHEAKIS, L. C., DIAS, J. M., CHOI, Y.-J., GONG, J., BRUCHEZ, M. P., LIU, J., WANG, E., Optical coding of mammalian cells using semiconductor quantum dots., *Anal. Biochem.* **2004**, 327, 200–208.

172 TKACHENKO, A. G., XIE, H., COLEMAN, D., GLOMM, W., RYAN, J., ANDERSON, M. F., FRANZEN, S., FELDHEIM, D. L., Multifunctional gold nanoparticle-peptide complexes for nuclear targeting., *J. Am. Chem. Soc.* **2003**, 125, 4700–4701.

173 JOSEPHSON, L., TUNG, C.-H., MOORE, A., WEISSLEDER, R., High-efficiency intracellular magnetic labeling with novel superparamagnetic-Tat peptide conjugates., *Bioconj. Chem.* **1999**, 10, 186–191.

174 Won, J., Kim, M., Yi, Y.-W., Kim, Y. H., Jung, N., Kim, T. K., A magnetic nanoprobe technology for detecting molecular interactions in live cells., *Science* **2005**, 309, 121–125.

175 Zhang, C., Ma, H., Nie, S., Ding, Y., Jin, L., Chen, D., Quantum dot-labeled trichosanthin., *Analyst* **2000**, 125, 1029–1031.

176 Hoshino, A., Fujioka, K., Oku, T., Nakamura, S., Suga, M., Yamaguchi, T., Suzuki, K., Yasuhara, M., Yamamoto, K., Quantum dots targeted to the assigned organelle in living cells., *Microbiol. Immunol.* **2004**, 48, 985–994.

177 Derfus, A. M., Chan, W. C. W., Bhatia, S. N., Intracellular delivery of quantum dots for live cell labeling and organelle tracking., *Adv. Mater.* **2004**, 16, 961–966.

178 Tkachenko, A. G., Xie, H., Liu, Y., Coleman, D., Ryan, J., Glomm, W. R., Shipton, M. K., Franzen, S., Feldheim, D. L., Cellular trajectories of peptide-modified gold particle complexes: Comparison of nuclear localization signals and peptide transduction domains., *Bioconj. Chem.* **2004**, 15, 482–490.

179 Dahan, M., Lévi, S., Luccardini, C., Rostaing, P., Riveau, B., Triller, A., Diffusion dynamics of glycine receptors revealed by single-quantum dot tracking., *Science* **2003**, 302, 442–445.

180 Chu, T. C., Shieh, F., Lavery, L. A., Levy, M., Richards-Kortum, R., Korgel, B. A., Ellington, A. D., Labeling tumor cells with fluorescent nanocrystal-aptamer bioconjugates., *Biosens. Bioelectron.* **2006**, 21, 1859–1866.

181 Gao, X., Nie, S., Quantum dot-encoded beads., *Methods Mol. Biol.* **2005**, 303, 61–71.

182 Pathak, S., Choi, S.-Y., Arnheim, N., Thompson, M., Hydroxylated quantum dots as luminescent probes for in situ hybridization., *J. Am. Chem. Soc.* **2001**, 123, 4103–4104.

183 Ferrara, D. E., Weiss, D., Carnell, P. H., Vito, R. P., Vega, D., Gao, X., Nie, S., Taylor, W. R., Quantitative 3D fluorescence technique for the analysis of en face preparations of arterial walls using quantum dot nanocrystals and two-photon excitation laser scanning microscopy., *Am. J. Physiol. Regul. Integr. Comp. Physiol.* **2006**, 290, R114–23.

184 Jaiswal, J. K., Simon, S. M., Potentials and pitfalls of fluorescent quantum dots for biological applications., *Trends Cell Biol.* **2004**, 14, 497–504.

185 Lewin, M., Carlesso, N., Tung, C.-H., Tang, X.-W., Cory, D., Scadden, D. T., Weissleder, R., Tat peptide-derivatized magnetic nanoparticles allow in vivo tracking and recovery of progenitor cells., *Nat. Biotechnol.* **2000**, 18, 410–414.

186 Moore, A., Josephson, L., Bhorade, R. M., Basilion, J. P., Weissleder, R., Human transferrin receptor gene as a marker gene for MR imaging., *Radiology* **2001**, 221, 244–250.

187 Ahmed, F., Discher, D., Self-porating polymersomes of PEG-PLA and PEG-PCL: Hydrolysis-triggered controlled release vesicles., *J. Controlled Release* **2004**, 96, 37–53.

188 Ahmed, F., Discher, D., Breaking down endolysosomal barriers for drug delivery with degradable polymersomes., Proceedings of American Institute of Chemical Engineers Annual Meeting, October 29–November 5, **2005**, Cincinnati, OH, available online at http://aiche.confex.com/aiche/2005/techprogram/P28864.htm.

189 Kunisawa, J., Masuda, T., Katayama, K., Yoshikawa, T., Tsutsumi, Y., Akashi, M., Mayumi, T., Nakagawa, S., Fusogenic liposome delivers encapsulated nanoparticles for cytosolic controlled gene release., *J. Controlled Release* **2005**, 105, 344–353.

190 Kam, N. W. S., O'Connell, M., Wisdom, J. A., Dai, H., Carbon nanotubes as multifunctional biological transporters and near-infrared agents for selective cancer cell destruction., *Proc. Natl. Acad. Sci. U.S.A.* **2005**, 102, 11 600–11 605.

191 PANTAROTTO, D., SINGH, R., MCCARTHY, D., ERHARDT, M., BRIAND, J.-P., PRATO, M., KOSTARELOS, K., BIANCO, A., Functionalized carbon nanotubes for plasmid DNA gene delivery., *Angew. Chem. Int. Ed.* **2004**, 32, 5242–5246.

192 CAI, D., MATARAZA, J. M., QIN, Z.-H., HUANG, Z., HUANG, J., CHILES, T. C., CARNAHAN, D., KEMPA, K., REN, Z., Highly efficient molecular delivery into mammalian cells using carbon nanotube spearing., *Nat. Meth.* **2005**, 2, 449–454.

193 STRYER, L., HAUGLAND, R. P., Energy transfer: A spectroscopic ruler., *Proc. Natl. Acad. Sci. U.S.A.* **1967**, 58, 719–726.

194 MAXWELL, D. J., TAYLOR, J. R., NIE, S., Self-assembled nanoparticle probes for recognition and detection of biomolecules., *J. Am. Chem. Soc.* **2002**, 124, 9606–9612.

195 MEDINTZ, I. L., CLAPP, A. R., MATTOUSSI, H., GOLDMAN, E. R., FISHER, B., MAURO, J. M., Self-assembled nano-scale biosensors based on quantum dot FRET donors., *Nat. Mater.* **2003**, 2, 630–638.

196 WANG, S., MAMEDOVA, N., KOTOV, N. A., CHEN, W., STUDER, J., Antigen/antibody immunocomplex from CdTe nanoparticle bioconjugates., *Nano Lett.* **2002**, 2, 817–822.

197 LEE, J., GOVOROV, A. O., KOTOV, N. A., Bioconjugated superstructures of CdTe nanowires and nanoparticles: Multistep cascade Forster resonance energy transfer and energy channeling., *Nano Lett.* **2005**, 5, 2063–2069.

198 KINBARA, K., AIDA, T., Toward intelligent molecular machines: Directed motions of biological and artificial molecules and assemblies., *Chem. Rev.* **2005**, 105, 1377–1400.

199 WARSHAW, D. M., KENNEDY, G. G., WORK, S. S., KREMENTSOVA, E. B., BECK, S., TRYBUS, K. M., Differential labeling of myosin V heads with quantum dots allows direct visualization of hand-over-hand processivity., *Biophys. J. Biophys. Lett.* **2005**, 88, L30–L32.

200 MÅNSSON, A., SUNDBERG, M., BALAZ, M., BUNK, R., NICHOLLS, I. A., OMLING, P., TÅGERUDA, S., MONTELIUS, L., In vitro sliding of actin filaments labeled with single quantum dots., *Biochem. Biophys. Res. Commun.* **2004**, 314, 529–534.

201 PATOLSKY, F., WIEZMANN, Y., WILLNER, I., Actin-based metallic nanowires as bio-nanotransporters., *Nat. Mater.* **2004**, 3, 692–695.

202 JIA, L., MOORJANI, S. G., JACKSON, T. N., HANCOCK, W. O., Microscale transport and sorting by kinesin molecular motors., *Biomed. Microdev.* **2004**, 6, 67–74.

203 VAN DEN HEUVEL, M. G. L., BUTCHER, C. L., LEMAY, S. G., DIEZ, S., DEKKER, C., Electrical docking of microtubules for kinesin-driven motility in nanostructures., *Nano Lett.* **2005**, 5, 235–241.

204 VAN DEN HEUVEL, M. G. L., BUTCHER, C. L., SMEETS, R. M. M., DIEZ, S., DEKKER, C., High rectifying efficiencies of microtubule motility on kinesin-coated gold nanostructures., *Nano Lett.* **2005**, 5, 1117–1122.

205 LIMBERIS, L., STEWART, R. J., Toward kinesin-powered microdevices., *Nanotechnology* **2000**, 11, 47–51.

206 BÖHM, K. J., STRACKE, R., MÜHLIG, P., UNGER, E., Motor protein-driven unidirectional transport of micrometer-sized cargoes across isopolar microtubule arrays., *Nanotechnology* **2001**, 12, 238–244.

207 BACHAND, G. D., RIVERA, S. B., BOAL, A. K., GAUDIOSO, J., LIU, J., BUNKER, B. C., Assembly and transport of nanocrystal CdSe quantum dot nanocomposites using microtubules and motor proteins., *Nano Lett.* **2004**, 4, 817–821.

208 RIEGLER, J., NICK, P., KIELMANN, U., NANN, T., Visualizing the self-assembly of tubulin with luminescent nanorods., *J. Nanosci. Nanotechnol.* **2003**, 3, 380–385.

209 HESS, H., CLEMMENS, J., QIN, D., HOWARD, J., VOGEL, V., Light-controlled molecular shuttles made from motor proteins carrying cargo on engineered surfaces., *Nano Lett.* **2001**, 1, 235–239.

210 DIEZ, S., REUTHER, C., DINU, C.,

Seidel, R., Mertig, M., Pompe, W., Howard, J., Stretching and transporting DNA molecules using motor proteins., *Nano Lett.* **2003**, 3, 1251–1254.

211 Weibel, D. B., Garsetcki, P., Ryan, D., DiLuzio, W. R., Mayer, M., Seto, J. E., Whitesides, G. M., Microoxen: Microorganisms to move microscale loads., *Proc. Natl. Acad. Sci. U.S.A.* **2005**, 102, 11 963–11 967.

212 Darnton, N., Turner, L., Breuer, K., Berg, H. C., Moving fluid with bacterial carpets., *Biophys. J.* **2004**, 86, 1863–1870.

213 Ferrari, M., Cancer nanotechnology: Opportunities and challenges., *Nat. Rev. Canc.* **2005**, 5, 161–171.

214 Moses, M. A., Brem, H., Langer, R., Advancing the field of drug delivery: Taking aim at cancer., *Cancer Cell.* **2003**, 4, 337–341.

215 Capaldi, B., Treatments and devices for future diabetes management., *Nurs. Times* **2005**, 101, 30–32.

216 Stahl, S. M., Applications of new drug delivery technologies to Parkinson's disease and dopaminergic agents., *Neural Transm. Suppl.* **1988**, 27, 123–132.

217 Harbaugh, R. E., Novel CNS-directed drug delivery systems in Alzheimer's disease and other neurological disorders., *Neurobiol. Aging* **1989**, 10, 623–629.

218 Langer, R., Peppas, N. A., Advances in biomaterials, drug delivery, and bionanotechnology., *AIChE J.* **2003**, 49, 2990–3006.

219 Müller, R. H., Keck, C. M., Drug delivery to the brain-realization by novel drug carriers., *J. Nanosci. Nanotechnol.* **2004**, 4, 471–483.

220 Johnson-Saliba, M., Jans, D. A., Gene therapy: Optimizing DNA delivery to the nucleus., *Curr. Drug Targ.* **2001**, 2, 371–399.

221 Plank, C., Schillinger, U., Scherer, F., Bergemann, C., Rémy, J.-S., Krötz, F., Anton, M., Lausier, J., Rosenecker, J., The magneto-transfection method: Using magnetic force to enhance gene delivery., *Biol. Chem.* **2003**, 384, 737–747.

222 Luo, D., Saltzmann, W. M., Enhancement of transfection by physical concentration of DNA at the cell surface., *Nat. Biotechnol.* **2000**, 18, 893–895.

223 Widder, K. J., Senyei, A. E., Scarpelli, D. G., Magnetic microspheres: A model system for site specific drug delivery in vivo., *Proc. Soc. Exp. Biol. Med.* **1978**, 58, 141–146.

224 Widder, K. J., Morris, R. M., Poore, G. A., Howard, D. P., Senyei, A. E., Selective targeting of magnetic albumin microspheres containing low-dose doxorubicin: Total remission in Yoshida sarcoma-bearing rats., *Eur. J. Cancer Clin. Oncol.* **1983**, 19, 135–139.

225 Swanson, N., Gershlick, A. H., Drug eluting stents in interventional cardiology – current evidence and emerging uses., *Curr. Drug Targets Cardiovasc. Haematol. Disord.* **2005**, 5, 313–321.

226 Huh, Y.-M., Jun, Y.-W., Song, H.-T., Kim, S., Choi, J.-S., Lee, J.-H., Yoon, S., Kim, K.-S., Shin, J.-S., Suh, J.-S., Cheon, J., In vivo magnetic resonance detection of cancer by using multifunctional magnetic nanocrystals., *J. Am. Chem. Soc.* **2005**, 127, 12 387–12 391.

227 Lübbe, A. S., Alexiou, C., Bergemann, C., Clinical applications of magnetic drug targeting., *J. Surg. Res.* **2000**, 95, 200–206.

228 Pankhurst, Q. A., Connolly, J., Jones, S. K., Dobson, J., Applications of magnetic particles in biomedicine., *J. Phys. D, Appl. Phys.* **2003**, 36, R167–R181.

229 Gallo, J. M., Varkonyi, P., Hassan, E. E., Groothius, D. R., Targeting anti-cancer drugs to the brain: II. Physiological pharmacokinetic model of oxantrazole following intraarterial administration to rat glioma-2 (RG-2) bearing rats., *J. Pharmacokinet. Biopharm.* **1993**, 21, 575–592.

230 Pulfer, S. K., Ciccotto, S. L., Gallo, J. M., Distribution of small magnetic particles in tumor-bearing rats., *J. Neurooncol.* **1999**, 41, 99–105.

231 Goodwin, S., Peterson, C., Hoh, C., Bittner, C., Targeting and retention of magnetic targeted carriers

enhancing intra-arterial chemo-therapy., *J. Magn. Magn. Mater.* **1999**, 194, 132–139.

232 ALEXIOU, C., ARNOLD, W., KLEIN, R. J., PARAK, F. G., HULIN, P., BERGEMANN, C., ERHARDT, W., WAGENPFEIL, S., LÜBBE, A. S., Locoregional cancer treatment with magnetic drug targeting., *Cancer Res.* **2000**, 60, 6641–6648.

233 KUBO, T., SUGITA, T., SHIMOSE, S., NITTA, Y., MARAKAMI, T., Targeted systemic chemotherapy using magnetic liposomes with incorporated adriamycin for osteosarcoma in hamsters., *Int. J. Oncol.* **2001**, 18, 121–125.

234 LÜBBE, A., BERGEMANN, C., RIESS, H., SCHRIEVER, F., REICHARDT, P., POSSINGER, K., MATTHIAS, M., DORKEN, B., HERRMANN, F., GURTLER, R., HOHENBERGER, P., HAAS, N., SOHR, R., SANDER, B., LEMKE, A. J., OHLENDORF, D., HUHNT, W., HUHN, D., Clinical experiences with magnetic drug targeting: A phase I study with 4′-epidoxorubicin in 14 patients with advanced solid tumors., *Cancer Res.* **1996**, 15, 4686–4693.

235 ZHANG, Y., KOHLER, N., ZHANG, M., Surface modification of superparamag-netic magnetite nanoparticles and their intracellular uptake., *Biomaterials* **2002**, 23, 1553–1561.

236 PETRI-FINK, A., CHASTELLAIN, M., JUILLERAT-JEANNERET, L., FERRARI, A., HOFMANN, H., Development of functionalized superparamagnetic iron oxide nanoparticles for interaction with human cancer cells., *Biomaterials* **2005**, 26, 2685–2694.

237 KOHLER, N., SUN, C., WANG, J., ZHANG, M., Methotrexate-modified superparamagnetic nanoparticles and their intracellular uptake into human cancer cells., *Langmuir* **2005**, 21, 8858–8864.

238 JAIN, T. K., MORALES, M. A., SAHOO, S. K., LESLIE-PELECKY, D. L., LABHASETWAR, V., Iron oxide nanoparticles for sustained delivery of anticancer agents., *Mol. Pharmacol.* **2005**, 2, 194–205.

239 PACIOTTI, G. F., MYER, L.,

WEINREICH, D., GOIA, D., PAVEL, N., MCLAUGHLIN, R. E., TAMARKIN, L., Colloidal gold: A novel nanoparticle vector for tumor directed drug delivery. *Drug Deliv.* **2004**, 11, 169–183.

240 CHEN, J.-F., DING, H.-M., WANG, J.-X., SHAO, L., Preparation and characterization of porous hollow silica nanoparticles for drug delivery applications., *Biomaterials* **2004**, 25, 723–727.

241 MOROZ, P., JONES, S. K., BRAY, B. N., Magnetically mediated hyperthermia: Current status and future directions., *Int. J. Hypertherm.* **2002**, 18, 267–284.

242 RAMACHANDRAN, N., MAZURUK, K., Magnetic microspheres and tissue model studies for therapeutic applications., *Ann. New York Acad. Sci.* **2004**, 1027, 99–109.

243 HERGT, R., ANDRÄ, W., d'AMBLY, C. G., HILGER, I., KAISER, W. A., RICHTER, U., SCHMIDT, H.-G., Physical limits of hyperthermia using magnetite fine particles., *IEEE Trans. Magn.* **1998**, 34, 3745–3754.

244 WUST, P., HILDEBRANDT, B., SREENIVASA, G., RAU, B., GELLERMAN, J., RIESS, H., FELIX, R., SCHLAG, P. M., Hyperthermia in combined treatment of cancer., *Lancet* **2002**, 3, 487–497.

245 GILCHRIST, R. K., MEDAL, R., SHOREY, W. D., HANSELMAN, R. C., PARROTT, J. C., TAYLOR, C. B., Selective induction heating of lymph nodes., *Ann. Surg.* **1957**, 146, 596–606.

246 CHAN, D. C. F., KIRPOTIN, D. B., BUNN, P. A., Synthesis and evaluation of colloidal magnetic iron oxides for the site-specific radiofrequency-induced hyperthermia of cancer., *J. Magn. Magn. Mater.* **1993**, 122, 374–378.

247 JORDAN, A., WUST, P., SCHOLZ, R., TESCHE, B., FAHLING, H., MITROVICS, T., VOGL, T., CERVOS-NAVARRO, J., FELIX, R., Cellular uptake of magnetic fluid particles and their effects on human adenocarcinoma cells exposed to AC magnetic fields *in vitro.*, *Int. J. Hypertherm.* **1996**, 12, 705–722.

248 HIRSCH, L. R., STAFFORD, R. J., BANKSON, J. A., SERSHEN, S. R., RIVERA, B., PRICE, R. E., HAZLE, J. D.,

HALAS, N. J., WEST, J. L., Nanoshell-mediated near-infrared thermal therapy of tumors under magnetic resonance guidance., *Proc. Natl. Acad. Sci. U.S.A.* **2003**, 100, 13 549–13 554.

249 O'NEAL, D. P., HIRSCH, L. R., HALAS, N. J., PAYNE, J. D., WEST, J. L., Photothermal tumor ablation in mice using near infrared-absorbing nanoparticles., *Cancer Lett.* **2004**, 209, 171–176.

250 LOO, C., LOWERY, A., HALAS, N., WEST, J., DREZEK, R., Immunotargeted nanoshells for integrated cancer imaging and therapy., *Nano Lett.* **2005**, 5, 709–711.

251 WANG, S., GAO, R., ZHOU, F., SELKE, M., Nanomaterials and singlet oxygen photosensitizers: Potential applications in photodynamic therapy., *J. Mater. Chem.* **2004**, 14, 487–493.

252 ROY, I., OHULCHANSKYY, T. Y., PUDAVAR, H. E., BERGEY, E. J., OSEROFF, A. R., MORGAN, J., DOUGHERTY, T. J., PRASAD, P. N., Ceramic-based nanoparticles entrapping water-insoluble photosensitizing anticancer drugs: A novel drug-carrier system for photodynamic therapy., *J. Am. Chem. Soc.* **2003**, 125, 7860–7865.

253 YAN, F., KOPELMAN, R., The embedding of meta-tetra(hydroxyphenyl)-chlorin into silica nanoparticle platforms for photodynamic therapy and their singlet oxygen production and pH-dependant optical properties., *Photochem. Photobiol.* **2003**, 78, 587–591.

254 HONE, D. C., WALKER, P. I., EVANS-GOWING, R., FITZGERALD, S., BEEBY, A., CHAMBRIER, I., COOK, M. J., RUSSELL, D. A., Generation of cytotoxic singlet oxygen via phthalocyanine-stabilized gold nanoparticles: A potential delivery vehicle for photodynamic therapy., *Langmuir* **2002**, 18, 2985–2987.

255 BAKALOVA, R., OHBA, H., ZHELEV, Z., ISHIKAWA, M., BABA, Y., Quantum dots as photosensitizers?, *Nat. Biotechnol.* **2004**, 22, 1360–1361.

256 SAMIA, A. C. S., CHEN, X., BURDA, C., Semiconductor quantum dots for photodynamic therapy., *J. Am. Chem. Soc.* **2003**, 125, 15 736–15 737.

257 BAKALOVA, R., OHBA, H., ZHELEV, Z., NAGASE, T., JOSE, R., ISHIKAWA, M., BABA, Y., Quantum dot anti-CD conjugates: Are they potential photosensitizers or potentiators of classical photosensitizing agents in photodynamic therapy of cancer?, *Nano Lett.* **2004**, 4, 1567–1573.

258 CSÖGÖR, ZS., NACKEN, M., SAMETI, M., LEHR, C.-M., SCHMIDT, H., Modified silica particles for gene delivery., *Mater. Sci. Eng. C* **2003**, 23, 93–97.

259 KNEUER, C., SAMETI, M., HALTNER, E. G., SCHIESTEL, T., SCHIRRA, H., SCHMIDT, H., LEHR, C. M., Silica nanoparticles modified with aminosilanes as carriers for plasmid DNA., *Int. J. Pharm.* **2000**, 196, 257–261.

260 BAHARALI, D. J., KLEJBOR, I., STACHOWIAK, E. K., DUTTA, P., ROY, I., KAUR, N., BERGEY, E. J., PRASAD, P. N., STACHOWIAK, M. K., Organically modified silica nanoparticles: A nonviral vector for in vivo gene delivery and expression in the brain., *Proc. Natl. Acad. Sci. U.S.A.* **2005**, 102, 11 539–11 544.

261 KNEUER, C., SAMETI, M., BAKOWSKY, U., SCHIESTEL, T., SCHIRRA, H., SCHMIDT, H., LEHR, C.-M., A nonviral DNA delivery system based on surface modified silica-nanoparticles can efficiently transfect cells in vitro., *Bioconj. Chem.* **2000**, 11, 926–932.

262 RADU, D. R., LAI, C.-Y., JEFTINIJA, K., ROWE, E. W., JEFTINIJA, S., LIN, V. S.-Y., A polyamidoamine dendrimer-capped mesoporous silica nanosphere based gene transfection reagent., *J. Am. Chem. Soc.* **2004**, 126, 13 216–13 217.

263 SANDHU, K. K., MCINTOSH, C. M., SIMARD, J. M., SMITH, S. W., ROTELLO, V. M., Gold nanoparticle-mediated transfection of mammalian cells., *Bioconj. Chem.* **2002**, 13, 3–6.

264 THOMAS, M., KLIBANOV, A. M., Conjugation to gold nanoparticles enhances polyethylenimine's transfer of plasmid DNA into mammalian cells., *Proc. Natl. Acad. Sci. U.S.A.* **2003**, 100, 9138–9143.

265 SCHERER, F., ANTON, M.,

SCHILLINGER, U., HENKE, J., BERGEMANN, C., KRÜGER, A., GÄNSBACHER, B., PLANK, C., Magnetotransfection: Enhancing and targeting gene delivery by magnetic force in vitro and in vivo., *Gene Ther.* **2002**, 9, 102–109.

266 HUGHES, C., GALEA-LAURI, J., FARZANEH, F., DARLING, D., Streptavidin paramagnetic particles provide a choice of three affinity-based capture and magnetic concentration strategies for retroviral vectors., *Mol. Ther.* **2001**, 3, 623–630.

267 KRÖTZ, F., SOHN, H.-Y., GLOE, T., PLANK, C., POHL, U., Magnetotransfection potentiates gene delivery to cultured endothelial cells., *Vasc. Res.* **2003**, 40, 425–434.

268 JIANG, H., ZHANG, T., SUN, X., Vascular endothelial growth factor gene delivery by magnetic DNA nanospheres ameliorates limb ischemia in rabbits., *J. Surg. Res.* **2005**, 126, 48–54.

269 TAN, S. T., WENDORFF, J. H., PIETZONKA, C., JIA, Z. H., WANG, G. Q., Biocompatible and biodegradable polymer nanofibers displaying superparamagnetic properties., *ChemPhysChem* **2005**, 6, 1461–1465.

270 SON, S. J., REICHEL, J., HE, B., SCHUCHMAN, M., LEE, S. B., Magnetic nanotubes for magnetic-field-assisted bioseparation, biointeraction, and drug delivery. *J. Am. Chem. Soc.* **2005**, 127, 7316–7317.

271 DONATH, E., SUKHORUKOV, G. B., CARUSO, F., DAVIS, S. A., MÖHWALD, H., Novel hollow polymer shells by colloid-templated assembly of polyelectrolytes., *Angew. Chem. Int. Ed.* **1998**, 37, 2201–2205.

272 SHENOY, D. B., ANTIPOV, A. A., SUKHORUKOV, G. B., MÖHWALD, H., Layer-by-layer engineering of biocompatible, decomposable core-shell structures., *Biomacromolecules* **2003**, 4, 265–272.

273 VOIGT, A., BUSKE, N., SUKHORUKOV, G. B., ANTIPOV, A. A., LEPORATTI, S., LICHTENFELD, H., BÄUMLER, H., DONATH, E., MÖHWALD, H., Novel polyelectrolyte multilayer micro- and nanocapsules as magnetic carriers., *J. Magn. Magn. Mater.* **2001**, 225, 59–66.

274 SHCHUKIN, D. G., RADTCHENKO, I. L., SUKHORUKOV, G. B., Micron-scale hollow polyelectrolyte capsules with nanosized magnetic Fe_3O_4 inside., *Mater. Lett.* **2003**, 57, 1743–1747.

275 GAPONIK, N., RADTCHENKO, I. L., SUKHORUKOV, G. B., WELLER, H., ROGACH, A. L., Toward encoding combinatorial libraries: Charge-driven microencapsulation of semiconductor nanocrystals luminescing in the visible and near IR., *Adv. Mater.* **2002**, 14, 879–882.

276 GAPONIK, N., RADTCHENKO, I. L., GERSTENBERGER, M. R., FEDUTIK, Y. A., SUKHORUKOV, G. B., ROGACH, A. L., Labeling of biocompatible polymer microcapsules with near-infrared emitting nanocrystals., *Nano Lett.* **2003**, 3, 369–372.

277 LU, Z., PROUTY, M. D., GUO, Z., GOLUB, V. O., KUMAR, C. S. S. R., LVOV, Y. M., Magnetic switch of permeability for polyelectrolyte microcapsules embedded with Co@Au nanoparticles., *Langmuir* **2005**, 21, 2042–2050.

278 LAI, C.-Y., TREWYN, B. G., JEFTINIJA, D. M., JEFTINIJA, K., XU, S., JEFTINIJA, S., LIN, V. S.-Y., A mesoporous silica nanosphere-based carrier system with chemically removable CdS nanoparticle caps for stimuli-responsive controlled release of neurotransmitters and drug molecules., *J. Am. Chem. Soc.* **2003**, 125, 4451–4459.

279 MORISHITA, N., NAKAGAMI, H., MORISHITA, R., TAKEDA, S.-I., MISHIMA, F., TERAZONO, B., NISHIJIMA, S., KANEDA, Y., TANAKA, N., Magnetic nanoparticles with surface modification enhanced gene delivery of HVJ-E vector., *Biochem. Biophys. Res. Commun.* **2005**, 334, 1121–1126.

280 SALEM, A. K., SEARSON, P. C., LEONG, K. W., Multifunctional nanorods for gene delivery., *Nat. Mater.* **2003**, 2, 668–671.

281 KUEHN, B. M., MRI reveals gene

activity in vivo. *J. Am. Med. Assoc.* **2005**, 21, 2584.

282 BHAKOO, K. K., BELL, J. D., COX, I. J., TAYLOR-ROBINSON, S. D., The application of magnetic resonance imaging and spectroscopy to gene therapy., *Meth. Enzymol.* **2004**, 386, 303–313.

283 GENOVE, G., DEMARCO, U., XU, H., GOINS, W. F., AHRENS, E. T., A new transgene reporter for in vivo magnetic resonance imaging., *Nat. Med.* **2005**, 11, 450–454.

284 ABOUL-FADL, T., Antisense oligonucleotides: The state of the art., *Curr. Med. Chem.* **2005**, 12, 2193–2214.

285 KRÖTZ, F., DE WIT, C., SOHN, H.-Y., ZAHLER, S., GLOE, T., POHL, U., PLANK, C., Magnetotransfection – a highly efficient tool for antisense oligonucleotide delivery in vitro and in vivo., *Mol. Ther.* **2003**, 7, 700–710.

286 HAMAD-SCHIFFERLI, K., SCHWARTZ, J. J., SANTOS, A. T., ZHANG, S., JACOBSON, J. M., Remote electronic control of DNA hybridization through inductive coupling to an attached metal nanocrystal antenna., *Nature* **2002**, 415, 152–155.

287 LUCKARIFT, H. R., SPAIN, J. C., NAIK, R. R., STONE, M. O., Enzyme immobilization in a biomimetic silica support., *Nat. Biotechnol.* **2004**, 22, 211–213.

288 LUNDQVIST, M., SETHSON, I., JONSSON, B.-H., Transient interaction with nanoparticles "freezes" a protein in an ensemble of metastable near-native conformations., *Biochemistry* **2005**, 44, 10 093–10 099.

289 SHARMA, R. K., DAS, S., MAITRA, A., Enzymes in the cavity of hollow silica nanoparticles., *J. Colloid Interface Sci.* **2005**, 284, 358–361.

290 LIU, H., RUSLING, J. F., HU, N., Electroactive core-shell nanocluster films of heme proteins, polyelectrolytes, and silica nanoparticles., *Langmuir* **2004**, 20, 10 700–10 705.

291 JI, Q., KAMIYA, S., JUNG, J.-H., SHIMIZU, T., Self-assembly of glycolipids on silica nanotube templates yielding hybrid nanotubes

with concentric organic and inorganic layers., *J. Mater. Chem.* **2005**, 15, 743–748.

292 WANG, Y., CARUSO, F., Mesoporous silica spheres as supports for enzyme immobilization and encapsulation., *Chem. Mater.* **2005**, 17, 953–961.

293 LVOV, Y., CARUSO, F., Biocolloids with ordered urease multilayer shells as enzymatic reactors., *Anal. Chem.* **2001**, 73, 4212–4217.

294 VILLALONGA, R., FRAGOSO, A., CAO, R., ORTIZ, P. D., VILLALONGA, M. L., DAMIAO, A. E., Supramolecular-mediated immobilization of trypsin on cyclodextran-modified gold nanospheres., *Supramol. Chem.* **2005**, 17, 387–391.

295 FISCHER, N. O., MCINTOSH, C. M., SIMARD, J. M., ROTELLO, V. M., Inhibition of chymotrypsin through surface binding using nanoparticle-based receptors., *Proc. Natl. Acad. Sci. U.S.A.* **2002**, 99, 5018–5023.

296 HONG, R., FISCHER, N. O., VERMA, A., GOODMAN, C. M., EMRICK, T., ROTELLO, V. M., Control of protein structure and function through surface recognition by tailored nanoparticle scaffolds., *J. Am. Chem. Soc.* **2004**, 126, 739–743.

297 WILLNER, I., KATZ, E., Magnetic control of electrocatalytic and bioelectrocatalytic processes., *Angew. Chem. Int. Ed.* **2003**, 42, 4576–4588.

298 ISHII, D., KINBARA, K., ISHIDA, Y., ISHII, N., OKOCHI, M., YOHDA, M., AIDA, T., Chaperonin-mediated stabilization and ATP-triggered release of semiconductor nanoparticles., *Nature* **2003**, 423, 628–632.

299 NOLTING, B., YU, J.-J., LIU, G.-Y., CHO, S.-J., KAUZLARICH, S., GERVAY-HAGUE, J., Synthesis of gold glyconanoparticles and biological evaluation of recombinant Gp120 interactions., *Langmuir* **2003**, 19, 6465–6473.

300 NIGGEL, J., SIGURDSON, W., SACHS, F., Mechanically induced calcium movements in astrocytes, bovine aortic endothelial cells, and C6 glioma cells., *Membr. Biol.* **2000**, 174, 121–134.

301 WINTER, J. O., GOMEZ, N., KORGEL,

B. A., SCHMIDT, C. E., Quantum dots for electrical stimulation of neural cells., *Nanobiophotonics and Biomedical Applications II*, CARTWRIGHT, A. N., OSINSKI, M. (Eds.), Proceedings of SPIE, **2005**, Vol. 5705, pp. 235–246.

302 BESANGER, T. R., EASWARAMOORTHY, B., BRENNAN, J. D., Entrapment of highly active membrane-bound receptors in macroporous sol-gel derived silica., *Anal. Chem.* **2004**, 76, 6470–6475.

303 LEE, S., EISENBERG, D., Seeded conversion of recombinant prion protein to a disulfide-bonded oligomer by a reduction-oxidation process., *Nat. Struct. Biol.* **2003**, 10, 725–730.

304 SCHEIBEL, T., PARTHASARATHY, R., SAWICKI, G., LIN, X. M., JAEGER, H., LINDQUIST, S. L., Conducting nanowires built by controlled self-assembly of amyloid fibers and selective metal deposition., *Proc. Natl. Acad. Sci. U.S.A.* **2003**, 100, 4527–4532.

305 KOÇER, A., WALKO, M., MEIJBERG, W., FERINGA, B. L., A light-activated nanovalve derived from a channel protein., *Science* **2005**, 309, 755–758.

306 HARRELL, C. C., KOHLI, P., SIWY, Z., MARTIN, C. R., DNA-nanotube artificial ion channels., *J. Am. Chem. Soc.* **2004**, 126, 15646–15647.

12
Nanoengineering of Biomaterial Surfaces

Ashwath Jayagopal and Venkatram Prasad Shastri

12.1
Introduction

Cells sample their environment through sensory elements that can discern minute changes in chemical composition and surrounding mechanical forces. These sensory elements are on the order of tens to hundreds of nanometers, and include lamellopodia that aid in cell locomotion, and receptor tyrosine kinases that can bind growth factors. Nanoscale interactions mediated by these elements lead to the activation of signaling pathways that influence cellular processes at different levels, ranging from mRNA synthesis and cell cycle progression to apoptosis. Therefore, it comes as no surprise that biomaterial surface characteristics, namely chemistry and topography, have a profound influence on the elicited cellular response of a tissue engineering construct. Early biomedical devices rarely incorporated nanoscale surface engineering as part of the design paradigm. However, this has changed in light of the growing biological evidence that has elucidated potential therapeutic avenues made possible by the nanoscale presentation of information. New approaches seek to reconstruct the natural tissue environment replete with biochemical and topological cues for the optimum restoration of function. In this chapter, we review the unique contributions and key examples of traditional and rapidly-emerging surface engineering techniques directed toward the progression of biomimetic tissue engineering. As various fundamentally different techniques are discussed, we have grouped biomaterial surface engineering strategies by their primary enabling features, from which some similarities can be drawn, such as the utilization of chemically-based or instrument-guided approaches. We discuss the amenability of these techniques for achieving micro- and nanoscale surface features, the applicability toward hard and soft materials and three-dimensional (3D) geometries, and the future implications of each technology concerning the development of clinically-relevant cellular and tissue engineering devices.

Nanotechnologies for the Life Sciences Vol. 9
Tissue, Cell and Organ Engineering. Edited by Challa S. S. R. Kumar
Copyright © 2006 WILEY-VCH Verlag GmbH & Co. KGaA, Weinheim
ISBN: 3-527-31389-3

12.2
Conventional Photolithography

The demand for miniaturization strategies in microelectronics long preceded the need for micro- and nanoscale features in tissue engineering devices prompted by the elucidation of extracellular matrix (ECM) proteins and signaling mechanisms, and thus many biomaterial surface engineering strategies are convenient adaptations of these well-established methods. An example is photolithography. The resolution limit of this technology is generally the wavelength of the light used for irradiation, so given the availability of UV excitation sources, photolithography has been frequently utilized to create organized cellular or biomolecular patterns upon nanoscale features engineered on planar surfaces. In the conventional scheme, a patterned photomask is used to control the light-induced decomposition of a spin-coated photoresist on the substrate, exposing defined regions of the substrate. These regions can be etched to create topographical features such as grooves, and can be chemisorbed with compounds such as organosilanes, which promote adhesion to proteins, as well as compounds that resist adsorption. The surface can then be utilized to site-specifically pattern cells and biomolecules. This photoresist-based photolithographic technique was employed for the micropatterned, co-cultivation of hepatocytes and fibroblasts, and cell placement was spatially-controlled by the patterning of collagen to the glass surface [1]. The topographical features produced by photolithographic processes have generated potent cellular responses, such as cell alignment and parallel cytoskeletal filament orientation along grooved features, as well as several other responses, as reviewed by Flemming et al. [2].

Photolithography-based surface engineering techniques can be modified such that the biomolecule itself, not the surface, is the photoactivatable conduit for biopatterning. Variations of photolithography include the use of photoreactive chemicals conjugated to a biomolecule, such that irradiation directly mediates site-specific biomolecular immobilization to the surface. Commonly used strategies based on this principle include arylazide, diazirine, benzophenone, and nitrobenzyl photochemistries [3, 4]. In the first three methods, UV irradiation results in active groups that insert themselves within chemical bonds, enabling photoimmobilization of the species attached to the photolabile moiety. The first two chemistries have been successfully applied to the attachment of cell adhesion peptides (RGD-containing) [5, 6] and/or proteins, such as biotin and antibodies [6, 7]. Benzophenone chemistry was applied to covalently attach a laminin peptide fragment [4]. Thus, these photochemical surface engineering techniques can be employed to recreate ECM-like environments that are most supportive of natural cellular proliferation and migration patterns. Nitrobenzyl chemistries have been successfully applied towards the photobiotinylation of polymer surfaces [8].

Photolithographic and photochemical techniques can reproducibly produce nanoscale features that are critical for influencing cellular behavior. The technology is well characterized, and facilitates the surface engineering of chemistry, through adsorption of compounds containing functional groups or biomolecules, and also provides a conduit for producing certain topographies. However, several elements of this technology are generally not satisfactory for tissue engineering applications. First, photolithography is not suitable for patterning nonplanar surfaces, and is associated with costly, specialized equipment, often using solvents not suitable for cell and protein patterning [9]. Thus, it is difficult to utilize photolithography for the ultimate goal of tissue engineering, which is the 3D biomimesis of the natural tissue environment. Furthermore, for photoreactive chemistries, extensive biomolecule conjugation to the photoactivatable adduct may be involved, UV irradiation must be used, which is damaging to biomolecules, and most processes cannot be carried out in aqueous media, all of which complicate tissue engineering device fabrication strategies that make use of this technology [10]. In addition, illumination is conventionally conducted through photomasks that do not alter the intensity of incoming light; thus, illumination is said to be "all-or-none", making it difficult to produce features of non-uniform height. This is a primary limitation in adapting photolithographic techniques for producing 3D features.

To address shortcomings of photoresist-based photolithographic techniques, several modifications in techniques and materials have been reported that have reinforced the utility of these methods in modern tissue engineering, in which biomolecule and cellular sensitivity to the processing environment is of critical importance. One approach has been the development of biocompatible photoresists. To expand on the applicability of photoresist-based techniques for protein and cell patterning, a "bioresist" based on a copolymer of methyl methacrylate and vinyl pyrrolidone was applied for the alignment of fibroblasts, obviating the need for potentially-denaturing solvent development steps used to remove photoresist [11]. Another biocompatible photoresist, based on poly(t-butyl acrylate), was utilized to pattern a polystyrene tubular support with antibodies, with only mild resist removal steps (specifically, 60 °C and exposure to basic media) that were tolerated by the biomolecules. Features of this technique were comparable with the high resolution of photolithographic techniques, with the patterning of 0.13 μm lines with protein arrangements <10 μm [12–14]. The feature of non-denaturing conventional photolithography is an important step in the transition from established microelectronics processes to tissue engineering micro and nanofabrication procedures that do not necessitate the use of biologically-incompatible solvents in processing steps. To address the all-or-none limitation inherent in the photomask used, several strategies involving what is called gray-scale photolithography have been utilized. In this approach, the mask itself contains tunable features, mainly varying light transmission levels, which control the site-specific degree of photoresist irradiation. In one approach, dye-filled microfluidic channels within an elastomeric mask consisting of varying concentrations of dye were successful in producing 3D features in resist, using standard equipment, with resolution in the single micron range [15].

Recent developments in photochemistry have introduced the possibility of photoconjugating biomolecules, and perhaps cells without the use of UV irradiation and dry environments. Furthermore, complex bioconjugation techniques used to attach photoactivatable compounds such as caging compounds to the bioactive agent are

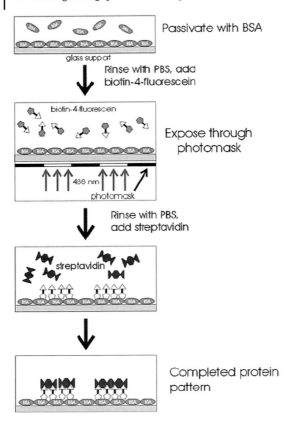

Fig. 12.1. Schematic for photoimmobilization of biotin-4-fluorescein to a bovine serum albumin (BSA)-coated glass surface using visible-light laser irradiation through a photomask. Photobleaching of the fluorescent dye initiates a singlet-oxygen-mediated binding reaction. (Reproduced with permission from Ref. [10].)

no longer the only option for utilizing light-guided patterning strategies. Holden et al. have reported a method for the fluorophore-mediated attachment of biomolecules such as enzymes to electron-rich surfaces in aqueous, neutral media, using visible light that is not harmful to proteins [10, 16] (Fig. 12.1). The technique uses the well-known fluorescein photobleaching reaction to immobilize biomolecules in patterns as arranged by photomasks and fluorescence illumination. Fluorescein and similar organic dyes used for patterning are commonly conjugated to proteins using simple, rapid laboratory techniques [17], thus making the technology accessible to most biological laboratories. One can envision the application of photochemical techniques such as this one for the patterning of biomolecules on complex geometries, simply by controlling the localization of laser irradiation in the *x*-*y*-*z* plane for site-specific radical-mediated immobilization, although this has not been explored to any significant degree. In another approach that obviates the need for UV-activatable groups, Luebke et al. described a Trp-containing peptide or protein crosslinking system based on blue excitation light and a ruthenium complex for specific cell patterning on a poly(ethylene glycol) (PEG)-coated glass surface, with single-cell resolution enabled by the use of digital micromirrors [18, 19]. These methods, like the advancements in biocompatible photoresists, also facilitate the use of non-denaturing chemistries in the photolithographically-mediated controlled patterning of tissue engineering scaffolds and implants.

By judicious selection of biomaterials, surface engineering of 3D cellular matrices can be accomplished by photolithographic and photochemical techniques, thus potentially addressing, at least in part, the difficulties of nonplanar patterning with conventional approaches. As an example, PEG-diacrylate hydrogels, known to photopolymerize in the presence of an initiator, were conjugated to nonspecific and specific cell adhesion peptides and exposed through a transparency-based photomask to create 3D layered hydrogel patterns, upon which dermal fibroblasts were specifically attached in specific cell adhesion peptide-defined regions [20] (Fig. 12.2). The study demonstrated that photolithographic methods can be used for the surface engineering of complex 3D cell and biomolecule spatial arrangements, with minimum expense, as the mask was printed on a transparency sheet by laser printer. A photochemistry-based technique for the 3D surface engineering of agarose hydrogels was demonstrated by Luo and coworkers, who conjugated

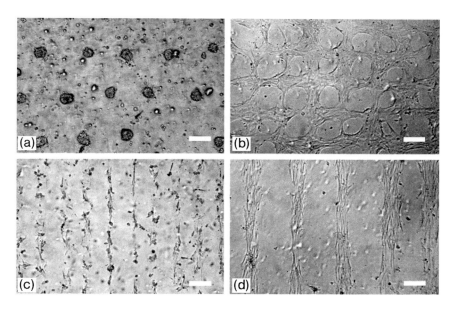

Fig. 12.2. Site-specific adhesion of human dermal fibroblasts to RGDS-immobilized PEG-diacrylate (PEGDA) hydrogels. (a)−(c) Hydrogel patterned with RGDS generic cell adhesion peptide. (d) Hydrogel patterned with RGDS and REDV (nonspecific) peptide. In all images, HDF-binding is limited to RGDS-patterned regions only. Scale bar = 250 μm. (Reproduced with permission from Ref. [20].)

photolabile nitrobenzyl-Cys groups to the hydrogel [21, 22]. Upon UV exposure, exposed sulfhydryl functional group channels were then exploited to site-specifically conjugate RGD-containing cell adhesion peptides, which were patterned to guide neurite outgrowth. In work by Revzin et al., 3D PEG hydrogels could be engineered to bear immobilized fluorescent proteins, supporting the possibility of incorporating ECM-signaling cues within hydrogel microstructures [23]. These findings demonstrate the potential of techniques based on conventional photolithographic and photochemical approaches to construct, with high-resolution and reliability, complex 3D environments based on tissue-like hydrogels to be potentially applied towards the controlled organization of functional tissue. Thus, photolithography and photochemistry, often suggested to be incompatible for tissue engineering applications, may actually be a potential route to the fabrication of 3D tissue constructs using a well-established technology.

12.3
Electron-beam Lithography

As previously discussed, we have found that nanoscale interactions are paramount to the optimization of a desired cellular response. To this end, nanofabrication techniques used in high-density storage media and electronics might be extended for purposes of patterning sub-10 nm features for tissue engineering applications, given the appropriate material and chemical conduits to enable such an approach. Electron beam lithography (EBL) is a technology capable of achieving such a high resolution, and initially has been shown to be potentially applicable to tissue engineering applications. In this technology, in a manner similar to photolithography, electron beams rather than light are used to irradiate polymer resists (e-beam resists), such as poly(methyl methacrylate) (PMMA). The resist can then be developed to produce several nanotopographies. The current resolution limits of this strategy are as low as 5 nm in the X-Y plane and 1 nm in the Z plane [24], thus warranting applications of EBL in tissue engineering device fabrication.

Given the advantage of high-resolution, biological applications of EBL have only recently emerged. While EBL is currently being utilized to first study in detail such nanoscale interactions between cells and their surrounding chemistry and topography [25–27], it seems plausible that such a technology could be also directed towards the precise fabrication of complex tissue engineering scaffolds. The potential of EBL to create high-resolution protein patterns on conventional substrates, such as silane-coated silicon wafers, has been demonstrated [28, 29]. In one example, EBL-patterned 1 μm PMMA grooves were seeded with collagen to control lung fibroblast orientation and patterning [30]. Hippocampal neuron perpendicular alignment was influenced heavily by EBL-etched nanoscale grooves [31, 32]. In another application, an EBL method capable of producing 20 nm features surface engineered with poly(caprolactone) (PCL) [33], a biodegradable polymer used in biomedical implants, was used to study the filopodial extensions of fibroblasts in response to different nanotopographies [34] (Fig. 12.3). Thus, as we elucidate the

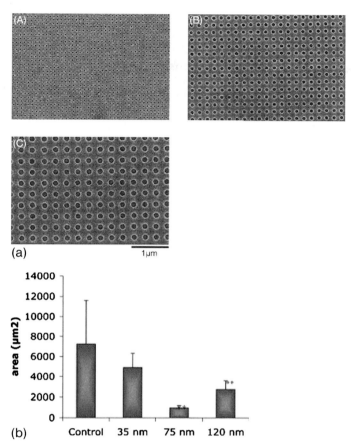

Fig. 12.3. (a) Scanning electron micrographs of nanopit topographies produced by electron-beam lithography, nickel die fabrication, and hot embossing. Diameters: (A) 35, (B) 40, (C) 120 nm. (b) Average number of filopodia per cell on planar and nanopit-containing surfaces. Results are mean ± SD; $P < 0.05*$, $P < 0.01**$ (student's t-test). (Reproduced with permission from Ref. [34].)

specific geometries capable of optimizing tissue organization and function, EBL, at least on a limited amount of materials, is a candidate for creating such features. Three-dimensional surface engineering features, while not explored in depth with EBL, have been at least demonstrated to be achievable, using EBL-enabled amino-functionalization of a gold surface templated with a self-assembled monolayer (SAM), upon which proteins are sequentially layered [35]. These preliminary findings suggest the potential utility of EBL in the surface engineering of tissue engineering devices, provided the materials and biological agents are compatible with the process.

Given the biocompatibility of PCL and PMMA substrates (discussed above for EBL applications) in tissue engineering, EBL has initially been demonstrated to be relevant to the surface engineering of biomaterials, and it is likely that, in conjunction with other fabrication techniques, EBL-based nanopatterns can be plated on other polymeric and metallic biomaterials as well. However, this has yet to be extensively investigated in a true cellular and protein surface engineering context, especially since most patterning for tissue engineering applications has been primarily explored on the microscale, where high-resolution EBL was not necessary. This might be addressed by investigating more biomaterials for their potential utility as e-beam resists. Furthermore, this highly-specialized technique can be both costly and time-consuming, with the serial drawing nature of the technology, and the extensive effort (computationally and physically) that is necessitated by patterns of increasing complexity. Similar to light-based techniques, EBL can require extensive optimization of process parameters, such as beam current, scan mode, and diameter, as well as resist properties [36]. Pattern reproducibility and resolution, as well as feature width, can be problems with this technology under various conditions. Nevertheless, the high resolution achievable by EBL warrants its further investigation concerning the surface engineering of biomaterials, and the method can still be utilized as an *in vitro* tool to study the fundamentals of cellular responses to varying topographies on the order of single nanometers, since that is the scale of cell plasma membrane features.

12.4
Soft Lithography

The enhanced specialization of biomaterial surfaces required for realistic functional tissue arrays has necessitated the development of new surface engineering techniques. Photolithography and EBL technologies, for instance, in general are not suitable for patterning on nonplanar surfaces. As a result, the diversity of biomaterials available to accommodate specific tissue structures may not be suitable fabrication platforms using these technologies. Furthermore, these techniques are time-consuming, can involve denaturing solvents, and require costly equipment and clean-rooms not accessible to biological laboratories [9, 37, 38]. For biomaterial surface engineering to complement advances in our knowledge base of optimal cellular environments, a family of micro- and nanofabrication technologies known as soft lithography have been developed that are suitable for patterning cellular and biomolecular arrays, with resolutions comparable to those of photolithography [9, 39–41].

Soft lithographic approaches, such as microcontact printing (uCP), microfluidic patterning [micromolding in capillaries (MIMIC)], membrane-based patterning (MEMPAT), and laminar flow patterning [38, 39, 41, 42], are related by the use of elastomeric stamps or channels for patterning, typically based on the silicone rubber poly(dimethylsiloxane) (PDMS). The polymer is cast against a patterned master mold generated by conventional photolithography to create a stamp with various features. PDMS stamps have been used effectively in the patterning of biomolecules and cells through uCP, in which the patterned stamp is coated or "inked" with a functional chemical substance, such as a self-assembled monolayer (SAM), normally consisting of alkanethiols, and transferred by brief contact to another surface, such as gold, glass, or polymeric materials [37]. Self-assembled films on the target surface can then be used to create patterns in the nanoscale regime [43–45]. MIMIC involves the use of elastomeric channels, a support (e.g., glass, gold, polymer, metal), and a liquid polymer of low viscosity, to form patterned layers by capillary action [42]. In MEMPAT and related techniques, a thin elastomeric lift-off membrane with circles and square-shaped features are used to constrain cell and protein seeding [39, 46, 47]. Laminar flow patterning takes advantage of the low Reynolds' number, parallel non-mixing fluid streams of molded capillary channels to selectively pattern cells and their biomolecular environment [48, 49]. Other novel techniques involve multilayering schemes for the development of 3D elastomeric devices such as microfluidic on/off valves [50]. The variety of soft lithographic techniques provides significant versatility in surface engineering applications.

Soft lithographic procedures incorporate versatile materials and procedures that are readily accessible to biologists [51]. The key component of these technologies, PDMS, is a biocompatible, inexpensive, oxygen-permeable material that can be utilized for cell culture [52], and, due to its elastomeric nature, it can readily assume the shape of nonplanar structures, and is a robust material that can be reused as a stamp for the same application, without leaving residual polymer on the targeting surface. Furthermore, while PDMS is hydrophobic and non-functional for bioconjugation, surface oxidation or polymer grafting procedures can be utilized to make the surface hydrophilic and surface functionalized [53–55]. A previous drawback in soft lithography was the photolithographic step used to generate the original stamp. However, this potentially time-consuming and costly step has been modified with the development of rapid prototyping (RP) steps that use less expensive materials and processing equipment, such as transparency-based photomasks [56] and inexpensive computer-aided design and printing tools [57]. Generally, soft lithography is a low-cost technique that does not necessitate specialized accommodations such as clean-rooms. Hence, the attractive features of soft lithography are versatility in the type of surface that can be patterned (including some 3D applications), low cost compared with other methods, micro- or nanoscale resolution appropriate for tissue engineering applications, and biocompatibility.

Soft lithography is a significant example of a technique that complements the need to fabricate complex cellular and biomolecular arrays for tissue engineering. The appeal of this technology for the engineering of surface chemistries and topologies on biomaterials can be illustrated through several examples. Nanoscale features of varying shape such as lines and circles can be achieved by uCP with certain modifications [58–60]. Alkanethiol SAMs terminated with fibronectin-adhesive and resistant groups were patterned by uCP on gold and silver surfaces to control endothelial cell adsorption on the microscale [61, 62]. uCP techniques have been used to pattern cells and proteins on gold, silver, glass and polymer substrates. Using SAMs with certain specificities toward a given biomolecule or cell, control is provided over cell–cell interactions, as well as cell confinement and

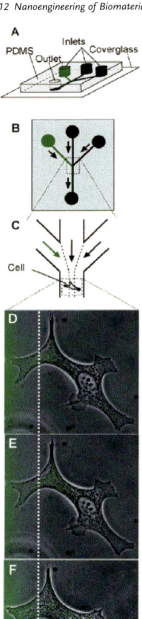

A
PDMS Inlets Coverglass
Outlet

B

C
Cell

D

E

F
50 μm

migration within a microscale area, which is critical to the goal of biomimesis. For example, by patterning biomolecules in a tight distribution on the nanoscale, subsequently a high density of protein-based information is created, which makes it possible to mimic the same density of information in living systems, such as integrin clustering. Findings that are a result of the patterning of cells using uCP techniques include the correlation of geometrical and biomolecular surface information with several biologically important events, such as cell growth and differentiation, cell cycle, cytoskeletal arrangement, and cell polarization [9, 37, 40]. For the fabrication of more complex structures, 3D channel networks can be harnessed by using MIMIC to fabricate multiple patterned arrays of protein gradients and cells, on a diverse pool of surfaces, such as biodegradable polymers (PLGA, PEG), glass, gold, and silicon [9, 37, 42]. Laminar flow patterning, due to its ability to deposit multiple cells or proteins in separate streams, adds significantly to this functionality, as streams containing distinct information (cells, growth factors, serum) can be rapidly patterned without the placement of physical implements to separate the streams. Furthermore, patterning directly over cell cultures, or even certain regions of cells, is possible with this technique, which is perhaps its major distinguishing feature. For example, Takayama et al. were able to demonstrate selective labeling of mitochondrial subpopulations, as well as the disruption of actin filaments in specific cell regions, using a technique known as partial treatment of cells using laminar flows (PARTCELL) [49, 63] (Fig. 12.4). Thus, soft lithographic patterning can be utilized to specifically influence cellular organization and function, with the added functionality of biochemical and physical manipulation on the single cell level, which could be useful for developing regions within a tissue construct, whereby different cells would require certain nutrients at distinct spatial and temporal intervals.

Biomaterial surface engineering techniques based on soft lithography have been shown to provide controlled cellular environments with various biochemical cues for optimum tissue function and organization. Fundamental information concerning the influence of such signals on cellular phenotype and function have been made possible by these methods, which can employ SAM and direct cellular and/or protein patterning strategies to present complex biochemical information with high resolution. To translate the past decade of patterning cells and biomolecules on mostly planar surfaces toward work to develop functional tissues and organs, increased attention is needed regarding the amenability of these approaches on other biocompatible substrates. The investigation of other elastomeric materials as

Fig. 12.4. (A)–(C) Schematic describing the partial treatment of cells using the laminar flow (PARTCELL) technique. A channel network molded in PDMS (A) is placed on a coverglass and consists of inlet and outlet ports. (B) The green inlet channel is injected with a mitochondrial staining probe (Mitotracker Green FM, Molecular Probes Invitrogen), while the other inlets are injected with cell media. (C) Illustration of non-mixing parallel streams. (D)–(F) Time-lapse fluorescence microscopy at 5, 11, and 35 min confirms the selective labeling of the mitochondria within the region of the cell exposed to the dye-injected port alone. (Reproduced with permission from Ref. [49].)

stamps with varying properties may enable new approaches. The most commonly used material currently, PDMS, is a hydrophobic polymer, and thus is susceptible to protein adsorption unless surface-modified, and can slowly recover hydrophobicity even in that case [64]. Also, in addition to the hydrophobic properties of the material, the ability to produce quality patterns on the stamp, as well as the quantity of different nanopatterns to be produced using the same stamp, are major factors affecting reproducibility of the pattern and the scope of application of the technique. In addition, the true nanoscale resolution limit of most of these technologies has not been ascertained. However, as there is no wavelength-imposed limit as in other technologies, it is likely this limit will be suitable for nanoscale feature patterning, if not already suitable. The usefulness of soft lithography for presenting topological information has not been explored for the most part, although PDMS is easy to mold and could be used in studies of cell–substrate interactions in such a context. In addition, soft lithography could potentially be combined with other nanofabrication techniques to produce desired features such as grooves if necessary. Even though some questions the technology may raise, this family of methods is still a low-cost method of producing quality patterns of cells and proteins at potentially nanoscale resolutions on nonplanar substrates. Furthermore, improvements have been demonstrated in extending the types of materials that can be used for pattern transfer. Patterned agarose hydrogel stamps have been used in inking/stamping procedures, such as in the method employed by Stevens et al. to transfer osteoblasts to hydroxyapatite scaffolds [65]. Thus, soft lithography pattern transfer can be accomplished with wet stamps which enhance cell viability, likely without sacrificing much control over patterns produced. Further work will likely improve the resolution provided by rapid prototyping methods, and continued efforts concerning multilayered, multifunctional surfaces that present structurally-favorable cellular microenvironments using these techniques, especially laminar flow patterning of biomolecules and cells using parallel nonmixing streams, are capable of facilitating the development of novel tissue engineering devices.

12.5
Polymer-demixed Nanotopographies

As discussed previously, highly specialized EBL techniques provide for the reproducible topological patterning of biomaterial surfaces, with biologically-relevant features such as grooves. However, such methods are time-consuming and expensive, and candidate surfaces and materials for EBL are limited. Accessible techniques for the high-resolution nanopatterning of biomaterial topographies at low cost are thus necessary to facilitate the progression of surface engineering strategies in tissue engineering fabrication. Recently, methods in surface engineering have been developed that exploit the bulk interactions between polymeric domains to produce a desired surface feature upon a substrate. By manipulating these phenomena, it has been demonstrated that surface *topologies*, in addition to surface chemistries, can be patterned upon a surface. Here we discuss a concept known as polymer demixing, whereby the inherent immiscibility of a binary polymer system, when cast upon a surface, results in phase separation-induced surface features [66]. Polymer demixing is a process tunable by polymer selection, relative polymer proportions and solubilities within the common solvent, and solvent selection, and can be utilized to produce nanoscale topographies such as pits and hills of tunable height and depth on the 10s of nanometers. Polymer blends initially explored with this strategy include poly(styrene) (PS) and poly(methyl methacrylate) (PMMA), PS and poly(4-bromostyrene) (PBrS), and poly(n-butyl methacrylate) (PnBMA) [66–69].

Surface engineering strategies of high tunability are highly desirable in the field for achieving specific, reproducible surface features. With polymer demixing, polymer fraction composition is a determinant of the type of topology produced, whereas polymer concentration in the solvent mixture is the parameter governing topology size. For example, by varying a 60/40 blend of PS/PBrS in total concentrations of 0.5–5% in toluene, nanoscale island features of heights from 13 to 95 nm were produced [67]. Interestingly, this study showed that positive-phenotyped human endothelial cells elicited nanoisland height-specific cytoskeletal and morphological responses, with the slightest grade (13 nm) eliciting the most rapid cellular spreading, relative to flat surfaces and higher features of the same polymer blend. Filopodia were observed to be in contact with the nanoislands, and cell shape was arcuate, in agreement with *in vivo* morphologies, suggesting that nanofeature-presenting surfaces may provide a more natural cell phenotype than flattened surfaces, such as the tissue culture polystyrene commonly employed. The 13 nm features produced by this method were also tested for fibroblast response, and again a greater cellular response was observed on the nanoisland-laden surface compared with the planar surface of the same composition. In addition, gene expression as assessed by microarrays was significantly enhanced for hundreds of signals in response to the nanoislands, with the timing of upregulation being consistent with focal adhesion formation upon the surface structures [69]. As an extension to tissue engineering approaches utilizing tubular constructs, polymer demixing was recently applied to the inner lining of nylon tubing to develop nanoislands of 40 and 90 nm, which had profound effects on cell spreading and cytoskeletal organization [70]. Thus, the application of polymer demixing could have implications for the development of neural growth cone elongation conduits and vascular grafts, for the control of cell adhesion and morphology, as well as gene expression. In these preliminary studies it is apparent that nanotopologies on the 10s of nanometers directly influence cell shape, proliferation, and signal transduction. Collectively, polymer demixing studies complement EBL studies on cell–biomaterial surface interactions in reinforcing the importance of nanoscale surface information on cellular response. Also, this work demonstrates that topography as the dominating feature, rather than chemistry as the primary mechanism, can be tuned to control cellular adhesion and signaling.

New strategies to present tunable nanoscale topological features will undoubtedly be useful in tissue engineering device fabrication as the knowledge base concerning optimum cell growth and function conditions is expanded to include in-depth nanoscale interactions. However, as technologies such as EBL and, in this case, polymer demixing are rather in their infancy in terms of biological applications there is considerable uncertainty of the extent of application of these technologies. The utility of polymer demixing in tissue engineering applications will depend on the number of biocompatible polymeric blends that can be induced by phase separation to form the desired structural features at the necessary resolution. If the ribbon and hill nanopatterns cannot be generated upon well-characterized biomaterials, it is unlikely that polymer demixing techniques can be utilized for tissue engineering, other than for the detailed study of cell–surface interactions in tissue engineering-like 3D constructs. However, phase separation techniques based on some different principles, such as *temperature*-induced phase separation (TIPS) [71], have been reported for developing relevant biomaterial constructs, in this case porous biodegradable polymer architectures. Therefore, additional parameters other than polymer–polymer and polymer–solvent interactions should be investigated for families of polymers to identify more relevant phase separations. In addition, future studies than address the combination of surface chemistries with these nanotopographies would be helpful and likely provide new insight on detailed nanoscale biochemical interactions, and also enable added features to this platform. For instance, while tall features were shown to be relatively inhibitory to cell adhesion, the grafting of cell adhesion peptides to the same feature might evoke a drastically different response in terms of cell spreading and growth rate. Nevertheless, polymer-demixed surfaces provide an interesting platform towards the study of nanotopography and its influence on cellular responses. The capability of reproducing *in vivo*-like cell phenotypes on an *in vitro* platform, as suggested by the arcuate morphology and structural changes of endothelial cells and the molecular expression profiles of fibroblasts on nanofeatures produced by this technology, could establish this technique as a correlative study tool to predict, for example, post-implantation responses. Of significance in this technology is the data gathered from cellular growth patterns on structures of varying height, which has some relevance should the same cells be patterned on 3D arrays in tissue engineering applications.

12.6
Star-shaped and other Novel Polymer Structures

In addition to phase separation strategies, novel properties resulting from alterations in the polymeric structural configurations themselves have been of significant benefit in surface engineering approaches. Since polymers can be tailored to present functional groups on their surface, the density of that information on a given surface can be of critical importance to the evoked tissue engineering response, in the form of cell adhesion or spreading, for example. Star polymers, for example, are such a structure. They consist of a central core from which multiple branches of varying physicochemical properties emanate. By incorporating star-shaped polymers onto biomaterial surfaces, the multivalency of the polymer can be exploited to present multiple surface chemistries, with a tunable number of branches and branch length, while the central core can impart its characteristic properties upon the whole surface.

Star-shaped architectures have been investigated in studies appropriate to tissue engineering applications. A promising application can be found in PEG-star architectures within hydrogel constructs. Initial studies by the Peppas laboratory concerning the utility of PEG star polymer hydrogels, for example, indicated that swelling properties of the construct could be adjusted by branch length, as well as the incorporation of acrylate moieties within the structure [72, 73], thus suggesting the utility of this approach in developing controlled-release tissue engineering surfaces, perhaps for the timely release of growth factors to encourage functional tissue development prior to or following implantation. In another example, star PEG coatings on titanium were explored for their utility of resisting nonspecific adhesion while promoting site-specific attachment of cells through a branch-conjugated RGD sequence [74], which illustrates the benefit of this approach in conferring added functionality upon titanium-coated implants in enhancing interactions and tissue integration with the surrounding environment, and reducing the adsorptive capacity of titanium for proteins, while also taking advantage of its mechanical stability in bioimplants. In other work to highly-tune cell adhesion on bioimplant surfaces, Griffith and coworkers investigated the utility of a "comb" copolymer consisting of a PMMA backbone with PEO branches, conjugated with varying RGD peptide densities. Surface film clustering of comb-presenting films either modified or unmodified with the peptide was demonstrated to be tunable for controlling cell–surface interactions [75–77]. The multifunctionality of star and comb polymers allows for incorporation of adsorption-resistant and adhesive surfaces within the same construct, and parameters such as the number and length of branches, and backbone/branch polymer composition, can be tuned, albeit with some difficulty, depending on the desired configuration, to develop complex surface arrangements of optimum density and spacing.

The degree of tunability of star-shaped polymer configurations appears to be very extensive, and the approaches used to make such modifications are diverse. In one example, such modifications are used to reversibly change the surface properties of a self-assembled film. Using a heteroarm star block copolymer of poly(2-vinylpyridine) and polystyrene, a self-assembled "brush" developed that could be tuned by solvent and pH-induced environments to present core–shell micelles with one of the two polymers on the surface. In addition to unimicellar interactions, lateral phase segregation was also observed for the monolayer, creating "dimple" and "ripple" morphologies, thus illustrating the development of an intelligent polymer coating with tunable, switchable surface chemistries and morphologies [78]. In another application, an amphiphilic diblock copolymer consisting of polystyrene and poly(ethylene oxide) (PEO) was synthesized in various star-shaped configurations, resulting in relative spreading and compaction of each polymer along the surface that were different that the linear form of the copolymer [79]. An exciting tuning approach highly relevant to star-shaped and dendrimer-branching polymers is the concept of molecular imprinting. In this process, the polymerization step is accompanied by a "templating" step, in which the polymer and template, such as a biomolecule or metabolite-binding probe, are crosslinked. Extraction of the template, for example by hydrolysis, creates highly-specific complementary binding sites for the templating structure. The presence of multiple surface functionality sites on star and other multibranched polymers thus facilitates this approach to developing highly-sensitive recognitive networks, and has been exploited in the specific imprinting of porphyrin and glucose-based recognition systems [80, 81]. These works are interesting precursors to the development of smart multifunctional biomaterial surface constructs that are responsive to the changing conditions of the microenvironment. While molecular imprinting is generally considered a technology for drug delivery and biosensor applications, it is conceivable that such imprinting technologies incorporated into hydrogels could serve as specific biomolecule-recruitment devices that sort out serum proteins to capture the proper growth factors for the cells that are entrapped within the construct. Alternatively, such a technique could be used to recruit specific endogenous stem cells within a construct to generate new tissue *in vivo*, although the technology is in its infancy. Nevertheless, the versatility of star and other hyperbranched, nonlinear polymers makes them promising candidates for several tissue engineering applications in which the presentation of surface chemistries on the nanoscale level is required. Manipulation of the biomaterial itself in this manner may avoid the time-consuming optimization of process parameters used in other methods to achieve desired surface chemistries within biomaterial constructs.

12.7
Vapor Deposition

In tissue engineering scaffolds or cellular and protein arrays for biosensor applications, patterns have been generated using the aforementioned photolithographic and soft lithographic techniques. These approaches may have limitations in the equipment and procedures involved, and/or the range of biomaterials that can be patterned. For instance, photolithographic equipment can be expensive, and soft lithographic fabrication techniques have been carried out primarily on gold and silicon substrates. Furthermore, PDMS, a common building block of microfluidic devices, nonspecifically adsorbs proteins via hydrophobic interactions, and does not express native functional groups for bioconjugation techniques [54, 55]. Thus, techniques adapted from the semiconductor industry have evolved that permit the thin film deposition of a wide variety of materials. Chemical vapor deposition (CVD) and related strategies, for example, are essentially a family of techniques that involve the chemical transformation of volatile gaseous precursors into thin films upon a substrate. The highlights of CVD that are relevant to biomaterial surface engineering are the capability of this technique to deposit various coatings (including multiple surface functionalities) on several substrates of varying composition and geometry. The process can be performed without catalysts and potentially-denaturing solvents at room temperature. These advantages have encouraged the exploration of CVD technologies in the patterning of defined chemistries on substrates of complex geometries that might be considered candidate tissue engineering constructs.

We now illustrate the versatility of CVD-based approaches for the patterning of surface chemistries on various biomaterial surfaces. Lahann and coworkers have employed CVD polymerization of poly(2-chloro-para-xylylene) to homogenously coat metallic stents, which permitted subsequent plasma treatments to significantly reduce platelet adhesion [82]. In addition, polymer deposition enabled the immobilization of a thrombin inhibitor on the strent to further reduce thrombogenicity [83]. This process addressed several shortcomings with previous attempts to coat Nitinol-based stents with polymers, such as the use of solvent-based defects, and the difficulty of evenly coating thin films below the micron scale. Thus, CVD is a potentially powerful technique for the coating of potentially thrombogenic or non-biocompatible metal surfaces with stable, evenly distributed functional polymer coatings for improving the performance and expanding the scope of application of existing materials. In other work by Lahann, CVD polymerization of functionalized poly(xylylenes) was applied for the thin film coating of steel, glass, gold, silicon, Teflon, and polyethylene surfaces, and, in conjunction with uCP, layer by layer (LbL) self-assembly was permitted for the controlled deposition of endothelial cells [84]. Therefore, the synergy of CVD with other patterning techniques for the spatial patterning of surface information is particularly of interest, as the benefits of soft lithographic pattern transfer techniques can be combined with the ability of CVD to pattern chemical functionalities on various surfaces. As further support that soft lithography and CVD polymerization can complement each other in the development of novel tissue engineering constructs, PDMS-based microfluidic devices themselves were functionalized on the luminal surface with the same functionalized paracyclophanes [85]. Reactive or nonreactive polymeric coatings of sub-100 nm thickness were homogenously-deposited within the luminal surface of both straight and curved PDMS microchannel geometries [86] (Fig. 12.5), suggesting the potential of this technique in the ultrathin patterning of biofunctional groups within complex tissue constructs.

In addition to the capability of CVD techniques to deposit a given material on various substrates, the properties of the materials and surface chemistries that can be patterned are diverse as well, and are highly-relevant to bioimplant and tissue engineering applications. As a first example, by the CVD-based deposition of a photoactivatable surface chemistry upon a microfluidic channel, a template was provided for the spatially-controlled functionalization of both hydrophilic (PEG) and protein coatings in discontinuous arrangements [87]. In a second example, CVD-based thin film formation of poly(2-hydroxyethyl methacrylate) (HEMA), a biocompatible polymer with extensive biomedical applications, was observed to be a swelling property-tunable process by adjusting the partial pressure of a cross-linking agent incorporated into the vapor phase [88]. As a result, the deposited film had tunable crosslinking properties and could function as a hydrogel. Recently, the utilization of CVD for the deposition of thin films with chemically-tunable top-

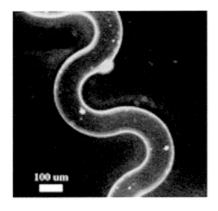

Fig. 12.5. Chemical vapor deposition (CVD) of straight and curved PDMS microchannel geometries of a hydrazide-reactive coating. The channels were exposed to biotin-hydrazide followed by TRITC-streptavidin. Channel depth is 75 μm, channel width is 100 μm. (Reproduced with permission from Ref. [86].)

ographies suitable for biological applications was demonstrated by the fabrication of aligned carbon nanotube arrays on a silicon substrate for the promotion of fibroblast growth [89]. This example suggests that CVD can serve as the patterning method for powerful, exciting nanotechnologies, to allow exploitation of their unique properties, and thus CVD can play an important role in the fabrication of complex 3D architectures for tissue engineering scaffolds.

CVD techniques provide a mechanism for the homogenous deposition of thin films presenting surface chemistries for the specific patterning of cells and biomolecules. While its application has primarily been directed towards *in vitro* diagnostic assays, the variety of compatible substrates and deposition candidate materials suggest the substantial applicability of CVD for biomaterial surface engineering for *in vivo* devices. While many of the polymer coatings investigated for CVD of stents and cellular patterning surfaces have not been associated with apparent cytotoxicity, more extensive *in vivo* studies would be welcomed concerning the biocompatibility of deposited polymer coatings, particularly with metallic stents, given the better characterized nature of that application. Given the well-deserved appeal of hydrogels in tissue engineering applications, it would also be desirable to attempt to expand the list of candidate polymer films that can be tuned to behave as hydrogels using CVD processes with suitable co-polymerization agents. The versatility of CVD for the solvent/catalyst-free deposition of multiple coatings and surface chemistries on various materials and geometries make it an attractive option for surface engineering applications, either alone or in conjunction with other known techniques such as photopatterning and soft lithography.

12.8
Self-assembly

We have discussed the utility of soft lithography in the patterning of cells and proteins on complex geometries. A technology that has facilitated this application (as well as photolithographic approaches) is the use of self-assembled monolayers (SAMs), namely alkanethiols, for patterning on gold and silver surfaces. These SAMs are known to pack onto a gold surface with nearly crystalline density. Thus, the properties of the SAM itself are assumed by the new surface, allowing for chemical tuning of the biomaterial. For example, through organic synthesis modifications, alkanethiols can be modified to present bioconjugation sites, such as carboxylic acid, or protein-resistant sites, based on the hydrophilic PEG moiety. Using uCP, alkanethiols have been patterned with resolutions of 200 nm [90]. uCP techniques have been routinely used to create alternating patterns of ECM proteins and cells, with adsorption-resistant surfaces in between them conferred by PEG groups. The surface properties of SAM-adsorbed surfaces can be analyzed by a host of techniques, such as ellipsometry, X-ray photoelectron spectroscopy, electron microscopy, and quartz crystal microbalance, thus enabling many standard quality control procedures in the development of biosensors and cell/protein arrays. In addition to nanoscale resolution, when combined with the molding properties of PDMS stamps, SAMs can be utilized in conjunction with soft lithography and other standard techniques to produce complex features of varying chemistry and topology [61].

Despite the highly-useful features provided by SAMs on these substrates, the overall utility in tissue engineering applications is unclear. While chemisorption of alkanethiols on gold is useful in studying fundamental interactions of proteins and cells on the nanoscale, there is limited application in using SAMs to produce patterns on other surfaces that are not noble metals. Furthermore, organic synthesis techniques are complex, with each functional group requiring its own set of synthesis protocols. Other SAMs, such as alkylsilanes on hydroxyl-presenting surfaces have been applied, but their synthesis is especially complex and its use is far less than alkanethiols for biological surface engineering applications. However, the biocompatible aspects of gold make it suitable in several tissue engineering applications, and should a host of biomaterials, such as titanium, be modified with gold surfaces where needed for SAM-assisted surface engineering, then SAM associated-strategies can conceivably be employed to create high-definition patterns for *in vivo* applications.

The nanoscale assembly of cells, proteins, and biomaterials such as polymers achieved by multiple nanofabrication processes has been also accomplished by the exploitation of alternating layers of materials containing opposing charges or hydrogen bond donor–acceptor pairs. Layer by layer (LbL) assembly techniques have emerged as low-cost nanopatterning strategies upon which various surface chemistries and topographies have been patterned. The method is simple, can be used to pattern stable films upon various hard and soft surfaces of varying geometries, and can bear biological information at varying heights within the film, without disturbing the overall assembly of the LbL network [91]. The use of electrostatically-complementary building blocks has also facilitated the use of other nanofabrication technologies, such as soft lithography, by enhancing pattern transfer from a stamped mold.

Several examples indicate the potential utility of LbL for the development of complex tissue engineering devices. In a co-culturing application, negatively-charged hyaluronic acid (HA), deposited by lithography, and the adhesive protein fibronectin were co-adsorbed to a glass substrate, followed by the specific attachment of cells to the fibronectin. A second adsorption medium consisted of positively-charged poly(L-lysine) (PLL), which bound electrostatically to the non-cell-coated HA layer to "switch" the originally non-adhesive chemistry to a cell-adhesive moiety that was permissive for the attachment of a secondary cell type, thus demonstrating the simplicity of LbL for patterned co-culture applications [92]. In another application, discrete, multiprotein regions were patterned within the same substrate using LbL in conjunction with photolithography, demonstrating the utility of LbL for facilitating hierarchical assemblies of distinct information [93]. Another LbL-based technology, colloidal lithography, has shown to be a promising, versatile application. In this technique, nanospheres are self-assembled on any number of surfaces, and can be the template for lithographic or vapor deposition processes, or themselves can contain functional groups for patterning. Since the nanoparticles are highly uniform and size-tunable, and can be tightly-packed if needed, they are highly-versatile materials for several surface engineering applications in a tissue engineering context. Additionally, the capability of this technology for the simultaneous presentation of surface texture and chemistry has been demonstrated. Shastri et al. investigated a functionalized nanoparticle-based system for simultaneously modifying surface chemistry and roughness (topology) without altering bulk substrate properties, on hard and soft materials such as titanium and polyurethane [94]. In another study involving microparticles electrostatically adsorbed on multilayer polyelectrolyte films, adjustments involving RGD peptide density on the microparticles and the packing density of the microparticles on the film surface, a function of the patterning area, were used to tune surface chemistries and roughness, with profound effects on cellular morphology [95].

LbL-based approaches obviate the need for substrate-specific bonding agents, such as SAMs based on alkanethiols and organosilanes [96]. The compatible materials, be it the thin film to be deposited or the substrate, are endless due to the availability of oppositely-charged polyelectrolytes, and with colloidal assemblies, both surface chemistry and surface texture can be presented to the cellular environment on microscale dimensions. This is a rare feature in biomaterial surface engineering techniques, which typically can present surface chemistry or surface topology, but not both simultaneously. The technique can be inexpensive compared with existing nanofabrication techniques, and/or can complement them. The size of nano/microparticles chosen for the application can be selected to modify thickness and surface area available for bioconjugation. Surface chemistries can be confined to different depths of a 3D multilayered construct, which enables the development of complex environments needed for tissue engineering applications as the library of matrix biology expands to define ever more complicated cell–ligand signaling processes.

The self-assembly phenomena observed for alkanethiols and polyelectrolytes on surfaces can also be observed for biomolecules. Here, the monomeric building blocks of the biomaterial surface chemistry and topology are peptides or proteins, which associate with each other through various non-covalent interactions. This has been very useful for biological surface engineering, since proteins do not have the same biocompatibility issues associated with polymers or other synthetic materials. Furthermore, protein structure and alignment is of great interest to the biological community, and the fact that this field has developed extensively will enable tissue engineering applications that take advantage of protein structural morphology to present high-resolution information to cells. Proteins themselves are the nanoscale messengers of information to cells, providing signals in the form of chemistry and topology, and thus techniques that directly apply this information to surfaces are desirable.

Nanoscale features from biomolecules can be created by hierarchical, irreversible systems, such as those based on avidin–biotin stacking systems, or interchangeable complexations. Such strategies have been used to develop beta-sheet monolayers [97, 98], as well as peptide-based nanotube architectures [99, 100]. Spontaneous formation of porous networks by a set of peptide families yielded filaments of 10–20 nm, with pores 50–100 nm in size, with complex geometries [101]. Thus, various nanoscale chemistries and topologies are achievable with this strategy. Self-assembling peptides can be patterned using uCP, to control cell–cell interactions by modulating the peptide sequence and surface density. The self-assembled peptides may be further conjugated to cell-signaling peptides, such as RGD for example, to allow for methods that are focused on presenting a surface topology and chemistry simultaneously. As an alternative to hydrogels based on synthetic polymers, Mi [102] and colleagues have developed RGD-incorporated hydrogels by the self-assembly of proteins with the included sequence, in a process dictated by temperature and pH. The RGD incorporation enabled the binding and spreading of fibroblasts on the surface. While this is a feat achievable by other approaches, this work illustrates that it is possible to encode biological signals within proteins that naturally assemble into modules, each consisting of similar or varying information. One can envision the modular construction of natural self-assembled polymeric architectures on a host of biocompatible surfaces to direct multiple processes such as cell proliferation, migration, and cytokine secretion, at the same time. Further tuning is possible, such as the mechanical properties of the self-assembled species, and the degree of self-assembly. For example, a peptide-amphiphile complex on the order of 6 nm developed by Hartgerink and coworkers have provided for a pH-tunable (and reversible) self-assembly mechanism, with reversible crosslinking of the complex tunable by disulfide bridges [103]. These features enabled the mineralization of hydroxyapatite on the complex surface in an orientation that resembled the collagen–hydroxyapatite alignment in bone. thus illustrating the advantage of using biomimetic structures based on biomolecule-based monomers.

As the enabling technology in this mode of self-assembly are the peptides themselves, then there is no particular substrate specificity. Any surface that allows the desired nanoscale self-assembly is then a candidate substrate for this approach, which makes the scope of application potentially quite extensive in tissue engineering applications. The ease with which biologists can synthesize recombinant proteins and peptides with functional domains, all without an appreciable reduction in normal protein function, makes self-assembled biomolecule arrays an exciting prospect that is also readily accessible.

In general, nanoengineering of surfaces based on self-assembly can be a very convenient approach. Ambient temperatures, no need for organic solvents, and the use of well-characterized films or biofilms for predictable, reproducible results are all potential features of this technique. As the number of bioactive agents that influence cellular and tissue function are elucidated, the number of potential biomolecular self-assembling candidates will likely expand as well. Future studies that investigate the synergistic effect of bioactive agents and surface topology (as initially achieved with colloidal-based approaches above) are warranted, as is the transition from protein patterning to viable tissue engineering constructs by extensive *in vivo* studies.

12.9
Particle Blasting

An important challenge of surface engineering strategies for biomaterials is the modification of a surface to achieve sufficient surface free energies that are required for adequate cell adhesion, motility, and proliferation. Surface engineering of implants, commonly in the orthopedic and dental implant arenas, often employ sandblasting technology for the attainment of certain tunable surface morphologies. By varying the physical properties of the blast material itself, and the pressure with which it is ejected, surface topographies can be achieved for modulating various biological responses similar to those observed for topographies produced by other methods. These techniques are generally applicable to a wide variety of biomaterials, and are relatively inexpensive.

Sandblasting techniques, while simple, can be readily applied to several well-characterized biomaterials and biomedical applications to modulate their biological response. An example can be seen with the sandblasting of PMMA, a widely-used material in bone cements and ocular applications, with alumina microparticles [104]. Compared to smooth PMMA, vascular and corneal cell adhesion was higher on roughened PMMA surfaces. The simplicity of sandblasting can also be applied toward the deposition of films in addition to surface texturing. By loading a sandblaster with hydroxyapatite powder, Ishiwaka et al., produced a high-strength bond between the powder and the titanium surface that was higher than titanium coated by electrochemical and electrophoretic deposition processes [105].

These examples illustrate the principle that surface roughness alone, whether patterned by the previously-mentioned techniques, or randomly distributed by simple sandblasting techniques, has a profound influence on the affinity of cells for biomaterial surfaces. Furthermore, to generate such topology to evoke the desired cellular response in terms of adhesion and organization, it may not be necessary to employ expensive, time-consuming steps, if simple and rapid techniques such as sandblasting can achieve the same effect. However, this technology is generally not suitable for advanced tissue engineering applications that require precise control of biomolecule and cell placement, since it does not provide for surface chemistry modifications, is random in topography distribution, and has not thus far offered fine nanoscale resolution. Furthermore, blasting particles can be embedded within the target surface, which may complicate biological applications. Nevertheless, there is utility of sandblasting for creating microscale surface topography at room temperature, without the need for chemical modifications (although acid etching steps often accompany it), and surface chemistry can also be deposited onto a surface using this technology. While its main applicability is directed toward orthopedic and implant dentistry techniques, and has not been extensively performed on most biomaterial surfaces not relevant to those fields, it is conceivable that particle blasting could be utilized for a broad range of tissue engineering devices. For example, controlled-release microspheres could be used as the blast material for a deformable hard substrate, to serve as a cell-supportive nutrient reservoir during cell seeding on titanium or polymer-coated implants, thus potentially providing both surface roughness and chemistry in the form of released biomolecules or surface functionalities (e.g., RGD).

12.10
Ion Beam and Plasma-guided Surface Engineering

These are surface engineering techniques that provide mechanical and chemical alteration of a biomaterial surface, without negatively affecting the overall bulk properties of the original material. For decades, ion beam-related technologies, such as ion implantation, plasma immersion ion implantation, and ion beam-assisted deposition, have been highly reproducible techniques for the modification of surface properties due to the degree of control provided over the process. Furthermore, these techniques and variations on these techniques have been utilized to modify the properties of many biomaterials in many ways, including the overall response of the biomaterial to *in vivo* environments. A plasma is generally defined as a mixture of gas in a reactor containing one or more charged and uncharged species, which can be energized/accelerated with an applied electric field. Ion implantation involves the bombardment of a surface with accelerated ions from an excited plasma. Superficial layers of the target material (generally <1 μm) are disrupted by the ions, and a host of physicochemical modifications occur as a result, depending on process parameters and intrinsic material and ion properties. The modifications generally do not "strip" material off the surface as may chemical treatment-based approaches or reactive ion etching, which makes this technique useful for enhancing biomaterials selected for their intrinsic properties. The ion–polymer interactions can strengthen the resultant material properties by the formation of crosslinks. Similarly, for metallic biomaterials, ion implantation can strengthen the original material without bulk modifications by the formation of alloyed species, such as that produced by titanium modification by nitrogen.

A main application of ion implantation strategies has been directed toward the improvement of titanium implants for integration with bone, a major tissue engineering arena. Titanium has long been utilized as a biomaterial due to its rigidity and biocompatibility. Several techniques such as those discussed previously have been carried out to enhance the bone–Ti interaction, such as with hydroxyapatite deposition. These endpoints can also be achieved with ion implantation methods. For example, implantation into Ti with sodium ions resulted in a complex that enhanced hydroxyapatite precipitation from a pretreatment solution, which led to improved cell densities relative to untreated surfaces [106]. An alternative route to hydroxyapatite deposition has been realized with ion implantation for enhancing osteointegration. Implantation of carbon (C^+) and carbon monoxide (CO^+) ions into titanium dental implants significantly improves bone integration, with stronger contacts between bone and implant made possible by covalent bonds not observed in the standard titanium oxide–bone interface [107]. Ion implantation techniques are also suited to the enhancement of other biomaterial interactions. Vascular smooth muscle cells seeded on fluorine-treated polystyrene was observed to have improved adhesion and survival characteristics. Ion implantation has also improved the endothelialization properties of polyurethane, polystyrene, and poly(ethersulfones) for the construction of vascular grafts [108–110].

Other related technologies are useful for surface engineering of biomaterials, by other mechanisms. Ion beam-assisted deposition (IBAD) directs the acceleration of ions toward a simultaneously-generated, vapor-deposited thin film on the surface. The deposited film then reacts with the ions, resulting in modified properties such as film density, texture, interfacial properties, and stress. This technique has been utilized to deposit bacteriostatic silver films on catheters, as well as sealant and electrode coatings on polymeric substrates [111]. In response to the "line-of-sight," serial modification limitations imposed by conventional ion beam technologies, a technique called plasma immersion ion implantation (PIII) was developed in the 1980s. This method enables the rapid implantation of complex geometries, and has been used to enhance the hydrophobicity of various polymers such as poly(tetrafluoroethylene) (PTFE), PDMS, and polyurethane, and to improve material properties of various metals [112–117]. Another technology, reactive ion etching (RIE), utilizes both ions and chemically-reactive species, to impart a chemical component to the primarily physical process. When the reactive species formed by the chemical component and the substrate surface is created, the byproduct can be removed by various means, such as the physical ejection of atoms from the substrate. For surface patterning applications, inductively-coupled plasma reactive ion etching (ICP-RIE) was utilized to etch 3D features in PMMA blocks for protein patterning and modification of the surface to resist nonspecific binding, thus obviating a nonspecific adsorption blocking step [118]. This variation on RIE allows for independent control of ion density and ion energy, to address the tradeoff between substrate damage and etching efficiency. RIE allows for tuning of both chemical and physical components, which are generally isotropic and anisotropic, respectively, to influence the detail of the etching process. Other plasma-based techniques for the etching, sterilization, and surface property modification of biomaterials are extensively reviewed elsewhere [119]. Of note are tetrafluoromethane plasmas for hydrophobic treatment of a surface, and oxygen and nitrogen plasmas for enhancing surface hydrophilicity.

Plasma-based techniques are a versatile method for the modification of many biomaterials, and are suited for developing new surface properties without substantially changing those of the background material, due in general to the low penetration of the bombardment species. Processes can be performed in dry conditions without harsh chemicals to tailor to a specific tissue engineering application, and the engineer has multiple tunable process parameters that can be modified to achieve desired surface properties. The technique is applicable to metals, ceramics, and polymers that may be used as biomaterials; in general, plasma treatments can enhance the biomaterial surface in terms of biocompatibility, biochemistry, tissue function, and mechanical properties, and are thus a key technology. However, caution must be exercised in process parameter optimization, such that undesired stripping of the surface and underlying layers does not occur. This is especially the case with new technologies such as deep RIE, with have etch rates far beyond single microns. In addition, these processes can be both expensive and hazardous, due to the equipment needed and the plasma compositions (e.g., Cl_2), respectively. Furthermore, ion interactions with a surface create a chemically-different species that must be independently evaluated (from the original polymer) for typical features, such as biocompatibility, etc. Nevertheless, this technology, due to the myriad of tunable parameters and candidate substrates, makes plasma-based treatments important for the development of biomaterial surfaces with novel properties.

12.11
Sol–Gel Technology

In addition to a biomaterial featuring certain metrics, such as satisfactory biocompatibility criteria and appropriate tissue interactions (cell adhesion, low thrombogenicity, etc.), in many cases the porosity of the structure within specific macroporous and mesoporous ranges are critical to the success of the bioimplant. Cells and tissues require specific porous architectures characteristic of their *in vivo* environment for tissue ingrowth and fully functional regeneration. For bone tissue regeneration, long a main focus of tissue engineering, specific porous architectures are essential for cellular migration and vascularization with a steady influx of nutrients and molecular cues to facilitate the process. Sol–gel technology has served as a promising route to achieving such structures for various tissue engineering applications. This processing technique can be utilized to enhance the biocompatibility of biomaterial surfaces, can readily coat a number of surfaces, and forms chemically-stable coatings in physiological environments. Of particular interest are the sol–gel derived bioactive glasses. Several incorporations of sol–gel processes are described elsewhere in detail, but they generally involve the gelation of a nanoparticulate suspension into a continuous network. Following an "aging" step, the gel is dried to produce a glassy solid. The process enables the synthesis of materials with various physical properties, made possible by allowing the mixture of organic and inorganic materials within the same process.

Several applications of bioactive glasses illustrate the importance of this technique in creating mechanically stable, highly-functional tissue engineering constructs. Scaffolds based on bioactive glasses were shown to have compressive strength similar to trabecular bone, with pore diameters within the microscale sufficient to allow vascularization of regenerated tissue, as well as cell migration [120]. Furthermore, it was determined that the sintering temperature of the process was a determinant of mesoscale pore size, which had profound effects on the dissolution rate and hydroxyapatite layer formation, as well as the compressive strength of the material. The tuning of surface area, and thus dissolution rates, in tissue engineering scaffolds for bone is highly powerful, as it allows for a mechanism of tuning the implant to the natural bone resorption rate in the body. In another example, sol–gel processing was utilized to develop bioactive foams presenting amine and mercaptan groups to allow for biomolecule incorporation with similar mesoscale and macroscale porosity as the previous application [121]. Thus, these networks supportive of natural tissue ingrowth and vascularization can also be biofunctionalized to allow for appropriate molecular signaling throughout the interconnected matrix. In other applications, sol–gel processing has provided a method of making conventional biomaterial surfaces more biocompatible. Titanium surfaces coated with hydroxyapatite sol–gel layers were shown to exhibit a higher activity of osteoblast-like cells compared with bare titanium, with tunable surface dissolution rates achievable by the incorporation of fluor-hydroxyapatite [122]. Other coatings, such as PLGA and titanium oxide, have also been patterned on titanium substrates using sol–gel processes, with cell adhesion on the sol–gel coated hydroxyapatite layer being higher than the plasma-sprayed substrate [123].

The low cost, favorable mechanical properties, and tunable nano and macroporous architectures made possible by sol–gel technology using a wide blend of organic and inorganic materials make it a promising surface engineering technique. With the added advantage of introducing chemical functionalities within the sol–gel processing steps, bone tissue engineering devices that utilize bioactive glasses and other sol–gel derived materials will benefit from the advances in bone tissue biology, which are already elucidating key molecules such as bone morphogenic protein (BMP-2), which have spatial and temporal distribution profiles that could be tuned by an implant for optimum tissue regeneration [124]. However, tissue engineering applications utilizing sol–gel processes would be enhanced by more extensive studies of the effects of process parameters on porosity of manufactured architectures for various composites. Further studies should also investigate the type of chemical functionalities that can be incorporated within porous architectures produced by sol–gel processes.

12.12
Nanolithography

A family of surface engineering technologies has been developed to produce high-contrast and high-resolution nanoscale patterns for the purpose of readily integrating this information within novel BioMEMS and tissue engineering applications. Collectively known as nanolithography, these strategies include soft lithography, which has been discussed previously, as well as dip-pen lithography (DPN) and nanoimprint, or nanoindentation lithography (NIL). In DPN, an atomic force microscope (AFM) scanning probe is "inked" analogous to a PDMS stamp with specific information, in the form of SAMs, peptides, oligonucleotides, or other biomolecules, and patterned upon a solid substrate by capillary transport, to produce features as low as 30 nm [125]. Substrates tested with DPN have primarily included silicon and gold, with multiple inks. The major advantage of DPN is not only its nanoscale resolution appropriate for biology, but also its direct-writing capability inherent in the probe-based technology, which allows for single cell and potentially single biomolecule manipulations.

DPN is not an old technology, and thus direct tissue engineering applications have not yet been reported at the time of this writing. Most studies are focused on determining the species that can be successfully patterned using DPN, and have been reviewed by Mirkin et al. [126]. Of particular interest is the development of SAMs that function as DPN-sensitive resists for forming features on solid substrates as low as 25 nm. Proteins have been successfully patterned with DPN, with features as small as 45 nm, and cells have specifically-adhered to patterned protein recognition sites [127, 128] (Fig. 12.6). To address the serial nature of the process inherent in other techniques, several studies are in place to develop mul-

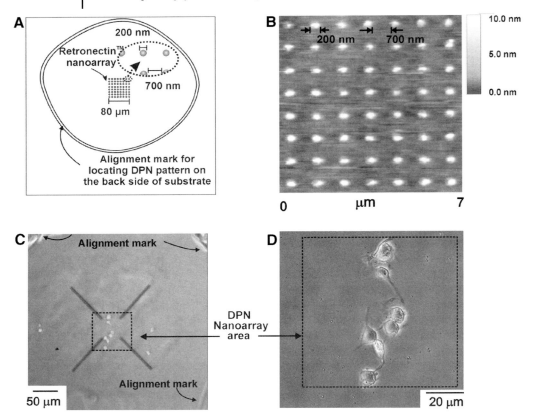

Fig. 12.6. Fibroblast adhesion to dip-pin lithography (DPN)-patterned retronectin (cell adhesion protein) arrays. (A) Fibroblasts were adsorbed onto a retronectin-patterned area of 6400 μm^2. (B) Retronectin protein array. (C, D) Optical images of intact cells adsorbed on retronectin-patterned area. (Reproduced with permission from Lee et al. [126].)

tiprobe arrays for higher throughput DPN. While these developments are exciting for semiconductor and biosensor applications, the applicability of this technique for the surface engineering of biomaterials is not clear. However, gold-plated biomaterials, such as Au pits within titanium implants, can be readily manufactured using previously discussed techniques, thus enabling AFM probe access if an application is found. Also, it seems likely, based on the variety of inks that can coat an AFM tip, that DPN could conceivably be utilized to pattern living tissues or tissue scaffolds to construct cell–protein arrays with nanoscale precision. If such a feat was achieved, then single biomolecules and cells could be manipulated and oriented toward appropriate functional locations within a tissue engineering construct, not unlike that achieved with LGDW as discussed previously. Regardless of whether such a transition is made, further investigations concerning the dynamics of inking and ink transfer to a substrate are warranted to optimize the technology. Various parameters govern this process, not limited to chemistry and geometry, which allow for some tunability in the process. However, the specialized equipment and conditions necessary to apply the technique will limit its accessibility to biological laboratories seeking patterning tools for hypothesis testing, although the embracing of DPN by industry carries with it the potential vision of developing high-throughput, accessible DPN methods for nanofabrication on order.

Similar to DPN technology, NIL is capable of resolutions down to 10 nm [129, 130]. In this technology, a silicon template generated by conventional patterning techniques serves as the template for a polymer (e.g., PMMA) elevated beyond its glass transition temperature, such that it is elastically deformable. Following a cooling step, the polymer assumes the pattern transferred by the mold. Several features make this technology very attractive over existing techniques, including DPN. The technology is not limited in resolution by beam parameters, requires no expensive equipment, and can be utilized in high-throughput processes for reproducible quality patterns. This was furthered by the development of air cushion press (ACP) technology to create more uniform pressures during mold imprinting (Nanonex Corp.). Furthermore, NIL is not as time-consuming as DPN and other beam-based serial technologies, and, like DPN, it is conceivable that single biomolecules could be manipulated with this technique. Thus, NIL is generally a user-friendly biocompatible technology that can be readily applied to surface engineering applications. Applications have included nanoscale protein patterning with preserved activity and high signal to noise ratios [131]. NIL has been combined with further processing steps such as self-assembly and RIE to create defined bioactive regions, thus demonstrating the synergy of NIL with various existing techniques [132]. Given the ultimate resolution of 10 nm for this technique, it could replace electron beam lithography for tissue engineering applications on a cost and biomaterial flexibility basis alone. However, not all materials can be manipulated by NIL, and until the full scope of candidate molding materials is investigated the true utility of the technique will not be fully realized. Furthermore, stamp fabrication can be a tedious process; in addition, the 10 nm resolution limit has not yet been realized for biological applications, and its suitability for most biomaterials has not been determined. Nevertheless, the NIL technique promises to be a flexible, low-cost, high-throughput technique for producing nanoscale features.

12.13
Laser-guided Strategies

The wide range of technologies discussed thus far have been associated with their own unique advantages and limitations, as there is no one size fits all approach in biomaterials surface engineering, in terms of patterning any surface with reproducible chemistries and topologies of perfect resolution for the application of interest. For example, EBL, plasma/ion beam, and photolithography are constrained by specialized equipment costs and lengthy procedures, whereas sandblasting techniques generally do not produce precisely repeatable features. We now discuss a family of technologies based on lasers, which is a promising alternative to the above techniques for the tuning of various biomaterial surface properties. Potential advantages of this approach include high reproducibility, the potential for automated guidance systems, and various sources to choose from of different excitation wavelengths and power densities, which immediately suggests the potential of generating patterns of varying intensity in a manner not unlike a "maskless" grayscale photolithography. The wide availability and familiarity with lasers is also an enticing aspect in incorporating this technology in tissue engineering applications, as well as the potential to incorporate into existing methods, recalling the method of Holden et al. for laser patterning of multiple dye-labeled biomolecules using photomasks [10]. Lasers have been successfully applied to the surface texturing and patterning of polymers, biomolecules, and even cells, to a host of material surfaces.

The generation of surface topologies is possible with multiple laser types, due mainly to the material properties of the target surface. First, since many substances absorb strongly in the ultraviolet range of the spectrum, UV-emitting lasers can thus be used to successfully modify biomaterial surfaces without altering bulk properties, since the penetration depth is limited by the high surface absorption of energy. This feature enables two surface modification routes: (a) the etching of material through photoablation and (b) the nanopatterning of topologies, made possible by the low UV wavelengths of the lasers. Photomasks can be incorporated in patterning steps if necessary to facilitate the formation of certain surface features as described elsewhere. Technically, this process could be performed on any geometry, since the only requirements of this technology is that the laser beam is in contact with the surface to be patterned, and that the surface has chemical bonds that are photolabile at the illuminating wavelengths. Excimer lasers have been used for the photoablation of polymeric surfaces for over two decades [133, 134], and thus an extensive library is available concerning candidate materials and their responses to texturing. Other (higher-emitting) lasers for the patterning of either non-polymeric or low UV-absorbing materials have also been employed. An example is the Nd:YAG laser texturing of titanium surfaces [135]. In general photoablation does not require chemicals or specialized equipment other than the laser to achieve the basic task, although the technology seems capable of incorporating multiple added functionalities, such as computer guidance (discussed below), complex masks, and incorporation into device fabrication assembly lines for rapid, simultaneous patterning of multiple features. Despite the scope of applicability of photoablative techniques, recent studies have demonstrated that they do not possess the resolution necessary for BioMEMs applications when compared with dry etching methods [136], and thus, as with any technique, the geometrical features required of the application must be assessed for suitability with the method used to pattern them.

The same ablative nature of laser irradiation of surfaces has also been exploited to deposit polymer and other biomaterial films, using pulsed laser deposition (PLD) techniques. Laser light is directed inside a vacuum chamber, and several process parameters such as wavelength, power, and duration can be tuned to control the vaporization of the surface material. Potential advantages of this technique are similar to those for texturing/etching using photoablation, such as the broad range of materials that can be ablated with laser irradiation. Furthermore, the tunability of process parameters and the high resolution (small spot size of irradiation) provides for an important control mechanism over the vaporization process. No chemicals or other catalysts (such as exogenous heating of substrate) are needed for PLD. This approach has been successfully applied to the layering of PTFE (Teflon) [137] and PMMA films [138, 139]. In addition, using protective detergent capsules, thin films of glucose oxidase were also formed by PLD, with preservation of activity [140]. While PLD provides for some biomaterial surface engineering applications, especially with polymer thin film deposition on complex geometries, it is difficult to extend its applications to keep up with the need for different scaffolds and bioimplants. Each PLD specimen requires optimized process parameters such that heating of the substance is not too substantial such that the polymer architecture is irreversibly damaged. Even more caution is required in the PLD-based manipulation of biomolecules and cells, and it is unlikely protective substances can safely encapsulate all such species while facilitating high-quality surface patterning. The lack of compatibility between biomolecules and UV irradiation is a limiting aspect. Nevertheless, the laser positioning outside of the vacuum chamber may provide for a favorable alternative to other film deposition techniques, in that the source orientation can be adjusted as needed, in addition to the other previously discussed process parameters that are user-tunable.

Recent approaches have been developed to circumvent some of the limitations of PLD. To improve laser-assisted film deposition methods, a technique called matrix-assisted pulsed laser evaporation (MAPLE) and its related direct-write counterpart (MAPLE DW) were devised, whereby the material to be deposited is dissolved in a judiciously-selected solvent and frozen; the complex is then irradiated to deposit the film of interest with removal of the solvent *in vacuo* [141]. With this system, exposure to laser irradiation is generally limited to the solvent capsule, which aids in preserving the integrity of the dissolved species. As an example, PEG films, which were damaged by direct UV irradiation, deposited using MAPLE were of favorable structural integrity for typical applications [142]. Interestingly, by encapsulating cells in a glycerol-based matrix, successful deposition with acceptable cell viability can be achieved with features <10 µm [141]. In a fundamentally different approach to enhance the biocompatibility of laser-based processes with cells without the use of UV light, the exploitation of optical forces have been used to laser-guide cells to specific locations on substrates. Near-infrared light irradiation can exert tiny forces on the cells, the net effect of which is for cells to move toward the center of the beam and away from the light source. This technique, called laser-guided direct writing (LGDW) is capable of single-cell resolution patterning, and has been employed for the co-patterning of endothelial cells and hepatocytes within Matrigel constructs to create self-assembled 3D constructs with similar morphology to hepatic sinusoids [143] (Fig. 12.7). This technique has the advantage of "point and shoot" control of single cells to create multicellular arrays without a compromise in cell viability, as the irradiation source is not ultraviolet. Further-

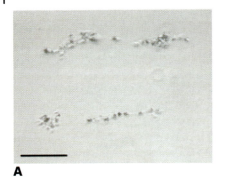
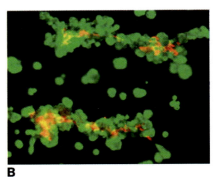

A   **B**

Fig. 12.7. Laser-guided direct writing (LGDW) of cell patterns approximating liver sinusoids *in vitro*. (A) HUVECs deposited by LGDW. Cell elongation along the pattern was observed. (B) 24 h post-deposition, added hepatocytes surrounded HUVECs to form sinusoid-like structures. (Green – albumin, red – CD31 staining; bar = 200 μm.) (Reproduced with permission from Ref. [141].)

more, optical trapping could likely pattern cells on multiple surface types, provided there is an aqueous medium to support light guidance. Although the method is time-consuming, involving a cell-by-cell manipulation technique, theoretically, multiple laser arrays could be utilized for the guidance of multiple cells at once to form hierarchical organization of tissue. This method could be incorporated into advanced schemes that seek to optimize tissue structural and functional properties, for instance by the enhanced migration of a slow-moving cell type by added optical guidance.

Future applications of lasers in biomaterial surface engineering could potentially be in areas in which scale-up for mass production is necessary. Laser-based approaches could be exploited for the capability of modifying surface texture with one laser type, and performing film and biomolecule deposition with other wavelengths, in an assembly line-style fashion. This readily-accessible technology has already been successfully applied to the surface engineering of biomaterials with promising results in producing nanoscale and microscale features. However, a much more detailed investigation of the interface between lasers and biological materials/environments is necessary to expedite such a transition.

12.14
Rapid Prototyping Techniques

The future of biomaterial surface engineering for tissue regeneration and bioimplant applications is highly dependent on the development of procedures for the nanofabrication of 3D cell-protein arrays that can be scaled-up for mass production. Early techniques that constituted the primary methods of scaffold fabrication, such as salt/porogen leaching, fiber bonding, and gas foaming, were not capable of high-throughput, reproducible construction of multilayered, highly interconnected porous architectures of consistent structural and morphological characteristics [144, 145]. To this end, several rapid prototyping (RP) or solid freeform fabrication (SFF) techniques, some of which have already been described (such as MAPLE DW), have emerged to meet the need for high-throughput, reproducible construction of functional tissue architectures. Various RP approaches are related by the use of computational design and fabrication guidance, using CAD-based software. The 3D model of interest is designed on computer using CAD software, and linked instrumentation translates this data into a physical model on a layer-by-layer basis. The building blocks can be powders, liquids, or crystalline materials, depending on the instrumentation employed. RP, if proven to be (a) flexible in the candidate biomaterials that can be used, (b) reproducible in the patterns produced, and (c) capable of achieving tissue engineering-relevant features, will undoubtedly constitute one of the most significant milestones in the field, as it will enable the application of numerous designs and concepts gathered from decades of multi-disciplinary research into clinical practice using high-throughput, reliable manufacturing processes. Here, we discuss four interesting RP technologies that in initial stages have demonstrated utility in their potential to recreate useful tissue architectures: stereolithography (SLA), selective laser sintering (SLS), three-dimensional printing (3DP), and fused deposition molding (FDM).

Computer-guided manufacturing of tissue engineering devices has been explored for some time as the need to not simply design devices but to manufacture them was anticipated by many groups. One route to this goal has been achieved through laser-mediated technologies. For example, Mikos et al. have adapted commercial stereolithographic (SLA) technology for the 3DP of a scaffold using servo-motor controlled, laser-based crosslinking of a polymer-photoinitiator composite layers upon a movable stage [146]. While the features produced were on the order of hundreds of microns, nevertheless the utility of CAD-CAM approaches for the manufacturing of clinically-useful biodegradable tissue scaffolds was demonstrated. SLA (and its improved-resolution derivative technology μSLA) has been applied with promising results in the manufacture of scaffolds incorporating polymers (e.g., PCL) [147] or ceramics [148] with improved microscale resolution, although the use of lasers presents potentially degradation problems as discussed below. Another laser-based commercial manufacturing technology applied to biomaterial surface modification is SLS. In this technology, a laser is directed upon a thin layer of powdered polymer. The heat generated at the surface, which is elevated beyond the polymer's glass transition, is capable of particle fusion, with some control over the pattern and microstructure generated. The 3D model within the CAD software contains slice-by-slice information that guides the laser rastering, scan speed, and intensity, while fresh powders are deposited above the fused layer for the formation of a new one. Early results concerning the use of SLS for polymeric scaffolds are mixed. SLS (as well as SLA) is a laser-based technique and, thus, the same issues concerning polymer degradation observed in laser applications discussed previously are equally applicable here. Degradative effects can include chain scission as well as photo-oxidation. Degradation was observed in the application of SLS for the manufacturing of polyethylene (PE)-based implants [149]. However, SLS has yielded useful results in the fabrication of ceramic and hydroxyapatite-based devices [150]. For example, SLS process parameters such as laser scan speed and laser intensity were tuned to successfully fabricate polyether-ketone (PEEK)–hydroxyapatite constructs for bone tissue engineering applications [151]. Thus, while the same precautions necessary in other laser-based nanoengineering strategies are applicable to SLA and SLS, the flexibility of the technique in accommodating a wide variety of materials and substrates is of particular significance, as well as the potential of the technique to be incorporated in computer-automated high-throughput and high-reliability manufacturing of tissue engineering scaffolds. For this technology to be successful in the surface engineering of biomaterials for clinical usage, work to study the influence of laser tuning parameters on the resulting morphology (e.g., porosity) and mechanical characteristics of the construct produced is necessary. The resolution of the technique will be limited by the wavelength of the laser used, and thus optimization of laser choice should be carried out to achieve the necessary spatial control during the production process. Additionally, the compatibility of these processes with existing methods to engineer surface chemistries is unclear, and should be addressed in future work. It seems feasible, however, to layer functionalized polymer microparticles as needed, or to possibly surface engineer functionalized moieties using other methods as each layer is constructed. The ability of the laser to not only crosslink polymers but to also initiate chemical reactions should be useful in developing SLS and SLA-based processes that incorporate biofunctionalization into scaffold construction. A potential advantage of this technique, which would allow such possibilities, is the amenability of this method for constructing tissues layer-by-layer, using local heating of the polymer in a room temperature environment, thus avoiding heating chambers, although "part bed" temperature is an adjustable parameter as needed [151]. Furthermore, organic solvents used in casting procedures are not necessary by this technique, which needs only a steady supply of deposited polymer microparticles as the starting material. The applicability of these techniques for manufacturing using the broad spectrum of biocompatible polymer-based biomaterials (i.e. those that can be synthesized in a powder form), followed by extensive *in vivo* investigations of scaffold performance, is certainly warranted given the promising results in preliminary work.

Another RP method that can be utilized in ambient temperatures is 3D printing (3DP), which uses essentially a specialized inkjet printer to deposit a binding solution layer-by-layer upon pre-deposited powder beds. Thus, any material that can be obtained in a powder form is a candidate for this technique if a compatible binder can be identified. Upon drying of the bed, unbound powder is removed. Key process control parameters in this technology are the size of the deposition nozzle and polymer microparticles (which dictate layer thickness), the type of binding solution utilized (biocompatibility, hydrophilicity), as well as the resolution step of the printer head. The binding material can be biocompatible, although several studies have used organic solvents as the binder [152]. Early work demonstrated the potential utility of 3DP for the manufacturing of biodegradable, functional polymeric scaffolds for cell adhesion [153]. A multilayered osteochondral scaffold intended for tissue engineering of articular cartilage based on PLGA, PLA, and CaP was fabricated using 3DP with bone and cartilage-like regions of suitable porosity and layer mechanical characteristics within acceptable ranges for hard and soft tissues [154]. This study demonstrates the benefits of a high-throughput multilayering approach for building hierarchically-organized hybrid materials, with sections delegated to specific functions. A problem with 3DP that is especially relevant to the creation of porous scaffolds has been a difficulty with removing internally-embedded compounds, and due to incomplete fusion of polymer particles during the process, undesired micropores can develop that compromise the resulting integrity of the construct [152]. However, future investigations into optimal binders and polymers will likely overcome these issues. The success of 3DP technologies for the surface engineering of biomaterials will depend on work to study the capability of the technique to accommodate micro- and nanoscale features as needed for multiple applications, such as bone tissue engineering (for ingrowth, cell migration, and vascularization), as well as elucidation of polymers (and binder materials) that are compatible with this technique in a tissue engineering context. As with SLS, traditional surface chemistry modifications could be performed after the deposition of each layer to impart bioactivity or mechanical characteristics within the model, but efforts to incorporate such steps during 3DP would be welcomed. Also, as with SLS, the polymers must be supplied in powder-form to be compatible with the process. However, in contrast to SLS, because the whole process is performed without heat, it is possible that bioactive molecules or cells could be incorporated within the constructs, although this has not been extensively investigated. The capability of cell or protein co-deposition will depend on the use of biocompatible binders. This young technology has been licensed to several companies (http://web.mit.edu/tdp), and will certainly yield therapeutically useful constructs of sufficient mechanical and biochemical properties should the scope of compatible biomaterials be expanded.

Reproducible, porous scaffold architectures have also been generated by a related nozzle-based technique known as fused deposition molding (FDM). In this technology, a thermoplastic material is pumped through a heated nozzle to fabricate the device layer-by-layer. A z-axis stepper facilitates the process, and the nozzle diameter as well as the polymer melt thickness influences thickness (z-resolution). An advantage of this technique is the ability to deposit architectures that are porous in three dimensions by adjusting the horizontal distance between deposited filaments (fill gap). Initial studies are supportive of the utility of the approach. FDM was used for the fabrication of 3D interconnected porous and resorbable PCL scaffolds, with satisfactory biocompatibility and mechanical properties [155, 156]. Furthermore, tissue formation was demonstrated with osteoblast-like cells and fibroblasts within the construct, and the mechanical properties and porosity were controllable by process parameters. Similar positive results have been obtained with PCL/CaP and other polymer hybrid systems [157, 158]. However, the scope of FDM is limited by the amount of biomaterials that can be deposited in the manner described. Compatible materials must have the proper melt characteristics that allow it to be extruded through the nozzle. Furthermore, FDM likely cannot be used to simultaneously deposit proteins or cells due to the melting process, as well potential viscosity modifications and nozzle clogging that would make the application very difficult. Z-resolution, as mentioned, is limited by the size of the extruded polymer filament. Nevertheless its application for PCL-based scaffolds warrants its further investigation, due to the favorable aspects of the polymer, such as its FDA-approved status and numerous tissue engineering applications that have already been investigated.

While each RP/SFF approach has its characteristic advantages and disadvantages, they each constitute major steps in the effort to rapidly and reliably produce functional tissue architectures. Given the wide variety of synthetic polymeric constructs that are compatible with these processes, the next step is to integrate these methods with natural polymers, such as ECM, which is underway in several laboratories. Furthermore, as many modifications such as surface chemistry are performed post-processing, the technology would greatly benefit from methods to incorporate such steps within the process itself when possible. Lastly, the further integration of SFF with established nanoengineering approaches discussed in this chapter will expand the capabilities of the constructs produced, especially if such nanoengineering can be performed following the deposition of each layer to develop highly-customized architectures. If these steps can be incorporated to some extent within these processes, the result will be a highly-tunable system for the development of "made to order" tissue engineering constructs that can be fabricated using CAD-based modeling that incorporates features specific to the patient's needs.

12.15
Conclusions

We have described both established and emerging biomaterial surface engineering technologies for the molecular-level presentation of information to cells. Each approach makes an overall contribution to the tissue engineering field by serving as a potential tool to tune the interfacial characteristics of a biomaterial for the optimal reconstruction of the tissue environment. The future will witness further advances in the effort to construct functional tissues with nanoscale precision through integrative strategies that control biochemical and topographical signaling, as nature-inspired developments in materials science continue to adapt to the criteria imposed by biology for recreating complex living systems.

Acknowledgments

This chapter was made possible by funding from the Vanderbilt University Institute for Integrative Biosystems Research and Education (VIIBRE), and a graduate research fellowship to AL through the Vanderbilt Vision Research Center Training Grant (NE1 T32 EY07135).

References

1 BHATIA, S. N., YARMUSH, M. L., TONER, M., Controlling cell interactions by micropatterning in co-cultures: Hepatocytes and 3T3 fibroblasts. *J. Biomed. Mater. Res.* **1997**, 34(2), 189–199.

2 FLEMMING, R. G., MURPHY, C. J., ABRAMS, G. A., GOODMAN, S. L., NEALEY, P. F., Effects of synthetic micro- and nano-structured surfaces on cell behavior. *Biomaterials* **1999**, 20(6), 573–588.

3 BLAWAS, A. S., REICHERT, W. M., Protein patterning. *Biomaterials* **1998**, 19(7–9), 595–609.

4 CLEMENCE, J. F., RANIERI, J. P., AEBISCHER, P., SIGRIST, H., Photoimmobilization of a bioactive laminin fragment and pattern-guided selective neuronal cell attachment. *Bioconj. Chem.* **1995**, 6(4), 411–417.

5 SUGAWARA, T., MATSUDA, T., Photochemical surface derivatization of a peptide-containing Arg-Gly-Asp (Rgd). *J. Biomed. Mater. Res.* **1995**, 29(9), 1047–1052.

6 SIGRIST, H., COLLIOUD, A., CLEMENCE, J. F., GAO, H., LUGINBUHL, R., SANGER, M., SUNDARABABU, G., Surface immobilization of biomolecules by light. *Opt. Eng.* **1995**, 34(8), 2339–2348.

7 PRITCHARD, D. J., MORGAN, H., COOPER, J. M., Patterning and regeneration of surfaces with antibodies. *Anal. Chem.* **1995**, 67(19), 3605–3607.

8 HENGSAKUL, M., CASS, A. E. G., Protein patterning with a photo-activatable derivative of biotin. *Bioconj. Chem.* **1996**, 7(2), 249–254.

9 KANE, R. S., TAKAYAMA, S., OSTUNI, E., INGBER, D. E., WHITESIDES, G. M., Patterning proteins and cells using soft lithography. *Biomaterials* **1999**, 20(23–24), 2363–2376.

10 HOLDEN, M. A., CREMER, P. S., Light activated patterning of dye-labeled molecules on surfaces. *J. Am. Chem. Soc.* **2003**, 125(27), 8074–8075.

11 HE, W., HALBERSTADT, C. R., GONSALVES, K. E., Lithography application of a novel photoresist for patterning of cells. *Biomaterials* **2004**, 25(11), 2055–2063.

12 DIAKOUMAKOS, C. D., DOUVAS, A., RAPTIS, I., KAKABAKOS, S., DIMOTIKALLI, D., TERZOUDI, G., ARGITIS, P., Dilute aqueous base developable resists for environmentally friendly and biocompatible processes. *Microelectron. Eng.* **2002**(61–62), 819–827.

13 DOUVAS, A., ARGITIS, P., MISIAKOS, K., DIMOTIKALI, D., PETROU, P. S., KAKABAKOS, S. E., Biocompatible photolithographic process for the patterning of biomolecules. *Biosens. Bioelectron.* **2002**, 17(4), 269–278.

14 DOUVAS, A., ARGITIS, P., DIAKOUMA-KOS, C. D., MISIAKOS, K., DIMOTIKALI, D., KAKABAKOS, S. E., Photolithographic patterning of proteins with photoresists processable under biocompatible conditions. *J. Vacuum Sci. Technol. B* **2001**, 19(6), 2820–2824.

15 CHEN, C. C., HIRDES, D., FOLCH, A., Gray-scale photolithography using microfluidic photomasks. *Proc. Natl. Acad. Sci. U.S.A.* **2003**, 100(4), 1499–1504.

16 HOLDEN, M. A., JUNG, S. Y., CREMER, P. S., Patterning enzymes inside microfluidic channels via photoattachment chemistry. *Anal. Chem.* **2004**, 76(7), 1838–1843.

17 HERMANSON, G. T., *Bioconjugate Techniques* Academic Press: San Diego, 1996, p. xxv, 785 pp.

18 LUEBKE, K. J., CARTER, D. E., GARNER, H. R., BROWN, K. C., Patterning adhesion of mammalian cells with visible light, tris(bipyridyl)ruthenium(II) chloride, and a digital micromirror array. *J. Biomed. Mater. Res. Part A* **2004**, 68A(4), 696–703.

19 SAMPSELL, J. B., Digital micromirror device and its application to projection displays. *J. Vacuum Sci. Technol. B* **1994**, 12(6), 3242–3246.

20 HAHN, M. S., TAITE, L. J., MOON, J. J., ROWLAND, M. C., RUFFINO, K. A.,

West, J. L., Photolithographic patterning of polyethylene glycol hydrogels. *Biomaterials* **2005**.

21 Luo, Y., Shoichet, M. S., Light-activated immobilization of biomolecules to agarose hydrogels for controlled cellular response. *Biomacromolecules* **2004**, 5(6), 2315–2323.

22 Luo, Y., Shoichet, M. S., A photolabile hydrogel for guided three-dimensional cell growth and migration. *Nature Materials* **2004**, 3(4), 249–253.

23 Revzin, A., Russell, R. J., Yadavalli, V. K., Koh, W. G., Deister, C., Hile, D. D., Mellott, M. B., Pishko, M. V., Fabrication of poly(ethylene glycol) hydrogel microstructures using photolithography. *Langmuir* **2001**, 17(18), 5440–5447.

24 Dalby, M. J., Riehle, M. O., Sutherland, D. S., Agheli, H., Curtis, A. S., Use of nanotopography to study mechanotransduction in fibroblasts – methods and perspectives. *Eur. J. Cell Biol.* **2004**, 83(4), 159–169.

25 Curtis, A. S. G., Wilkinson, C. D., Reactions of cells to topography. *J. Biomater. Sci.-Polym. Ed.* **1998**, 9(12), 1313–1329.

26 Wilkinson, C. D. W., Riehle, M., Wood, M., Gallagher, J., Curtis, A. S. G., The use of materials patterned on a nano- and micro-metric scale in cellular engineering. *Mater. Sci. Eng. C-Biomimetic Supramol. Systems* **2002**, 19(1–2), 263–269.

27 Curtis, A. S. G., Gadegaard, N., Dalby, M. J., Riehle, M. O., Wilkinson, C. D. W., Aitchison, G., Cells react to nanoscale order and symmetry in their surroundings. *IEEE Trans. Nanobiosci.* **2004**, 3(1), 61–65.

28 Harnett, C. K., Satyalakshmi, K. M., Coates, G. W., Craighead, H. G., Direct electron-beam patterning of surface coatings and sacrificial layers for micro-total analysis systems. *J. Photopolym. Sci. Technol.* **2002**, 15(3), 493–496.

29 Harnett, C. K., Satyalakshmi, K. M., Craighead, H. G., Low-energy electron-beam patterning of amine-functionalized self-assembled monolayers. *Appl. Phys. Lett.* **2000**, 76(17), 2466–2468.

30 Dupont-Gillain, C. C., Alaerts, J. A., Dewez, J. L., Rouxhet, P. G., Patterned layers of adsorbed extracellular matrix proteins: Influence on mammalian cell adhesion. *Bio-Med. Mater. Eng.* **2004**, 14(3), 281–291.

31 Rajnicek, A. M., Britland, S., McCaig, C. D., Contact guidance of CNS neurites on grooved quartz: Influence of groove dimensions, neuronal age and cell type. *J. Cell Sci.* **1997**, 110, 2905–2913.

32 Rajnicek, A. M., McCaig, C. D., Guidance of CNS growth cones by substratum grooves and ridges: effects of inhibitors of the cytoskeleton, calcium channels and signal transduction pathways. *J. Cell Sci.* **1997**, 110, 2915–2924.

33 Gadegaard, N., Thoms, S., Macintyre, D. S., McGhee, K., Gallagher, J., Casey, B., Wilkinson, C. D. W., Arrays of nano-dots for cellular engineering. *Microelectron. Eng.* **2003**(67–68), 162–168.

34 Dalby, M. J., Gadegaard, N., Riehle, M. O., Wilkinson, C. D. W., Curtis, A. S. G., Investigating filopodia sensing using arrays of defined nano-pits down to 35 nm diameter in size. *Int. J. Biochem. Cell Biol.* **2004**, 36(10), 2005–2015.

35 Biebricher, A., Paul, A., Tinnefeld, P., Golzhauser, A., Sauer, M., Controlled three-dimensional immobilization of biomolecules on chemically patterned surfaces. *J. Biotechnol.* **2004**, 112(1–2), 97–107.

36 Petronis, S., Gretzer, C., Kasemo, B., Gold, J., Model porous surfaces for systematic studies of material-cell interactions. *J. Biomed. Mater. Res. Part A* **2003**, 66A(3), 707–721.

37 Shim, J., Bersano-Begey, T. F., Zhu, X. Y., Tkaczyk, A. H., Linderman, J. J., Takayama, S., Micro- and nanotechnologies for studying cellular function. *Curr. Top. Med. Chem.* **2003**, 3(6), 687–703.

38 Xia, Y. N., Whitesides, G. M., Soft

lithography. *Angew. Chem.-Int. Ed.* **1998**, 37(5), 551–575.

39 OSTUNI, E., KANE, R., CHEN, C. S., INGBER, D. E., WHITESIDES, G. M., Patterning mammalian cells using elastomeric membranes. *Langmuir* **2000**, 16(20), 7811–7819.

40 ZHAO, X. M., XIA, Y. N., WHITESIDES, G. M., Soft lithographic methods for nano-fabrication. *J. Mater. Chem.* **1997**, 7(7), 1069–1074.

41 WHITESIDES, G. M., OSTUNI, E., TAKAYAMA, S., JIANG, X. Y., INGBER, D. E., Soft lithography in biology and biochemistry. *Annu. Rev. Biomed. Eng.* **2001**, 3, 335–373.

42 KIM, E., XIA, Y. N., WHITESIDES, G. M., Polymer microstructures formed by molding in capillaries. *Nature* **1995**, 376(6541), 581–584.

43 ODOM, T. W., THALLADI, V. R., LOVE, J. C., WHITESIDES, G. M., Generation of 30–50 nm structures using easily fabricated, composite PDMS masks. *J. Am. Chem. Soc.* **2002**, 124(41), 12 112–12 113.

44 ODOM, T. W., LOVE, J. C., WOLFE, D. B., PAUL, K. E., WHITESIDES, G. M., Improved pattern transfer in soft lithography using composite stamps. *Langmuir* **2002**, 18(13), 5314–5320.

45 SCHMID, H., MICHEL, B., Siloxane polymers for high-resolution, high-accuracy soft lithography. *Macromolecules* **2000**, 33(8), 3042–3049.

46 FOLCH, A., TONER, M., Cellular micropatterns on biocompatible materials. *Biotechnol. Progr.* **1998**, 14(3), 388–392.

47 FOLCH, A., JO, B. H., HURTADO, O., BEEBE, D. J., TONER, M., Microfabricated elastomeric stencils for micropatterning cell cultures. *J. Biomed. Mater. Res.* **2000**, 52(2), 346–353.

48 TAKAYAMA, S., McDONALD, J. C., OSTUNI, E., LIANG, M. N., KENIS, P. J. A., ISMAGILOV, R. F., WHITESIDES, G. M., Patterning cells and their environments using multiple laminar fluid flows in capillary networks. *Proc. Natl. Acad. Sci. U.S.A.* **1999**, 96(10), 5545–5548.

49 TAKAYAMA, S., OSTUNI, E., LEDUC, P.,

NARUSE, K., INGBER, D. E., WHITESIDES, G. M., Selective chemical treatment of cellular microdomains using multiple laminar streams. *Chem. Biol.* **2003**, 10(2), 123–130.

50 UNGER, M. A., CHOU, H. P., THORSEN, T., SCHERER, A., QUAKE, S. R., Monolithic microfabricated valves and pumps by multilayer soft lithography. *Science* **2000**, 288(5463), 113–116.

51 McDONALD, J. C., DUFFY, D. C., ANDERSON, J. R., CHIU, D. T., WU, H. K., SCHUELLER, O. J. A., WHITESIDES, G. M., Fabrication of microfluidic systems in poly(dimethylsiloxane). *Electrophoresis* **2000**, 21(1), 27–40.

52 PINO, C. J., HASELTON, F. R., CHANG, M. S., Seeding of corneal wounds by epithelial cell transfer from micropatterned PDMS contact lenses. *Cell Transplant.* **2005**, 14(8), 565–571.

53 BOWDEN, N., HUCK, W. T. S., PAUL, K. E., WHITESIDES, G. M., The controlled formation of ordered, sinusoidal structures by plasma oxidation of an elastomeric polymer. *Appl. Phys. Lett.* **1999**, 75(17), 2557–2559.

54 HU, S. W., REN, X. Q., BACHMAN, M., SIMS, C. E., LI, G. P., ALLBRITTON, N., Surface modification of poly(dimethylsiloxane) microfluidic devices by ultraviolet polymer grafting. *Anal. Chem.* **2002**, 74(16), 4117–4123.

55 HU, S. W., REN, X. Q., BACHMAN, M., SIMS, C. E., LI, G. P., ALLBRITTON, N. L., Tailoring the surface properties of poly(dimethylsiloxane) microfluidic devices. *Langmuir* **2004**, 20(13), 5569–5574.

56 DUFFY, D. C., McDONALD, J. C., SCHUELLER, O. J. A., WHITESIDES, G. M., Rapid prototyping of microfluidic systems in poly(dimethylsiloxane). *Anal. Chem.* **1998**, 70(23), 4974–4984.

57 QIN, D., XIA, Y. N., WHITESIDES, G. M., Rapid prototyping of complex structures with feature sizes larger than 20 μm. *Adv. Mater.* **1996**, 8(11), 917.

58 BURGIN, T., CHOONG, V. E., MARACAS,

G., Large area submicrometer contact printing using a contact aligner. *Langmuir* **2000**, 16(12), 5371–5375.

59 ROGERS, J. A., PAUL, K. E., JACKMAN, R. J., WHITESIDES, G. M., Generating similar to 90 nanometer features using near-field contact-mode photolithography with an elastomeric phase mask. *J. Vacuum Sci. Technol. B* **1998**, 16(1), 59–68.

60 ROGERS, J. A., PAUL, K. E., WHITESIDES, G. M., Quantifying distortions in soft lithography. *J. Vacuum Sci. Technol. B* **1998**, 16(1), 88–97.

61 MRKSICH, M., CHEN, C. S., XIA, Y. N., DIKE, L. E., INGBER, D. E., WHITESIDES, G. M., Controlling cell attachment on contoured surfaces with self-assembled monolayers of alkanethiolates on gold. *Proc. Natl. Acad. Sci. U.S.A.* **1996**, 93(20), 10 775–10 778.

62 MRKSICH, M., DIKE, L. E., TIEN, J., INGBER, D. E., WHITESIDES, G. M., Using microcontact printing to pattern the attachment of mammalian cells to self-assembled monolayers of alkanethiolates on transparent films of gold and silver. *Exp. Cell Res.* **1997**, 235(2), 305–313.

63 TAKAYAMA, S., OSTUNI, E., LeDuc, P., NARUSE, K., INGBER, D. E., WHITESIDES, G. M., Laminar flows – Subcellular positioning of small molecules. *Nature* **2001**, 411(6841), 1016–1016.

64 LINDER, V., VERPOORTE, E., THORMANN, W., DE ROOIJ, N. F., SIGRIST, M., Surface biopassivation of replicated poly(dimethylsiloxane) microfluidic channels and application to heterogeneous immunoreaction with on-chip fluorescence detection. *Anal. Chem.* **2001**, 73(17), 4181–4189.

65 STEVENS, M. M., MAYER, M., ANDERSON, D. G., WEIBEL, D. B., WHITESIDES, G. M., LANGER, R., Direct patterning of mammalian cells onto porous tissue engineering substrates using agarose stamps. *Biomaterials* **2005**, 26(36), 7636–7641.

66 WALHEIM, S., BOLTAU, M., MLYNEK, J., KRAUSCH, G., STEINER, U., Structure

formation via polymer demixing in spin-cast films. *Macromolecules* **1997**, 30(17), 4995–5003.

67 DALBY, M. J., RIEHLE, M. O., JOHNSTONE, H., AFFROSSMAN, S., CURTIS, A. S. G., In vitro reaction of endothelial cells to polymer demixed nanotopography. *Biomaterials* **2002**, 23(14), 2945–2954.

68 DALBY, M. J., RIEHLE, M. O., JOHNSTONE, H. J. H., AFFROSSMAN, S., CURTIS, A. S. G., Nonadhesive nanotopography: Fibroblast response to poly(n-butyl methacrylate)-poly(styrene) demixed surface features. *J. Biomed. Mater. Res. Part A* **2003**, 67A(3), 1025–1032.

69 DALBY, M. J., YARWOOD, S. J., RIEHLE, M. O., JOHNSTONE, H. J., AFFROSSMAN, S., CURTIS, A. S., Increasing fibroblast response to materials using nanotopography: Morphological and genetic measurements of cell response to 13-nm-high polymer demixed islands. *Exp. Cell Res.* **2002**, 276(1), 1–9.

70 BERRY, C. C., DALBY, M. J., McCLOY, D., AFFROSSMAN, S., The fibroblast response to tubes exhibiting internal nanotopography. *Biomaterials* **2005**, 26(24), 4985–4992.

71 NAM, Y. S., PARK, T. G., Porous biodegradable polymeric scaffolds prepared by thermally induced phase separation. *J. Biomed. Mater. Res.* **1999**, 47(1), 8–17.

72 PEPPAS, N. A., KEYS, K. B., TORRES-LUGO, M., LOWMAN, A. M., Poly(ethylene glycol)-containing hydrogels in drug delivery. *J. Controlled Release* **1999**, 62(1–2), 81–87.

73 KEYS, K. B., ANDREOPOULOS, F. M., PEPPAS, N. A., Poly(ethylene glycol) star polymer hydrogels. *Macromolecules* **1998**, 31(23), 8149–8156.

74 GROLL, J., FIEDLER, J., ENGELHARD, E., AMERINGER, T., TUGULU, S., KLOK, H. A., BRENNER, R. E., MOELLER, M., A novel star PEG-derived surface coating for specific cell adhesion. *J. Biomed. Mater. Res. Part A* **2005**, 74A(4), 607–617.

75 BANERJEE, P., IRVINE, D. J., MAYES, A. M., GRIFFITH, L. G., Polymer

latexes for cell-resistant and cell-interactive surfaces. *J. Biomed. Mater. Res.* **2000**, 50(3), 331–339.

76 IRVINE, D. J., RUZETTE, A. V. G., MAYES, A. M., GRIFFITH, L. G., Nanoscale clustering of RGD peptides at surfaces using comb polymers. 2. Surface segregation of comb polymers in polylactide. *Biomacromolecules* **2001**, 2(2), 545–556.

77 IRVINE, D. J., MAYES, A. M., GRIFFITH, L. G., Nanoscale clustering of RGD peptides at surfaces using comb polymers. 1. Synthesis and characterization of comb thin films. *Biomacromolecules* **2001**, 2(1), 85–94.

78 LUPITSKYY, R., ROITER, Y., TSITSILIANIS, C., MINKO, S., From smart polymer molecules to responsive nanostructured surfaces. *Langmuir* **2005**, 21(19), 8591–8593.

79 LOGAN, J., MASSE, P., SKOLNIK, A., FRANCIS, R., ANGOT, S., TATON, D., GNANOU, Y., DURAN, R. S., Novel dendrimer-like and star architectures of poly(ethylene oxide)-block-polystyrene copolymers as monolayers ath the air/water interface. *Abstr. Papers Am. Chem. Soc.* **2003**, 225, U610–U610.

80 ZIMMERMAN, S. C., WENDLAND, M. S., RAKOW, N. A., ZHAROV, I., SUSLICK, K. S., Synthetic hosts by monomolecular imprinting inside dendrimers. *Nature* **2002**, 418(6896), 399–403.

81 ORAL, E., PEPPAS, N. A., Responsive and recognitive hydrogels using star polymers. *J. Biomed. Mater. Res. Part A* **2004**, 68A(3), 439–447.

82 LAHANN, J., KLEE, D., THELEN, H., BIENERT, H., VORWERK, D., HOCKER, H., Improvement of haemocompatibility of metallic stents by polymer coating. *J. Mater. Sci. Mater. Med.* **1999**, 10(7), 443–448.

83 LAHANN, J., KLEE, D., PLUESTER, W., HOECKER, H., Bioactive immobilization of r-hirudin on CVD-coated metallic implant devices. *Biomaterials* **2001**, 22(8), 817–826.

84 LAHANN, J., BALCELLS, M., RODON, T., LEE, J., CHOI, I. S., JENSEN, K. F., LANGER, R., Reactive polymer coatings: A platform for patterning proteins and mammalian cells onto a broad range of materials. *Langmuir* **2002**, 18(9), 3632–3638.

85 LAHANN, J., BALCELLS, M., LU, H., RODON, T., JENSEN, K. F., LANGER, R., Reactive polymer coatings: a first step toward surface engineering of microfluidic devices. *Anal. Chem.* **2003**, 75(9), 2117–2122.

86 CHEN, H. Y., ELKASABI, Y., LAHANN, J., Surface modification of confined microgeometries via vapor-deposited polymer coatings. *J. Am. Chem. Soc.* **2006**, 128(1), 374–380.

87 CHEN, H. Y., LAHANN, J., Fabrication of discontinuous surface patterns within microfluidic channels using photodefinable vapor-based polymer coatings. *Anal. Chem.* **2005**, 77(21), 6909–6914.

88 CHAN, K., GLEASON, K. K., Initiated chemical vapor deposition of linear and cross-linked poly(2-hydroxyethyl methacrylate) for use as thin-film hydrogels. *Langmuir* **2005**, 21(19), 8930–8939.

89 CORREA-DUARTE, M. A., WAGNER, N., ROJAS-CHAPANA, J., MORSCZECK, C., THIE, M., GIERSIG, M., Fabrication and biocompatibility of carbon nanotube-based 3D networks as scaffolds for cell seeding and growth. *Nano Lett.* **2004**, 4(11), 2233–2236.

90 WILBUR, J. L., KIM, E., XIN, Y. N., WHITESIDES, G. M., Lithographic molding – a convenient route to structures with submicrometer dimensions. *Adv. Mater.* **1995**, 7(7), 649–652.

91 HAMMOND, P. T., Form and function in multilayer assembly: New applications at the nanoscale. *Adv. Mater.* **2004**, 16(15), 1271–1293.

92 KHADEMHOSSEINI, A., SUH, K. Y., YANG, J. M., ENG, G., YEH, J., LEVENBERG, S., LANGER, R., Layer-by-layer deposition of hyaluronic acid and poly-L-lysine for patterned cell co-cultures. *Biomaterials* **2004**, 25(17), 3583–3592.

93 MOHAMMED, J. S., DECOSTER, M. A., MCSHANE, M. J., Micropatterning of nanoengineered surfaces to study neuronal cell attachment in vitro. *Biomacromolecules* **2004**, 5(5), 1745–1755.

94 SHASTRI, V. P., LIPSKI, A. M., SY, J. C., ZNIDARSIC, W., CHOI, H., CHEN, I.-W., Functionalized Nanoparticles as Versatile Tools for the Introduction of Biomimetics on Surfaces in Nano-engineered Nanofibrous Materials, GUCERI, S. et al. (ed.), Kluwer Academic Publishers, **2004**, 257–264.

95 ZHENG, H. P., BERG, M. C., RUBNER, M. F., HAMMOND, P. T., Controlling cell attachment selectively onto biological polymer-colloid templates using polymer-on-polymer stamping. *Langmuir* **2004**, 20(17), 7215–7222.

96 JIANG, X. P., ZHENG, H. P., GOURDIN, S., HAMMOND, P. T., Polymer-on-polymer stamping: Universal approaches to chemically patterned surfaces. *Langmuir* **2002**, 18(7), 2607–2615.

97 SNEER, R., WEYGAND, M. J., KJAER, K., TIRRELL, D. A., RAPAPORT, H., Parallel beta-sheet assemblies at interfaces. *ChemPhysChem* **2004**, 5(5), 747–750.

98 RAPAPORT, H., MOLLER, G., KNOBLER, C. M., JENSEN, T. R., KJAER, K., LEISEROWITZ, L., TIRRELL, D. A., Assembly of triple-stranded beta-sheet peptides at interfaces. *J. Am. Chem. Soc.* **2002**, 124(32), 9342–9343.

99 RECHES, M., GAZIT, E., Casting metal nanowires within discrete self-assembled peptide nanotubes. *Science* **2003**, 300(5619), 625–627.

100 RANGANATHAN, D., SAMANT, M. P., KARLE, I. L., Self-assembling, cystine-derived, fused nanotubes based on spirane architecture: Design, synthesis, and crystal structure of cystinospiranes. *J. Am. Chem. Soc.* **2001**, 123(24), 5619–5624.

101 ZHANG, S. G., ALTMAN, M., Peptide self-assembly in functional polymer science and engineering. *Reactive Funct. Polym.* **1999**, 41(1–3), 91–102.

102 MI, L., FISCHER, S., CHUNG, B., SUNDELACRUZ, S., HARDEN, J. L., Self-assembling protein hydrogels with modular integrin binding domains, *Biomacromol* **2006**, 7, 38–47.

103 HARTGERINK, J. D., BENIASH, E., STUPP, S. I., Self-assembly and mineralization of peptide-amphiphile nanofibers. *Science* **2001**, 294(5547), 1684–1688.

104 LAMPIN, M., WAROCQUIER CLEROUT, R., LEGRIS, C., DEGRANGE, M., SIGOT LUIZARD, M. F., Correlation between substratum roughness and wettability, cell adhesion, and cell migration. *J. Biomed. Mater. Res.* **1997**, 36(1), 99–108.

105 ISHIKAWA, K., MIYAMOTO, Y., NAGAYAMA, M., ASAOKA, K., Blast coating method: New method of coating titanium surface with hydroxyapatite at room temperature. *J. Biomed. Mater. Res.* **1997**, 38(2), 129–134.

106 MAITZ, M. F., PHAM, M. T., MATZ, W., REUTHER, H., STEINER, G., RICHTER, E., Ion beam treatment of titanium surfaces for enhancing deposition of hydroxyapatite from solution. *Biomol. Eng.* **2002**, 19(2–6), 269–272.

107 DE MAEZTU, M. A., ALAVA, J. I., GAY-ESCODA, C., Ion implantation: Surface treatment for improving the bone integration of titanium and Ti6Al4V dental implants. *Clin. Oral Implants Res.* **2003**, 14(1), 57–62.

108 LEE, J. S., KAIBARA, M., IWAKI, M., SASABE, H., SUZUKI, Y., KUSAKABE, M., Selective adhesion and proliferation of cells on ion-implanted polymer domains. *Biomaterials* **1993**, 14(12), 958–960.

109 BACAKOVA, L., MARES, V., BOTTONE, M. G., PELLICCIARI, C., LISA, V., SVORCIK, V., Fluorine ion-implanted polystyrene improves growth and viability of vascular smooth muscle cells in culture. *J. Biomed. Mater. Res.* **2000**, 49(3), 369–379.

110 PIGNATARO, B., CONTE, E., SCANDURRA, A., MARLETTA, G., Improved cell adhesion to ion beam-irradiated polymer surfaces. *Biomaterials* **1997**, 18(22), 1461–1470.

111 SIOSHANSI, P., TOBIN, E. J., Surface treatment of biomaterials by ion beam processes. *Surf. Coatings Technol.* **1996**, 83(1–3), 175–182.

112 RANGEL, E. C., BENTO, W. C. A., KAYAMA, M. E., SCHREINER, W. H.,

CRUZ, N. C., Enhancement of polymer hydrophobicity by SF6 plasma treatment and argon plasma immersion ion implantation. *Surf. Interface Anal.* **2003**, 35(2), 179–183.

113 THORWARTH, G., MANDL, S., RAUSCHENBACH, B., Plasma immersion ion implantation using titanium and oxygen ions. *Surf. Coatings Technol.* **2000**, 128, 116–120.

114 THORWARTH, G., MANDL, S., RAUSCHENBACH, B., Plasma immersion ion implantation of cold-work steel. *Surf. Coatings Technol.* **2000**, 125(1–3), 94–99.

115 MANOVA, D., HUBER, P., MANDL, S., RAUSCHENBACH, B., Surface modification of aluminium by plasma immersion ion implantation. *Surf. Coatings Technol.* **2000**, 128, 249–255.

116 MANDL, S., RAUSCHENBACH, B., Plasma immersion ion implantation. New technology for homogeneous modification of the surface of medical implants of complex shapes. *Biomedizinische Technik* **2000**, 45(7–8), 193–198.

117 MANDL, S., KRAUSE, D., THORWARTH, G., SADER, R., ZEILHOFER, F., HORCH, H. H., RAUSCHENBACH, B., Plasma immersion ion implantation treatment of medical implants. *Surf. Coatings Technol.* **2001**, 142, 1046–1050.

118 RUCKER, V. C., HAVENSTRITE, K. L., SIMMONS, B. A., SICKAFOOSE, S. M., HERR, A. E., SHEDIAC, R., Functional antibody immobilization on 3-dimensional polymeric surfaces generated by reactive ion etching. *Langmuir* **2005**, 21(17), 7621–7625.

119 PONCIN-EPAILLARD, F., LEGEAY, G., Surface engineering of biomaterials with plasma techniques. *J. Biomater. Sci.-Polym. Ed.* **2003**, 14(10), 1005–1028.

120 JONES, J. R., EHRENFRIED, L. M., HENCH, L. L., Optimising bioactive glass scaffolds for bone tissue engineering. *Biomaterials* **2006**, 27(7), 964–973.

121 LENZA, R. F. S., VASCONCELOS, W. L., JONES, J. R., HENCH, L. L., Surface-modified 3D scaffolds for tissue engineering. *J. Mater. Sci.-Mater. Med.* **2002**, 13(9), 837–842.

122 KIM, H. W., KIM, H. E., KNOWLES, J. C., Fluor-hydroxyapatite sol-gel coating on titanium substrate for hard tissue implants. *Biomaterials* **2004**, 25(17), 3351–3358.

123 SATO, M., SLAMOVICH, E. B., WEBSTER, T. J., Enhanced osteoblast adhesion on hydrothermally treated hydroxyapatite/titania/poly(lactide-co-glycolide) sol-gel titanium coatings. *Biomaterials* **2005**, 26(12), 1349–1357.

124 DERNER, R., ANDERSON, A. C., The bone morphogenic protein. *Clin. Podiatr. Med. Surg. North Am.* **2005**, 22(4), 607–618, vii.

125 PINER, R. D., ZHU, J., XU, F., HONG, S. H., MIRKIN, C. A., "Dip-pen" nanolithography. *Science* **1999**, 283(5402), 661–663.

126 GINGER, D. S., ZHANG, H., MIRKIN, C. A., The evolution of dip-pen nanolithography. *Angew. Chem.-Int. Ed.* **2004**, 43(1), 30–45.

127 LEE, K. B., LIM, J. H., MIRKIN, C. A., Protein nanostructures formed via direct-write dip-pen nanolithography. *J. Am. Chem. Soc.* **2003**, 125(19), 5588–5589.

128 LEE, K. B., PARK, S. J., MIRKIN, C. A., SMITH, J. C., MRKSICH, M., Protein nanoarrays generated by dip-pen nanolithography. *Science* **2002**, 295(5560), 1702–1705.

129 CHOU, S. Y., KRAUSS, P. R., RENSTROM, P. J., Imprint of sub-25 nm vias and trenches in polymers. *Appl. Phys. Lett.* **1995**, 67(21), 3114–3116.

130 CHOU, S. Y., KRAUSS, P. R., RENSTROM, P. J., Nanoimprint lithography. *J. Vacuum Sci. Technol. B* **1996**, 14(6), 4129–4133.

131 HOFF, J. D., CHENG, L. J., MEYHOFER, E., GUO, L. J., HUNT, A. J., Nanoscale protein patterning by imprint lithography. *Nano Lett.* **2004**, 4(5), 853–857.

132 FALCONNET, D., PASQUI, D., PARK, S., ECKERT, R., SCHIFT, H., GOBRECHT, J., BARBUCCI, R., TEXTOR, M., A novel approach to produce protein nanopatterns by combining nanoimprint lithography and molecular self-assembly. *Nano Lett.* **2004**, 4(10), 1909–1914.

133 DYER, P. E., JENKINS, S. D., SIDHU, J., Novel method for measuring excimer laser ablation thresholds of polymers. *Appl. Phys. Lett.* **1988**, 52(22), 1880–1882.

134 DYER, P. E., SIDHU, J., Excimer laser ablation and thermal coupling efficiency to polymer-films. *J. Appl. Phys.* **1985**, 57(4), 1420–1422.

135 ROMANOS, G. E., EVERTS, H., NENTWIG, G. H., Effects of diode and Nd: YAG laser irradiation on titanium discs: A scanning electron microscope examination. *J. Periodontol.* **2000**, 71(5), 810–815.

136 MELLO, A. P., BARI, M. A., PRENDERGAST, P. J., A comparison of excimer laser etching and dry etching process for surface fabrication of biomaterials. *J. Mater. Processing Technol.* **2002**, 124(3), 284–292.

137 SMAUSZ, T., HOPP, B., KRESZ, N., Pulsed laser deposition of compact high adhesion polytetrafluoroethylene thin films. *J. Phys. D-Appl. Phys.* **2002**, 35(15), 1859–1863.

138 CRISTESCU, R., SOCOL, G., MIHAILESCU, I. N., POPESCU, M., SAVA, F., ION, E., MOROSANU, C. O., STAMATIN, I., New results in pulsed laser deposition of poly-methyl-methacrylate thin films. *Appl. Surf. Sci.* **2003**, 208, 645–650.

139 SAVA, F., CRISTESCU, R., SOCOL, G., RADVAN, R., SAVASTRU, R., SAVASTRU, D., Structure of bulk and thin films of poly-methyl-methacrylate (PMMA) polymer prepared by pulsed laser deposition. *J. Optoelectronics Adv. Mater.* **2002**, 4(4), 965–970.

140 PHADKE, R. S., AGARWAL, G., Laser-assisted deposition of preformed mesoscopic systems. *Mater. Sc. Eng. C-Biomim. Supramol. Systems* **1998**, 5(3–4), 237–241.

141 WU, P. K., RINGEISEN, B. R., CALLAHAN, J., BROOKS, M., BUBB, D. M., WU, H. D., PIQUE, A., SPARGO, B., McGILL, R. A., CHRISEY, D. B., The deposition, structure, pattern deposition, and activity of biomaterial thin-films by matrix-assisted pulsed-laser evaporation (MAPLE) and MAPLE direct write. *Thin Solid Films* **2001**, 398, 607–614.

142 BUBB, D. M., RINGEISEN, B. R., CALLAHAN, J. H., GALICIA, M., VERTES, A., HORWITZ, J. S., McGILL, R. A., HOUSER, E. J., WU, P. K., PIQUE, A., CHRISEY, D. B., Vapor deposition of intact polyethylene glycol thin films. *Appl. Phys. a-Mater. Sci. Processing* **2001**, 73(1), 121–123.

143 NAHMIAS, Y., SCHWARTZ, R. E., VERFAILLIE, C. M., ODDE, D. J., Laser-guided direct writing for three-dimensional tissue engineering. *Biotechnol. Bioeng.* **2005**, 92(2), 129–136.

144 CHEN, G. P., USHIDA, T., TATEISHI, T., Scaffold design for tissue engineering. *Macromol. Biosci.* **2002**, 2(2), 67–77.

145 TSANG, V. L., BHATIA, S. N., Three-dimensional tissue fabrication. *Adv. Drug Delivery Rev.* **2004**, 56(11), 1635–1647.

146 COOKE, M. N., FISHER, J. P., DEAN, D., RIMNAC, C., MIKOS, A. G., Use of stereolithography to manufacture critical-sized 3D biodegradable scaffolds for bone ingrowth. *J. Biomed. Mater. Res. Part B-Appl. Biomater.* **2003**, 64B(2), 65–69.

147 MATSUDA, T., MIZUTANI, M., Liquid acrylate-endcapped biodegradable poly(epsilon-caprolactone-co-trimethylene carbonate). II. Computer-aided stereolithographic microarchitectural surface photoconstructs. *J. Biomed. Mater. Res.* **2002**, 62(3), 395–403.

148 ZHANG, X., JIANG, X. N., SUN, C., Micro-stereolithography of polymeric and ceramic microstructures. *Sensors Actuators A-Phys.* **1999**, 77(2), 149–156.

149 RIMELL, J. T., MARQUIS, P. M., Selective laser sintering of ultra high molecular weight polyethylene for clinical applications. *J. Biomed. Mater. Res.* **2000**, 53(4), 414–420.

150 HUTMACHER, D. W., SITTINGER, M., RISBUD, M. V., Scaffold-based tissue engineering: Rationale for computer-aided design and solid free-form fabrication systems. *Trends Biotechnol.* **2004**, 22(7), 354–362.

151 TAN, K. H., CHUA, C. K., LEONG, K. F.,

CHEAH, C. M., CHEANG, P., ABU BAKAR, M. S., CHA, S. W., Scaffold development using selective laser sintering of polyetheretherketone-hydroxyapatite biocomposite blends. *Biomaterials* **2003**, 24(18), 3115–3123.

152 LAM, C. X. F., MO, X. M., TEOH, S. H., HUTMACHER, D. W., Scaffold development using 3D printing with a starch-based polymer. *Mater. Sci. Eng. C-Biomimetic Supramol. Systems* **2002**, 20(1–2), 49–56.

153 PARK, A., WU, B., GRIFFITH, L. G., Integration of surface modification and 3D fabrication techniques to prepare patterned poly(L-lactide) substrates allowing regionally selective cell adhesion. *J. Biomater. Sci.-Polym. Ed.* **1998**, 9(2), 89–110.

154 SHERWOOD, J. K., RILEY, S. L., PALAZZOLO, R., BROWN, S. C., MONKHOUSE, D. C., COATES, M., GRIFFITH, L. G., LANDEEN, L. K., RATCLIFFE, A., A three-dimensional osteochondral composite scaffold for articular cartilage repair. *Biomaterials* **2002**, 23(24), 4739–4751.

155 HUTMACHER, D. W., SCHANTZ, T.,

ZEIN, I., NG, K. W., TEOH, S. H., TAN, K. C., Mechanical properties and cell cultural response of polycaprolactone scaffolds designed and fabricated via fused deposition modeling. *J. Biomed. Mater. Res.* **2001**, 55(2), 203–216.

156 ZEIN, I., HUTMACHER, D. W., TAN, K. C., TEOH, S. H., Fused deposition modeling of novel scaffold architectures for tissue engineering applications. *Biomaterials* **2002**, 23(4), 1169–1185.

157 ENDRES, M., HUTMACHER, D. W., SALGADO, A. J., KAPS, C., RINGE, J., REIS, R. L., SITTINGER, M., BRANDWOOD, A., SCHANTZ, J. T., Osteogenic induction of human bone marrow-derived mesenchymal progenitor cells in novel synthetic polymer-hydrogel matrices. *Tissue Eng.* **2003**, 9(4), 689–702.

158 RAI, B., TEOH, S. H., HO, K. H., HUTMACHER, D. W., CAO, T., CHEN, F., YACOB, K., The effect of rhBMP-2 on canine osteoblasts seeded onto 3D bioactive polycaprolactone scaffolds. *Biomaterials* **2004**, 25(24), 5499–5506.

Index

Nanotechnologies for the Life Sciences Vol. 9
Tissue, Cell and Organ Engineering. Edited by Challa S. S. R. Kumar
Copyright © 2006 WILEY-VCH Verlag GmbH & Co. KGaA, Weinheim
ISBN: 3-527-31389-3